Praise for *The Woman Who Smashed Codes*

"It's unsurprising that the name Elizebeth Friedman doesn't ring a bell for most Americans, given how much of her work was classified during the war. . . . Still, this Quaker-born poet from Indiana was the grandmother of the National Security Agency and virtually created the modern code-breaking profession. Trust us on this one."

—*Forbes*

"[Fagone] records the pair's accomplishments, trials, and love affair, taking care to ensure that Elizebeth finally receives the recognition she deserves. The impressive endnotes will prove useful to researchers who wish to further explore the contributions of female codebreakers. Fans of Margot Lee Shetterly's *Hidden Figures* and Andrew Hodges's *Alan Turing: The Enigma* will enjoy this carefully researched story of a smart and loyal but overlooked woman."

—*Library Journal* (starred review)

"Fagone is a superb writer and has created a fascinating tale of a woman who brought down Prohibition-era smugglers, Nazis, counterfeiters, gangsters, and more. "

—Ben Rothke, RSA Conference

"A powerful love story, a story of war, and a fascinating biography, *The Woman Who Smashed Codes* is a magnificent work of literary nonfiction that sheds light on an important hidden figure. You will devour this book."

—Karen Abbott, *New York Times*
bestselling author of *Sin in the Second City*

"Deeply reported and stunningly original, *The Woman Who Smashed Codes* is a riveting narrative about one of the most overlooked figures in American history—a figure whose remarkable story was essentially ignored for more than seventy years simply because she was a woman."

—Stefan Fatsis, bestselling author of *Word Freak*

"Jason Fagone is a master storyteller—and he's telling one damn good story about a long-forgotten American heroine. It is, among many things, the compulsively readable history of the national security state in its infancy. His book is filled with memorable villains, intrigue, and love."

—Franklin Foer, *New York Times* bestselling author of *How Soccer Explains the World* and the forthcoming *World Without Mind*

"Jason Fagone's stunning narrative unearths an intimate and unexpected history of code breaking. This remarkable tale reveals the fundamental role cryptology has played in our past, and the untold story of the pioneering woman behind its evolution. It is a treasure of a book."

—Nathalia Holt, *New York Times* bestselling author of *Rise of the Rocket Girls: The Women Who Propelled Us, from Missiles to the Moon to Mars*

"In *The Woman Who Smashed Codes*, Jason Fagone rights a historical wrong, unshrouding an unsung heroine and revealing the love story at the root of the modern world's spy games. But this book's true revelation is the author's talent: sure-handed, thrilling, and lyrical."

—Benjamin Wallace, *New York Times* bestselling author of *The Billionaire's Vinegar*

THE

WOMAN

WHO

SMASHED

CODES

DEY ST.
An Imprint of WILLIAM MORROW

THE WOMAN WHO SMASHED CODES

A True Story of Love, Spies,

and the Unlikely Heroine Who Outwitted

America's Enemies

JASON FAGONE

Grateful acknowledgment is made to the following for the use of the images that appear throughout the text: George C. Marshall Research Foundation (pages xi, 21, 37, 93, 110, 127, 168, 177, and 327); New York Public Library, Manuscripts and Archives Division (pages 33 and 45); Hathi Trust Digital Library (page 88); National Archives of the United Kingdom (page 223); and U.S. National Archives and Records Administration (page 249).

HarperCollins books may be purchased for educational, business, or sales promotional use. For information please e-mail the Special Markets Department at SPsales@harpercollins.com.

A hardcover edition of this book was published in 2017 by Dey Street Books, an imprint of William Morrow.

FIRST DEY STREET BOOKS PAPERBACK EDITION PUBLISHED 2018.

Designed by Paula Russell Szafranski

Frontispiece © Jonathan Weiss/Shutterstock, Inc.

Library of Congress Cataloging-in-Publication Data has been applied for.

ISBN 978-0-06-243051-9

20 21 22 DIX/LSC 10

The king hath note of all that they intend,
By interception which they dream not of.
—SHAKESPEARE, *HENRY V*, 1599

Knowledge itself is power.
—FRANCIS BACON, *SACRED MEDITATIONS*, 1597

CONTENTS

Umol-huun   tah-tiyal

William Frederick

yetel

Elizebeth Smith Friedman

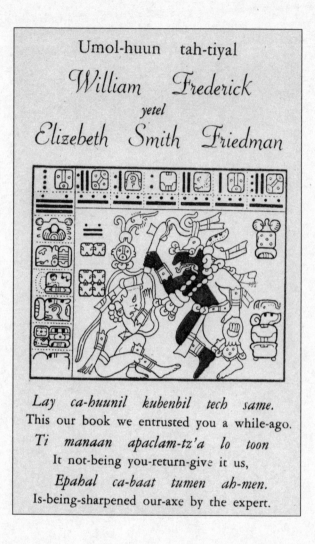

Lay ca-huunil kubenbil tech same.
This our book we entrusted you a while-ago.

Ti manaan apaclam-tz'a lo toon
It not-being you-return-give it us,

Epahal ca-baat tumen ah-men.
Is-being-sharpened our-axe by the expert.

Prying Eyes

This is a love story.

In 1916, during the First World War, two young Americans met by chance on a mysterious and now-forgotten estate near Chicago. At first they seemed to have little in common. She was Elizebeth Smith, a Quaker schoolteacher who found joy in poetry. He was William Friedman, a Jewish plant biologist from a poor family. But they fell for each other. Within a year they were married. They went on to change history together, in ways that still mark our lives today. They taught themselves to be spies—of a new and vital kind.

What they learned to do, better than anyone in the world, was reveal the written secrets of others. They were codebreakers, people who solve secret messages without knowing the key. Puzzle solvers. In a time when there were only a handful of experienced codebreakers in the entire country, the two lovers became a sort of family codebreaking bureau, a husband-and-wife duo unlike any that existed before or has since. Computers didn't exist, so they used pencil, paper, and their brains.

Over the course of thirty years, while raising two children, Elizebeth and William Friedman unscrambled thousands of messages spanning two world wars, prying loose secrets about smuggling networks, gangsters, organized crime, foreign armies, and fascism. They also invented new techniques that transformed the science of secret writing, known as cryptology. Today the insights of this one couple lurk at the base of everything from huge government agencies to the smallest fluctuations of our online lives. And the Friedmans did it all despite having little to no training in mathematics. The basic unit of their life was not the equation but the word. At heart they were people who loved words—words kneaded and pulled and torn, words flipped and arranged in grids and squares and strips and in lines marching down the pale sheet of scratch paper.

In the decades since the Second World War, the husband, William Friedman, has become a revered figure to intelligence historians. He is called "the world's greatest cryptologist" by the eminent chronicler of secret writing David Kahn: "Singlehandedly," Kahn writes, "he made his country preeminent in his field." William Friedman is also widely considered to be the father of the National Security Agency, the part of the U.S. government that intercepts foreign communications and sifts them for information—"signals intelligence." He wrote the definitive textbooks that trained generations of NSA analysts who are still working today. In 1975 the agency named its main auditorium after William Friedman, at its headquarters in Fort Meade, Maryland, and a bronze bust of William's head still stands guard there, above a plaque that reads CRYPTOLOGIC PIONEER AND INVENTOR, FOUNDER OF THE SCIENCE OF MODERN AMERICAN CRYPTOLOGY.

Today Elizebeth isn't nearly as famous, despite her talent and contributions. Early on she worked side by side with William and collaborated on several of their groundbreaking scientific papers; she was considered by some of their friends to be the more brilliant of the pair; she ultimately carved out a spectacular career of her own; and by 1945 the government considered *both* Friedmans to be pioneers of their field. A then-secret document said of Elizebeth, "She and her husband are among the founders of American military cryptanalysis"—cryptanalysis is another word for codebreaking—and a federal prosecutor told the FBI that "Mrs. Friedman and her husband . . . are recognized as the leading authorities in the country." Yet in the canonical books about twentieth-century code-breaking, Elizebeth is treated as the dutiful, slightly colorful wife of a great man, a digression from the main narrative, if not a footnote. Her victories are all but forgotten.

I started reading about the Friedmans in 2014, after Edward Snowden shocked the world by revealing that the NSA was gathering the phone records of millions of ordinary Americans. Curious to know more about Elizebeth, I found a brief bio on the website of a Virginia library, along with a set of pictures. There she was, a petite woman in a white dress, standing on a patch of grass almost one hundred years ago, skin porcelain, head cocked at the photographer, smiling and squinting slightly in what must have been a blinding sun.

The library held the Friedmans' personal papers. One morning I drove down to Virginia and asked the chief archivist to show me what Elizebeth had left. In the back of an office, he unlocked a solid gray metal door and an inner door of silver metal bars, led me into a darkened, humidity-controlled vault, and pointed to multiple shelves of gray archival boxes, twenty-two boxes in all. "We try to tell people that Elizebeth's stuff is amazing," the archivist said, but usually researchers want to see William's papers.

You get these moments sometimes as a journalist, if you're lucky. You hear a voice that bursts from a body or a page with beauty or urgency or insight. Elizebeth's boxes contained hundreds of her letters. Love letters. Letters to her kids *written in code.* Handwritten diaries. A partial, unpublished autobiography. I'm not a mathematician, and I'll never be an expert on codes and ciphers, but Elizebeth's descriptions of her work gave me a sense of what it must have felt like to be her—the excitement of solving the kind of puzzle that could save a life or nudge a war. She liked to say that codes are all around us: in children's report cards, in slang, in headlines and movies and songs. Codebreaking is about noticing and manipulating patterns. Humans do this without thinking. We're wired to see patterns. Codebreakers train themselves to see more deeply.

As rich as Elizebeth's papers were, they struck me as incomplete. The records trailed off around 1940. What was she doing in the Second World War? No one seemed to know.

It took me almost two years to find the answer. She spent the war catching Nazi spies, among other little-known feats. Working with an elite codebreaking unit that she founded in 1931 and collaborating closely with both British and U.S. intelligence, Elizebeth became a secret detective, a Sherlock Holmes on the trail of fascist agents infiltrating the Western Hemisphere. She tracked and exposed them, smashing the spy rings, ruining Nazi dreams.

In a broader sense, she filled gaps in agencies that weren't prepared for the battle of wits that now faced them, a pattern that repeated throughout her entire career. The FBI, the CIA, the NSA—to different degrees Elizebeth pressed her thumb into the clay of all these agencies when the clay was still wet. She helped to shape them and she battled them, too, a woman hammering herself into the history of what we now call the "intelligence community." But when powerful men started telling the story, they left her out of it. In 1945, Elizebeth's spy files were stamped with classification tags and entombed in government archives, and officials made her swear an oath of secrecy about her work in the war. So she had to sit silent and watch others seize credit for her accomplishments, particularly J. Edgar Hoover. A gifted salesman, Hoover successfully portrayed the FBI as the major hero in the Nazi spy hunt. Public gratitude flowed to Hoover, increasing his already considerable power, making him an American icon, virtually untouchable until his death in 1972.

It's not quite true that history is written by the winners. It's written by the best publicists on the winning team.

What follows is my attempt to put back together a puzzle that was fragmented by secrecy, sexism, and time. I relied on the Friedmans' letters and papers, declassified U.S. and British government files, Freedom of Information Act requests, and my own interviews. Anything between quotation marks in this book is from a letter or other primary source document.

In these files I found a story of a true American adventure. A young woman with no money or connections is hired during the First World War by a millionaire to probe an odd theory about the works of Shakespeare. Through the millionaire's sleight of hand and the urgencies of war, this eccentric literary project turns into a life-or-death hunt for actual enemy secrets, one that spawns a completely new science of codebreaking. The woman goes on in the 1930s to become one of the world's most famous codebreakers, a front-page celebrity, before the government recruits her for one of the most closely guarded missions of the Second World War. And through it all she serves as muse and colleague to her husband, a troubled genius who lays the foundation of modern surveillance.

All democracies ride the line between security and transparency, secrecy and disclosure. What do people have a right to know? What must stay secret and why? The Friedmans lived these tensions more deeply than most. Their journey took them to great heights in the service of their country—and also to the depths of paranoia, poverty, and madness.

Jason Fagone
PHILADELPHIA

Terminology: A Cheat Sheet

You don't need to know math to enjoy this book, just a bit of lingo.

A *code* is a fixed relationship between one set of symbols or ideas and another. It can be a very ordinary and everyday thing. Slang is code, emoji are code. Think of Paul Revere hanging lamps in the Boston steeple to signal the route of the British invasion: one if by land, two if by sea. That's a code.

A *cipher* is a rule for altering the letters in a message. Usually it involves a one-to-one exchange: one letter gets replaced with one other letter, or a digit. For instance, if A=B, B=C, and so on, SMASH becomes TNBTI.

A *cryptogram* is a catchall term for a string of garbled text, solution unknown. It can be generated by a code or a cipher.

You can think of codes and ciphers as different sorts of locks that protect words, like a padlock protects money in a safe. In this analogy, the security professionals who make the locks and the keys are called *cryptographers,* and the thieves who try to pick the locks without having the keys are called either *codebreakers* or *cryptanalysts,* two terms that mean exactly the same thing.

The broad science of codes and ciphers—making them, breaking them, studying them, writing about them—is *cryptology.*

At different points in their careers, Elizabeth and William were asked to make codes, and they were good at this task, but their most significant feats involved codebreaking. They snuck into vaults of text, sometimes alone, sometimes together, feeling for the click of the bolt. Their lives became a series of increasingly spectacular and improbable heists. They used science to steal truth.

PART I

RIVERBANK

1916–1920

CHAPTER I

Fabyan

Sixty years after she got her first job in codebreaking, when Elizebeth was an old woman, the National Security Agency sent a female representative to her apartment in Washington, D.C. The NSA woman had a tape recorder and a list of questions. Elizebeth suddenly craved a cigarette.

It had been several days since she smoked.

"Do you want a cigarette, by the way?" Elizebeth asked her guest, then realized she was all out.

"No, do you smoke?"

Elizebeth was embarrassed. "No, no!" Then she admitted that she did smoke and just didn't want a cigarette badly enough to leave the apartment.

The woman offered to go get some.

Oh, don't worry, Elizebeth said, the liquor store was two blocks away, it wasn't worth the trouble.

They started. The date was November 11, 1976, nine days after the election of Jimmy Carter. The wheels of the tape recorder spun. The agency was documenting Elizebeth's responses for its classified history files. The interviewer, an NSA linguist named Virginia Valaki, wanted to know about certain events in the development of American codebreaking and intelligence, particularly in the early days, before the NSA and the CIA existed, and the FBI was a mere embryo—these mighty empires that grew to shocking size from nothing at all, like planets from grains of dust, and not so long ago.

Elizabeth had never given an interview to the NSA. She had always been wary of the agency, for reasons the agency knew well—reasons woven into her story and into theirs. But the interviewer was kind and respectful, and Elizabeth was eighty-four years old, and what did anything matter anymore? So she got to talking.

Her recall was impressive. Only one or two questions gave her trouble. Other things she remembered perfectly but couldn't explain because the events remained mysterious in her own mind. "Nobody would believe it unless you had been there," she said, and laughed.

The interviewer returned again and again to the topic of Riverbank Laboratories, a bizarre institution now abandoned, a place that helped create the modern NSA but which the NSA knew little about. Elizabeth and her future husband, William Friedman, had lived there when they were young, between 1916 and 1920, when they discovered a series of techniques and patterns that changed cryptology forever. Valaki wanted to know: What in the world happened at Riverbank? And how did two know-nothings in their early twenties turn into the best code-breakers the United States had ever seen—seemingly overnight? "I'd be grateful for any information you can give on Riverbank," Valaki said. "You see, I don't know enough to . . . even to ask the first questions."

Over the course of several hours, Valaki kept pushing Elizebeth to peel back the layers of various Riverbank discoveries, to describe how the solution to puzzle A became new method B that pointed to the dawn of C, but Elizebeth lingered instead on descriptions of people and places. History had smoothed out all the weird edges. She figured she was the last person alive who might remember the crags of things, the moments of uncertainty and luck, the wild accelerations. The analyst asked about one particular scientific leap six different times; the old woman gave six slightly different answers, some meandering, some brief, including one that is written in the NSA transcript as "Hah! ((Laughs.))"

Toward the end of the conversation, Elizebeth asked if she had thought to tell the story of how she ended up at Riverbank in the first place, working for the man who built it, a man named George Fabyan. It was a story she had told a few times over the years, a memory outlined in black. Valaki said no, Elizebeth hadn't already told this part. "Well, I better give you that," Elizebeth said. "It's not only very, very amusing, but it's actually true syllable by syllable."

"Alright."

"You want me to do that now?" Elizebeth said.

"Absolutely."

The first time she saw George Fabyan, in June 1916, he was climbing out of a chauffeured limousine in front of the Newberry Library in Chicago, a tall stout man being expelled from the vehicle like a clog from a pipe.

She had gone to the library alone to look at a rare volume of Shakespeare and to ask if the librarians knew of any jobs in the literature or research fields. Within minutes, to her confusion and mystification, a limousine was pulling up to the curb.

Elizebeth Smith was twenty-three years old, five foot three, and between 110 and 120 pounds, with short dark-brown curls and hazel eyes. Her clothes gave her away as a country girl on an adventure. She wore a crisp gray dress of ribbed fabric, its white cuffs and high pilgrim collar imparting a severe appearance to her small body as she stood in the lobby and watched Fabyan through the library's glass front doors.

He entered and stormed toward her, a huge man with blazing blue eyes. His clothes were more haggard than Elizebeth would have expected for a person of his apparent wealth. He wore an enormous and slightly tattered cutaway coat and striped trousers. His mustache and beard were iron gray, and his uncombed hair was the same shade. His breath shook the hairs of his beard.

Fabyan approached. The height differential between them

was more than a foot; he dwarfed her across every dimension. With an abrupt motion he stepped closer, frowning. She had the impression of a windmill or a pyramid being tipped down over her.

"Will you come to Riverbank and spend the night with me?" Fabyan said.

Elizebeth didn't understand any part of this sentence. She didn't know what he meant by spending the night or what Riverbank was. She struggled for a response, finally stammering a few words. "Oh, sir, I don't have anything with me to spend the night away from my room."

"That's all right," Fabyan said. "We'll furnish you anything you want. Anything you need, we have it. Come on!"

Then, to her surprise, Fabyan grabbed Elizebeth under one elbow, practically lifting her by the arm. Her body stiffened in response. He marched her out of the library and swept her into the waiting limousine.

People often guessed that she was meek because she was small. She hated this, the assumption that she was harmless, ordinary. She despised her own last name for the same reason; it seemed to give people an excuse to forget her.

"The odious name of Smith," she called it once, in a diary she began keeping at age twenty. "It seems that when I am introduced to a stranger by this most meaningless of phrases, plain 'Miss Smith,' that I shall be forever in that stranger's estimation, eliminated from any category even approaching anything interesting or at all uncommon." There was nothing to be done: changing her name would cause horrendous insult to blood relations, and complaining provided no satisfaction, because whenever she did, people asked why she didn't just change her name, a response so "inanely disgusting" that it made her feel violent. "I feel like snipping out the tongues of any and all who indulge in such common, senseless, and inane pleasantries."

Her family members had never shared this fear of being ordinary. They were midwestern people of modest means, Quakers from Huntington, Indiana, a rural town known for its rock quarries. Her father, John Marion Smith, traced his lineage to an English Quaker who sailed to America in 1682 on the same boat as William Penn. In Huntington he worked as a farmer and served in local government as a Republican. ("My Indiana family," Elizebeth later wrote, "were hide-bound Republicans who had never under any instances voted for any other ticket.") Her mother, Sopha Strock, a housewife, delivered ten children to John, the first when she was only seventeen. One died in infancy; nine survived. Elizebeth was the last of the nine, and by the time she was born, on August 26, 1892, most of her brothers and sisters had already grown up and scattered. She got along with only two or three, particularly a sister named Edna, two years her elder, a practical girl who later married a dentist and moved to Detroit.

Sopha had decided to spell "Elizebeth" in a nonstandard way, with *ze* instead of the usual *za*, perhaps sensing that her ninth child named Smith would want something to set her apart in the world. But Elizebeth didn't need the hitch in her first name to know she was different. Prone to recurring fits of nausea that began in adolescence and plagued her for years, she had trouble sitting still and keeping her tongue. She clashed with her father, a pragmatic, stubborn man who ordered his children around and believed women should marry young. She questioned her parents' faith. John and Sopha, though not devout, were part of a Quaker community and believed what Quakers do: that war is wrong, silence concentrates goodness, and direct contact with God is possible. Elizebeth's God was more diffuse: "We call a lot of things luck that are but the outcome of our own bad endeavor," she wrote in the diary, "but there is undoubtedly something outside ourselves that sometimes wins for us, or loses, irrespective of ourselves. What is it? Is it God?"

Her father didn't want Elizebeth to go to college. She defied him and sent applications to multiple schools, vowing to pay her own tuition; a friend later recalled that she was full of "determination and energy to get a college education with no help or encouragement from her father." (John Smith did end up loaning her some money—at 4 percent interest.) After being rejected from

Swarthmore College in Pennsylvania, a top Quaker school, she settled on Wooster College in Ohio, studying Greek and English literature there between 1911 and 1913. Then her mother fell ill with cancer and Elizebeth transferred to another small liberal arts school, Hillsdale College in Michigan, to be closer to home. At both schools she earned tuition money as a seamstress for hire. Her dorm rooms were always cluttered with dresses in progress and stray ribbons of chiffon.

College took Elizebeth's innate tendency to doubt and gave it a structure, a justification. At Wooster and Hillsdale she discovered poetry and philosophy, two methods of exploring the unknown, two scalpels for carving up fact and thought. She studied the works of Shakespeare and Alfred, Lord Tennyson, carrying books of their poems and plays around campus, annotating and underlining the pages until the leaves separated from the bindings. A course on philosophy introduced her to a new hero, the Renaissance scholar Erasmus, who "believed in one aristocracy— the aristocracy of intellect," she wrote in a paper. "He had one faith—faith in the power of thought, in the supremacy of ideas." Elizebeth, a smart person from a working-class family, found this concept liberating: the measure of a person was her ideas, not her wealth or her command of religious texts. She wrote a poem about this epiphany:

> I sit stunned, nerveless, amid the ruins
> Of my fallen idols. The iconoclast Philosophy
> Has shattered for me
> My God . . .
> But through the confusing ruins, Faith, still hoping,
> Somehow raises her hands and bids me—
> Yearn on! Finally
> Through the mazes of error and doubt and mistrust
> You will come, weary heart
> To the final conclusion upon which you will build anew.
> You will find triumphant
> The Working Hypothesis,
> The Solid Rock.

In addition to the well-worn volumes of Shakespeare and Tennyson, she lugged her own diary from place to place, a book with a soft black binding that said "Record" on the cover in silver script. The round-cornered pages were lined. She wrote in wet black ink with a quill pen, in a slanted cursive hand that was not too beautiful, about the importance of choosing the right words for things, even if those words offended people. She didn't like it when she heard a friend say that a person who had died had "passed away" or that a staggering drunk at a party was "a bit indisposed." It was more important to be honest. "We glide over the offensiveness of names and calm down our consciences by eulogistic mellifluous terms, until our very moral senses are dulled," she wrote. "Let things be shown, let them come forth in their real colors, and humanity will not be so prone to a sin which is glossed over by a dainty public!"

Sometimes Elizebeth had trouble channeling these energies and frustrations into cogent work. Her professors found her intensely bright, yet unfocused and argumentative. More than one told her, she said, that "I have marvelous abilities, yet do not use them." A philosophy professor wrote on the back of her Erasmus paper, "Very suggestive, with lots of good ideas and phrases. Also novel. But the style is choppy and the ideas are not in proper sequence." Next to these words Elizebeth scribbled a defiant note, dismissing the criticism on the grounds that she had recently won second place in a state oratorical contest.

She found herself attracted to male artists. Attending a choral concert one night, "my musical heart was carried completely away by a baritone," she wrote. "He loved the very act of singing—it could be seen in his eyes, in his mouth, in his very hands, as they irrepressibly moved in half-gestures. It made me want to be able to sing well myself, so badly that—well, I just couldn't sit still with the desire of it." At Hillsdale she dated a poet named Harold Van Kirk, called Van by his friends, handsome and athletic. He typed French sonnets for her and later joined the army and moved to New York. Van's roommate, Carleton Brooks Miller, wooed her when the relationship with Van fell apart, urging Elizebeth to read James Branch Cabell's erotic science fiction novel *Jurgen* because "it reveals the naked man-soul as it is." Carleton joined the army, too, then became the minister of a Congregational church near the college, writing to Elizebeth a few years later that he was still looking for a mate.

When graduation came around in the spring of 1915, Elizebeth still felt like "a quivering, keenly alive, restless, mental question mark." She had no sense of where to go or what she wanted to do with her life. That fall she accepted a position as the substitute principal of a county high school twenty miles west of her childhood home. The landscape of small-town Indiana was depressingly familiar to Elizebeth, and while she enjoyed parts of her job—she taught classes in addition to running the school—she also felt trapped. For an educated American woman in 1915, teaching at the high school level or below was what you did. Almost 90 percent of professors at public universities were male; only 939 women in the country received master's degrees in 1915, and 62 women earned Ph.D.s. Elizebeth had arrived at the last stop on a dreary train. There was no path from teaching that led anywhere else she might want to go. A woman taught, had kids, retired, died.

All her life, Elizebeth assumed that her restlessness was a defect that adulthood would somehow remove. She had called it "this little, elusive, buried splinter" and hoped for it to be "pricked from my mind." But she was learning to see the splinter as a permanent piece of her, impossible to remove. "I am never quite so gleeful as when I am doing something labeled as an 'ought not.' Why is it? Am I abnormal? Why should something with a risk in it give me an exuberant feeling inside me? I don't know what it is unless it is that characteristic which makes so many people remark that I should have been born a man."

Wanting something more, and ready to take a risk, Elizebeth quit her job at the Indiana high school in the spring of 1916 and moved back in with her parents to think about what was next. She soon remembered how unpleasant it was to live with her father. She reached her limit and packed a suitcase in early June. Nervous, but forcing herself to be brave, she boarded a train for

Chicago, hoping to find a new job there, or at least a new direction.

That month, the war in Europe—the First World War, then called the "Great War"—was two years old. America had not yet joined the battle. Woodrow Wilson was finishing his first term as president and campaigning for reelection in November on a platform of peace. More than a thousand Republican delegates had just kicked off their national convention to nominate a challenger to Wilson. They were gathered in the same city that now lured Elizebeth: Chicago, the young capital of the Midwest, an upstart empire of stockyards and skyscrapers.

The scale of the city jangled her. Pedestrians brushed past each other on sidewalks that cut mazes through the downtown office buildings, banks, apartments, hotels. It rained most every day, a cold, miserable rain, sheets of fat, icy drops that saturated the wool coats of the political delegates and swamped the grass at the baseball parks, canceling Cubs and White Sox games. Elizebeth stayed in the apartment of a friend on the South Side and ventured out each morning in search of work, visiting job agencies and presenting her qualifications. She told the receptionists she would like to work in literature or research. She pictured herself at a desk in a room of desks, taking notes with a sharp pencil. Not clerical work but something that required the brain. The people at the job agencies said they were sorry, but they didn't have anything like that.

She had no other cards to play. No money or connections, no means of bending Chicago to her will. She felt small and anonymous. After a week she decided to return home.

Before boarding the train, though, she wanted to make one more stop in the city, at a place she had heard about, the Newberry Library, which owned a rare copy of the First Folio of William Shakespeare, a book whose backstory had intrigued her when she learned it in college. The Bard's plays were never collected and printed in one place during his lifetime, because the culture in which he worked, Queen Elizabeth I's England, revered the spoken word above the written. It wasn't until 1623, seven years after his death, when a group of admirers gathered thirty-six of Shakespeare's comedies and tragedies in a single hefty volume that came to be known as the First Folio. Simply publishing the book was a radical act, a statement that the phrases of a playwright deserved to be documented with the same care as the Gospels. A team of London artisans produced about a thousand copies, each typeset and bound by hand. Five men memorized portions of the plays to help them set type faster, stacking metal letters one by one into words and sentences.

Over the centuries, most of the copies were lost or destroyed. The Newberry had one of the few on display in America. So, on what she thought would be her final day in Chicago, Elizebeth made her way to the library.

The library was an odd institution, created by a dead man's will and a quirk of fate. A rich merchant named Walter Newberry died on a steamship in 1868. The crew preserved his body for the remainder of the voyage in an empty rum cask before returning it to his beloved city, where lawyers discovered that Newberry had left behind almost $2,150,000 for the construction of a public library.

According to his will, the library had to be free to use, and it had to be located in North Chicago. These were the only conditions. The library didn't even have books to start with, because three years after Newberry's death, his own hoard of rare volumes was destroyed in the Great Chicago Fire of 1871.

The slate of the library was blank. Now the library's trustees wrote their status anxieties upon it. These were wealthy Chicago businessmen who felt they lived in one of the finest cities in the world and were painfully aware that the world did not agree. For all of Chicago's sudden material success, its skyscrapers and factories and department store empires and slaughterhouses, it lacked the institutions of art and music and science that elevated New York and Boston and Paris in traditional measures of civic greatness but omitted Chicago and made its large men feel small.

They wished to prove that they were men of culture and refinement, and they were willing to spend whatever it took.

This was the same insecurity that drove the fathers of Chicago to raise the dreamlike White City, the temporary pavilions of the Chicago World's Fair that soared along the southern edge of the lake in the summer of 1893. The White City exhibited the future in prototype, pieces of an unfinished puzzle. On August 26, 1893, a day of demonstrations at the Palace of Mechanical Arts, a building twice as large as the U.S. Capitol, the palace rumbled and whirred with machines that turned raw sugar into candy, made sausages and horseshoe nails and bricks, and sewed ten thousand button holes per hour. All day long one hundred thousand people wandered the sprawling aisles, eardrums split by machine roar, drinking lemonade that spurted from a fountain. An entire newspaper was printed in exactly sixty-three minutes starting with raw planks of wood that were pulped as people watched. "Everywhere was a demonstration of the almost irresistible power of mind when matter is set to do its bidding," the *Tribune* reported. The world's tallest man, Colonel H. C. Thurston of Texas, eight foot one and a half in boots, mingled with the throngs, and in the afternoon fifty thousand gathered outdoors to watch a fat man dive for a bologna sausage that dangled from a pole above the lake's lagoon.

This was the day Elizebeth turned one year old in Indiana. And as crowds of Americans roamed the White City in awe, builders completed construction of the Newberry Library, ten miles north of the noisy fairgrounds, and the first patrons entered the library in reverent silence.

Unlike the White City, a spectacle for the masses, the library was designed as "a select affair" for "the better and cleaner classes," the *Chicago Times* wrote with approval when the Newberry first opened. It was an imposing five-story building of tan granite blocks. All visitors had to fill out a slip stating the purpose of their research and they were turned away if they could not specify a topic. The books, available for reference only, were shelved in reading rooms modeled after the home libraries of wealthy gentlemen, cozy and intimate spaces containing the rarest and most sophisticated books that vulgar Chicago money could buy. During the library's first decades, the masters of the Newberry acquired books with the single-mindedness of hog merchants. They bought hundreds of incunabula, printed volumes from before 1501, written by monks. They bought fragile, faded books written by hand on unusual materials, on leather and wood and parchment and vellum. They bought mysterious books of disputed patrimony, books whose past lives they did not know and could not explain. One book on the Newberry's shelves featured Arabic script and a supple, leathery binding. Inside were two inscriptions. The first said that the book had been found "in the palace of the king of Delhi, September 21st, 1857," seven days after a mutiny. The second inscription said, "Bound in human skin."

In one especially significant transaction, the library acquired six thousand books from a Cincinnati hardware merchandiser, a haul that included a Fourth Folio of Shakespeare from 1685, a Second Folio from 1632, and most exceptional of all, the First Folio of 1623, the original printing of Shakespeare's plays.

This is the book that Elizebeth Smith was determined to see in June 1916, when she was twenty-three.

Opening the glass front door of the Newberry, she walked through a small vestibule into a magnificent Romanesque lobby. A librarian at a desk stopped her and sized her up. Normally Elizebeth would have been required to fill out the form with her research topic, but she had gotten lucky. The year 1916 happened to be the three hundredth anniversary of Shakespeare's death, and libraries around the country, including the Newberry, were mounting exhibitions in celebration.

Elizebeth said she was here to see the First Folio. The librarian said it was part of the exhibition and pointed to a room on the first floor, to the left. Elizebeth approached. The Folio was on display under glass.

The book was large and dense, about 13 inches tall and 8 inches wide, and almost dictionary-thick, running to nine hundred pages. The binding was red and made of highly polished goatskin, with a large grain. The pages had gilded edges. It was opened to a pair of pages in the front, the light gray paper tinged with yellow due to age. She saw an engraving of a man in an Elizabethan-era collar and jacket, his head mostly bald except for two neatly combed hanks of hair that ended at his ears. The text said:

MR. WILLIAM SHAKESPEARES
COMEDIES,
HISTORIES, &
TRAGEDIES.
Publifhed according to the True Originall Copies.
LONDON
Printed by Ifaac Iaggard, and Ed. Blount. 1623.

Elizebeth later wrote that seeing the Folio gave her the same feeling "that an archaeologist has, when he suddenly realizes that he has discovered a tomb of a great pharaoh."

One of the librarians, a young woman, must have noticed the expression of entrancement on her face, because now she walked over to Elizebeth and asked if she was interested in Shakespeare. They got to talking and realized they had a lot in common. The librarian had grown up in Richmond, Indiana, not far from Elizebeth's hometown, and they were both from Quaker families.

Elizebeth felt comfortable enough to mention that she was looking for a job in literature or research. "I would like something unusual," she said.

The librarian thought for a second. Yes, that reminded her of Mr. Fabyan. She pronounced the name with a long *a*, like "Faybe-yin."

Elizebeth had never heard the name, so the librarian explained. George Fabyan was a wealthy Chicago businessman who often visited the library to examine the First Folio. He said he believed the book contained secret messages written in cipher, and he had made it known that he wished to hire an assistant, preferably a "young, personable, attractive college graduate who knew English literature," to further this research. Would Elizebeth be interested in a position like that?

Elizebeth was too startled to know what to say.

"Shall I call him up?" the librarian asked.

"Well, yes, I wish you would, please," Elizebeth said.

The librarian went off for a few moments, then signaled to Elizebeth. Mr. Fabyan would be right over, she said.

Elizebeth thought: What?

Yes, Mr. Fabyan happened to be in Chicago today. He would be here any minute.

Sure enough, Fabyan soon arrived in his limousine. He burst into the library, asked Elizabeth the question that so bewildered and stunned her—"Will you come to Riverbank and spend the night with me?"—and led her by the arm to the waiting vehicle.

"This is Bert," he growled, nodding at his chauffeur, Bert Williams. Fabyan climbed in with Elizabeth in the back.

From the Newberry, the chauffeur drove them south and west for twenty blocks until they arrived at the soaring Roman columns of the Chicago & North Western Terminal, one of the busiest of the city's five railway stations. Fabyan hurried her out of the limo, up the steps, between the columns, and into the nine-hundred-foot-long train shed, a vast, darkened shaft of platforms and train cars and people rushing every which way. She asked Fabyan if she could send a message to her family at the telegraph office in the station, letting them know her whereabouts. Fabyan said no, that wasn't necessary, and there wasn't any time.

She followed him toward a Union Pacific car. Fabyan and Elizebeth climbed aboard at the back end. Fabyan walked her all the way to the front of the car and told her to sit in the frontmost seat, by the window. Then he went galumphing back through the car saying hello to the other passengers, seeming to recognize several, gossiping with them about this and that, and joking with the conductor in a matey voice while Elizebeth waited in her window seat and the train did not move. It sat there, and sat there, and sat there, and a bubble of panic suddenly popped in her stomach, the hot acid rising to her throat.

"Where am I?" she thought to herself. "*Who* am I? Where am I going? I may be on the other side of the world tonight." She wondered if she should get up, right that second, while Fabyan had his back turned, and run.

But she remained still until Fabyan had finished talking to the other passengers and came tramping back to the front of the car. He packed his big body into the seat opposite hers. She smiled at him, trying to be proper and polite, like she had been taught, and not wanting to offend a millionaire; she had grown up in modest enough circumstances to be wary of the rich and their power.

Then Fabyan did something she would remember all her life. He rocked forward, jabbed his reddened face to within inches of hers, fixed his blue eyes on her hazel ones, and thundered, loud enough for everyone in the car to hear, "Well, WHAT IN HELL DO YOU KNOW?"

Elizebeth leaned away from Fabyan and his question. It inflamed something stubborn in her. She turned her head away in a gesture of disrespect, resting her cheek against the window to create some distance. The pilgrim collar of her dress touched the cold glass. From that position she shot Fabyan a sphinxy, sidelong gaze.

"That remains, sir, for you to find out," she said.

It occurred to her afterward that this was the most immoral remark she had ever made in her life. Fabyan loved it. He leaned way back, making the seat squeak with his weight, and unloosed a great roaring laugh that slammed through the train car and caromed off the thin steel walls.

Then his facial muscles slackened into an expression clearly meant to convey deep thought, and as the train lurched forward, finally leaving the station, he began to talk of Shakespeare, the reason he had sought her out.

Hamlet, he said. *Julius Caesar, Romeo and Juliet, The Tempest,* the sonnets—the most famous written works in the world. Countless millions had read them, quoted them, memorized them, performed them, used pieces of them in everyday speech without even knowing. Yet all those readers had missed something. A hidden order, a secret of indescribable magnitude.

Out the train window, the grid of Chicago gave way to the silos and pale yellow vistas of the prairie. Each second she was getting pulled more deeply into the scheme of this stranger, destination unknown.

The First Folio, he continued. The Shakespeare book at the Newberry Library. It wasn't what it seemed. The words on the page, which appeared to be describing the wounds and treacheries of lovers and kings, in fact told a completely different story, a secret story, using an ingenious system of secret writing. The messages revealed that the author of the plays was not William Shakespeare. The true author, and the man who had concealed the messages, was in fact Francis Bacon, the pioneering scientist and philosopher-king of Elizabethan England.

Elizebeth looked at the rich man. She could tell he believed what he was saying.

Fabyan went on. He said that a brilliant female scholar who worked for him, Mrs. Elizabeth Wells Gallup, had already succeeded in unweaving the plays and isolating Bacon's hidden threads. But for reasons that would become clear, Mrs. Gallup needed an assistant with youthful energy and sharp eyes. This is why Fabyan wanted Elizebeth to join him and Mrs. Gallup at Riverbank—his private home, his 350-acre estate, but also so much more.

Genius scientists lived there, on his payroll, working in laboratories unlike any on earth. Celebrities made pilgrimages to get a glimpse of projects under way. Teddy Roosevelt, his personal friend. P. T. Barnum. Famous actresses. Riverbank was a place of wonders. She would see.

After they'd been riding west for ninety minutes or so, traveling thirty-five miles across the plain, the train began to slow, hissing as it came to a full stop. Fabyan opened the door and he and Elizebeth walked down the length of a platform and emerged into a handsome waiting room of dark enameled brick with terra-cotta flourishes. They continued out the front door, into the main street of Geneva, Illinois, a village of two thousand. Originally settled by a Pennsylvania whiskey distiller, Geneva had swelled with foreign immigrants in recent years, Irish and Italians and Swedes leaving crowded Chicago for the open spaces of the prairie. Whiskey still accounted for a good portion of Geneva's commerce, grain from the fields mixing with the sweet water of the Fox River, which bisected the town north to south.

To Elizebeth's amazement, a limousine was waiting for her at Geneva Station—not the one she'd ridden in an hour ago in Chicago but a second limousine with a second chauffeur. She climbed in with Fabyan and was carried south along a local road known as the Lincoln Highway for a bit more than a mile, until a long, high stone wall appeared to the left. Then a gate.

The limousine slowed. It pulled off the highway, to the right, across from the wall and the gate, and came to a stop in front of a two-story farmhouse with a wide front porch.

The Lodge, Fabyan announced. Elizebeth would be staying here tonight.

CHAPTER 2

Unbelievable, Yet It Was There

Elizebeth Smith and George Fabyan
at Riverbank, summer 1916.

A naked woman was living in a cottage at Riverbank. This was the story going around town in Geneva. The woman was said to be young, in her late teens or early twenties. Above the entrance of the cottage hung a sign that read *Fabyan*.

The story mutated as it passed from teller to teller. The cottage at Riverbank was stocked with attractive women, kept by Fabyan to satisfy his lust. They had been seen disrobing. Five women, ten.

Rumors about Fabyan and his strange laboratory were always spreading through the small farm towns surrounding the estate. The grounds were private and only open to the public at particular times. A stone wall protected part of the 350 acres, patrolled by Fabyan's guards, and at night the lighthouse on the island in the river broadcast a continual warning to intruders in code: two white flashes followed by three red ones, signifying "23-skidoo," meaning "keep out." Sometimes, on Sundays, he opened Riverbank to local residents as a gesture of goodwill, a benevolent king allowing his people to roam the castle grounds. The electric trolley operated by the Aurora-Elgin and Fox River Electric Company, which usually ran past the estate without stopping, was permitted to stop by the river. People tumbled out and wandered in awe through an elaborate Japanese Garden. And then Monday came, and the trolley did not stop at Riverbank anymore, and people once again had to guess from afar what might be happening there.

They heard loud noises from the direction of the estate, things that sounded like bombs exploding. They saw what looked like warplanes buzzing around the buildings and making an incredible racket. The press often called him "Colonel Fabyan" or simply "The Colonel." It seemed obvious that Fabyan was performing military research, but the townspeople did not know exactly what. They gleaned clues from newspapers and magazines. Fabyan was always inviting journalists and professors to tour the laboratories, under controlled conditions, and their reports spoke of Riverbank as a wonderland, a place almost beyond earthly reckoning. Visitors called Riverbank, variously:

A Garden of Eden on Fox River
Fabyan's colony
a wonder-working laboratory near Chicago
one of the strangest and, at the same time, most beautiful
    country estates in America

As for George Fabyan himself, visitors described him as:

one of the greatest cipher experts of the world
one who has achieved triple success in three distinct fields of
    activity, those of business, letters, and science
the man of a thousand interests
the lord and master
Chicago inventor
multi-millionaire country gentleman
the seer of Riverbank
the caliph on a grand scale

Guests of Riverbank went away telling two main types of stories. On the one hand, the visitors spread bizarre rumors and anecdotes of Fabyan's personal behaviors, portraying him as a mad king: "Credible persons," one newspaper reported, "say that a pair of sprightly, highly groomed zebras dash down with a station wagon to the Geneva station . . . to meet him mornings and evenings." These tales of bacchanalia were mingled with incredible stories of scientific experimentation at the laboratories, hints of anatomical investigations, and tales of secret cipher messages divined from old books.

Before he built the laboratories, Fabyan had often appeared in the Chicago newspapers in connection with more conventional tycoon activities: donations to political figures, board meetings of the stock exchange. People thought they knew his story. The black sheep of a prosperous New England family, he had dropped out of boarding school at age sixteen after repeated clashes with his father. He ran away from home and wandered the West for several years in the 1880s, making a living by selling lumber and railroad ties. Later, moving to Chicago, he reconciled with his father, and when the old man died, George inherited his $3 million fortune—equal to almost $100 million today—along with the reins of the family business, Bliss Fabyan & Company, one of the largest fabric companies in America. George used his gift for salesmanship to grow the company. After a Bliss Fabyan textile mill in Maine started making a type of striped seersucker cloth, he christened it

"Ripplette," a wonder fabric that required no ironing and resisted stain, undyed white bedspreads staying white after repeated washings, "white and clear as the driven snow . . . the name 'Ripplette' on a bedspread is the only sure indication of Ripplette quality. . . ."

Fabyan never claimed to be an altruist. "I ain't no angel," he said once, "and there are no angels in the New England cotton textile business, and if there are, they will all be broke." But in one part of his life he did strive toward some kind of greater good, and he wanted people to know it. In his free time, for his own amusement, he had made himself into a man of science. The steel magnates of Pittsburgh collected paintings, old and contemporary masterpieces. Newspaper tycoon William Randolph Hearst would soon build a 165-room castle in California full of marble statues. Fabyan was thinking bigger. "Some rich men go in for art collections, gay times on the Riviera, or extravagant living, but they all get satiated," he said. "That's why I stick to scientific experiments, spending money to discover valuable things that universities can't afford. You can never get sick of too much knowledge."

The atom had not been split in 1916. The structure of DNA was undescribed. There were no antibiotics. Aspirin, vitamins, blood types, and the medical uses of X-rays had all been discovered in the last twenty-one years. Einstein's theory of general relativity was only a year old. According to Einstein, space and time were one and the same, related by the universal force of gravity, and people did not know what to make of it. They came to Riverbank knowing that major scientific discoveries had emerged from the private laboratories of Thomas Edison and Nikola Tesla, so they were primed to believe that a new age of wonders was just over the next hill. And Fabyan gave them a peek. He paraded them from lab to lab, marvel to marvel, a former teen runaway and dropout showing off his Eden of science.

There seemed to be experiments happening everywhere, even inside his own house, known as the Villa. A man from the *Chicago Herald* was shocked to notice a swarm of bees flying through the Villa's open window. Fabyan laughed. "Those bees are just going into the music room to deposit their honey," he said. "You see I didn't trust that particular bunch of bees, so I had their hive placed inside the [Villa] and had it glassed in so we could watch them and see that they didn't cheat. . . . It's made honest bees out of them—this constant supervision."

A correspondent from the *Chicago Daily News* visited on a clear spring morning. Fabyan asked him, "Do you ever think? No, I don't think you do. Ninety-nine percent of the people don't, so why should you? I can make you think. We're all thinkers out here. Yes-siree, every one of the 150 souls in the Riverbank community." Fabyan, wearing a bowler hat, a lavender scarf, a tailored vest, and a frock coat with his French Légion d'Honneur rosette pinned to the lapel, added that he himself was "just a worker" like all the others, that there were no bosses at Riverbank, no time clocks, no iron regulations. Then he removed a gold-tipped cigarette from a cigarette case, snapped it in two, and lowered the halves toward a nearby monkey enclosure. The monkeys took the cigarette halves from Fabyan's outstretched hand, peeled off the paper, and jammed the tobacco into their mouths.

"Yes," Fabyan continued, "a community of thinkers." He took the reporter to the farm and explained how his scientists were taking cows and pigs and sheep and freezing them with rocks of ice and then slicing them thin as salami, to study their anatomy; he showed off the statues of the duck and the Egyptian thrones next to the Villa and pointed out with glee that they weren't made of marble or stone but of concrete, which lasts longer than stone and can be carved like stone; he pointed to the Dutch windmill and bragged that it was fully functional, that he used the mill to grind flour and bake fresh loaves of bread for the workers. He invited the reporter into the Acoustics Laboratory, built around an ultraquiet test chamber where the buzz of a stray mosquito seemed as loud as an air siren, and a pencil writing on paper sounded like a dozen people coughing. Fabyan said that experiments here would someday make cities more livable by eliminating the "racket ogre" of machines and crowds.

"Look through this telescope thing," boomed Colonel Fabyan proudly. He struck a tuning fork. The visitor squinted, and saw a flickering light, like a gas flame in a wind. "That's the sound made by this tuning fork! Sure, you're seeing it!"

And all through the tour, Fabyan kept circling back to the primary mission of the laboratories, the glimmering idea at the bottom of it all: immortality. Extending human life. Each person could live to be one hundred or more, he said. The thinkers of Riverbank had sequestered themselves in this lush, remote location to learn how not to die.

"Over there in that hothouse, they're trying genetics on nasturtiums, orchids, roses, and tulips," Fabyan told the Chicago correspondent, jabbing a finger in the direction of Riverbank's greenhouse. "What for? Why, look at the average human being. A mighty pitiful contraption of flesh and bones. If we of the Riverbank community can improve the human race by experimenting first with flowers and plants—say, won't that be a wonderful thing?"

Some experiments veered into ethically dubious territory. A journalist visiting from Philadelphia stopped and asked for directions from "a pretty girl, clad in blue overalls," with a "slim young figure—one of Colonel Fabyan's colony crowned with a head of bobbed blonde hair." Fabyan told the reporter that he had enrolled a number of young women in a series of studies to correct their defective posture. He recruited these human subjects from a boarding school adjacent to the Riverbank property, the Illinois State Training School for Delinquent and Dependent Girls at Geneva, really a low-security juvenile prison in the countryside, a place where judges across Illinois sent "wayward girls" deemed mentally deficient or sexually promiscuous. The founder of the Training School ordered the girls beaten with rawhide whips and thought society should force them to be sterilized: "When they begin to grow and attain some size the blood that runs in their veins will begin to tell and the incorrigible girl is the result." The school housed a rotating population of five hundred women ages ten to eighteen, and some lived in a cottage built with a donation from George and Nelle Fabyan. This was the cottage that townspeople gossiped about. The donation explained why the Fabyan sign hung above the door, and the posture experiments explained the rumor of nudity; the girls were required to undress for physical examinations. "The results of our experiments on the girls at Geneva have been marvelous," George Fabyan boasted. "Their so-called 'debutante slouch' has disappeared. They are learning to stand erect and not like anthropoid apes just learning to walk. I am trying to improve the human race, to discover what's wrong with the female figure. What will the next generation be like if all the women have hollow chests?"

The Philadelphia reporter also revealed that "in his effort to impress on the young women the terrors of crooked spines," Fabyan maintained a laboratory at Riverbank that he called "The Chamber of Horrors," containing actual human skeletons with grotesquely deformed spines, procured through methods that Fabyan never explained. Multiple Riverbank employees later told an Illinois historian that Fabyan "collected from hospitals and cemeteries numerous unclaimed cadavers that his scientists would radiate, cut, probe, and dissect, and then bury the remains in secret graves around the estate." At night in the laboratories "the beams would creak and the chairs would seem to move," and several staffers "recalled looking out the windows into the dark yard and seeing running girls with flowing white trains." Opinions about these visions differed: some staffers thought they were seeing "wayward girls" from the Training School escaping momentarily, and others believed in ghosts.

One summer a science journalist visited. Austin Lescarboura was a professional debunker, a man who had once partnered with Houdini to prove that fortune-tellers were liars and frauds. George Fabyan led Lescarboura into a darkened room in one of the laboratories. "The staff in charge moved about like so many Egyptian priests of old guarding the darkest secrets," Lescarboura later reported in *Scientific American*.

To deepen the mystery still further, a pretty girl was brought in. We were ushered into a small booth with dull black curtains for walls. It reminded us strongly of our psychic experiments back in New York, when we exposed one of the leading mediums after three sittings. At the command of the Colonel, the demonstration got under way. In a few minutes, we were astounded by what we were witnessing. It seemed unbelievable, yet it was there, in plain black and white. We had been brought face to face with certain facts regarding the human mechanism which we would hardly dared to have surmised in the absence of such a convincing demonstration. We were shown how—well, at this point we can go no further. Colonel Fabyan made us promise that nothing would be said about the nature of this investigation until some later date, when the experiments have progressed further.

What he was seeing was a woman standing behind an X-ray screen, the structure of her bones illuminated by the penetrating energies of $750,000 worth of radium. X-rays had been discovered in 1895, so they were hardly new technology by the time Lescarboura arrived at Riverbank, but the aura of mystique at Riverbank was so thick, the range of scientific experiments so wide, that even a trained skeptic like Lescarboura could not necessarily distinguish between the real and the fantastic. "Every so often the world reaches a point bordering on stagnation, because everything seems to be fully developed," he wrote. "But the scientist, pegging away at the secrets of nature, sooner or later breaks down existing barriers, opens the way to a new field, and we are soon confronted with brand new opportunities for exploration."

Twenty-three-year-old Elizebeth Smith climbed the stairs of the Lodge to the porch, opened the front door, and found herself in a warm, spacious drawing room. The walls were lined with double-paned casement windows that looked out across a grassy field on one side and back toward the road. There were people milling about.

In a brusque, hurried way, Fabyan introduced Elizebeth to a pair of magnificently dressed women, then disappeared, leaving her with these strangers.

The aristocratic appearance of the women was so incongruous that Elizebeth looked them over a few times to be sure they were real. They were sisters. The first, an older woman, wore a dark dress and a necklace that glittered with jewels. Her gray hair was tied in a bun and escaped in wisps that framed a delicate face. It seemed as if a French duchess had been teleported to the prairie, and her voice dripped with learning. Her name, she said, was Mrs. Elizabeth Wells Gallup. She ran the Riverbank Cipher School. The other woman was her younger, darker-haired sister, Miss Kate Wells.

Young Elizebeth gathered from this brief conversation that Mrs. Gallup and Miss Wells lived and worked here at Riverbank in this very building, which also contained quarters for the cooks and servants who fed and catered to the sisters and to the other scholars and scientists who worked on the estate.

The sisters informed Elizebeth that dinner would soon be served here at the Lodge, and that she would be dining with the two of them and some of the scientists as well. Mrs. Gallup and Miss Wells suggested that she head upstairs and freshen up in a certain spare bedroom where she would be sleeping. Elizebeth did as asked, and some minutes later, when she descended the stairs, she saw that Fabyan had returned, in striking new clothes: riding pants, a billowing shirt with a riding collar, and a big, broad cowboy hat. He looked ready to jump on a horse and gallop away. It didn't make sense to her at the time, given that he was about to eat dinner; later she would realize that Fabyan simply enjoyed dressing up as the ideal of a country squire. She would never once see him wearing a traditional business suit at Riverbank.

People began streaming into the Lodge in ones and twos, walking up the steps to the porch in the fading prairie light. Elizebeth sat on the bannister of the staircase, looking out across the Lincoln Highway, listening to crickets chirp and cicadas sing, watching the guests arrive. They all wore semi-formal clothes with a country feel, except for a slim man in a pinstriped shirt and pants, a neat bow tie, and sparkling white buck shoes. He had short, dark hair parted in the exact middle of his head and pomaded to each side, and his ears were pointy; he seemed like the youngest of the arriving guests, and by far the best dressed, as if he were attending a society dinner at some mansion in the city. The young man reminded Elizebeth of Beau Brummell, the eighteenth- and nineteenth-century British fashion icon who polished his boots with champagne and peach marmalade.

They sat down to dinner at a long communal table covered with fine china and linens. Swedish and Danish servants appeared in crisp white uniforms and brought heaping plates of meat and vegetables from Riverbank's working farm. To keep food costs low and to satisfy his own taste for meat, Fabyan kept the farm stocked with chickens, ducks, sheep, and turkeys, and his wife, Nelle, bred prize-winning livestock here. Mrs. Gallup sat at the head of the table, flanked by the other guests, all of them lured here by Fabyan to investigate different pieces of the world. Elizebeth spoke little and tried to get a sense of who these people were and what they were doing here. A sweet-seeming man in his fifties introduced himself as J. A. Powell, president of the University of Chicago Press and the top public relations man at that university; his job there was to "cause the University of Chicago to be known as well in Peking as in Peoria," the *Tribune* once put it. Another dinner guest was Bert Eisenhour, Riverbank's chief engineer and builder of structures, a short man with a ruddy complexion who struck Elizebeth as a country bumpkin.

Then there was the well-dressed man with the white buck shoes. He smiled shyly at Elizebeth and introduced himself as William Friedman, head of the Genetics Department at Riverbank. He worked here studying seeds and plants, breeding new strains of corn, wheat, and other crops, trying to infuse them with desirable properties.

Altogether it was a curious bunch of characters. Elizebeth couldn't see an obvious thread that connected them. Literary scholars, an engineer, a geneticist. Perhaps Fabyan was the kind of rich person who collected people in addition to banknotes and stocks.

The dominant personality that night was Mrs. Gallup. As the smell of meat and the noise of clinking silverware filled the room, and the servants whisked away the empty plates, Mrs. Gallup told stories of her travels while researching Francis Bacon and Shakespeare, staying in the homes of wealthy patrons around the world, in France and in England, who believed in her theories and had sponsored her work. When she spoke about the details of her investigations and findings, no one interrupted her to ask skeptical questions. It was obvious to Elizebeth that people here were used to treating Mrs. Gallup with great deference, that she was an important person at Riverbank, and that dinner conversations like this had probably happened many times before, with Mrs. Gallup holding court and the others nodding and smiling. Elizabeth got the sense that "Mrs. Gallup had dwelt only among those who agreed with her premise and that she had little personal contact with the viewpoint of those who did not believe."

After dinner, the guests went their separate ways. Fabyan gave Elizabeth a set of men's pajamas to wear to bed, telling her they would have more to discuss in the morning, and wishing her good night. She went upstairs to her room and found that a pitcher of ice water had been left on her bedside table along with an enormous bowl of fresh fruit, plus knives to carve it up.

On Elizabeth's second day at Riverbank, after she woke in the Lodge and got dressed, Fabyan found her and said she ought to see the rest of the estate. He assigned an employee to give her a short tour.

Fifty or sixty yards along the highway from the Lodge was a smattering of buildings known collectively as Riverbank Laboratories, where many of the scientists worked. Elizabeth was told that a new laboratory was under construction for the study of sound waves, designed by the top acoustics expert in the country, Professor Wallace Sabine of Harvard University, who would move to Riverbank when the new lab was complete. Adjacent to Riverbank Laboratories was the ordnance building, a low concrete hut where Fabyan and several scientists tested bombs and mortars for potential use by the U.S. military.

Elizabeth wasn't shown inside these buildings but instead was led across the highway to an iron gate she had seen yesterday while riding in the limo. She walked through it. A short, curving driveway led down a gentle slope to Fabyan's personal residence, known as the Villa, a long, low two-story house in a cruciform shape with a heavy roof and thin clapboard siding that seemed to press the house downward into the hill. Originally a far smaller farmhouse, it had been expanded in 1907 by the famous architect Frank Lloyd Wright, who produced a mansion for Fabyan that looked like a peaceful part of the countryside. Strange objects dotted the lawn: a concrete wading pool, a concrete table and concrete semicircular bench carved with elaborate Egyptian hieroglyphs, a concrete chair whose front legs were sphinxes, and a concrete duck the size of a human eight-year-old.

Inside the Villa the walls were paneled with squares of dark walnut, and the sun shone through a series of thin slats that Wright's builders had carved in the hill-facing wall and that decorated the opposite wall with rhombuses of light. Elizebeth was amazed to see that the chairs and divans in the living room and drawing room and even the beds upstairs were suspended from the ceiling on chains—no chair or bed legs anywhere to be seen. She had no idea what to make of this. Fabyan and Nelle each had their own private bedchamber. It was unclear whether they slept in the same room. On a wide veranda that looked down to the river, another piece of furniture swung on chains, a wicker chair with arms woven from thick reeds.

Taxidermized animals stared out from walls and glass display cases inside the house, beasts that Fabyan or his wealthy friends had killed and stuffed: a buck, an alligator, a Gila monster, a shark, birds of all kinds (grouses, owls, hawks), and hundreds of bird eggs, speckled with blue, yellow, and pink. There was also a valuable work of art in the Villa that had once been displayed to millions at the White City: a life-size marble statue of a naked woman petting a lion, her right hand falling across the lion's mane, the lion looking calmly to the side. The statue was called *Diana and the Lion, or Intellect Dominating Brute Force.*

The mastery of Nature. This appeared to be Fabyan's preoccupation.

Back outside, Elizebeth walked down the steepening hill toward the Fox River until it leveled out a hundred yards from the water. A curving path took her through a Shinto arch of wood and into a pristine garden ringed by buildings, benches, and lanterns of Japanese design. She was told it had all been devised by one of the emperor's own personal gardeners. Flowering trees were aflame with pink and red and blue and orange blossoms that breathed their reflections onto a circular pool at the garden's center, the surface of the water like a painter's palette smeared with color. A half-moon footbridge spanned the pool. Every leaf and flower, every plank of wood and drop of water seemed designed for maximum tranquility, except for the low concrete structure to the right of the pool, shaped like a hexagon, protected by heavy black iron bars. It was a bear cage.

Fabyan kept two pet grizzly bears inside. Their names were Tom and Jerry.

The river lay beyond, a placid silver width flowing southward, from left to right, away from the center of Geneva. Elizabeth could see a small island in the middle of the river, connected to the near shore by two bridges, and on the far bank, an impressive Dutch windmill, a giant X spinning against the sky. It was explained to her that Fabyan had bought the windmill in Holland and transported it here, piece by piece.

Later that day, after the tour, Elizabeth sat down with Mrs. Gallup in the Lodge to discuss the work they might do together if Elizabeth were to accept the research position. They talked for two or three hours, not long enough for Elizabeth to grasp the full nature of the project but sufficient to get a sense of Mrs. Gallup's immediate needs and her personality.

Unlike Fabyan, Mrs. Gallup spoke in a restrained, careful manner, the tones of a scholar. There was nothing hucksterish about her at all. She illustrated her explanations with oversize sheets of paper that were curled up like scrolls. She rolled them out to their full length to show Elizabeth, and placed weights on the ends to prevent them from curling up again. The sheets were beautiful and full of hand-drawn letters of the alphabet in subtle variations, lowercase and uppercase, roman and italic:

Mrs. Gallup said she had drawn these letterforms from photographic enlargements of the Newberry Library's First Folio of Shakespeare, and the drawings had helped reveal the secrets that Francis Bacon had woven into Shakespeare's plays. In some way Elizabeth didn't understand yet, the hidden messages were embedded in the shapes of the letters themselves, in small variations between an *f* on one page of the Folio and an *f* on another.

According to Mrs. Gallup, she had already discovered these messages. She knew what they said; she was certain they existed. The problem, as she saw it, was that some literary experts disputed her method and doubted that the messages were really there. So Elizebeth's job at Riverbank would be twofold. First she would use Mrs. Gallup's method to reproduce her existing results, providing scientific confirmation and silencing the critics. Then Elizebeth would assist Mrs. Gallup with new investigations. Mrs. Gallup believed that in addition to authoring Shakespeare, Bacon also secretly wrote works commonly attributed to Christopher Marlowe, Ben Jonson, and other major figures of the age. Together Elizebeth and Mrs. Gallup would rewrite the history of seventeenth-century England—and by extension, the history of all English literature.

George Fabyan popped in briefly, to see how the women were getting on, rolling out one of Mrs. Gallup's scrolls and regarding it with evident satisfaction.

It was a lot for Elizebeth to process. Over the last twenty-four hours she had been accosted by an eccentric tycoon, dragged to his hall-of-wonders laboratory in the countryside, introduced to a merry team of scientific experts, told about a secret cipher embedded in the heart of the First Folio, and invited to assist them in turning history upside down.

That evening Elizebeth returned to her room in the Lodge to find a vase of flowers and another bowl of fresh fruit by the bed, filled to abundance. She lay awake for a time, thinking about all she had seen and heard, stunned and a bit jangled by the weirdness of Fabyan's kingdom, yet impressed with Mrs. Gallup's erudition and quiet confidence. To be sure, their theory was unconventional, but what if it was correct? If there was even a chance that Fabyan and Mrs. Gallup were on to something here, how could Elizebeth pass up a chance to join an effort of such magnitude? The next day, when she rode the Union Pacific back to Chicago, she was buzzing with "a mixture of astonishment, incredulity, and curiosity."

On the morning of June 7, around the time Elizebeth Smith had first arrived in the city to look for a job, five thousand women marched toward the Republican National Convention, being held at the Coliseum, to demand the right to vote. The wind and rain shoved the women this way and that by the handles of their increasingly useless umbrellas. Dye from their yellow sashes streamed down their legs. Reaching the convention hall, the women surged through the entryway. Water poured from straw hats, hems, and sleeves, pooling at their feet in a spreading puddle. Many held rain-blurred signs. "We want to be citizens. Do we look desirable?" The protesters demanded that the GOP support a constitutional amendment granting women the vote, but after debating the issue, the delegates decided that an amendment would violate "the right of each state to settle this question for itself."

Elizebeth Smith wasn't involved in the suffrage movement or any other. Her views on women's rights at age twenty-three were complicated. She idolized the suffrage pioneers but doubted that men would give up their power without a vicious fight. Earlier that year she had found herself riding a crowded bus when a heavyset woman looked straight at her and then used her rump to shove her out of the way. Elizebeth fumed in her diary, "No woman's rights was adequate to the situation then! I wanted genuine masculine title; if I had been a man she'd never have dared do it."

There in the city, she reviewed her options. Should she take the job that George Fabyan had offered, or go back home to Indiana? She couldn't decide. Fabyan scared her. But she had wanted an unusual job, and in all her life she had never seen a place so unusual as Fabyan's estate.

She was running out of clean clothes in her suitcase. She was out of time.

Elizebeth made her way to the Chicago & North Western train station. At the ticket counter she asked, in her firm, polite voice, for a fare to Geneva, Illinois.

When she arrived once again at Riverbank, Fabyan and Mrs. Gallup were glad to see her. They wasted no time, and began teaching her how to dive for what Francis Bacon had left behind: a sunken treasure of words, a ship of gold at the bottom of the sea.

CHAPTER 3

Bacon's Ghost

The Bacon-Shakespeare investigators at Riverbank, 1916.
Mrs. Gallup is seated in the front row on the left, and
Elizebeth Smith is third from the left standing in the middle row.

Mrs. Gallup had to know if her new assistant, this Elizebeth Smith, could be trained to see. So this is where they began—with a deciphering test.

In the Lodge, Mrs. Gallup placed several pages in front of Elizebeth. One was a worksheet of white paper with some typing on it, eight lines of text from the Shakespeare Folio. The text was broken into five-letter blocks:

```
TheWo rkeso fWill iamSh akesp earec ontai ninga llhis Comed
iesHi stori esand Trage diesT ruely setfo...
```

When Elizabeth read the text and skipped over the spaces, she could make sense of it as English:

The workes of William Shakespeare containing all his Comedies Histories and Tragedies Truely set . . .

She recognized these words from one of the early pages of the First Folio, "The Names of the Principall Actors."

According to Mrs. Gallup, Francis Bacon had concealed a message on this page. She already knew the secret, but needed to know if Elizebeth could find it, too.

Mrs. Gallup always said that as a devout Christian she was appalled when she first discovered the secret messages of Francis Bacon. She did not traffic in such matters as Bacon discussed: deception, blackmail, adultery, the insatiable lusts of queens and earls. "Surprise followed surprise," Mrs. Gallup wrote, "as the hidden messages were disclosed, and disappointment as well was not infrequently encountered. Some of the disclosures are of a nature repugnant, in many respects, to my very soul." However, her own moral beliefs were irrelevant. "The sole question is—what are the facts? These cannot be determined by slight and imperfect examinations, preconceived ideas, abstract contemplation, or vigor of denunciation."

She was not the first person who claimed that Shakespeare was really Francis Bacon in disguise. This idea, known as the "Baconian" theory, enjoyed broad appeal and made a certain sense. Francis Bacon and Shakespeare had lived in the same country in the same era, the England of Queen Elizabeth I, and of the two men, Bacon was by far the more distinguished, a child prodigy who graduated from Trinity College at age fifteen, studied law, served in Parliament, became lord chancellor, won the lofty titles of Baron Verulam and Viscount St. Albans, and wrote manifestoes that heralded the dawn of the scientific age and inspired generations of inventors and revolutionaries. Charles Darwin idolized Francis Bacon; Thomas Jefferson thought Bacon was among the two or three greatest men who ever lived; Teddy Roosevelt's love of Bacon's writings encouraged him to create America's system of national parks.

The radical idea that made Bacon a legend is one of the epigraphs of this book: "Knowledge itself is power" (his admirers often shortened it to "knowledge is power"). What people called science in Bacon's day was more like philosophy or logic: the thinking of beautiful thoughts. Bacon said no, science is about *physical evidence.* Knowledge is found not in the skull but in contact with Nature. And Bacon made it his mission to collect and classify all forms of knowledge, arguing that if enough knowledge was gathered and sorted and pinned to the page, there was nothing men could not achieve. In an unfinished utopian novel, *The New Atlantis,* Bacon imagined a lush, remote island ruled by superintelligent scientists. The people spend their days studying the native beasts and plants, running experiments in towers, caves, artificial lakes, and specially constructed laboratories. The island is a like a cross between a research university and a nature preserve, a place devoted to the investigation of light, acoustics, perfumes, engines, furnaces, mammals, fishes, flowers, seeds, geometry, illusions, deceptions, and, above all, methods of extending human life, of achieving immortality. He thought humans might learn to live forever, be immortal, become like gods.

All in all, Bacon was such an impressive person that it seemed perfectly plausible to writers and scholars in the late nineteenth and early twentieth centuries that Bacon might have written Shakespeare's plays under a pseudonym. Mark Twain believed it. So did Nathaniel Hawthorne. Was there proof? Had Bacon left a hidden signature? Men and women romped through the grassy fields of the texts and scraped at the individual letters with every kind of tool imaginable. Anagrams: whereby existing letters are rearranged to create new words and phrases. (The phrase "Maister William Shakespeare" in the 1623 Folio can be anagrammed into "I maske as a writer I spelle Ham.") Numerology: whereby letters are converted into numbers that seem significant. (If A=1 and B=2, the name "Bacon" is 2+1+3+14+13, equaling 33; the

appearance of 33 in any count of Shakespeare's words is a signature of Bacon.) One man, Orville Ward Owen, a physician
from Detroit, invented a machine he called a "wheel," two large
wooden spools stretched with one thousand feet of canvas on
which he had printed thousands of pages of different Elizabethan
texts. Dr. Owen and his team of assistants would spin the wheel,
look for instances of four "code words" they believed were important (FORTUNE, HONOUR, NATURE, and REPUTA-
TION), write down words that appeared next to those four
words, and arrange them into sentences.

What made Mrs. Gallup different from these other investigators was that she presented herself first and foremost as a scientist,
and her system for finding the messages was the most scientific
and plausible yet. It wasn't something that came to her in a dream.
The method had been demonstrated by Francis Bacon himself, in
his book *De Augmentis Scientarium,* published the same year as
Shakespeare's First Folio, 1623.

Bacon revealed that year that he had invented a new type of cipher, a method to signify "omnia per omnia": anything by means
of anything. It possessed what he said were the three virtues of
a good cipher: it was "easy and not laborious to write," it was
"safe," and it did not raise suspicion—that is, an enciphered message would not appear, at first glance, to be in cipher at all. These
are still sound principles today. His insight was that all letters
of the alphabet can be represented with only two letters, if the
two letters are combined in different permutations of five-letter
blocks. The letters *i* and *j,* and *u* and *v,* were interchangeable in
Bacon's time, so, choosing *a* and *b* for the two letters that represent all the rest, the new alphabet looks like this:

A	B	C	D	E	F
aaaaa	aaaab	aaaba	aaabb	aabaa	aabab

G	H	I,J	K	L	M
aabba	aabbb	abaaa	abaab	ababa	ababb

N	O	P	Q	R	S
abbaa	abbab	abbba	abbbb	baaaa	baaab

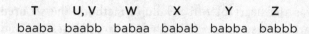

T	U, V	W	X	Y	Z
baaba	baabb	babaa	babab	babba	babbb

Each letter becomes five, so a word like *Riverbank*, written in this cipher, grows five times as long: *baaaa abaaa baabb aabaa baaaa aaaab aaaaa abbaa abaab.*

This is exactly like binary code, the language at the root of computers, and Morse code as well. In all of these systems, just two symbols, arranged in different combinations, can stand for many others. Binary code uses 0s and 1s, Morse code dots and dashes. Francis Bacon discovered the basic principle in 1623.

Crucially to Mrs. Gallup, he also showed the flexibility and power of his cipher *by example.* Bacon pointed out that the two letters that represent the others in his cipher don't have to be *a* and *b.* They can be *c* and *d,* or *x* and *y.* They can be physical objects, like apples and oranges arranged on a table; they can be sounds, like the alternating and audibly distinct shots of a musket and a cannon. In Bacon's cipher, the plaintext for ♪♪♫♩♪♩♪♫♪♩♩♪♪♪♪♪♫ is "deaf." ♥♥♥♡♡♥♡♥♥♥♥♥♡♥♥ means "die." All that's required is a "biliteral alphabet," an alphabet made of any two forms that are recognizably different. Write a manifesto with candies, send a love letter with bullets. As long as you specify an *a*-form and *b*-form, you can make anything stand for anything else. *Omnia per omnia.*

You can even camouflage a secret message in plain sight.

A message that reads *aaaba abbab aaabb aabaa* is obviously written in cipher, and anyone who intercepts it will know it contains a secret. Bacon suggested creating a "bi-formed alphabet" to overcome this problem—an alphabet with two slightly different versions of each letter, an *a*-form and a *b*-form. For example, an italic letter might be the *a*-form, and a normal, "roman" letter the *b*-form. A string of text like

knowl*edge* is power

might translate to

run

This was the heart of Mrs. Gallup's method. She scoured photo enlargements of pages from Shakespeare's First Folio and other Elizabethan books, looking for minute differences in the shapes of letterforms to discover the "biformed alphabet" she believed Bacon had planted in the text—the two alphabets with letters of different shapes. Then she drew charts of the *a*-form letters and the *b*-form letters. Then she went back through the original texts of the old books and compared each letter to the drawings of the letters on her charts, deciding if a letter was an *a*-form or a *b*-form. After classifying five letters in this manner, she was able to check Bacon's key (*aaaaa*=A, *aaaab*=B, *aaaba*=C) and write one letter of the final message. And that was when she found the secrets that troubled her Christian conscience.

Queene Elizabeth is my true mother and I am the lawfull heire to the throne. Finde the cypher storie my bookes containe. It tells great secrets, every one of which, if imparted openly, would forfeit my life. —F. Bacon.

Francis of Verulam is author of all the plays heretofore published by Marlowe, Greene, Peele, Shakespeare, and of the two and twenty now put out for the first time. Some are alter'd to continue this history.

Francis St. Alban, descended from the mighty heroes of Troy, loving and revering these noble ancestors, hid in his writings Homer's *Iliad* and *Odyssey* (in cipher), with the *Aeneid* of the noble Virgil, prince of Latin poets, inscribing the letters to Elizabeth. . . . He in this way, and in his Cypher workes, gives full directions, in a great many places, for finding and unfolding of severall weightie secrets, hidden from those who would persecute the betrayer.

You will either finde the guides or be lost in the labyrinth.

—Fr. St. Alban.

First published in her 1899 book *The Biliteral Cypher of Sir Francis Bacon Discovered in His Works and Deciphered by Mrs. Elizabeth Wells Gallup,* the messages told an alternate history of Elizabethan England that riveted journalists and divided scholars. According to Mrs. Gallup's decipherments, Francis Bacon wasn't just the great genius of his age. He was a secret king: the bastard son of Queen Elizabeth, known for her indiscretions, and the Earl of Leicester. During his own lifetime, Bacon was afraid that if he claimed his royal blood, he would be killed to suppress a scandal, so he had found a way to smuggle the truth into history, using his cipher to conceal messages in "great dramaticall works" he wrote under pseudonyms (Shakespeare, Marlowe) and also in "workes of science" published under his own name. He conspired with printers to sneak the cipher into books without anyone catching on, and he taught the cipher to a clandestine society of engineers, the Rosicrucian Society of England, who conducted scientific experiments in secret, fearing accusations of witchcraft. Using the cipher, Bacon and the Rosicrucians were able to exchange dangerous knowledge without fear of discovery and design technologically advanced machines.

Her work unleashed a furor. Mrs. Gallup seemed to come out of nowhere with an impressive scientific procedure and reams of proof. "Here are 360 pages of deciphered matter," one journalist wrote in a typical review, "with sufficient means of proof to satisfy any investigator." Skeptics questioned the veracity of the messages and said Mrs. Gallup must be imagining them; she savaged her critics in icy pamphlets and letters to the editor, writing that her style of analysis was "impossible to those who are not possessed of an eyesight of the keenest and most perfect accuracy of vision in distinguishing minute differences in form, lines, angles and curves in the printed letters. Other things absolutely essential are unlimited time and patience, and aptitude, love for overcoming puzzling difficulties, and, I sometimes think inspiration." She argued that if other people could not replicate her findings, it was their own fault—they had poor eyesight, they were lazy, they were uninspired.

She traveled to Oxford, England, and won converts in the literary community there. She produced testimony from researchers in England and America who swore that they had been able to replicate her decipherments: Mrs. Gertrude Horsford Fiske, Mrs. Henry Pott, Mr. Henry Seymour, Mrs. D. J. Kindersley, Mr. James Phinney Baxter. And of all her supporters, no one had more faith than George Fabyan. He invited Mrs. Gallup and her sister to Riverbank in 1912 and gave them carte blanche to pursue their investigation to its ultimate end. There was nothing he would not buy or build to support her work, no mode of investigation too outlandish or expensive. After establishing herself at Riverbank, Mrs. Gallup reported that she had deciphered a message from Bacon describing an "acoustical levitation device," an antigravity machine he apparently invented in the seventeenth century. It used the vibrations of musical strings to lift a rapidly rotating cylinder off the ground. Fabyan ordered his chief engineer, Bert Eisenhour, to build the machine out of wood. The result looked like a water wheel. Eisenhour couldn't get it to work. Something about the tuning of the strings. Fabyan was undaunted, saying, "The inheritance which the world received from Mrs. Gallup's work is the greatest that has ever been given to posterity."

He wrote that in 1916, the same year Elizebeth Smith arrived at Riverbank and was handed her first deciphering test by Mrs. Gallup.

Elizebeth looked at the test worksheet:

TheWo rkeso fWill iamSh akesp earec ontai ninga llhis Comed iesHi stori esand Trage diesT ruely setfo...

Along with the typed worksheet, Mrs. Gallup had provided a photo enlargement of the actual page from the First Folio on which these words appeared. There was a copy of the biformed alphabet that Mrs. Gallup had already extracted from this part of the Folio—a list of all the *a*- and *b*-forms apparently inserted by Bacon. Mrs. Gallup also gave Elizebeth a looking glass of her own, and the key to the biliteral cipher: *aaaaa* means *A, aaaab* means *B,* and so on. To find the secret message, Elizebeth would need to squint at the Folio page through the glass, decide if each letter was an *a*-form or a *b*-form, and write a dash or a slash on the typed worksheet above the corresponding letter: a dash for *a*-form, a slash for *b*-form. Once Elizebeth made five dashes or slashes, she should check the key and write one letter of the final message. For instance, if her pattern of dashes and slashes looked like

--//-

then she would write the letter *G*.

Elizebeth knew nothing about secret writing at this point. She had never studied codes and ciphers. She had never even been particularly fond of puzzles. She was as fresh to the whole subject as any person off the street. But Mrs. Gallup had given her the rules of the game, and now she tried to follow them.

TheWo rkeso fWill iamSh akesp earec ontai ninga llhis Comed iesHi stori esand Trage diesT ruely setfo...

She started looking back and forth between the Folio page and the biformed alphabet, trying to tell if the letters were *a*- or *b*-forms. It was slow going. She got stuck on the first couple of letters, staring and staring through the glass, unable to decide if a letter was an *a*- or *b*-form. The variations were subtle: a slight wobble in the stem of an *H*, a tilt in the ovals of a *g*. It was like trying to sort blueberries by color, or beach pebbles by smoothness. Ultimately she needed Mrs. Gallup's help to get the answer, and it still took her eight hours to produce the twenty-four-word plaintext translation: "As I sometimes place rules and directions in other ciphers, you must seeke for the others, soone to aide in writing. Fr. of Ve (Francis of Verona)." She signed it on the bottom:

It went like this with the tests that followed. Mrs. Gallup handed Elizebeth a new Folio passage to decipher, and Elizebeth struggled for hours, solving it only with her boss's intervention. From time to time Elizebeth carried her materials over to Gallup's desk and set them down; Gallup pressed her eye to her looking glass, made some sharp pencil marks on Elizebeth's sheet, and handed it back. Impressed, Elizebeth always asked Mrs. Gallup how she succeeded when Elizebeth failed—had she modified the list of *a*- and *b*-forms, tweaking the alphabet to get the "right" answer? No, she hadn't: Elizebeth, being a novice, had failed to see the subtleties in the letters, had overlooked a little angle or an accent or a tiny shift in the position of the dot above an *i*.

At first she didn't worry that she struggled with Mrs. Gallup's system. Elizebeth awoke each morning in a dreamland. She had arrived to begin her new job during Riverbank's sweetest season, the moment of peak summer pleasure, the colors and smells dialed all the way up, the food most plentiful: breakfasts of fresh eggs from the chickens on the farm, dinners of meats and fruits prepared by the nimble Danish and Swedish cooks. She took walks along the river. Wild orchids grew on the banks. Shapes of sunlight twitched on the water like spinning coins. Ragtime music played from somewhere. She turned her head searching for the source. Fabyan had installed a series of loudspeakers across the property, operable from a single control panel in the Villa, so that he and Nelle might listen to music at any spot on the estate, and the songs changed throughout the day, switching directions, coming now from the garden and now from the veranda, ragtime shifting to jazz, then to a Beethoven symphony, the boss, intense and restless, wanting to hear every kind of music all at once and never getting his fill.

Fabyan assigned her a bedroom in a two-story building called Engledew Cottage, named after a local florist and friend of the Fabyans', one of the larger of the many cottages spread across the estate where "brain workers" lived. Engledew stood down the road from the Lodge a few hundred yards to the south, next to the farm with its big barn and Nelle Fabyan's prize cows. There were shared work areas in Engledew Cottage as well as the Lodge, and during the day men and women walked back and forth between the cottages along the highway, carrying papers and books, as horse-drawn carriages and automobiles drove past.

Elizebeth wasn't the only young woman assigned to cipher research at Riverbank. When she arrived there were at least two others, sisters from Chicago, barely out of high school. Fabyan tended to hire women out of clerical pools because it was convenient, but he had come to believe that in many ways they were better than men at analyzing ciphers. Women had the stamina and patience to look at text all day, and complained less. "Our experience at Riverbank," Fabyan wrote, "has demonstrated that women are particularly adapted for this kind of work."

After a few weeks Elizabeth fell into a routine, adjusting to the rhythm of her new job. Mrs. Gallup often worked with her assistants in the Lodge's spacious living room, with its tall casement windows that looked east across the highway, toward the Villa and the river. The work atmosphere in the Lodge was a bit like how she imagined a museum of natural history or a lepidopterist's lab to be, a place where people analyzed delicate objects, pinning dead butterflies to pages, drawing pictures. Mrs. Gallup sat at a handsome wooden desk, peering through an oblong looking glass at photo enlargements of pages from old books. The enlargements were made by William Friedman, the geneticist with the white buck shoes who had caught Elizebeth's eye at her first dinner in the Lodge. Because William happened to be handy with a camera and a darkroom, Fabyan had roped him into the cipher project, even though it wasn't his job, and he often visited the Lodge to drop off new prints for Mrs. Gallup.

The woman would raise the looking glass, lower it, write a few words in a notebook, raise it and lower it again, and write some more, hour after hour. When Elizebeth asked the other girls what Mrs. Gallup was doing, they said she was attempting to complete Bacon's unfinished science fiction novel *The New Atlantis,* to recover the remainder of the text, which Mrs. Gallup believed was woven throughout Bacon's works.

The cursive line of her pen was exceedingly fine. Each page of her notebook resembled a piece of art. She kept images of Bacon close at hand, for inspiration: an engraving of Bacon in his prime, a handsome youth with curls and a ruff; a picture of Gorhambury House, Bacon's mansion outside London. She filled small wooden boxes with news clippings about her own research and that of her competitors, ultimately pasting the clips into scrapbooks.

The women worked long hours, into the evening, sun dipping

low, flies swarming on the porch. "We lived hard and fast," Elizebeth later recalled to the NSA's Valaki, then paused, embarrassed. No, she did not mean to imply anything salacious. "I mean, there was absolutely no carousing, no parties, no nothing. Fabyan had use for only one kind of worker, and that was one that knew his business and worked at it damned hard." He paid the codebreakers and scientists tiny salaries but promised to take care of them in all other ways. Food, lodging, recreation: they would live like the "minor idle rich" as long as they stayed under his wing at Riverbank.

And who would ever want to leave? On weekends Elizebeth put on her bathing suit, skidded down the hill to the river's edge, and walked across the bridge to the island, which Fabyan called "Isle of View" because he liked that it sounded similar to "I love you." The lighthouse rose above the northern bank of the island, and the southern bank was crowned with tall trees and a magnificent swimming pool built by Fabyan for the enjoyment of his staff, lit at night by floodlamps and lined with soaring Roman columns. Swimming there made Elizebeth feel like an Italian princess or an actress in a movie. The cool water licked off the sweat and she dried herself in the sun, talking and laughing with Mrs. Gallup's other female assistants.

In August 1916 she turned twenty-four.

The men of Riverbank noticed Elizebeth early on, particularly the brain workers. They smoked pipes stuffed with cheap tobacco. They asked around about her story, tried to figure out if she was single, looked for ways to get her attention. Bert Eisenhour, the carpenter and engineer, pulled strings to borrow Fabyan's Stutz Bearcat, a roadster with a four-stroke engine, and invited Elizebeth to climb in; seconds later she was ripping along the Lincoln Highway at 60 or 70 miles per hour with the top off. In an era when most roads were dirt and gravel, the highway was paved, a result of cooperation between Fabyan and other local business owners; as a result, "no billiard ball on the smoothest billiard table ever made could have more pleasure in motion than that enjoyed by the ordinary Illinoisan skimming in a powerful car over that gleaming, winding stretch of concrete," a visitor once wrote. In the passenger seat of the Bearcat, Elizebeth raced past barns and silos, her body inches from the ground, wind plastering her curls to her head, engine roaring in her ears. She didn't know if her head would blow off.

Another man who crossed her path during free hours was William Friedman, the geneticist and Mrs. Gallup's photo assistant. He didn't like to swim, because he was afraid he'd catch cold, but he enjoyed bicycling, and he and Elizebeth took leisurely rides together around the estate, stopping to picnic on the grass, sandhill cranes and red hawks circling above.

At twenty-five he was one of the younger male scientists, closest to her own age, so it felt natural to spend time with him. She appreciated his shy, precise way of speaking, his soft, halting voice that seemed to encode its own refutation, as if he were constantly checking a mental ticker tape of his words for correctness. One day he showed her where he lived on the estate. It was a working windmill—not the big showy Dutch contraption on the other side of the river but a smaller windmill on the same side of the road as the Lodge and the ordnance lab. It was two stories tall. William opened the door and she walked into an old, creaky structure, damp and warm, with a powerful smell of soil. She saw that the ground floor contained some microscopes and work shelves. An interior door led to the greenhouse that William managed, which is where Fabyan had him breeding new strains of crops and flowers, violets and wheat, and a type of corn with no cob.

Upstairs, he said, was his sleeping quarters, and down here was a little laboratory where he ran genetics experiments with living fruit flies, *Drosophila melanogaster*. Elizebeth could see his bottles full of the teensy-weensy flies. Each bottle was about the size of a coffee mug, only thinner, and was smeared with some overripe banana that the flies ate. William explained that geneticists like to use fruit flies in experiments because they reproduce very quickly, then die. If you marry a normal fruit fly with a fly that has yellow eyes, say—a genetic mutation, an alteration in the biological code—they will produce children in three weeks, and you can look at the children to see if they have yellow eyes, showing they inherited the yellow-eye gene. There was something incongruous and surreal about seeing this good-looking young man in a crisply pressed white shirt and bow tie working in a rustic prairie windmill that smelled of banana and the sweetish decay of plant matter. Elizebeth used to watch him there, mating the flies, carefully pouring one bottle of flies into another, getting them to exchange their codes.

The size and scope of Riverbank was dawning on her. What had appeared on her first visit to be a sparsely populated stretch of land now revealed itself to be a small self-contained village, a community of 150 workers, some of whom had been with Fabyan for more than decade: the caretaker of the Japanese garden, Susumu Kobayashi; the boathouse manager, Jack "the Sailor" Wilhelmson, a happy and well-built Norwegian; Fabyan's personal secretary, Belle Cumming, originally from Scotland, who kept Riverbank's financial records in black folders and hurled torrents of profanity at guests she felt were overstepping their bounds; Silvio Silvestri, Fabyan's personal sculptor. Fabyan hired them on whims. He trusted his own impressions of people instead of their accomplishments or educations. He brought people to Riverbank if he decided they were spectacular. He was always saying that to Elizebeth and everyone else: "Achieve success! Be spectacular! Then things break your way."

To entice spectacular individuals to stay, he welcomed their spouses and children. Every child born on the estate received a sum of money from Fabyan, placed in a bank account to grow and pay for future schooling. This was another aspect of the place that made Elizebeth marvel: there were families here, boys and girls growing up at Riverbank. Fabyan seemed to genuinely love children. He handed out shiny dimes he kept in his coat pocket. He stopped whatever he was doing to answer their questions about Riverbank's zoo creatures and explain the curious behaviors of the animals, to remove a snake from a cage with his own hands and demonstrate how a snake was able to disengage its jaw in order to swallow an egg.

Elizebeth realized that everything that appeared so hallucinatory to her about Riverbank must seem perfectly normal to these children. It was normal for them to live where two monkeys roamed outdoors wearing red diapers, one a kleptomaniac with a habit of stealing men's keys. It was normal for Jack the Sailor to sing sea shanties to the children, dance the jig on their command, and teach them how to tie knots. It was normal to be out-side playing and see a famous actress walk by, or Teddy Roosevelt, who liked to stroll the grounds with Fabyan and talk about crops, genetics, and Francis Bacon. In summer Jack the Sailor always wove a gigantic spiderweb out of rope that spanned two elm trees; squirrels climbed it, and children tried, and so did Lillie Langtry, a stage and vaudeville actress and an adventurous horsewoman. Other celebrities vacationed at Riverbank: the curly-haired ac-tress and pilot Billie Dove; the aviator and polar explorer Richard Byrd; Broadway producer Flo Ziegfeld and his actress wife, the elegant Billie Burke, who would later play the role of Glenda the Good Witch in *The Wizard of Oz;* the titans of the Chicago Stock Exchange, of which Fabyan was a member. They ate dinner with the Fabyans, George and Nelle, then drank and smoked around a campfire.

Elizebeth wasn't impressed by the celebrities. She met and talked with Lillie Langtry, one of the most famous women in America at the time, and only mentioned it later in passing. Eliz-ebeth was proud not to be "afflicted with the star-complex and hero-worship," she would write later. "Whatever quality it is which is possessed by those who love the adulation and star wor-ship seems to be, in my case, supplanted by an intense reach for freedom from observation—and for privacy."

This is one reason Elizebeth became wary of George Fabyan almost immediately after starting her job: He seemed interested in prying into every part of her life. Soon after she arrived and settled in, Fabyan told her that the modest white and gray dresses she liked to wear were inadequate and she needed to buy a new wardrobe at Marshall Field's in Chicago. Frugal by nature, Eliz-ebeth resisted paying a premium for a name brand, but when she raised her voice to complain, Fabyan told her to hush. "That's so typically Fabyan," Elizebeth recalled: if you told him he was wrong, "[t]he next thing you know there'll be a gun rammed down your throat." He sent her into the city with one of his sec-retaries, who accompanied her to the department store and made sure she bought Fabyan-approved items.

She figured out within a week or two that she was dealing with a half-crazy individual of unlimited funds and a split personality.

There was a side of him that was authoritarian, that craved order and ceremony, which explained the bugler who played reveille in the morning and taps at night, and the American flag that was raised each morning and lowered in the evening, folded into a triangle as a cannon fired a ceremonial ball into the prairie dusk. The staff called him "Colonel" or "the Colonel." Elizabeth was told to address Fabyan as "Colonel Fabyan," and she did, assuming he must have served in the military; it was only later when she learned the truth, that the title of Colonel was an honorary one, bestowed by the governor out of gratitude after Fabyan allowed the Illinois National Guard to use the estate as a training ground. The governor even named a group of cavalry scouts the Fabyan Scouts. Chest leaping with pride, Fabyan recruited some local farmhands to join a militia he called the Fox Valley Guards, as if he aimed to build a personal army.

He liked to sit in the wicker chair that hung on chains from the tree next to his villa. Elizabeth heard people call it the "hell chair" and soon understood why. If Fabyan was angry at an employee or a guest, he brought that person over to the hell chair and sat there, screaming at the offender, giving them hell, while he swung to and fro, chains creaking. He sat there at night sometimes, in the hell chair, stoking the coals of a campfire in the dark.

The other side of Fabyan loved chaos and ripped through the days under power of impulse and inspiration. He had a habit of buying supplies sight unseen from train boxcars—a skyscraper's worth of steel I-beams; seventy-five plows—and storing them in a warehouse next to the Dutch windmill that he called the Temple de Junk. He seemed to glory in randomness to the point of mocking the foundations of his world. He published a book, *What I Know About the Future of Cotton and Domestic Goods,* by George Fabyan, and kept copies in his office. Visitors grabbed the book with sweaty hands and flipped through, hoping for a stock tip from a wealthy cotton magnate. Inside were one hundred blank pages. It was Fabyan's joke about the riddle of commerce, the arbitrary American system that kinged him with enough money that he didn't have to care about money anymore.

He liked to dress up as a horseman or hunter, in riding coats and knee-high boots, but no one ever saw him riding a horse or shooting wild game, and he liked to dress up as a yachtsman, in a white sweater and jaunty blue cap, but no one ever saw him sailing a boat. He spilled across his kingdom on foot, thundering from place to place, dropping heavy ideas and moving on, letting others do the lifting. One day he walked past the swimming pool and saw a little girl, Sumiko Kobayashi, the daughter of his head gardener, resisting her first swimming lesson, crying because she was afraid of the water. Fabyan commanded an adult to throw her in the pool and let her *learn by doing*. He watched the terrified girl thrash for her life in the water, then walked away, satisfied with his solution, while the adults dove in and saved poor Sumiko from drowning.

He may have been a monster. But he was no idiot. To underestimate his intelligence was dangerous, Elizebeth sensed. She considered him to be, despite his lack of formal education, "a very bright man" with a cunning mind and a proven ability to predict how people and institutions would react to moments of stress and crisis. He could get anyone to listen to him. He didn't read scientific papers; Elizebeth never saw him read anything longer than a newspaper headline. But he had been blessed with a near-photographic memory, and whatever his scientists told him, he could repeat back verbatim. This skill for mimicry, combined with his innate abilities as a salesman, made Fabyan seem like a credible prophet of science even when he was talking about things that science said were impossible. He pursued schemes for perpetual motion, infinite energy from nothing, and once showed Elizebeth a prototype of a perpetual-motion machine that he kept behind one of the labs. She was unimpressed: "I remember going, looking at it for quite a while, and it just seemed to me like a great, huge, metal something-or-other." He argued that common human ailments could be traced to the fact that our primate ancestors crawled on their stomachs and humans have never properly learned to walk. And he wasn't selling these ideas cynically; he really believed them. He was good at blurring the line between fantasy and reality because he didn't believe any such line existed. As he once told William Friedman, *I have seen impractical and improbable things accomplished.* All it took to achieve improbable things was an optimistic attitude and a refusal to give up.

"We play the game day-to-day as best we can," he was fond of saying.

Of all the investigations at Riverbank, Fabyan sold the Bacon cipher project the hardest. Though he gave every visitor at least a taste of the cipher work, presenting it as one element of the general package of wonders, he organized separate junkets to persuade hesitant or openly hostile academics that Riverbank had found the answer.

In the late summer of 1916 he began to lean on Elizebeth for help. He had already realized that when she spoke, even though she was only twenty-four, people listened to her—her good looks caught the eye of men and her precision and earnest intelligence held attention. He started to let her know that Professor So-and-So from Such-and-Such College was coming to learn about the cipher discoveries and Elizebeth needed to persuade this person that Riverbank's approach was correct. "We'll get along fine," Fabyan told Elizebeth. "We'll see if we can induce him to stay."

The academic would come, all expenses paid. There would be a lot of food, a lot of wine. Fabyan usually delivered a presentation on the ciphers using lantern slides, square photographic negatives printed by William Friedman and projected onto the wall of a darkened room through a curved piece of glass. He had contempt for what he saw as the timid and conformist mind-set of literary intellectuals while at the same time wanting to win them over, and took pains to present himself as the sort of careful and factual man he felt they would be likely to respect. While displaying slides of the Folio and Mrs. Gallup's lovely drawings of biformed alphabets she claimed to find within, Fabyan explained that he did not care about the "useless Bacon-Shakespeare controversy" of who wrote Shakespeare; that he only cared about getting to the bottom of the cipher contained in the plays; that he and his Riverbank colleagues had no use for anything but "hard, cold facts"; that the existence of the cipher was such a fact; that it had passed careful tests; that no one at Riverbank was making any money from these investigations; and that they were doing it for the benefit of humanity, committed to sharing their discoveries with the world. Who could object? The combination of his gravelly voice in the shadows and the delicate letters on the wall tended to disorient the guest and lull him into a state of increased charity toward the Riverbank view. Heads, ever so slightly, began to nod. Jaws to relax. And Elizebeth played her part. If a visitor grew sick of listening to Fabyan and turned to Elizebeth, asking what she thought, she said she was convinced that the work was solid, that the messages were really there.

Privately, though, she was beginning to doubt. Skeptical visitors made arguments difficult to refute. The head of the English department at the University of Chicago, John Matthews Manly, an authority on Chaucer and an amateur cryptologist, stayed at Riverbank for a time and concluded that it was all bunk. Manly was already famous in his field, didn't need money or anything else Fabyan could offer, and he took delight in pointing out the holes in Mrs. Gallup's method, like a boy in a roomful of red balloons, stomping them flat one by one. Fabyan asked Elizebeth to "wrassle" with Manly for a weekend, and she found him a pompous ass. At one point during an argument, Manly's voice rising sharply, Elizebeth staying calm but arguing back, Manly pushed her on the shoulder, baffled and upset that anybody might challenge the great John M. Manly—she never forgot it. "Oh, my! That was too much to take. Ahhh!"

But there was substance to what he and other skeptics were saying, a stubborn logic that tugged at the hem. Mrs. Gallup's technique depended on discerning small yet consistent fluctuations in letterforms in books made long ago, with the technology of a more primitive era. It strained credulity to think that the printers, setting the type by hand in 1623, could have duplicated these minute fluctuations across hundreds of copies of the First Folio, and in fact the variations between different Folio copies were sometimes larger than the variations Mrs. Gallup thought she saw in a single book.

Another skeptical argument moved Elizebeth. It was the literary case against Bacon's secret messages. There was no kind way to put it: the messages were badly written. *Francis St. Alban, descended from the mighty heroes of Troy, loving and revering these noble ancestors*—was this tedious author the same one who gave such light and supple voice to Romeo's desire for Juliet? *See how she leans her cheek upon her hand. O that I were a glove upon that hand, that I might touch that cheek.* To believe Mrs. Gallup's theory, you had to believe that the plays, these warm-blooded, ravishing beasts, had been conceived almost as an afterthought, as mere envelopes for a stilted memoir about a guy whose mom was the queen. It made no sense. It would be like God creating a galaxy simply to tell a knock-knock joke to some distant deity, enciphered in the shapes of stars.

The big question then became: If the secret messages discovered by Mrs. Gallup weren't really there, what was she seeing?

Elizebeth never once suspected that Mrs. Gallup was a fraud. Deception was not in her. The only possibility was that she had been somehow deceiving *herself.* Humans are so good at seeing patterns that we are often able to see patterns even when they aren't really there. Mrs. Gallup must have been altering the rules of her method to fit the desired result, changing the all-important assignment of letters to the two baskets (*a*-form and *b*-form) until she saw words that made sense. Decades later Elizebeth and William would describe what they thought was happening with Mrs. Gallup, in a book they wrote as coauthors:

> She could go through the texts extracting from them what she unconsciously wished to see in them. . . . With each successive letter deciphered she had a choice—limited but definite—of possibilities; and so, as she went on, there would be a kind of collaboration between the decipherer and the text, each influencing the other. Hence perhaps the curious maundering wordy character of the extracted messages, very like the communications of the spirit world: with some sense but no real mind behind them, just a sort of drifting intention, taking occasional sudden whimsical turns when the text momentarily mastered the decipherer.

This was the clarity of hindsight. In the moment, at Riverbank, Elizebeth didn't know what to do with her doubt. She saw how her bosses responded to criticism. Mrs. Gallup restated her conclusions in combative letters and articles, denying that it was all a figment of her imagination and comparing herself to Galileo: "The idea that the earth moves, was once thought an illusion."

And Fabyan doubled down on publicity. He released a picture book for children, *Ciphers for the Little Ones,* that taught the story of Bacon and his biliteral cipher. He printed business cards alleging that Bacon was the bastard son of Queen Elizabeth and added at the bottom:

ALL INQUIRIES REGARDING THE SOURCE OF, AND
AUTHORITY FOR, THESE HISTORICAL, BUT HITHERTO
UNKNOWN FACTS, WILL BE PROMPTLY ANSWERED FROM

RIVERBANK LABORATORIES
Geneva, Illinois.

When people did inquire, Fabyan replied with a form letter describing the cipher project:

> *Riverbank Laboratories are a group of serious, earnest researchers, digging for facts. It is supported by Colonel Fabyan at his country home in Geneva, for his own information and amusement. . . .*

A pressure was building in Elizabeth's chest. It was the old scalding sensation she remembered from college when she realized people valued politeness more than truth. For now she kept her doubts to herself. She doubted her doubt. Who was she to declare that she was right and everyone else was wrong? Was it her vanity telling her that? How would she prove her case if she did speak up? Would she lose her job? Would anyone stick up for her? She was twenty-four. She was a nobody here. She was a nobody anywhere.

During conversations in the Lodge she looked around the room at the faces of her colleagues, trying to tell if they really believed or if they were just pretending. She sometimes met the eye of William Friedman. She wondered what he was thinking.

Lately they'd been talking more and more. William carried his camera everywhere, a black box that hung from his neck. Elizebeth was becoming his favorite photo subject. He would ask her to stand in a garden or on a square of grass, and he would hold the camera at chest level and look down at the image of her face in the glass.

She was learning more about him, where he came from and how he got here. His family was Jewish, originally from a town in Russia called Kishinev, where he was born with a different name, Wolfe. His parents changed it to William when they sailed to America a year after his birth, escaping a famine in Russia and the anti-Jewish laws of the Czar. They settled in Pittsburgh.

William said his father was a serious, bookish man, fluent in eight languages, a student of the Talmud. In Russia he had been a postal clerk but had trouble finding a good job in America and resorted to selling Singer sewing machines door-to-door. His mother worked as a peddler for a clothing company. So William and his four siblings grew up poor in Pittsburgh, poorer than Elizebeth's family in Indiana.

He went to Cornell on a scholarship and chose to study genetics because it was a young field that "seemed to offer great possibilities for research and ingenuity." He earned his degree and stayed on to teach a few courses as an untenured lecturer. That was when the unsolicited letter arrived from Fabyan in the general mailbox of the biology department. William didn't know who Fabyan was. He said he ran a private research facility in Illinois and needed an expert in heredity to launch a genetics department and supervise experiments in the breeding of crops and fruit flies. William wrote back, introducing himself, and over the next three months, Fabyan courted him by mail, promising a life of intellectual freedom and adventure: "I am not looking for an agricultural expert, the woods are full of them; and I am not looking for a man to duplicate work that is being done at every agricultural station in the country, and at every advanced school and university. . . . If I should hear of something anywhere this side of Hell that I thought would do us any good, I might want you to go there and find out about it; in other words, I don't want to go backwards."

William replied with deference, formality, and gratitude: "I realize the value of the opportunity you are giving me to make good and I hope that our future relations will be mutually agreeable and profitable."

There were hints in this exchange that Fabyan would be difficult to work with. William, cautious by nature, asked about salary. Fabyan responded with a vague, long-winded riff: "I want to get some practical level-headed fellows that will carry themselves, and a community which is asking no favors and yet having the best there is, where people will have to come for what we have." William asked what he raised on the farm at Riverbank. Fabyan said he raised hell. His analogies were beautiful and bizarre. Writing of his desire to breed a new strain of wheat that would thrive in dry climates and help feed the hungry, Fabyan told William, "Here is a problem that has come up in my mind, that I want you to work on. I want the father of wheat, and I want a wife for him, so that the child will grow in an arid country." He added that "one of my wealthy Jewish friends" was also working on the problem, but "if I can beat him to it, he will foot the bills, and be damned glad to. . . . This may seem impractical and improbable, but I have seen impractical and improbable things accomplished."

Eventually Fabyan offered to pay William a hundred dollars a month, on top of free lodging, and William accepted. It happened to be an old yearning of his to live on a farm, a dream wrapped up in his Jewish identity. As a kid he'd heard stories from his parents of the pogroms in Russia, mob violence against Jews that they barely escaped, and by the time he got to high school he realized that anti-Semitism was spreading in the United States, too. Popular American magazines portrayed Jewish immigration from Eastern Europe as a "Jewish Invasion," a threat to the jobs of whites, with the Russian Jew said to be especially conniving thanks to his "nervous, restless ambition." Concerned and wanting to protect themselves, William and some of his high school classmates fell under the spell of the "back-to-the-soil" movement, a homespun brand of Zionism that encouraged Jewish kids in America to resist anti-Semitism by tilling the land, making themselves strong and self-sufficient. William took this idea seriously enough to enroll in some courses at a Michigan agricultural college. When he actually tried farming, he realized that everything about it, from the physical labor to the grit in his clothes, made him miserable. He went to Cornell instead.

Now, at Riverbank, he found himself on a farm again. Of sorts.

Elizabeth had never gone for shy men. But she liked William, and so did her elder sister, Edna, who visited Riverbank to see how Elizebeth was getting along. Edna's dentist husband had recently died, leaving her a widow, and the dapper geneticist left an impression. She wrote William two flirty letters. Edna informed him that her sister was growing fond—"I think E[lizebeth] cares a very great deal more for you than she lets herself or anybody else believe"—but also implied that perhaps she, Edna, the mature and responsible sister, might make a better mate for William than younger, flakier Elizebeth. Edna wrote to William, "My idea of real love-making is sort of the Lochinvar kind"—Lochinvar, the hero of an old Scottish poem: *So faithful in love, and so dauntless in war / There never was knight like the young Lochinvar.* "Him riding up furiously, sweeping his bride up before him with one hand and riding away."

Although Elizebeth found William attractive, she was drawn at first to his way of carrying himself, his scrupulous precision about words and facts and clothes, his modesty—qualities that made him George Fabyan's opposite. She liked checking in with William after spending hours in the blast zone of Fabyan's hype cannon. It felt healing, like drinking a glass of cold lemonade after a long walk. And what a mind he had! Talking to most people, Elizebeth felt like she could see the rough carpentry of their thoughts, the joints and tenons that never quite fit, but with William, ideas emerged smooth and whole, as if from a workshop. And he was so *playful* about it all, unlike Fabyan and Mrs. Gallup. Science to them was about results: defeating gravity, rewriting literary history, finding the secret to eternal life. Huge, epic, shocking, revolutionary ends. William never used such words. He didn't care about the answers so much as the questions. He enjoyed science because it was an interesting way of being alive.

He had a feel for ciphers thanks to his work with Mrs. Gallup, and also a youthful fascination with Edgar Allan Poe's short story "The Gold-Bug." The plot of the story revolves around a cryptogram whose solution points to a buried treasure chest full of diamonds, rubies, emeralds, sapphires, and gold coins, placed there by a murderer. Poe wrote articles about codes and ciphers, bragging that he could solve any cryptogram and daring readers to stump him, and for decades to come, Americans associated codebreakers with the sunken-cheeked, disreputable figure of Poe. William took a more whimsical approach to ciphers. He liked to blend them with his knowledge of botany to make jokes and works of art. At Riverbank he drew a sketch of a long-stemmed plant with many fine veins in its leaves; although from a distance it looked like an ordinary botanical illustration, closer examination revealed patterns of notches in the roots and leaves and petals that spelled out the words "Bacon" and "Shakespeare" in the biliteral cipher. He captioned the drawing, "CIPHER BACONIS GALLUP," "A MOST INTERESTING AND PECULIAR PLANT, PROPAGATED AT RIVERBANK RESEARCH LABORATORIES."

The autumn weeks burned away. The temperature dropped and Elizebeth experienced her first Riverbank winter, a gray duration of pitiless wind that scraped across the plains unbroken and slammed into the estate. The sky threw down a tarp of pale blue light. Your breath crystallized in the air like clouds of cigar smoke, and the cold groped into your lungs. The cottages and labs burned coal all day and the dark gray smoke rose from the chimneys. Elizebeth and William were growing closer. She didn't know what to call it—more than friends, less than lovers. William would perch in a rocking chair sometimes and she would sit on his lap as he pushed the chair forward and back, his thin arms around her thin waist, the chair creaking in a steady rhythm, neither of them saying much at all.

It took a while for her to get up the courage to share her doubts about Mrs. Gallup's work; she worried that William would look at her strangely, would think she was wrong and think less of her. But Elizebeth's mind wouldn't let it rest, and eventually, she asked what he thought. Wasn't it strange how Mrs. Gallup could see these things that no one else could see?

To her enormous relief, William said he had been wondering the same. Sometimes a thought floated to the front of his mind, the deepest heresy at Riverbank: *There are no hidden messages in Shakespeare.*

The idea rang in the air between them like a broken chime. Ugly, dissonant notes. Elizabeth and William exchanged a look. For the first time, but not for the last, each gathered strength from the other, and the notes resolved into a chord: *There are no hidden messages in Shakespeare.*

What if everyone involved in the Bacon work was crazy, except for the two of them?

CHAPTER 4

He Who Fears Is Half Dead

and then begins *step step leap*
she continues these leaps
scramble the code scramble uphill scramble eggs
and without premeditation but in full arc if possible
have a good time.
—ANNE CARSON

The intercepted and decoded telegram burned its way from hand to hand, from junior diplomat to senior diplomat, first in London and then in Washington, producing involuntary noises of surprise and bulging eyes. It was obvious that the president himself needed to see it. At 11 A.M. on February 27, 1917, the U.S. secretary of state, Robert Lansing, carried a copy of the intercepted telegram to the White House and showed it to Woodrow Wilson. The president read it and grew uncharacteristically angry: "Good Lord!" he said. "Good Lord!"

The telegram had been sent from Germany to Mexico on January 16, traveling by three separate telegraph routes and encoded as a series of number blocks: 130 13042 13401 8501 115 3528 416 17214. The British had intercepted the message, and a small team of civilian codebreakers toiled for a month in a secret office inside Whitehall to scrub away the grime of code and make the plaintext visible. What they saw, to their shock, was nothing less than a conspiracy plot against the United States.

Written by Germany's foreign minister, Arthur Zimmermann, the telegram proposed an alliance between Germany and Mexico: "We intend to begin unrestricted submarine warfare on the first of February. We shall endeavor in spite of this to keep the United States neutral. In the event of this not succeeding, we make Mexico a proposal of alliance on the following basis: make war together, make peace together, generous financial support, and an understanding on our part that Mexico is to reconquer the lost territory in Texas, New Mexico, and Arizona. The settlement in detail is left to you."

The Zimmermann Telegram, as it came to be known, was indisputable proof of a German plot against America, "clear as a knife in the back and near as next door," as the historian Barbara Tuchman has put it. Residents of Texas were particularly displeased to learn that the Kaiser was trying to give them to Mexico, but outrage against Germany was general across the States. The telegram sped up history. It pushed America into war with Germany, whether America was ready for war or not.

It was not.

And this is how the telegram changed the destinies of Elizebeth Smith and William Friedman: as American codebreakers, they happened to possess an extraordinarily rare and suddenly indispensable set of skills.

Elizebeth got word in January 1917 that her mother, Sopha, long ill from cancer, was on the verge of death, and Elizebeth should come to Indiana to say goodbye. She packed a bag and rode the train back to Huntington and her childhood home. Her father was there, and her sister, Edna. The two sisters consoled each other as physicians prowled through the old house, murmuring about a growth. Sopha was in a lot of pain and vomited violently. A doctor turned her on her belly and spread iodine across her back. He used cocaine to numb a particular spot and tapped a metal rod into the skin, removing what Elizebeth felt was a horrifying quantity of pinkish fluid.

She had brought some cipher materials with her, hoping to get work done. "My book-bag lies here unopened," she wrote to William at Riverbank. "I try to make myself work, but I cannot.

I sit a moment, then spend the hours pacing back and forth from Mother's bed, in the vain hope that there is something I can do. It is so awful—Billy Boy—to look on the face of death like that—the beckoning face—Do you know it makes me think a lot about posterity, and responsibility, and all that?" She wasn't sure what to call William in these letters, or to call herself in relation, so she mostly kept things platonic, signing her letters "yours, Elsbeth," and thanking William for being "one of the truest friends I've ever had," although she did admit that she missed William's "rocking," his comforting way with a rocking chair, and in one letter she slipped in something stronger: "I love you / Elsbeth."

When Sopha died, in February 1917, Edna stayed behind to arrange the funeral, while Elizebeth returned to Riverbank, seized by a new impatience. She had no desire to spend any more time on the Bacon ciphers. Life was too short to waste on fruitless quests. When she reunited with William, he said he felt the same. They both agreed they had to remove themselves from the project. The question was how.

Confronting Mrs. Gallup seemed a little cruel. She had worked too many years in a single direction to admit her compass was broken. She had treated both of them with kindness. They tried talking to Fabyan instead. On a few occasions the two youngsters buttonholed him and tried to get him to listen. Mrs. Gallup's theory was unsound, they said. Fabyan's money might be better spent on other projects. He shouted them down, as they expected. Fabyan said he wasn't paying them to question the theory, only to persuade the academy that it was correct.

But by now, even if he didn't want to admit it, a new scheme was diverting Fabyan's attention from the literary ciphers. Shakespeare, Bacon, Mrs. Gallup, old books, dead men—it paled in urgency to the world of the living.

For months now Fabyan had been advertising his patriotism and his willingness to place Riverbank at the disposal of the flag. He had ordered his groundskeepers to expand the network of model trenches next to the Lodge, and after months of digging by a team of mud-spattered workers the trenches reached a total length of three miles, enough to be useful for the Fox Valley Guards to conduct infantry drills complete with live mortar rounds. And Fabyan had told officials in Washington that if they needed help with codebreaking, Riverbank stood ready to serve.

"Gentlemen," he wrote to Washington on March 15, 1917, "I offer anything I have to the government, and if you care to have any of your local men call on me, and see the work that is being done, I should be very glad to show it to them." He described his interest in old ciphers, especially the biliteral cipher of Francis Bacon, and added: "To avoid any possibility of being considered a crank, or a theorist, I respectfully call your attention to the fact that I was the business partner of the late Cornelius N. Bliss, formerly Secretary of the Interior, whom most of the older men in Washington remember with a great deal of respect and admiration."

Military officials were of course reluctant to give any power or responsibility to a fake colonel in Illinois, but they had little choice but to accept Fabyan's offer. They were desperate for codebreakers because of the way radio and wireless technology was changing the art of war.

In earlier conflicts, codebreakers had mattered less; fewer military and diplomatic messages were encrypted because the messages were harder to intercept. If you wanted to steal an enemy message, you had to capture a messenger on horseback, or open an envelope at a post office, or install a tap on a telegraph wire. But with radio, all it took to intercept a message was an antenna. The air was suddenly full of messages in Morse code, dots and dashes that registered as audible pings and whines. You could pluck them out of the sky. So to protect their secrets, armies had begun encrypting their wireless messages before sending them over the wireless in Morse.

This simple fact transformed codebreakers from disreputable freaks into potential superheroes, wizards with power over life itself. Now the air was full of encrypted information of enormous tactical significance and the utmost stakes. The routes of ships at sea. Troop movements on the ground. Airplane sightings. Diplomatic negotiations and gossip. Reports of spies. Thousands upon thousands of puzzles zipping through the atmosphere, any one of which, if decrypted, might win or lose a battle, wipe out a regiment, sink a ship. In this new world, a competent codebreaker was suddenly a person of the highest military value—a savior, a warrior, a destroyer of worlds.

And yet, as Elizebeth would later write, "There were possibly three or at most four persons" in the whole United States who knew the slightest thing about codes and ciphers. She was one of them, William another.

The government lacked the capacity to reliably intercept foreign messages, much less break the codes and read them. The CIA didn't exist in 1917. There was no NSA, and the FBI was a crumb of its future self, a nine-year-old organization known as the Bureau of Investigation, which fielded only three hundred agents, on a total budget of less than half a million dollars. There simply was no intelligence community as we think of it today. The Department of Defense was called the War Department then, which operated the army, and though the War Department did contain an intelligence-gathering section, the Military Intelligence Division (MID), it was tiny and underfunded. On the day Congress declared war, April 6, 1917, the MID employed just seventeen officers. The officer in charge of the MID, Major Ralph van Deman, considered the government's ignorance of codes and ciphers an "emergency."

So, in the second week of April, the War Department dispatched an emissary to Riverbank, an army colonel named Joseph Mauborgne, to check out the place and report back on its suitability.

Mauborgne was one of the three or four people in America who knew something about codebreaking. In 1912, while stationed at the Army Signal Corps School in Kansas, a bare-bones airfield and laboratory to probe radio technology, Mauborgne had made history by figuring out how to send a radio signal from a plane to the ground for the first time, and in 1914 he became the first American to break the British army's field cipher, known as the Playfair Cipher, based on a table of letters arranged in a five-by-five grid.

When Mauborgne arrived at Riverbank, Fabyan greeted him with the usual overwhelming gusto and brought him to the second floor of the laboratory building, declaring with a flourish that the Riverbank Department of Ciphers was open for business. The office appeared busier and more crowded than it had ever been. In anticipation of the army man's visit, Fabyan had gone out and hired a dozen clerks, stenographers, and translators fluent in German and Spanish, to provide support for Elizebeth and William. Fabyan hoped the two young people would be able to lead the effort, to break codes for the government, while Mrs. Gallup continued her long labors on the Bacon ciphers. Superficially, the office looked like the picture in Fabyan's imagination, the pitch he had sold to Washington. It looked like a codebreaking agency on the prairie.

There in the new Department of Ciphers, Elizebeth and William introduced themselves to Mauborgne. They clicked with him immediately. He was thirty-six and big—big body, big voice, big brain, with perfectly round, black glasses. He was the only man they had ever seen stand eye to eye with Fabyan and not seem intimidated. Mauborgne liked Elizebeth and William, too. He saw a spark in the pair of young codebreakers. (He would later call them "the two greatest people I have ever known.") They had little formal training but were bright and eager. Fabyan made him wary—a mess of a man, lunging wildly from promise to promise—but it was undeniable that Riverbank had excellent security from a military standpoint. Aside from the virtue of being in the middle of nowhere, safe from enemy attack, it was protected by the lighthouse, and Fabyan also had the Fox Valley Guards nearby—his own private army. If all else failed and the Germans invaded, Fabyan said he would open the bear and wolf cages in the garden and sic the beasts on the intruders.

On April 11, Mauborgne informed his commanders that Riverbank was ready. He urged the army and also the Justice Department "to take immediate advantage of Colonel Fabyan's offer to decipher captured messages," owing to "the mass of data" in his private library of cipher books, the security of his compound, and the quality of his employees, "a force of eight or ten cipher experts who spend their time delving into the works of antiquity, discovering historical facts hidden away."

After reading Mauborgne's enthusiastic report, Van Deman of the MID wrote to Fabyan with gratitude, thanking him for "your exceedingly kind and patriotic offer of assistance," and soon encrypted messages started arriving at Riverbank from Washington. They came in the mail and by telegram, sent by different parts of the government: the War Department, the navy, the Department of State, the Department of Justice. The messages had been intercepted by covert means, mostly from various telegraph and cable offices across the country.

Fabyan had gotten his wish: for the foreseeable future, Riverbank would become ground zero for military codebreaking in America, a de facto government agency. He had drafted Elizebeth and William into the war, assuming they would be able to handle what was coming. But when they looked at the messages, the fresh piles of gobbledygook spilling from the mail sacks onto their desks, they weren't sure that he was right.

A woman and a man are sitting side by side in a busy room. People come and go and the door opens and closes and there is the sound of typewriter keys smacking ink into pages. Outside the window, hawks fly and cows moo and a bear scratches himself in an iron cage and a parrot sings and a river runs and there are also monkeys in diapers for some reason.

The two people, Elizebeth and William, notice nothing except what is in front of them on a slab of desk. They are looking down at a sheet of paper. All of their intensity is shining down at the paper, a bright beam of desire to understand the text that is typed there.

It looks like nothing. It is clearly not written in the biliteral cipher of Francis Bacon that they are familiar with. It is something else, a new level of mystery. They must understand it. But they don't know what they are looking at.

BGVKX	TLXWB	SHSFW	KWGRI	KZTZG
RKZFE	YDIWT	KOFOB	GUHGD	SFVRE
UIUQX	HSLDS	OHSRM	HTWKY	VHUIK
BJDUH	VSART	BGVNG	VBAFO	AZOXG
PQPMJ	DRODW	RCNML	MTMXL	SSVAR

A hiss of symbols, a raw block of babble. A cryptogram. Someone wrote and sent it for a purpose, and someone else intercepted it, and now it is here on your desk. These letters contain meaning. How to unlock it without knowing the key?

The basic task of codebreaking might seem impossible if you think about how many different ways a message might be encrypted. Each human language has its own quirks and curiosities. Then, within each language, a cryptographer can choose from among dozens of varieties of locks—codes and ciphers. And each lock will accommodate only one of a vast array of possible keys.

For instance, one of the simplest kinds of ciphers, called a mono-alphabetic substitution cipher, or MASC, swaps out one set of letters for another. Perhaps A=B, B=C, and C=D, or perhaps A=X, B=G, and C=K—or any other map between the 26 letters of the plaintext alphabet and 26 different letters in the ciphertext. A MASC is a very basic method of encrypting a message. But even here, there are 403,291,461,126,605,635,584,000,000 potential alphabets: 403 *septillion*. A thousand computers, each testing one million alphabets per second, would take more than a billion years to exhaust the possibilities.

And yet anyone who has ever solved a cryptogram on a newspaper puzzle page has conquered the 403 septillion possibilities, because, of course, there are shortcuts, ways of taming the task by grabbing on to certain patterns in the text.

This is the essence of codebreaking, finding patterns, and because it's such a basic human function, codebreakers have always emerged from unexpected places. They pop up from strange corners. Codebreakers tend to be oddballs, outsiders. The most important trait is not pure math skill but a deeper ability to pay attention. Monks, librarians, linguists, pianists and flutists, diplomats, scribes, postal clerks, astrologers, alchemists, players of games, lotharios, revolutionaries in coffee shops, kings and queens: these are the ones who built the field across the centuries and pushed the boundaries forward, stubborn individuals with a lot of time to sit and think and not give up.

Most were men who did not believe women intellectually or morally capable of breaking codes; some were women who took advantage of this prejudice to steal secrets in the shadows. One of the more cunning and effective codebreakers of the seventeenth century was a Belgian countess named Alexandrine, who upon the death of her husband in 1628 took over the management of an influential post office, the Chamber of the Thurn and Taxis, which routed mail all throughout Europe. The countess had a taste for espionage and transformed the Chamber into a brazen spy organization, employing a team of agents, scribes, forgers, and codebreakers who melted the wax seals of letters, copied their contents, broke any codes, and resealed the letters. This was an early example of what the French would later call a *cabinet noir,* or black chamber, a secret spy room in a post office. The countess's male contemporaries were slow to discover her true occupation because they couldn't imagine that a woman was capable of such deceptions. "What if this countess does not merely open our letters but is also capable of deciphering their contents?" one diplomat wrote in panic to another. "God knows what she is capable of doing to us!"

The two most prominent codebreakers in America when Elizebeth and William started were a married couple, Parker Hitt and Genevieve Hitt. Parker was a tall, dashing Texan with weathered skin, an army infantry commander in his thirties who had gotten interested in cryptology after volunteering to fight in the Spanish-American War and trying his hand with messages intercepted from the Mexican army. His wife, Genevieve, a proper southern girl, had scandalized her family by falling in love with a man they saw as a cowboy. She also studied cryptology, eventually becoming chief of the code operation for the War Department's Southern Division, based in San Antonio. "This is a man's size job," she wrote to her mother-in-law, "but I seem to be getting away with it, and I am going to see it through. . . . I am getting a great deal out of it, discipline, concentration (for it takes concentration, and a lot of it, to do this work, with machines pounding away on every side of you and two or three men talking at once)." Parker supported Genevieve and was proud of her: "Good work, old girl," he wrote to her in one letter.

Parker was the only American to have written a serious book about cryptology. Aimed at army units with no prior training, *Manual for the Solution of Military Ciphers* showed how to set up a quick-and-dirty deciphering office in the field with five or six soldiers, some radio equipment to intercept enemy signals, and a day or two of study. Hitt went over the basics of military cryptography and explained, accurately, that the methods of the world's armies had not changed much in hundreds of years. Just like there are millions of chicken recipes in the world but only several basic methods to cook the bird (roasting, frying, poaching, boiling), there are countless ciphers but only a handful of common types. Then he laid out some basic steps for solving a cryptogram written in cipher. Today a computer could do any of these steps in picoseconds, but in 1917 it all had to be done by hand, with a pencil and paper.

The first step was usually very simple: count the letters in the cryptogram. In English, the most frequently used letter is E, the most frequent two-letter group is TH, and the most frequent three-letter group is THE. So if you count the letters in the ciphertext and the most common letter is B, it might stand for E, and if the most common three-letter group is NXB, it might stand for THE.

You can count other things in a cryptogram, like the total number of vowels and consonants, and how often particular letters or groups of letters appear before or after other letters. All of these counts give hints to the hidden structure. A frequency count can also reveal if the plaintext was written in English, German, French, Spanish, or some other language, because the frequency of letters in a language is like a unique signature. The most common six letters in German, starting with the most common, are E, N, I, R, T, S. In French, E, A, N, R, S, I. In Spanish, E, A, O, R, S, N.

It's best to do the counting in a systematic way. You might start by drawing a thing called a frequency table. You chop the cryptogram into its component parts and sort them according to the letters by which they're surrounded. It looks like this:

A	B	C	D	E	F	G	H	I	J	K	L	M	N	O	P	Q	R	S	T	U	V	W	X	Y	Z
IP	HF		NR	PX	BA	JN	DJ	UJ	IG	NR	RJ		GD	JW	RE	AD	DP	FQ	AR	—I		OR	EY	XJ	
OQ	FR		QH		WS	PR	UN	RA	YO	PR	AI		HH	RA	AG	PJ	WI	JF		JH		JF	RP	JQ	
FT	FN		HR		JB	JJ	NB	WQ	HU				BG	WR	RQ	SW	GO			RI		QI	QP		
GL	PR				SB	NA	RD	QR	QW				QQ	RJ	XQ	=	TP					RO			
IP	HP						NB	LQ	LF				PH	IJ	XQ	PJ	IL								
							JR	PN	QG				IK		RB	IX	BH								
								QR	GS						RN	PN	DX								
								UA	OY						BI	NR	QP								
								QR	OH						AK	RI	BW								
								QO								YI	OR								
																RI	RP								
																	KQ								
																	IU								
																	KO								
																	IQ								
																	H—								

Though it may look like gibberish, it's a powerful tool—"the Real Stuff," in Elizebeth's phrase—because with a quick glance down the columns, you can identify the most frequent letter groups in the cryptogram and the letters that come before and after them. Letters in a given language are like children in a kindergarten class; they have affinities, cluster in cliques. In the lunch line, one kid likes to walk behind a second kid and in front of a third kid while a fourth sits off in the corner, eating from a paper bag. What you're really looking at when you look at this frequency table is a picture of "certain internal relations in the English alphabet," as Elizebeth and William would put it. You're looking at the structure of the underlying language itself.

Now you have some grip on the puzzle. You can begin to peel back the skin of the message, to see familiar shapes in the strangeness. Like with a crossword puzzle, there's no direct, guaranteed route to solving a cryptogram. The solver has to make educated guesses, plug in letters and see if they lead to recognizable words, backtrack and erase if a guess is wrong, try a new letter.

Elizebeth quickly got the hang of it, plowing through messages and counting letters, although it felt completely new and weird to her, a totally different way of looking at language than what she was used to. All her life she had celebrated the improbable bigness of language, the long-lunged galaxy that exploded out from the small dense point of the alphabet, the twenty-six humble letters. In college she trained herself to hear the rhythms of playwrights and poets, the syllables that slip from the tongue in patterns. Tennyson:

There lives more faith in honest doubt,
Believe me, than in half the creeds.
There LIVES more FAITH in HON-est DOUBT,
Be-LIEVE me, than in HALF the CREEDS.

But before, she had gone no further than chopping lines into meters. She left the words in their boxes, intact. Codebreaking required more drastic measures. Now Elizebeth had to shake the words until they spilled their letters. To rip, rupture, puncture, chisel, scissor, smash, and scoop up the rubble in her arms. To chip off flakes from the smooth rock of the message and place them in piles and ask questions about them. It involved a kind of hard-hearted analytic violence that she had never contemplated before. It was reaching into the red body of the text until the hands dripped with blood.

Ahhhh!

The first few messages she broke, real military messages, had been intercepted from the Mexican army. Like most military cipher messages, they were written in blocks of five letters, like TZYTV RGFQF MQFHC, in order to fuzz out the original lengths of the words and therefore make the messages harder for adversaries to break. Elizabeth counted the letters, drew her frequency tables, consulted materials on the frequencies of various letters and letter combinations in the Spanish language, scratched her guesses into the graph paper, and there, *right there*—she saw things that started to look like words. A lovely shape pried out of the murk, glistening.

The process gave her a sensation of power that was electric and new and made her want to keep going. It was nothing like working on the biliteral ciphers with Mrs. Gallup. Here there was no mystery, no squinting through a looking glass at the curls of italic letters and trying to sort them into categories based on vague criteria never fully explained. Here the method was sharp and clear, a series of small and logical steps that built toward a goal. "The thrill of your life," Elizabeth said later, describing how it felt to solve a message. "The skeletons of words leap out, and make you jump."

And she was never alone. That was the other thrill. She and William operated as a team. During the day they were never more than a few feet apart, handing papers back and forth, checking each other's work, asking questions when stuck, keeping up a friendly patter, "calling out" letters on their sheets in the "word-equivalent" alphabet commonly used by the U.S. Army: Able for the letter A, Boy for B, Cast for C, Dock, Easy, Fox, George, Have, Item, Jig, King, Love. If Elizabeth needed to read the ciphertext FVGEQ, she would call out, "Fox! Vice! George! Easy! Quack!"

It was demanding work. Each solution had to be checked. Errors corrected. A single miscopied letter could wreck hours of effort. You got tired and needed a friend to look at your page while you rubbed your eyes. Each learned to recognize signs of fatigue in the other and knew when to suggest that it was time for a break. The less you had to think about, the better and more accurate your work. Elizebeth and William used the same kind of pencil, the same kind of paper, and never deviated from these choices. They liked pencils with soft lead and big erasers, the eraser end seeing as much action as the lead end.

Cast! Easy! Jig! King! Opal! They called out letters all day long like teachers taking attendance at a strange school. *Pup! X-ray! Vice! Love! Sail!* The pencils at Riverbank were plentiful and free, black with white erasers, and doubled as advertising tools; a cipher alphabet was printed on each pencil in white letters, along with RIVERBANK LABORATORIES—GENEVA, ILLINOIS.

Mike, she called out, a smile playing at the corners of her mouth. *Zed. Rush. Fox. Zed.*

Watch, he called out, grinning. *Dock. Yoke. Pup. Easy.*

The paper they used was graph paper with a grid of quarter-inch squares. One letter per square. They never threw anything out. "Work sheets SHOULD NOT BE DESTROYED," the pair would soon write in one of several scientific papers about their discoveries. Worksheets "form a necessary part of the record pertaining to the solution of the problem. No work is too insignificant to discard, therefore it should be done well from the start."

Tare. Yoke. George. George. Able.

Unit. King. Nan. Zed. Boy.

No one told them how to set up this workflow, and no one told them they had to collaborate. They simply found, by trial and error, that collaboration made things go faster, that "a group of two operators, working harmoniously as a unit, can accomplish more than four operators working singly. Different minds, centered on the same problem, will supplement and check each other; errors will be found quickly; interchange of ideas will bring results rapidly. In short, two minds, 'with but a single thought,' bring to bear upon a given subject that concentration of effort and facility of treatment which is not possible for one mind alone."

Although William and Elizebeth solved their first batches of military messages using techniques they learned from Hitt's manual, they soon exhausted its teachings. Elizebeth filled the margins with her own notes in blue pen, comments about sentences she found imprecise, things Hitt had gotten slightly wrong, things he could have explained better or had left unexplored. (She underlined a sentence on page 85 and wrote next to it, "This is poorly expressed.") She and William had reached the cordon of what was known, the edge of the map. From here on, they would need to invent new techniques—to become scientists, explorers, pushing into a wild land.

One way of thinking about science is that it's a check against the natural human tendency to see patterns that might not be there. It's a way of knowing when a pattern is real and when it's a trick of your mind. Elizebeth and William had begun at Riverbank by looking for the false patterns of Mrs. Gallup. But now, over the next several years, they found ways of seeing true patterns. It was as if they had been tossed into a raging river of delusion without knowing how to swim and figured out how to save themselves from drowning, clinging to each other the whole time. This struggle made them stronger than they could have ever imagined. They climbed out of the river transformed, with new powers, shaking the water from their backs, and then took off at speed, racing across the mountains and through the swamps of an undiscovered continent.

Between 1917 and 1920, George Fabyan used Riverbank's vanity press to publish eight pamphlets that described new kinds of codebreaking strategies. These were little books with unassuming titles on plain white covers. Today they are considered to be the foundation stones of the modern science of cryptology. Known as the Riverbank Publications, they "rise up like a landmark in the history of cryptology," writes the historian David Kahn. "Nearly all of them broke new ground, and mastery of the information they first set forth is still regarded as the prerequisite for a higher cryptologic education."

The eight Riverbank Publications are commonly attributed to William alone, with two exceptions. Inside his personal copy of one paper, Riverbank No. 21, *Methods for the Reconstruction of Primary Alphabets,* William wrote in black ink beneath the title, "By Elizebeth S. Friedman and William F. Friedman." A second paper, *Methods for the Solution of Running Key Ciphers,* never included her name, but she and William always told colleagues it was a joint effort.

However, there's evidence that Elizebeth was involved with more than just the two papers. The original typewritten and hand-edited drafts of the Riverbank Publications are now held by the manuscript division of the New York Public Library, and her handwriting is all over them. William seems to have written a lot of the technical sections, with the drafts marked up by both of them, Elizebeth's comments interspersed with his, while Elizebeth wrote and researched the historical sections, which he edited in a similar fashion.

They worked as a team in most matters and the soon-to-be-legendary papers were no different. In a 1918 letter to Elizebeth, William referred to the early Riverbank Publications as "our pamphlets"—*our*, not *my*. And other Riverbank workers contributed as well: men and women, codebreakers and translators. The publications were "a piece of work that was done by the staff," Elizebeth said later. "No one person was mentioned as the sole conqueror or anything like that. Everybody worked together." This is as far as she ever went in claiming a piece of the credit. Today it's hard to know exactly what she did, because she wanted it that way. "Mrs. Friedman had a tendency to see that the record made little or no mention of her contribution to a number of their joint efforts," the custodian of the Friedmans' personal papers wrote in 1981 to a researcher interested in Elizebeth. "And therefore it will be difficult to get a clear picture of her exact role."

Why hide her role? Partly it was expected at the time, that the man was the scientist and the woman the helpmate, but Riverbank was a bubble world where the usual rules didn't apply. Fabyan had no trouble championing the work of women, as he proved with his zealous promotion of Mrs. Gallup. A more likely explanation is that Elizebeth was trying to help William win a battle with Fabyan over the copyrights of the Riverbank Publications. At first, Fabyan didn't even let William place his name on the covers of the pamphlets, only on the inside pages, and Fabyan registered the copyright under his own name. He said he saw no ethical problem because he had paid for the research. "It may be egotism on my part," Fabyan told William, "but so long as I pay the fiddler, I am going to have the privilege of selecting a few of the tunes."

It was hard enough for William—a credentialed scientist, a genetics Ph.D.—to get credit for the work. He and Elizebeth may have decided it would be doubly hard to convince Fabyan to share credit with her, too.

Whatever the case, the Riverbank Publications, and the breakthroughs they describe, still seem incredible today. Seven of the eight pamphlets were written in the space of two years, in a little cottage in the middle of Illinois, the cryptologic equivalent of Albert Einstein's annus mirabilis, when Einstein rewrote the language of light, mass, and time in the space of a single year, at age twenty-six, while working as a patent clerk in Switzerland, staring out the window of his office and bouncing ideas off a fellow clerk. This is the achievement that the NSA interviewer in 1976, Virginia Valaki, kept begging Elizebeth to explain: *How?* Elizebeth gave unsatisfying answers, noted in the transcript:

"That World War I leapt on, and so many things happened so fast. . . ."

"Nothing was ever as carefully executed as that. It was sort of on a day-to-day basis. You did what you could with what you had to do it with."

"I don't think I remember offhand. I was too busy either getting on this swing or getting off that one. ((Laughs.))"

"I feel no confidence whatever to speak on that point; wouldn't have the faintest idea what to say."

The likely truth: it only looked improbable in retrospect. At the time they didn't know what was supposed to be hard, and there was no one around to tell them. They didn't see themselves inventing a new science. They were playing the game day-to-day as best they could, as Fabyan always said. They were just trying to solve messages as they poured in and not get stuck.

The mail from Washington contained a frothy mix of messages from all over, a zoo of alphabets that had to be studied and classified. There were two main animal kingdoms of cipher, "transposition" and "substitution." A transposition cipher was like Scrabble, a jumbling of the same letters into a new order. A substitution cipher was a swapping of letters. Each kingdom contained a diverse multitude of beasts that had to be tamed in different ways, and there was always a time crunch, someone demanding a quick answer. Invention under pressure.

One day in early 1917 a heavyset man showed up at Riverbank on a mission all the way from Scotland Yard, the police headquarters in London. He had been referred to Riverbank by the U.S. Department of Justice. Fabyan barked an introduction at Elizebeth and William, and the detective opened a briefcase. Stacks of messages spilled out. He said the messages had been intercepted by British postal censors, and the recipients included as many as two hundred individuals in India, then a British colony. Scotland Yard suspected an attempt by Germany to spark a revolution among Hindu separatists, but no one knew for sure. All the detective knew were the names of a few of the suspects, which he told to Elizebeth and William.

The young codebreakers looked at the messages and found them "quite baffling." They were written in numbers, grouped together in shorter or longer blocks:

```
38425   24736   47575   93826
97-2-14
35-1-17
73-5-3
82-4-3
```

Elizebeth and William assumed from these groupings that the separatists were using three different codes. The blocks of five numbers looked like a simple type of codes based on a rectangular grid of letters:

	1	2	3	4	5	6	7
1	A	B	C	D	E	F	G
2	H	I	J	K	L	M	N
3	O	P	Q	R	S	T	U
4	V	W	X	Y	Z		

The grid turns a letter into a number (C is 13), which then has a number added to it, based on a prearranged key word. If the word is LAMP, the value of L (25) might be added to C (13), making 38. Elizebeth and William deduced the key word and solved the messages by analyzing frequent numbers and the intervals between them.

Thinking about the second set of numbers (97-2-14), which were confined to a single long message in the detective's trunk, Elizebeth and William noticed that the middle number was always either a 1 or a 2. This was a clue that the conspirators were using a specific, yet-unknown book to encrypt their messages and that the book had two columns of type, like a dictionary. The numbers likely pointed to words at certain locations in this mysterious book. For instance, in a sequence like 97-2-14, 97 meant the page number, 2 meant the right-hand column, and 14 meant the four-teenth word in the column. Applying similar logic to the third set of numbers in the detective's messages (73-5-3, 82-4-3), the young codebreakers deduced that the numbers pointed to *individual let-ters* within a different book possessed by the conspirators. The recipient of the letter, seeing the number 73-5-3, would turn to the seventy-third page of the book, go to the fifth line, and write down the third letter in the line.

Of course, this wasn't enough information to solve the mes-sages, and the young codebreakers at Riverbank weren't sure they could go further: If the letters and words had been selected from specific books, and Elizebeth and William didn't know the names of the books or have copies of them, what was the use?

"For a time," William wrote, "it looked like an insurmount-able task."

But they wrote down all the numbers in order, searched for repetitions, thought about it some more, and found a foothold. A Harvard professor had recently counted the words in a long English text, and the prairie codebreakers read his study. Of 100,000 total words, only 10,161 were unique, and just 10 words accounted for 26,677 of the 100,000: "the," "of," "and," "to," "a," "in," "that," "it," "is," "I." "You can't convey much intelligence using only these words," William wrote, "and yet you can't con-struct a long, intelligible, unambiguous message without using them over and over again."

Turning to the numbers that seemed to come from a dictionary, the codebreakers reasoned that a word at the beginning of
the alphabet, like "and," would correspond to a code group beginning with a lower number (1, 2, 3, 4) than a word that came
higher in the alphabet, like "the." This insight helped them solve
the most frequent words in the messages, and from there it was
possible to work out others: If 97-2-14 was YOU, then 99-2-17
must be a word close to "you" in the dictionary, perhaps "your,"
"young," or "youth." Elizebeth and William ended up solving 95
percent of the message in this manner, without ever seeing a copy
of the conspirators' dictionary. As for the last set of numbers, the
ones that seemed to refer to letters in an unknown book, they
used a similar process, matching frequent code groups with frequent letters and pairs of letters, and reverse-engineering the text
of the book as they went. Whenever they discovered a new letter
of plaintext, it told them more about the content of the book, and
whenever they pieced together a new line in the book, it told them
more about the plaintexts of the messages. One of the conspirators' notes read, in part:

> I challenge anybody who dares ignore the solid work done
> through our agencies. . . . Our men worked, suffered. Still
> suffering. . . . We have succeeded in laying foundation for
> future work . . .

The two young codebreakers ended up solving the whole trunkful of messages for Scotland Yard, revealing an intricate separatist
plot by Hindu activists living in New York to ship weapons and
bombs to India with the help of German funds and assistance:
dates, times, places, names. Several conspirators were charged in
San Francisco, and prosecutors summoned William to testify in
open court about how he broke the codes. Elizebeth wasn't asked.
She hated staying in Illinois while William went on an exciting
trip to the West Coast—she thought she deserved to be called as
a co-witness—"but someone had to stay behind and sort of oil the
machinery at Riverbank." She didn't speculate about why William
was chosen instead of her, perhaps because the answer was obvious: prosecutors thought the jury would more easily believe a male

expert. As it turned out, the trial erupted in spectacle: Before William had a chance to say his piece on the witness stand, an Indian man in the gallery stood up, pulled a handgun from his vest, and shot one of the defendants in the chest. He yelled a single word—"Traitor!" Then a U.S. marshal fired at the gunman over the heads of the shocked spectators, killing him. The shooter apparently thought the defendant had snitched to the government, betraying his friends by revealing the code. He didn't know about William, Elizebeth, and the science of codebreaking.

For the first eight months of the war, as incredible as it sounds, William and Elizebeth, and their team at Riverbank, did *all of the codebreaking* for every part of the U.S. government: for the State Department, the War Department (army), the navy, and the Department of Justice. And the broader scientific insights of the Riverbank Publications emerged directly from these day-to-day puzzles solved under wartime pressure. The pair would solve a cryptogram and realize they may have stumbled onto some more general method. Then they would test the method on additional examples, trying to see where it broke, what its limits were, aiming to strengthen a one-time solution into a universal principle and to share that knowledge with others.

It had long been known that the frequencies of letters in a cryptogram provide clues to its solution. Knowing this, cryptographers had invented many ways of obscuring the letter frequencies, making messages harder for adversaries to break. It was possible to encrypt a message with multiple cipher alphabets instead of just one (a poly-alphabetic cipher). It was possible to rely on a secretly chosen novel or dictionary to generate a code message, in the style of the Hindu revolutionaries. If the sender and recipient happened to select a key in their book that was exactly the same length as the message—known as a "running key"—the message became even harder to break. The War Department considered running-key messages to be indecipherable. And these methods could be mixed and matched to further frustrate the codebreaker. For instance, a plaintext word like "strawberry" could be turned into a block of code like WUBCW, then those letters transformed with a cipher into LWJIJ—a process of "enciphered code."

Each of these techniques placed a wall between the message and the codebreaker—sometimes a pane of frosted glass, sometimes a sheet of metal or stone. Elizebeth and William invented new tools for destroying these walls—hammers, corrosive acids, explosives. They learned to identify and solve several different kinds of substitution ciphers: straight alphabets, direct alphabets, reversed alphabets, poly alphabets, mixed alphabets. They developed general techniques of solving book ciphers without needing a copy of the book. They taught themselves to solve messages enciphered with running keys. Together in Engledew Cottage they strolled through cities of text with their wrecking kit, swinging hammers with glee, blowing up brick, melting steel, the sound of breaking glass echoing out into the prairie. Then they wrote the scientific papers, the Riverbank Publications, documenting exactly how they did it, and how other people could do it, too, if they followed the same steps.

This part was crucial. The test of a scientific discovery is if others can replicate it and get the same results. Mrs. Gallup had never passed this test. Elizebeth and William wanted to pass. They later wrote, "What Colonel Fabyan failed to realize, throughout his campaign to 'sell' Mrs. Gallup's decipherments, was that no demonstration, however good, can take the place of experiments which can be repeated and will produce identical results."

To drive home this point, William even invented a new word: "cryptanalysis," synonymous with "codebreaking." The new Riverbank methods were not magic but a species of *analysis,* similar to the analysis performed by a chemist or an astronomer or an engineer designing a bridge.

Serendipity still played a role in codebreaking. "Many times," the pair wrote, "the greatest ally the mind has is that indefinable, intangible something, which we would forever pursue if we could—luck." Epiphanies happened. Insights that seemed to come from nowhere, bolts from the blue, guesses that made more progress on a problem than days of dreary labor. Mrs. Gallup had always called this "inspiration." Elizebeth and William preferred to speak of "flexibility of mind" or "intuitive powers" because these phrases sounded less magical. Intuition, to them, was like a hard-earned internal compass, a grooved-in sense of how to move forward that came from patience, skill, and experience. It could be cultivated.

Starting here at Riverbank and continuing throughout their lives, people tended to describe the brains of Elizebeth and William in gendered ways, as if her style of solving puzzles had a starkly different texture than his. Elizebeth's was usually said to be the more intuitive mind, William's the more mathematical. He was supposed to be better with machines and she with languages— Elizebeth was rapidly picking up German and Spanish, and learning pieces of other tongues. There may have been some truth in it. But the reality is that they were both mathematical neophytes, even William. A future colleague of William's, Lambros Callimahos, a classical flutist and trained mathematician, idolized William to the point of copying his personal habits; upon learning that William liked to use tobacco snuff, Callimahos took up snuffing. But Callimahos recognized that whatever made William good had little to do with math. He described William as a man "cursed by luck," writing, "Even if he computed odds incorrectly, it didn't make any difference because he would forge ahead in his blissful ignorance and solve the problem anyway. On several occasions he told me that if he had had more of a mathematical background, he might not have been able to solve some of the things he did." If William had been older or better trained, "he could have been ruined. His definition of a cryptogram was simply a secret message that was meant to be solved, just that."

To those who had a chance to watch them both work, the minds of William and Elizebeth appeared equally amazing and equally incomprehensible. Their brains were Easter Island statues, stony and imposing. Colleagues resorted to mystical analogies. William was like a latter-day King Midas: "Everything he touched turned to plaintext." Elizebeth's gift for puzzles was "God-given," "an effect without a discernible cause." Who was the better codebreaker, William or Elizebeth? People gave up trying to figure it out. A raffish young army officer from Virginia, J. Rives Childs, met William and Elizebeth at Riverbank in November 1917; they taught him the science of codebreaking, and he went on to serve with distinction in the war. Childs found it impossible to tell if William was smarter than Elizebeth or if it was the other way around: "I was never able to decide which was the superior."

Elizabeth and William sometimes played into the stereotype that he was the mechanical male thinker and she the sensitive female thinker. It was a helpful shorthand for explaining the inexplicable.

There's a now-famous story that encapsulates how they thought about their own differing brains. One day during the war, a series of five short messages arrived at Riverbank from Washington. It was a test of sorts. The messages had been encrypted with a small hand-operated device recently invented by the British army to make their field communications more secure. The device was a kind of cipher disc, with two alphabets printed on rings that rotated with respect to each other, but with a twist: while the outer ring had the usual 26 letters, the inner ring had 27. The extra letter introduced a degree of irregularity, making it harder for a codebreaker to visualize the alphabets sliding against each other. The device also allowed the cryptographer to change the alphabets quickly and easily.

The British had already concluded that the device was unbreakable. So had experts in France and a few in America. But to be certain, an official in Washington had used the device to encrypt five test messages, using two alphabets of his choosing. He then sent the messages to the Colonel, to see if Riverbank could solve them.

William looked at the messages. He had been given a description of the device that produced them, but not alphabets. His only chance to solve the messages was to reverse-engineer the alphabets the Washington official had used.

He began with the assumption that the official wasn't an expert cryptographer: a safe assumption, because almost no one is. Therefore the official might have made any number of common blunders that people often make when trying to communicate securely. The strength of a cryptographic system usually has less to do with its design than with the way people tend to use it. Humans are the weak link. Instead of changing keys or passwords at regular intervals, we use the same ones over and over, for weeks or months or years. We repeat the same words (such as "secret") at the start of multiple messages, or repeat entire messages multiple times, giving codebreakers a foothold. We choose key phrases that are easy to guess: words related to where we live or work, our occupation, or to whatever project we're working on at the moment. A couple of human mistakes can bring the safest cryptographic system in the world to its knees.

It struck William that the Washington official, in preparing this important test of a cryptographic device, might have used key words related to the practice of cryptography. So William tried words like "cipher," "alphabet," "indecipherable," "solution," "system," and "method." After two hours of intense focus, he was able to piece together what he thought was the alphabet on the outer disc, which seemed to use the key phrase *cipher*. William now assumed that if the official had been careless enough to use a guessable key phrase in one alphabet, he had probably used a similar key phrase in the second alphabet too, on the inner disc. But this one proved tougher to crack. William tested all sorts of key phrases; nothing worked. He turned to Elizebeth for help.

"I was sitting across the room from him," she recalled, "busily engaged on another message":

> He asked me to lean back in my chair, close my eyes and make my mind blank, at least as blank as possible. Then he would propound to me a question to which I was not to consider the reply to any degree, not even for one second, but instantly to come forth with the word which his question aroused in my mind. I proceeded as he directed. He spoke the word cipher, and I instantaneously responded, "machine." And, in a few moments Bill said I had made a lucky guess.

Later, in writing and interviews, Elizebeth would try to explain the "springlike elasticity" of her mind in this moment. What led her to blurt out the word "machine"? Where did it come from? All she would say is that because the British device was small and hand operated, "it did not occur to [William's] meticulous mind to use the word machine. But to me it was a machine." Thanks to Elizebeth's guess, she and William were able to solve the five test messages in less than three hours.

William attributed Elizebeth's insight in this case to the fact that she was a woman. He later said in a lecture, "The female mind is, as you know, a thing apart." He appears to have made a joke about sex as well. He and Elizabeth had just gotten married when this story took place. William recalled in his lecture to a roomful of men, "I came to the end of my rope and said to the new Mrs. Friedman: 'Elizabeth, I want you to stop what you are doing and do something for me. Now make yourself comfortable'— whereupon she took out her lipstick and made a few passes with it." Imagine laughter.

To hear Elizabeth tell it, William was the brighter one. Before they were married and before they were a courting couple, she was already starting to praise his abilities in a way that minimized or overlooked her own, setting the pattern for the rest of her life, the moments when she would describe William to friends and to reporters as a man of history and destiny, "a wonderfully warm man, with the broadest of minds and intelligence," and even "the smartest man who ever lived." All the same, she was competitive by nature, and at times the two of them indulged a cheerful rivalry.

Once, in a dusty 1896 issue of the literary magazine *Pall Mall*, William and Elizabeth discovered an article about ciphers used by anarchist opponents of the old Russian czars. The article included a brief cryptogram at the end. In general, the shorter the cryptogram, the harder it is to solve, the same way a song is harder to identify by three notes of its melody instead of twenty. This "Nihilist" cryptogram consisted of only a few numerals and two question marks:

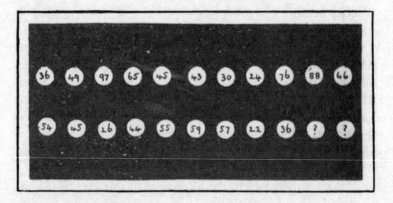

No solution was given. "The meaning of the cipher which now follows will never be solved by any one," the author wrote, concluding that the lock "has now closed and firmly shut its fastenings."

Naturally, William grabbed a pencil and began trying to pick the lock. "Well," Elizebeth writes, when William "met up with that message, he took the challenge and set his teeth into the tough nut with a snap. And would you believe it, he deciphered the message, short as it was, *and the key,* in 15 minutes!" The key relied on a single repeated word: "courage." The plaintext read: "He who fears is half dead."

It's a convincing piece of testimony to his greatness: William Friedman, the smartest man who ever lived. But instead of ending the story here, Elizebeth goes on: "Of course, when I learned that, I too had to try my hand" at the cryptogram. "I unlocked the forever-to-be-hidden secret in 17 minutes."

By the spring of 1917, William was in agony. He had known Elizebeth for eight or nine months now and he wanted her all the time. He was afraid to say it out loud because he didn't know if she would reciprocate. She had never called him anything but a friend. But he was sure he loved her. It was getting hard to sit with her all day and pretend to be thinking about work. In the moments when he appeared to be scratching away at a puzzle, he was really wondering what her hair would look like if he reached behind her neck and removed the pin, her beautiful loosened hair if he pulled her close.

He imagined a life with her, a house, children, and at the same time he could not imagine it. He knew that his Jewish family and friends in Pittsburgh would not approve of him marrying a non-Jew. The community there had always seen marriage between Jews and Gentiles as a kind of betrayal, a weakening of Jewish resistance to a hostile and bigoted American culture. When William was growing up, Pittsburgh's *Jewish Criterion* weekly newspaper made the case against intermarriage in repeated articles and editorials:

The glacial undercurrents of racial antipathy between Jew and non-Jew cannot be tepified by even the hottest fiercest rays of the sun of love! Statistics and the divorce courts prove this.

A part cannot become merged into a whole without ceasing to be a part. The Jews don't want to merge.

WILL THE JEWS COMMIT SUICIDE THROUGH MIXED MARRIAGES?

He feared what his people would say. But desire trampled the fear.

Soon enough, but not yet, he would tell Elizebeth what he thought of her: that "your soul and spirit and heart are as fine, sweet, and pure as your body is beautiful"; that the perfection of her body was matched only by the clockwork of her mind, "brilliant and quick and clever"; that she had him "skinned to a frazzle" in the brains department. He marveled at her ability to escape the bounds of a problem, to strike the flint of her thoughts against different rocks, against history and math and logic and against William, too, shooting out sparks, ribbons of flame. "You're lots smarter than I am in ciphers. You can soar away into the clouds and still remain planted firmly upon solid ground and reason. You can dream and be practical." He would tell her, over and over, that he couldn't express what was in his heart: "Oh Divine Fire Mine, I adore you, how futile are words!"

Divine Fire. A nod to the Bible of his father. The devouring God of the Old Testament, *whose fire is in Zion, and his furnace in Jerusalem.*

One of the mysteries of falling in love is that it makes you inarticulate and eloquent at the same time. You lose the ability to speak and write in normal ways (*How futile are words!*) even as you develop, with this person you love, assuming this person loves you back, a shorthand of glances and gestures. At first it seems like your beloved is "speaking in code"; later, maybe, it's like the two of you are *sharing* a secret code.

This feeling may have a deep scientific explanation. In the 1930s and '40s, before the digital computer was invented, a young scientist from rural Michigan named Claude Shannon wrote two papers that were like magic beans for the computing revolution, growing the great beanstalks of IBM, Apple, Silicon Valley, the Internet. As a graduate student at the Massachusetts Institute of Technology, Shannon realized that electronic circuits could be arranged to solve logic problems and make decisions, and that 0s and 1s could encode all the world's information, from a song to a Van Gogh. He didn't create the first computer, but he was one of the first to grasp the immensity of what digital computers could do.

Shannon, who would later work with William Friedman and other cryptologists on secret NSA projects, also enjoyed thinking about codes and ciphers. While employed at Bell Labs, he came up with the insight that the problem of communicating through a noisy system, like a phone wire, is almost identical to the process of enciphering and deciphering a message. In other words, according to Shannon, making yourself understood to another person is essentially a problem of cryptology. You reduce the noise of the channel between you (instead of noise, Shannon called it "information entropy") in a way that can be quantified. And the method for reducing the noise—for recovering messages that would otherwise be lost or garbled—is decryption.

Viewed through Shannon's theory, intimate communication is a cryptologic process. When you fall in love, you develop a compact encoding to share mental states more efficiently, cut noise, and bring your beloved closer. All lovers, in this light, are codebreakers. And with America going to war, the two young codebreakers at Riverbank were about to become lovers.

CHAPTER 5

The Escape Plot

Elizebeth with William in his army
lieutenant's uniform, around 1918.

To be your North Star—Billy Boy—I'd like to be!"
She wrote "North Star" on the page of the letter without
knowing what it meant or how she wanted him to respond. It was
just a pair of words that captured a tug of attraction toward William Friedman, a chemistry that made her curious: a small, persis-
tent tilt in his direction, like a plant bending toward a patch of sun.

Elizabeth first noticed it when she was home in Indiana watching her mother die, and writing to him at Riverbank, telling him—what? I don't love you but "I miss you infinitely." I am not sure I *love-you* love you but "I shall work for you" if you ask. I have dreamed about you but I don't remember what the dream was. "Anyway, Billy Boy, like me just a little bit always. I want you for the dear good friend you are, if nothing more. I want, oh, so much, for us both to 'achieve.' "

Before any feelings of passion, this is what Elizabeth expressed to William: a vague desire for the both of them to *win*. "Work hard on the letter tests—for my sake! You must win—because I want you to!" She saw his talent and was starting to understand the size of her own abilities. She sensed that the two of them were more powerful together than they were alone. *That* excited her, if not the thought of romance, a relationship, making love. When they kissed for the first time it didn't do much for her, as she would recall later in a poem:

> There was a time when for my love I did not care
> The hot wooing, the passionate kisses
> left me cold.
> I yielded to him
> because he was good to me.
> And compassion led me to return his kisses
> when his longing eyes and eager heart spoke,
> "I wish you cared as I."

She wasn't sure what to do when he started talking about marriage in early 1917. The ice cracked on the Fox River, spring picked the lock of winter, the water moved, and William Friedman spoke to her about the pros and cons of a potential union, obstacles and advantages, in a careful, unemotional tone, as if discussing a job opportunity. (He confessed later that he was simply trying to hold back the flood of his feelings, lest the dam burst and he embarrass himself.) He did not get down on his knees and propose, and his hesitation gave her the room to respond in kind, to examine the idea with detachment, to let the possibility of marrying him drift past at a distance like a harmless puff of cloud.

She could see a case for marrying him and a case against. If she did choose to marry, her family and neighbors back home would likely be confused—nice Quaker girls didn't marry Jewish boys in Indiana—but she had made more of a break with her family than William had. She had fought hard for the freedom to choose her own path.

Most of the time, talking it over that spring, Elizebeth and William agreed that getting married would be silly. There were too many barriers: family differences, religion, money. Incredibly, Fabyan was still paying them both the same salary as when they started, thirty dollars a month, despite their massive new responsibilities, and some months he didn't pay at all. If they did decide to get married, where would they live? In William's windmill? The prospect of being a married couple at Riverbank seemed absurd to both of them, as much of a fantasy as the perpetual motion machine and the messages from Bacon's ghost.

At the same time, they weren't sure they wanted to stay at Riverbank anyway.

The longer they lived here, the more concerned Elizebeth and William became about Fabyan's dark side, his need to control the people around him. He seemed to take special pleasure in humiliating William. Once, while both men were traveling in Washington, Fabyan demanded that William fetch a newspaper from a street vendor. William pointed out that papers were available at the hotel desk; Fabyan bellowed that he wanted one from the street. William obliged. Later, when William showed up to dinner in a freshly pressed evening suit, Fabyan, who was dressed shabbily, forced William to change into more casual clothes to match. "It just didn't go down to be treated like chattels," Elizebeth said. "We were sick and tired of Fabyan's scheming and dishonesty. Fabyan always came out ahead, and we always came out at the other end."

Fundamentally Elizebeth and William were two ambitious people. *Why should something with a risk in it give me an exuberant feeling?* She wanted to live a daring life, and he wanted to "make a mark in something," he told her. "Perhaps it will be Genetics." He expressed to Elizebeth "my ambition to know one little thing better than any other person, to be a pioneer in that field and to blaze new trails for the rest to follow. Why I feel that way, I don't know—it's just in me and will have to come out in some form or other." Riverbank had launched them on an incredible adventure, but now it was holding them back, and they sensed that if they were ever going to escape, they needed to do it together.

On a cool, rainy Monday in May 1917, they went missing from Riverbank.

It wasn't like them to skip work. The hours ticked away without them, the cows eating grass in the field, the fruit flies multiplying in jars. When the pair returned that evening, William was dressed in a dark blazer, light-colored striped pants, and a striped tie, and Elizebeth wore a simple gown of white lace. Colleagues gathered around and the couple shared the happy news: They had gone to Chicago and gotten married. A rabbi named Hersh performed the ceremony.

The wedding announcement ran on the front page of the May 23 *Geneva Republican*, next to a story about a Selective Service bill just passed by Congress, requiring men ages twenty-one to thirty-one to register for the military draft. "Mr. Friedman came to Riverbank soon after his graduation from Cornell University and was employed for some time on experimental work in the Riverbank greenhouses," the paper wrote of William. "He later took up the work in connection with the Bacon research studies." As for Elizebeth: "Miss Smith's home is in Indiana. She is a college graduate and a splendid reader. She and Mr. Friedman have lectured on the Bacon cipher before colleges, schools and clubs." The article didn't mention that Elizebeth was instrumental in convincing him to "take up the work."

She married him without being in love. She admitted this later in her diary, after picking up a novel called *The Prairie Wife* and reading the opening line of the book's woman narrator: "Splash! . . . That's me falling plump into the pool of matrimony before I've had time to fall in love!" Elizabeth recognized that sentence as one "I might myself have spoken." She married William because he was a good person, and he wanted it with such overwhelming intensity, and she trusted that the rest would come soon, certainly soon, because so much had already happened in the shortest time. It had been almost exactly a year since the Colonel met Elizabeth in Chicago and whisked her off to this patch of prairie. "I am learning to take things as they come," she wrote in the diary soon after the marriage ceremony, taking stock of how her life had been transformed in a flash:

> To be mangled and torn and castigated and macerated in soul—to wish passionately day after day only to die . . . and then to be brought, by a miracle, to a new place—to work that is absorbing, fascinating—to a place where I forget and find peace, glorious peace—and oh, miracle of miracles, to Love! Ah, Heaven is good! Truly, truth is stranger than fiction! I could not have believed it possible, but here am I. Is it possible I am to have them after all— Youth and Love, and Life?

The reaction to their marriage in Pittsburgh was just as they had feared. When William traveled back home briefly to tell his parents, his mother collapsed at the news that her son had married a shiksa. He told Elizabeth about it in a wire to Riverbank. She read it and felt sick. "I am cast into a whirl of remorse, pain, and sorrow for you," Elizabeth wrote to him. "Oh, Billy, Billy, what have we done?" She told her family in later years that when she visited her in-laws in Pittsburgh, William's mother would sit and weep. "You would have thought that Bill had committed murder," Max Friedman, one of William's brothers, later recalled. "If he had still been living in Pittsburgh he would have been ostracized."

But William didn't live in Pittsburgh anymore. He lived in Illinois, in a rich man's windmill. And now Elizabeth did, too. She moved from Engledew Cottage into the windmill. He made room for her journals and papers and books and brought her up the steps to the second floor. It was humid and cramped, and it smelled of soil, but it was theirs.

That summer the military was sending teams around the country to recruit volunteers. Fabyan invited the army to Riverbank in July. He ordered his employees to build a wooden stage at the highest point of the lawn next to the Villa, and a recruiting tent next to the stage. Three thousand people came from Geneva and surrounding towns, clogging the roads with horse-drawn buggies and automobiles, a prairie traffic jam.

A U.S. Army captain stood on the stage, tall and clean-shaven, with a hank of brown hair gelled to a stiff peak, wearing a uniform of olive drab wool with gold piping and black dress shoes. "Better to go and die than not to go," he cried. "Women, plead with your sons and brothers and sweethearts." The captain said that anyone who spoke of peace should be shot as a traitor. At the end of the speech, a boy from Elgin stood and walked to the recruiting tent. The crowd cheered, and more boys stood and followed him. A bit later, Fabyan invited the guests to tour his model trenches for a fee of twenty-five cents per person, to be donated to the Red Cross. He raised more than three hundred and fifty dollars. Men in bowler hats and women holding parasols stood at the edge of the trenches, peering down. Children ventured inside and played in the mud.

William did not volunteer for the army that day, but he was starting to think about it, partly out of guilt—he was a healthy male in wartime—but also out of concern for himself and Elizebeth, their future together. He wondered if he could use his codebreaking skills to get commissioned as an officer. The army paid more than Fabyan. It laid out clear paths to promotion. And there were army bases all over the country and the world. When the time came for the couple to leave Riverbank, they could leave with good prospects.

He began to pester Fabyan about it, asking the boss to reach out to his contacts in Washington and recommend him to the Army Signal Corps, Joseph Mauborgne's section. William said he wanted to go to France and apply his code and cipher knowledge closer to the fighting. Fabyan always waved him off. William was needed in his present position. He should forget about the army and concentrate on his work.

Frustrated, William took matters into his own hands, writing to Joseph Mauborgne and asking if the army had any use for his abilities. At the same time Elizebeth wrote to the navy to inquire about codebreaking positions. They waited for replies. Months passed. Nothing.

It wasn't until later that the Friedmans learned the truth. They heard it from Mauborgne and others who had been desperately trying to reach them the whole time. Fabyan was intercepting the Friedmans' mail. He had taken the job offers that arrived for them from Washington, put them in a drawer, and responded himself, informing Washington that the Friedmans were unavailable.

Also, one army officer who visited Riverbank for cryptologic training told William he discovered secret listening devices in the classrooms. Bugs. It seemed obvious that Fabyan didn't place the bugs to spy on the students. The students didn't know anything. It would have been pointless. The only logical explanation was that Fabyan had been spying on the Friedmans, in order to anticipate their movements and prevent them from ever leaving his Garden of Eden. *It's made honest bees out of them, this constant supervision:* Fabyan was surveilling his young employees as if they were two honeybees in his colony, under glass.

A tiny slip of paper fluttered down to Elizebeth. She was outdoors at Riverbank with William and Mr. Powell, the gentle University of Chicago publicity agent, the three of them working in the grass, the fresh air. She picked up the paper and saw a line of cursive written in light pencil. It was from William. "My dearest, I sit here studying your features. You are perfectly beautiful!! B.B." Billy Boy. She hid the note so Mr. Powell wouldn't see it, later pressing it between two pages of her diary. "My heart sang," she wrote there, "carolling bursts of ecstasy."

She wasn't pretending anymore or yielding to William out of kindness. She was the one throwing her arms around him in the cottage when Mrs. Gallup wasn't looking and pulling him into a kiss. "My Lover-Husband," she called him now:

TONIGHT MY LOVER-HUSBAND *and I made a tryst with the
future.*
THE GOAL IS *set; will we win? We planned it all—cheek to
cheek—facing the swelling power of the new moon—*
"WONDER-GIRL," HE SAID, *"It shall be all for You—only for
You!"*
AS I HELD *him close and caught my breath in the intensity of
hope, he said—"Dear Heart! You are not crying?"*
AND I REPLIED—*"No dear, only praying." And this was my
prayer:*
"OH SPIRIT WITHOUT *and Within, keep me sweet! Keep me
working on & on & keep me well—keep the Fire Burning!"*

Their work started to dry up in the summer and fall of 1917. Each
new parcel from Washington was lighter, containing fewer intercepted messages to solve. Something had changed. Fabyan paced
and fretted. He raged in the hell chair.

It turned out that the War Department had recently launched
a codebreaking unit of its own, under the command of a twenty-
eight-year-old lieutenant named Herbert O. Yardley, a scrawny
Indiana native who had become entranced with cryptology after reading library books that told about the old black chambers
of Europe. "Why did America have no bureau for the reading
of secret diplomatic code and cipher telegrams of foreign governments?" Yardley asked himself. "Perhaps I too, like the foreign cryptographer, could open the secrets of the capitals of the
world."

Fearless and charming, and a shark at poker, Yardley considered his new bureau to be an American black chamber, and he
had no trouble convincing the War Department to let him solve
messages that would have otherwise been shipped to Riverbank,
seven hundred miles away. Known officially as MI-8, and based at
the Army War College in Washington, Yardley's bureau had shattered Fabyan's near monopoly on American codebreaking.

Fabyan, aware that he was losing influence and power by the
day, now came up with a plan to win it back. He knew the military needed many more codebreakers than it could locate and
train quickly, both to work for Yardley's Washington bureau

and for the American Expeditionary Forces (AEF) in France. America needed a codebreaking school, and here was Riverbank, already set up as a university of sorts. He invited the army to send men to Riverbank for training. The army took him up on it.

The first students arrived in November 1917, four young lieutenants destined for the war front. They knew nothing about codes and ciphers. They were, as an NSA historian would later say, "as dumb as anyone just off the street." Fabyan asked Elizabeth and William to teach them.

The Friedmans had never taught a class before. They had no lesson plans and a grand total of one year of codebreaking experience. There was nothing to do but to do it.

"What was taught was taught," Elizabeth later recalled, "and we taught it with what we had."

The first four students soon became eighty young officers in training sent from the Army War College, many accompanied by their wives. There wasn't enough space to house the visitors at Riverbank, so the Colonel booked the largest hotel in the nearby town of Aurora, and William and Elizabeth taught class there every day, lecturing in the morning and correcting problem sets in the afternoon. They started with the biliteral cipher of Francis Bacon and moved on to more contemporary methods of encryption and decryption, using actual messages from the Spanish-American War and German intercepts from the first two years of the Great War. Mrs. Gallup sat off to the side during the classes, observing but not teaching.

The students knew that after graduating, many of them would be going straight to war, deployed to France as code and cipher officers with the AEF, and others would be assigned to similar work in Washington. This awareness of ordeals ahead made Riverbank seem that much sweeter. It was like being stationed in paradise. Fabyan provided the students with daily box lunches with fresh food from the farm, organized outings into the countryside, and threw parties where the single men could mingle with local girls, including a lavish military ball that ushered the golden-haired daughters of Geneva into the arms of the uniformed officers. At least four of the officers' wives took the classes and completed the course. The Colonel commended the women for their "excellent work" in a letter to the War Department, though he did not list their own names but instead the names of their husbands.

On the last day of the course, in late February 1918, the students and instructors gathered outside the hotel for a photo, lining up in two rows that stretched from one end of the building to the other. In the photo, William, Elizebeth, and Fabyan sit front and center, William looking off to the right; the army men take stiff poses, some of them angling their heads 45 degrees to either side, some looking straight on. The significance of this curious feature of the photograph escaped almost all who viewed it at the time: *Each person stood for a letter of ciphertext in the biliteral cipher.* The ones looking to the side were the *b*-form of the cipher. The ones looking straight on were the *a*-form. United they spelled out the motto of Francis Bacon, the phrase chiseled into stone by the Colonel above the Acoustics Lab door: KNOWLEDGE IS POWER.

For the rest of his years, William would keep an enlargement of this photo beneath the glass surface of his work desk, glancing at it most every day of his life. It was a reminder of a more innocent moment, a time before two dark and interrelated forces began to draw boxes around his days, shaping his path and Elizebeth's, too. War was one. Justifiable paranoia the other.

William finally raised enough of a stink with Fabyan that the boss said he could leave Riverbank on the condition he return to Riverbank when the war was over. He entered the army as a first lieutenant in the signal corps, an officer but a low-ranking one. He was headed to France to ply his code and cipher abilities with the AEF. An army photographer snapped his official portrait in a darkened room with a lamp to the left, illuminating the left half of his face and body. In the picture, he looks serious and delicately handsome. His ears seem to stick out more than usual. He liked it and gave Elizebeth a framed copy as a gift. In May 1918 they said goodbye, Elizebeth smiling through her tears, and Lieutenant Friedman boarded a train to Chicago on the way to his destination, American General Headquarters (GHQ), in the farming town of Chaumont, France.

Elizabeth wanted to go with him. She saw no reason why she should not be allowed to serve in France as an AEF cryptologist. But the army told Elizabeth that "I, a mere woman, could not follow to pursue my 'trade,'" so she stayed behind at Riverbank, continuing to break codes in the Lodge with the other brain workers of the estate. In her diary she wrote original poems about war, exploring "the heartache of separation from the Dear One overseas" and recognizing that she needed to take care of herself, to preserve an inner mental space that was solid and clear ("a calm Whole, a unified peace"), and all the while reading the time-lagged stream of letters that arrived from William weeks or even months after he sent them, the envelopes stamped with the red mark of the AEF censor before crossing the ocean to the prairie.

She could tell from his letters that he missed being able to talk to her about a puzzle when he got stuck. For security reasons, he had to speak vaguely and omit all technical details. "The work is so hard," he wrote, "and the results so very, very meagre. Sometimes I fear that I haven't got it in me at all. I cannot explain to you—but just imagine yourself at work absolutely in the dark, up against the most baffling problem, with no data to base speculation upon, no guiding generalizations, except the most vague and unreliable—Oh, I tell you Honey, it's going to be an awfully hard task to make good."

At the same time it was clear from the letters that William's reputation was growing in the army, that some days solutions appeared to him "out of the clear blue," startling his colleagues. "On Saturday Col. M brought around a visitor, some Col—I don't remember his name. When he came to my desk he introduced me and said, 'He is our wizard on Code.' Dearest, I was quite embarrassed, and didn't know what to say." Col. M was his commander, Frank Moorman, and William made it a point to tell Elizabeth that Moorman admired the Riverbank Publications, hoping she would feel proud of her work on them. "Love-girl," he wrote, "yesterday at conference Major M passed around our R.K. pamphlet"— Riverbank No. 16, *Methods for the Solution of Running-Key Ciphers*—"and said that he went through it with much interest and that it was the best thing he had seen on the subject."

William told her not to worry about his physical safety. Chaumont was far enough from the trenches and the firefights that he felt there was no danger of needing the .45 pistol that he carried on his hip.

Like the other officers, William lived in a billet, the private home of a French woman he called Madame. It was so dark in the French countryside at night that during his first weeks he had trouble finding his billet and had to use cigarettes as torches. In the day he worked in a building behind GHQ known as the Glass House, surrounded by the other AEF codebreakers and radio operators. The Germans used a field cipher based on six letters: A, D, F, G, V, and X. A message might look like FAXDF ADDDS DGFFF, or DDFAX SDGVV AFAFX. William spent a lot of time fiddling with these six-letter nonsense messages, groping for the light cord in the darkened room of the cryptograms. The men in his section preferred to work alone, which he found baffling. "You know how much 'group work' counts in our business," he wrote to Elizebeth. "What can one person alone accomplish?"

He tried drinking French wine and didn't care for the taste. Lemonade was more to his liking. At the officers' club he kept to himself, nursing highballs in a plush chair in front of a roaring fire, except when the men dragged him into a poker game, which he always regretted, losing money each time he played. The Midas of codes had no talent for reading human faces. In this he was the opposite of Herbert Yardley, leader of the MI-8 codebreaking unit in Washington, who also spent time in France during the war. William met Yardley there for the first time and thought he seemed fake. "I must confess to considerable distaste for Y. Frankly, I didn't like him at all," he wrote to Elizebeth. "He acted like a wooden Indian."

Feeling lost and out of place, and wishing the army had permitted his wife to serve in Chaumont, he walked back to his billet in the dark at the end of the night and spent hours writing to her. He had placed a picture of her next to an oil lamp and each time he struck a match to light the lamp he looked at the picture and said, out loud, "Hello, you darling! Hello, Rita Bita Girl," then lit the flame and started to write, imagining he was back in bed with her at Riverbank, stroking her hair, talking in baby talk. He fantasized about spanking her. "Do you miss your Biwy Boy, my darling? Have you been naughty? Do you need to be spanked? You little 'imp.'" He said that was as far as he dared to go, with the censor reading every word, and promised that someday he would cable her "some real stuff which may burn the insulation of the wires."

During these months in Chaumont, four thousand miles from home, William became tormented by feelings of inferiority and romantic inadequacy that would never completely go away, gnawing at him for the rest of his life. He worried he was too unprosperous for Elizebeth, too interested in science instead of money, too effeminate. He apologized for "the many imperfections in my makeup." He asked for reassurance that he was a "good lover": "You have told me that, haven't you?" One day he happened to meet Colonel Parker Hitt, the Texas codebreaker now posted to France, and needed to crane his neck up to look at him: "He actually towered above me." He went to sleep at night with the windows open and often dreamed of his wife, a recurring dream where she was leaving him because she didn't love him anymore, then he woke up in a sweat and lunged for his pen and a piece of paper to copy it down: "You didn't yike me at all," he wrote, baby talking, "and I was all broken up." He begged her forgiveness for leaving her alone at Riverbank with "no money and a lot of debts" and promised to pay more attention to her happiness. "When we were together," he wrote, "I was particularly mean not to take you out more often, even though we couldn't 'afford' it, or even though there was lots of work to do. The Spring Time of love was ours—and I failed to make it all that it should have been for you. I owe it to you—and you shall have it all a thousandfold over when we are together once again. 'Afford it' or not."

Against these anxieties and regrets William possessed only one weapon: language, wordplay. Every day he felt he was losing a little more of his wife and every day he felt he must fight to win her back, so he labored over his letters, making corrections, emendations, fixing rare grammatical mistakes, turning the pages 90 degrees and adding sentences at right angles in the margins, trying to find the magic incantation of symbols to crush the globe flat and cheat the distance between them. He filled the pages with encoded messages of devotion he had every reason to believe she would understand. "This cable will read apparently harmless— but each letter and punctuation mark on it is but a group standing for a whole phrase which I wish I had the power to vary—but have not. The phrase is 'I love you!' It has two alternates—'I adore you!' and 'I worship you!' So if those tiny flashes of electricity can talk to you they will whisper to you over and over again, with an infinite permutation of expression my message of love for you." It was a lover's code:

A = I love you! / I adore you! / I worship you!

B = I love you! / I adore you! / I worship you!

C = I love you! / I adore you! / I worship you!

! = I love you! / I adore you! / I worship you!

. = I love you! / I adore you! / I worship you!

Elizebeth wrote long letters back to him. In one envelope she enclosed a lock of her hair.

Her letters don't exist today. It's likely she destroyed them after the war. Still, they left traces in William's: sentences of hers that he quoted, questions she asked that he answered.

A frequent topic was their future at Riverbank. Should they stay there or leave? What should they do about George Fabyan? The man was relentless; all through William's deployment, Fabyan had been writing him in Chaumont, asking that he return to Riverbank at the soonest opportunity. The Friedmans discussed this issue with caution, abbreviating Riverbank as R., George Fabyan as G.F., and the Bacon Cipher project as B.C. It occurred to them that Fabyan might have friends in the censor's office and they didn't want the rich man prying into the conversation any more than he already had.

Elizebeth tried to tell William in her letters that she no longer felt safe at Riverbank. She made a vague reference to Fabyan's "excesses," causing William to say he didn't understand: "How and where did you learn of these?" When William told Elizebeth he wanted to have children with her, she replied that it wasn't safe to have children at Riverbank. "You are perfectly right," he agreed. "When we are 'safe,' the children."

On September 21, 1918, she revealed something to him in a letter. All that survives is William's reaction: "Honey, I could have committed several crimes after reading what it had to say about that old nameless rascal. I was upset all day as a result. To think that he would do such a thing after all we have done." Elizebeth later confided to friends that Fabyan made sexual advances while William was in France.

William encouraged her to leave Riverbank if she was unhappy: "Honey, don't be afraid to take a step. You have ability and more brains than any other woman I've known. You can fill any job a woman can and many jobs that men fill."

The German lines collapsed in October 1918, British and American troops advancing and seizing territory. The roads around Chaumont began to fill with convoys of emaciated German prisoners of war. On November 10 at GHQ, a group of American soldiers huddled around a newspaper, laughing and shouting: The Kaiser had abdicated his throne. The war was over, the Allies had won. Three miles away the men of the AEF Gas Defense School blew up bombs and fired rockets in celebration, thunderclaps disturbing the sky. The dazed citizens of Chaumont wandered into the street and hung lanterns in their windows.

As William's colleagues drank and sang, he stayed indoors at his billet and wrote to Elizebeth, vomiting a great pent-up mass of insecurities and dreams onto the page. "Dearest Woman in the Universe," he began, "This is surely a fateful day." Then he made a series of promises, talking about what their lives would be like when he got home. He said he didn't want her to be consumed by housework. As a child he had seen his mother exhausted by her cleaning duties. "Home does not entail a spotless kitchen and a faultless parlor," William wrote. "Home does entail the presence of hearts that beat in unison—whether the shelter be a hovel or a palace." He was offering her the same freedom to pursue her intellectual ambitions that she had always extended to him—but did she really mean that? In her private heart did Elizebeth wish that her husband had more of a bank account and less of a brain? "Elsbeth, my Dearest, when you say that you want me to go on with my research work—blaze the trail and all that—do you realize that those chaps, poor fortunate-unfortunates, are usually not bank presidents? I should be happy, I think, with a fair share of the comforts and goods of this world, if I could continue with my studies, and unless I am seriously mistaken—and I don't think I am—you are not the woman to be hankering after life's luxuries and fineries. If you were, we would never have been attracted to one another."

That night, after 11 P.M., the oil in his lamp burned out and he went to bed. In the morning he learned of the Armistice and added a line to the bottom of the letter: "Honey, it's all over now."

At Riverbank she heard it from the news well before she received his letter, and she began a letter of her own, which William then quoted back to her. "The signing of the Armistice had one result—my indulging in thoughts, last night . . . dear, intimate things that burn one up with a fire of longing and ache of wanting you," Elizabeth wrote. "I must not again."

William replied, "What shall I say of the thrills that took possession of me on reading those words? I, too, have indulged in thoughts. . . . Ah, Dear One, when shall we too live them over again?"

He then broke some bad news: The army wasn't releasing him yet. He had to stay in Chaumont to write a secret history of the code and cipher work as a technical reference for future army use. He might be there for months.

This is when Elizabeth finally decided to leave Riverbank. She packed a bag in stealth without telling Fabyan and slipped onto a train for Indiana, reasoning that with the war over and no urgent messages to decipher, Fabyan could do without her for a bit, whether he liked it or not.

To pass the time in Huntington, she got a temporary job in the local library, a two-story building of limestone with a special room of materials about railroad engineering. She helped farmers find books and opened letters from the men in her life.

Some of these letters were job offers, eager replies to inquiries she had already sent. The Office of Naval Intelligence in Washington wanted Elizabeth to join its Code and Signal Section. Also, an officer in the War Department thought her Riverbank training would be "of the greatest value" in the MID; he was none other than John Manly, the Chaucer expert who used to argue with Elizebeth about the Bacon Ciphers and once shoved her on the shoulder. Manly now worked with Herbert Yardley. "Most of the work handled here necessitates a thorough knowledge of Spanish or German," Manly wrote. "Women who can *think* in either of these languages are needed as cryptographers at $1400 per annum."

William wrote to her, of course. He seemed as effusive and insecure as ever. He asked her if she knew how small an electron is, using that as the basis of an extended riff about the incomprehensible size of his love for her. He said he had gotten her a piece of lingerie in Paris, a silk teddy, custom sewn, with the help of an army captain who told him what measurements to use ("Can't two perfectly 'spectable married men get together on designing a perfectly proper—even if private—piece of woman's apparel?"), and he ruminated on their postwar future. Fabyan had demanded that William return to Riverbank at once: "You have had a long enough vacation," Fabyan wrote, "your salary has been going on and I do want you to get back at the earliest possible moment." But William worried that if he and Elizebeth did resume working at the estate, and Fabyan forced them to continue probing the Bacon Ciphers, it would destroy their credibility as cryptologists and make it hard to find other jobs. "I refuse to have anything to do with the B.C.," William wrote. "I think that whole business would be an excellent experiment for a psychologist. . . . Furthermore, I shall keep you away from it too. Nothing but unhappiness, and accusations, and unfruitfulness have ever come out of the whole business. Aside from our deep, and perfect love, the greatest treasure which life holds, we have found little else at R. but heartache, and argument, and unhappiness."

By the end of the letter he came around to the idea that they should leave Riverbank forever. "I don't want to flatter ourselves, but [Fabyan] is going to have one fine time trying to replace the Friedman Combination!" He signed off one letter with a love note in cipher, written using a type of transposition cipher called a "rail fence":

several times. She
has a blouse exactly
like yours, so you are
members of the same
"lingerie sorority."
Mrs. R. Owen,
Madison, Wisconsin

$$I \quad E \quad R \quad O \quad U \quad !$$
$$L \quad V \quad Y \quad H \quad L \quad !$$
$$O \quad U \quad M \quad S \quad D \quad O$$
$$V \quad O \quad U \quad I \quad S \quad S$$
$$E \quad Y \quad C \quad H \quad A \quad Y$$

Bury.

To find the hidden message, start on the upper left, with I. Read down that column, ILOVE, then start at the bottom of the next column to the right and read up, YOUVE, and then down again, up again, down, up.

George Fabyan also sent Elizebeth letters while she was in Indiana. She dug her nail into the wax and his black words uncoiled.

"I am wondering how you are and what you are doing," he wrote with a strained politeness that barely masked his fury, "and if your vacation has not been long enough to suit you." He liked to sign his name in a flourish of blue colored pencil, making the widest line possible, the tip round and blunt. Elizebeth didn't understand how anyone could bear writing with an unsharp pencil. It was barbarism.

He asked her in several different ways to come back, alternating charm with threats. He attempted a strategy of divide-and-conquer, suggesting that if Elizebeth committed to returning to Riverbank, it wouldn't necessarily bind William to do the same. (When Elizebeth relayed this to her husband in France, William was enraged: "Does he suppose you'd live at R. and I at Chicago!! ... any man who attempts to sow dissension and create unharmony between man and wife, in that manner, is a scoundrel.") He tried to impress Elizebeth with his power, recounting a recent conversation with the head of MID in Washington, who had offered to hire all the women codebreakers at Riverbank, including her. "I told him that I would see them in hell before any girl whom I was interested in went to Washington under existing conditions," Fabyan wrote.

Finally, he tried being reasonable, addressing the Friedmans as a couple, through Elizebeth. "I am an old man going down hill; you are young people climbing up and it is for you to decide whether your opportunity lies at Riverbank or elsewhere." He said he wanted to talk it over with Elizebeth in person, in Chicago, given the "rather unsatisfactory" and slow nature of mail.

Elizebeth shot back in a letter, "I am inclined to agree with you that in most cases, correspondence is rather unsatisfactory. But with you I confess it has some advantages—for, you see, in conversation you insist on doing all the talking! Now I suppose you are going to retort, 'This, from a woman?'"

She finally got the letter she'd been waiting for in early February 1919: William was coming home. The army was done with him in France. "Won't our reunion be better than any honeymoon you can think of? I love you! I love you! I love you, love you love you!!"

He arrived in New York City two months later, in April, on a ship with other returning troops. Elizebeth went to New York to meet him, and they saw each other for the first time in a year.

They stayed in the East for a few weeks in rented rooms, thinking about what to do next.

Elizebeth knew she couldn't go back to Riverbank. Her husband, for reasons she didn't quite understand, had always found Fabyan at least slightly entertaining, but to her, the man was a scoundrel. As for William, he didn't want to be in the military anymore. He hated knowing that the army could send him anywhere in the world on a whim, separating him once again from his wife. The little he had seen of war convinced him there was no glory in it, once telling Elizebeth in a letter, "The War will not make better men or women out of us." If he could choose his own path, he confessed, he would unwind the last few years of his professional life and return to his first love, genetics. Maybe he could continue his plant and fruit-fly experiments at a university, or failing that, join a corporation and make some money.

Elizebeth agreed that William's "extraordinary gift of scientific analysis" should be properly appreciated and rewarded, and she encouraged him to get his discharge from the army, in April. After that they traveled so William could meet with potential employers. Elizebeth figured she would find work of her own wherever he landed. She noticed that the corporate executives who interviewed him were invariably amazed at his knowledge of codes and ciphers: "Everybody said, 'But where has this been all these years? Here we have this wonderful science all opened up for us now and where has it been hiding?'"

Strangely, however, no company offered William a job. Wherever the Friedmans went, a telegram would arrive from Fabyan, commanding them to give up the search: "Come back to Riverbank, your salary is still going on." The only way he could have known their whereabouts was if he had dispatched a spy. It was logical to conclude that Fabyan was threatening William's potential employers. "He had us followed," Elizebeth said. "He opened our mail."

Feeling defeated and not seeing any other options, the Friedmans told the Colonel they would return to Riverbank if he let them live in their own house in Geneva, gave them both a raise, and allowed them to question Mrs. Gallup's theory based on hard evidence. He agreed and welcomed them back home.

Fabyan didn't keep his promises. The raises never materialized. He continued to ignore and even suppress criticism of the Bacon Ciphers. When a famous type designer wrote a report showing how Mrs. Gallup misunderstood the printing practices of Shakespeare's era, Fabyan shoved the report into a drawer, even though he himself had paid for it to be written. (The Friedmans stumbled across the mothballed report years later in the Library of Congress.) Worst of all, he maneuvered to deny William credit for a crowning scientific achievement, his paper "The Index of Coincidence and Its Applications in Cryptography," written in 1920.

William had noticed that in any piece of English text, a letter sometimes appears directly atop the same letter in the line below—*d* on top of *d*, *w* on *w*, *q* on *q*. William discovered that the frequency of this "coincidence" could be measured, and it was distinct for each language, a kind of signature. In English, a coincidence happens exactly 6.67 percent of the time. Seven columns out of 100 contain an alignment. This insight married modern statistics with cryptology for the first time, and by doing so, kicked open a door that couldn't be closed. "The Index of Coincidence" and its offspring would lead directly to important feats of codebreaking in the Second World War. Instead of putting William's name on the cover, Fabyan had the paper published first in France; people got the idea that a French cryptologist was the author.

William and Elizebeth were enraged by "Fabyan's skullduggery"—Elizebeth scrawled this and other angry phrases in the margins of Fabyan's letters to William, annotating his duplicity, keeping a file of his lies—and within a year of returning to Riverbank they felt desperate to escape, reaching out to associates in Washington. This time they sent and received letters at their own address in Geneva, avoiding Fabyan's surveillance net. Joseph Mauborgne of the army leapt at the chance to hire the Friedmans and promised to create positions for both of them at once. "We feel that it would be a great misfortune if the Friedman family were to retire to some other kind of a job," Mauborgne wrote.

The Friedmans accepted the offer in December 1920. William was afraid to tell Fabyan, and he worried what the Colonel might do to his friend Mauborgne, too. William warned Mauborgne, "He is as powerful as he is ruthless." They all dreaded the rich man's reaction. "I expect a lively row when the news breaks upon

Colonel Fabyan's portly frame and expect that no little of his fury
will be vented upon me," Mauborgne wrote to William. "Perhaps
you had better fix up that side of it—if you can."

Elizebeth wanted to leave in the middle of the night without
telling Fabyan. William found this overly cruel. She begged him
to reconsider: If they wanted to escape, she said, they had to be
just as tricky as Fabyan.

So together the Friedmans planned a clandestine operation:
"our secret plot to be able to get away without getting our throats
cut," Elizebeth called it. One morning they loaded all of their
possessions into a car they had managed to borrow, cleaned out
the house they'd been renting in Geneva, locked all the doors,
drove to Riverbank, and located the Colonel. They showed him
the car with all their luggage. They said they were leaving on the
three o'clock train and their decision was final.

They thought he'd explode, turn red and scream, maybe try to
restrain them. Instead, with an eerie calmness, Fabyan smiled and
wished them well. It was so out of character that William assumed
he must have already decided to seek his revenge at a future moment of his choosing.

There would be time to worry about that later. For now, they
were giddy as they traveled east to Washington. They thought
they were free. William believed that "after a very limited number
of years," Riverbank "will disappear from the Earth and be but a
black memory." He was grateful to escape and eager to work with
his wife in a new city, keeping the Friedman Combination intact,
as he had talked about during the war: "Oh, you are some partner,
I'll say! . . . and to think you love poor me!"

Elizebeth's professional intentions were a little different. She
thought leaving Riverbank might unshackle her from William to
some degree. At times there, working at his side, she had felt pressure to compete, but the war had inflated William's renown to
the point where she couldn't possibly keep up, and she figured
that no one in Washington would expect her to match him. There
was relief in that. As she put it later, "By the end of the war I was
more or less known as a military cipher expert, but I was better
known as the wife of my husband," who had "made a reputation

so startling that I regarded the task of catching up to him as being altogether hopeless." But if she believed she would never again rival her husband, she was wrong. In the nation's capital, she was about to carve her own path, her own name, with a set of blades that would one day turn out to be just the right shape for dismembering the plans of Nazis.

PART II

TARGET PRACTICE

1921–1938

To work in this field, you have to become devious yourself. You have to think like a malicious attacker to find weaknesses in your own work. . . . Cryptographers are professional paranoids. It is important to separate your professional paranoia from your real-world life so as not to go completely crazy.

—CRYPTOGRAPHY ENGINEERING,
FERGUSON, SCHNEIER & KOHNO, 2010

American cryptology was in a shambles after the First World War, a perilous mess. The codes used by the troops were already out of date, "completely inadequate" going forward, in William's opinion.

The problem was twofold: speed and security. Old pencil-and-paper methods of generating cipher messages were too slow compared to the speed with which dots and dashes of Morse code could travel by radio. "Military, naval, air, and diplomatic cryptographic communications had to be sped up." And because more vital messages were increasingly transmitted by radio and could therefore be intercepted and studied by the enemy, new techniques were required to protect that information from prying eyes. Security and intelligence are two sides of the same coin. Without one, the other is no good. You can intercept and decrypt the juiciest enemy secrets in the world, but if your own codes and ciphers aren't secure, you are defeating yourself, filling a leaky bucket at the top while secrets spill out the bottom.

Every powerful government on earth realized this now: the urgent need to invent new kinds of machines to make cipher messages, machines that were faster, easier to use, and dramatically more secure, all at once. "All the countries of the world were trying to develop something that nobody else could read and make sense out of," Elizebeth later said. "They were all playing with machines."

This epiphany marked the dawn of a peacetime arms race that would draw the battle lines of the next world war, and the Friedmans were plunged into the heart of it as soon as they arrived in Washington in the last days of 1920, fresh from their traumas at Riverbank, with no time to relax and take a breath.

Elizebeth and William both went to work for the government on January 3, 1921, three days after the first dry New Year's Eve in U.S. history, when 1,400 newly hired Prohibition agents ("dry agents") across the country surveilled the midnight parties, and the restaurateurs of Washington, D.C., spared no effort "to make the celebration, well, as festive as it could ever be under modern circumstances," one hotelier told the *Washington Herald*. That following Monday the Friedmans reported to the Munitions Building in Washington, a low hulk of concrete on the National Mall. Hastily constructed during the Great War, it teemed with fourteen thousand army and navy workers, including the staff of the Army Signal Corps, the part responsible for communications. The chief of the signal corps was Joseph Mauborgne, the kindly cryptologist, and the Friedmans now became his junior colleagues.

It was the first proper job they'd had in years, with regular paychecks and a nonderanged boss, and they felt happy, hopeful, and profoundly relieved. They had escaped from the clutches of a paranoid millionaire without getting their throats cut. Now they could live without fear.

William was placed in the army reserves at his prior rank of lieutenant and worked for the signal corps in that capacity, while Elizebeth worked as a civilian. He was twenty-nine, she twenty-eight. His starting salary at the army was $4,500 and Elizebeth's was $2,200, equal to $58,000 and $28,000 in today's dollars. This seemed like a lot of money to both of them after the poverty wages of Riverbank. And they were finally in a real city, which the young couple thought was just as nice as getting paid.

They could see friends, go to the theater, movies, the orchestra. Their first Washington apartment was a piano studio above a bakery. They woke each morning to the smell of fresh bread, and when they left for work, the owner of the apartment taught piano students there. Some evenings, when it was warm, Joe Mauborgne visited with his cello, and the three of them opened the windows and played together, Elizabeth on piano and William on violin, pedestrians stopping on the sidewalk outside to listen, not knowing that the musicians inside were three of the most experienced cryptologists in the country—almost the *only* experienced cryptologists.

The world of American cryptology was still tiny. There were only three codebreaking units in government, with fewer than fifty employees among them. The largest and best-funded unit, with two dozen people, was run by the former army lieutenant Herbert Yardley. After the war he had won funding from the State Department to launch a codebreaking bureau in New York City, in a four-story town house off Lexington Avenue. He considered it a modern version of the cabinets noirs of old Europe, the secret rooms in post offices where clandestine agents melted the wax seals of letters—an American black chamber. Yardley and his wife, Hazel, lived in an apartment on the top floor, and Herbert and his employees worked on floors below, reading the mail of foreign diplomats. They dealt in paper ciphers, and had some successes breaking the messages of Japanese diplomats, though Herbert wasn't skilled enough to go further, into the era of machines. The Friedmans knew the Yardleys socially and had dinner with them sometimes in New York, the two men chatting amiably across a deep chasm of professional rivalry and personal incompatibility. William was shy and monogamous, Yardley a boozy raconteur who during the war had kept a Paris apartment for his mistress. Although Herbert respected Elizabeth's intelligence, he once told William that she had "an edge on her."

The other two American codebreaking units were in Washington, one at the navy and one at the army, each smaller and more poorly funded than Yardley's black chamber.

The army is where the Friedmans started out, in a small windowless office in the Munitions Building. William smoked a pipe, Elizabeth smoked cigarettes. By the end of the afternoon the room resembled an industrial city seen from a distance, the couple's bodies like churches poking out of the haze. Together they produced the first scientifically constructed set of pencil-and-paper codes and ciphers in army history. There was still a place for "hand" or "paper" ciphers as opposed to machine ciphers. The best paper system was more secure than a weak machine, or a strong machine improperly handled, which is why inventors of cipher machines in the 1920s were struggling to make their prototypes easy to use, almost idiot-proof.

At the request of the army, William started tinkering with these machines, analyzing them, flipping them all around, searching for their soft and vulnerable places. He was the first Friedman to confront the machine era of cryptology, although Elizabeth would one day take on the machines as well, in the most spectacular way.

"There's all the difference in the world between machine cipher and paper cipher," Elizabeth explained later. When trying to break a paper cipher, with pencil and paper and brain alone, you had to depend on finding repetitions, patterns in the messages. But the new breed of machines generated what seemed to be patternless winds of letters. "You can start from here and go to the end of the world and never have a repetition," Elizabeth said. In theory, the only way to read a message was to know the starting configuration of the machine's internal parts—the "key"—which only the sender and recipient would possess. What's more, the machines were designed to survive capture and study. Even if you got your hands on a copy and took it apart like a broken clock, examining each gear for as long as you pleased, you would still not be able to read messages produced by another such machine.

Many inventors of cipher machines were private citizens, and a quirky bunch at that: hucksters, dreamers, engineers, lotharios, thieves. William met Edward Hebern, a burly Californian who had built a machine based on alphabet wheels that turned with the application of electrical current. Called rotors, these electrified wheels represented an important advance that would find more sophisticated expression in the German Enigma machine; rotors could be easily removed, swapped, and linked in a chain. William asked Hebern how he happened to think of this elegant concept of a wired rotor. Hebern replied, "Well, you see, I was in jail." William asked what for. Hebern said, "Horse thievery." William asked if he was guilty. Hebern said, "The jury thought so."

Hebern believed his machine was unbreakable and was trying to sell it to the navy. William placed the machine on his desk and thought about it. He stared at the small box for six straight weeks in 1923. He told Elizebeth he was "discouraged to the point of blackout." Then the solution occurred to him one night when he and Elizebeth were getting dressed for a party: "As I was tying my black tie, it suddenly came to me." It was the first-ever solution of a wired rotor machine. William mentioned the accomplishment to George Fabyan in a letter; Fabyan had let him borrow one of his cryptologic curios—a nineteenth-century cipher device—for a lecture, and William was returning the device. "P.S.," he scrawled at the bottom of the letter, "I busted a beautiful 'indecipherable cipher' machine recently and it is extremely important in many respects. . . . Gave the Navy an awful jolt! Best piece of analysis I ever did. B."

In contrast to Elizebeth, who never wanted to talk to Fabyan again and didn't write to him after leaving Riverbank, William kept a connection alive. He wrote regular letters to the tycoon from his signal corps office, sharing bits of personal news and asking for Fabyan to send him pieces of cryptologic literature from his unrivaled library at Riverbank. He did it partly out of professional fear; it was best to maintain cordial relations with such a powerful figure. But William also coveted the documents that Fabyan still controlled, the records of the Riverbank Cipher Division. He realized that the records could be used to write a true history of cryptology, a history that did not exist. William wrote to Fabyan, "It's a striking paradox that a subject which forms the basis of material which is exchanged every hour of the day over the whole face of the globe, under its waters, and through its ether—material that touches directly or indirectly every human being—lacks an authentic, detailed history. Somebody, somewhere, is going to write that history." He hoped it would be him, and he needed access to Fabyan's papers to do it.

After busting the Hebern machine, William moved on to the next supposedly unbreakable device, invented in 1924 by a German named Alexander von Kryha, who committed suicide in 1955. The Kryha cipher machine was shaped like a half-moon and contained two discs of alphabets, one a fixed semicircle, the second a circle that rotated against it. According to the inventor, it could encipher a message in 2.29 x 10^{82} ways, a number larger than the number of atoms in the observable universe. William was not impressed: "The number of permutations and combinations which a given machine affords, like the birdies that sing in the spring, often have nothing or little to do with the case." Other features were more important, like the method of selecting alphabets and the motions of the wheels. William conquered the Kryha and later demonstrated his mastery by solving a two-hundred-word message prepared by a lawyer in New York who believed the Kryha was unbreakable. William found the solution in a mere three hours and thirty-one minutes, including a fifty-minute lunch break.

In this new era of machines, William was showing that the human was still king. Until the invention of digital ciphers in the 1960s, the field of cryptology would be defined by heroic human attacks on physical cipher machines. These attacks would often be *aided* by machines specially built to speed the attacks—like the famous electromechanical "bombes" designed by the British codebreaker Alan Turing, and some of the world's first computers, monstrosities of wires and vacuum tubes that occupied entire rooms—but not necessarily. It was still possible at this point to defeat a machine with mere pencil and paper. The human brain could beat the machines, if it was the right brain, and if the owner of the brain was willing to accept the cost of victory.

The common saying about cryptologists, as William phrased it, was that "it is not necessary" to be insane, "but it helps." There was uncomfortable truth in the joke. To operate at the highest level of the field seemed to require the kind of pitiless attention and focus that turned some otherwise pleasant and well-adjusted people into zombies who stumbled down the stairs. Mental breakdown was a hazard of the job. During the coming war, several American cryptologists would crack under the strain. Captain Joe Rochefort, one of the navy's top codebreakers from 1925 until the end of the Second World War, suffered from ulcers. A slender, high-strung man, Rochefort recalled later that he would come from work three days out of four and lie in bed for two hours, unable to eat, because he felt so much pressure. The pressure didn't come from superior officers, or even from the urgencies of war; it came from within. "Here is a bunch of messages and I can't read them," Rochefort said. "Now what's wrong? It was this sort of a thing, you see. And this was sort of standard. You'll find people like this who maybe go into a trance, what looks like a trance when you're talking to them, and the first thing you notice is that they are not paying any attention to you at all, and their mind is on this other problem which they brought home from the office." Rochefort had to stop breaking codes for two years in the late 1920s for health reasons, but the navy pressed him back into duty. He knew William Friedman well. "There is no one that could compare with Friedman," Rochefort said, "no one at all."

There was one other cipher machine that William studied briefly: Enigma, invented by a young German engineer in 1918 and available at that time on the open market. It looked a bit like a typewriter. Under the cover of the Enigma were three rotors, wired wheels similar to the ones in the Hebern machine but invented independently and capable of more intricate movements, the electrical current crossing and doubling back in mazes.

William was impressed with the Enigma and its "clever inventor" but didn't make a serious effort to solve the machine. It was just a curiosity at this point, not a haunted piece of technology cloaking the schemes of fanatical killers. The Nazi Party had just been founded in 1920. Hitler was speaking to modest crowds in beer halls. William Friedman had no way of predicting the fatal battles around Enigma yet to come, and no way to know that Enigma loomed larger in the destiny of his wife than his own: that one day Elizebeth would conquer multiple Enigmas with pencil and paper.

By this point, she had quit her job. She wasn't working as a cryptologist anymore. After about a year at the army, Elizebeth resigned in the spring of 1922, saying she planned "to stay home and write some books." William had encouraged her to quit. For one thing, he was excited about her books—he admired good writers and thought highly of her as a prose stylist—and in another sense it was just the path of least resistance, the expected arrangement, for a young Washington wife to stay home. The decision seemed to clear the way for Elizebeth's ambitions and for the couple to start raising children. But as soon as she quit, William missed her at the office, complaining to a friend that Elizebeth was home "and I am all alone." Then she left Washington for a five-week vacation through the Midwest, visiting family and friends, and his loneliness shaded into panic. She sent William letters about how much fun she was having. "I drove 36 miles to Lake Erie and then swam and broiled ham on the shore for supper. After it got dark we told stories and sang around the fire—it was a glorious night with the moon making fairyland of the lake."

She mentioned that a certain midwestern stranger had been following her to social events, expressing romantic intentions. William tried to make her jealous in his replies. He described going for an evening walk with the attractive wife of a friend, an outing cut short by clouds of mosquitoes. Elizebeth wrote back, "My dear, I'm proud of you! If fate had only been gentle with you and spared the chiggers, what a nice Memory you would have." She added, "Hold as many hands as you can! Life grows short!" In another letter she asked, "Am I wicked to be glad you are missing me?" and signed off with a love note in a rail-fence cipher:

It's 11:00 now. The girls
want to go to bed.

60 *stillion* kisses
to my sweetestheart!

E T O R O N R I
J A D E M M A !

Your
alsbeth

Please see that there
isn't any food left any-
where in cupboards or
icebox to smell, since
you aren't eating home
and if you aren't using
garbage can, better put
it out back.

So glad you had a nice
party — why not do it some
more? and be sure to
get in your fishing?

Got these pictures at Krausi's.
Jean is tall and a little beauty.

Plaintext: JE T'ADORE MON MARI, "I love you my husband" in French.

Not long after Elizebeth returned from her trip, the Friedmans
decided to move out of the city, to the pine forest of what is now
Bethesda, Maryland. They rented a house with a wraparound
porch that looked out to pines, apple trees, a tulip grove, and a
garden spattered with irises of purple, blue, red, yellow, and white.
The interior was comfortable and a bit creaky, the forest damp-
ness making the paint peel from walls. They got a dog and called
him Krypto, after *kryptos*, the Greek root of the word "cryptol-

ogy," meaning secret or hidden. Krypto was an Airedale terrier, tan on both ends and black in the middle, with the alert eyes of a hunter and a wiry coat. They added a cat to the mix, Pinklepurr, named after the A. A. Milne poem about "a little black nothing of feet and fur."

William rode the trolley to the Munitions Building in the morning, an hour spent in his head, thinking about alphabets. After Elizebeth resigned, the army had given him an assistant, a former boxer with cauliflower ears. The man's only skill was typing. If you would like to imagine the birth of the mighty National Security Agency, please visualize two men in a small room, one with a pug nose, pecking at a typewriter, the other a dandy in a suit and bow tie, smoking a pipe, wondering what his wife was up to at home, and if she was missing him.

"Now," Elizebeth wrote on her typewriter, "I wouldn't be a bit surprised to learn that there are some persons reading this book who if suddenly awakened from a sound sleep would not be able to recite the alphabet, so here it is:

A B C D E F G H I J K L M N O P Q R S T U V W X Y Z

"It is composed of 26 letters and their order is fixed. I couldn't for the life of me tell you why A comes first, B second, and so on, nor can anybody else tell you—and don't let them kid you into thinking they can, either."

Sitting with her typewriter by a crackling fire, Elizebeth worked on a book about codebreaking aimed at teens and curious adults, a "little book" to "afford you some amusement for leisure moments." The idea had been in the back of her head since Riverbank. She wanted to write about codes and ciphers with a light and whimsical touch nowhere to be found in other literature. She thought up playful analogies to explain cryptologic concepts. "Miss Transposition Merely Turns Her Clothes Around"; Miss Substitution changes into a new outfit. Adopting the tone of a mischievous schoolteacher, stern but kind, Elizebeth walked readers through sample problems and cheered them along: "You're just eating 'em up." "Bravo!" Meanwhile, she wrote a draft of a *second* book, a children's history of the alphabet, illustrated with her own drawings of hieroglyphs and cuneiform tablets. She had begun working on it at Riverbank. The alphabet, part of the backdrop of our lives, like the sky or electricity or advertising, but the one tool that makes all the others possible—she wanted kids to know that the alphabet is a miracle.

The goal with both books was the same: to share, explain, demystify, get people excited about the possibilities of words. This was the opposite of what government seemed to be about, in her experience—the dreary, smoky vault at the Munitions Building. She was having fun, developing a voice of her own, and she might have been perfectly happy here, writing books by a fire for the rest of her life, if men from the government had not begun to knock on her door, asking her to solve puzzles for America again.

Her skill was the force that pulled them. There were just so few cryptologists of her ability, or William's. They were like a binary star system in a void, twin suns rotating around each other, drawing lesser bodies by their light.

Having to choose between the two Friedmans, people almost always approached William first. He was the man, the one with his name on publications, the one who had served in France during the war. A retired astronomy professor asked William to analyze radio signals collected from a device that printed the waves as black lines on a thirty-foot-long piece of film. He theorized that if aliens existed on Mars, they might be transmitting; William found no pattern in the signals (they were probably interference). He consulted on criminal investigations. A man sent a bomb to Huey Long, the Louisiana politician, along with a note full of hieroglyphic markings. William solved it. The warden of an Ohio penitentiary sent him a cryptogram smuggled into prison by the mother of a bank robber. William sent back the plaintext, which revealed a plot to help the inmate escape by placing bombs along a prison wall and exploding them on a Sunday during church.

In 1923 William gave expert advice to a congressional committee that was investigating political bribes written in coded letters. Top officials in the Warren Harding administration had taken cash from oil tycoons—the "Teapot Dome" affair, the biggest corruption scandal in U.S. history. His testimony caught the eye of a twenty-eight-year-old J. Edgar Hoover, then an FBI agent working the Teapot Dome case, and in the years that followed, after Hoover was named director, he asked William to consult on a few FBI cases. The bureau had no cryptanalytic section, no skilled codebreakers of its own. It had to rely on outside experts. When Hoover's agents arrested several of John Dillinger's gang members, notes scrawled with code were found in the pockets of the machine gunners. Hoover sent the notes to William. He solved them.

A wealthy businessman, *Washington Post* owner Edward McLean, hired William in 1924 to design a code for his personal correspondence, then refused to pay him. This was a man who had bought the Hope Diamond as a gift for his wife. She wore it around her neck at parties she threw for wounded veterans. And McLean stiffed William on his fee.

Elizebeth urged her husband to protest, but he let the matter drop, fearing the stereotype of the "money-grubbing Jew." America was growing more anti-Semitic in the 1920s, the successes of Jewish immigrants provoking ugly responses. The president of Harvard changed admissions rules to keep Jews out. Henry Ford launched an anti-Semitic weekly newspaper with a declaration that "the Jew is the world's enigma." Officers in the Military Intelligence Division of the War Department tracked intelligence reports about Jewish activities on index cards labeled "Jews: Race" and kept a central dossier on the "Jewish Question" that included documents like "The Power and Aims of International Jewry." The MID men were William's colleagues.

So he treaded carefully in Washington, lest he provoke an anti-Semitic reaction that seemed to always be near. He tried to get along, said yes a lot, loaned out pieces of his brain. Inevitably there was only so much of him to go around, and requests for his time spilled over to Elizebeth. "When they couldn't get him, I'd be offered a job," she said later. "That's the story of my life. Somebody asks for my husband and they can't get him, so they take me."

On the one hand, she found this insulting. It was like people were trying to use William's brain "second-hand" by hiring the woman who presumably enjoyed intimate access to it. "Sad for me," she said. But it also gave Elizebeth a way to demonstrate that she was a master codebreaker in her own right.

First it was the navy that wanted to hire her, in late 1922. They had lost a civilian cryptologist, Agnes Meyer Driscoll, a woman with a mathematics Ph.D. who had left for the private sector, partnering with the horse thief Hebern to launch a cipher machine factory in California. "I didn't want to work for the Navy," Elizebeth said, "but they were just sitting on my doorstep all the time and the only way to get rid of them was to go there for a little while until they found someone else." She took Driscoll's place and filled in for a time, designing codes for sailors. Then she became pregnant. Elizebeth left the navy after five months, again thinking she would never go back to the government, and gave birth to the Friedmans' first child, a girl they named Barbara, in 1923.

Elizebeth had always been hesitant to bring a baby into the world, her own childhood having been less than warm. She had watched her mother, Sopha, raise nine kids, exhausting herself in the process and dying of cancer, suppressing her own desires and dreams to the extent that she never said what they might have been. Elizebeth didn't want to sacrifice herself like that. Ever since she married in 1917 it had been her husband who pushed the issue of children and she who pushed back, saying she wanted to wait. William confessed in letters from France that he wished they had begun having children at Riverbank, before he deployed to France with the AEF. "Often I feel that in many ways we made a mistake," William wrote. "Our love will only be complete when there is a third—flesh and blood of yours and mine." She lacked this conviction; when Elizebeth did think about having children, it was an impulse that gripped her and then quickly passed. "Sometimes I wish I were going to have a child," she wrote in one letter to France. William replied that "a queer sensation" came over him when he read those words. A child, "the wonder of all mysteries!" Whatever Elizebeth wrote in response, she destroyed.

When children finally arrived, Elizebeth didn't put aside her ambitions, although the delivery of Barbara was a difficult one, leaving the mother laid up in bed for months with spinal pain. William made the mistake of mentioning Elizebeth's pain in a letter to Fabyan, who offered to prescribe some back exercises, spinal alignment being his long obsession. "If she were here I am quite sure we could help her," Fabyan wrote to William. "The general idea is to get the anterior curve in the back near the hips out, and one way of doing it is to lie on your back on a table and stick up your knees, bringing the heels as near the buttocks as you can." William replied that he thought she would be all right.

She hired a nanny, a black woman named Cassie (her last name is not recorded), and continued to work on the two book projects from home, sharing the cooking duties with Cassie and taking breaks to play with Barbara, the firstborn, followed by a second child, John Ramsay, three years later, in 1926.

The daughter and the son had opposite personalities and Elizebeth felt constantly perplexed by the size of the gap. Barbara was an affectionate, verbally prolific infant with blond ringlets who charmed all of the Friedmans' friends, loved books, and shared her dolls with Krypto the dog. John Ramsay didn't talk much, didn't like to be sung to, didn't seek out books, and swung wildly between moods, "his mirth excited by so little, his hilarity so whole-souled," Elizebeth wrote. "And then in a flash, fury and rage takes its turn." The only music he seemed to enjoy was whatever Elizebeth played at top volume on the Victrola.

She took a hands-off approach to parenting, hewing to a doctrine of no doctrines, in agreement with William. The Friedmans were determined not to consciously teach their kids anything or tell them what to believe, only to create a comfortable environment, pour in vitamins, "and let the rest take care of itself," as Elizebeth put it. She respected her offspring as autonomous, independent creatures and was captivated by their efforts to learn how to communicate. Some of Barbara's noises sounded like verbalized cryptograms. Elizebeth, always looking for insight into language, transcribed her daughter's babble inside a book Fabyan had given her, *What I Know About the Future of Cotton and Domestic Goods,* the one that contained only blank pages. "She strings together consonants impossible to an English tongue," Elizebeth observed. "Within the last week she says I don't know in answer to questions. When I first told her father of it, he scoffed saying the concept of not knowing a thing was impossible at her age. But as I pointed out, she does not use the separate words—it is rather one elided sound—IDONTKNOW, a perfect imitation." She copied Barbara's ejections verbatim and analyzed the text with a codebreaker's curiosity. *Pfnr-*

pfnh-hnwhwp. It seemed clandestine. There was structure in it, a pattern at the edge of legibility: *pfnr-pfnh-hnwhwp*. What did it mean? *IDONTKNOW*.

After two years living in the forest of Bethesda, the Friedmans decided to move back into the city, closer to William's job, and bought their first house together, at 3932 Military Road in northwest Washington, a newly constructed home with an elephant-shaped door knocker. The neighborhood was full of families and young oak trees, two hundred yards from the Maryland state line. Fathers tended to work for the military or the government. Lieutenants sat on porches, reading the newspaper. This is the house where Barbara and John Ramsay grew up—the Friedmans would stay at 3932 Military Road for more than two decades—and it's where government agents once again came knocking for Elizebeth, hoping to use her husband's brain secondhand.

One day in 1925, when her daughter was two, she answered the door to find a man in his forties with a boyish face and large ears. He wore a double-breasted navy coat and a shiny navy cap. He said his name was Charles Root, he was from the U.S. Coast Guard, and he needed her help.

Root might have had the most thankless job in America at the time. The coast guard was responsible for patrolling American waters, trying to catch the nimble "rum-running" boats that flouted Prohibition law and smuggled bootleg liquor from sea to shore. It was a cat-and-mouse game that the mice were winning. From the start of Prohibition, the coast guard's task had been laughably difficult—they owned just 203 slow small patrol boats to police five thousand miles of coastline—but recently the rum-runners had deployed shortwave radios and sophisticated codes to conceal their movements, giving them the decisive advantage.

One year earlier Root had launched an intelligence division within the coast guard, the first of its kind. He told Elizebeth he was desperate to get a handle on the rumrunners' communications systems. The coast guard happened to be years ahead of other government agencies in radio prowess. It had built several large radio towers along the East Coast to aid in search-and-rescue operations at sea, to help save boats and sailors in storms, and these same posts were able to intercept radio messages sent by the bootleggers. Still, no one in his office knew how to break the codes in the intercepted messages, and over the last several years hundreds of messages had piled up, unsolved and unread. Root said he wanted Elizebeth to come in and tackle the backlog. He asked her to consider breaking codes for America again. Sensing her reluctance to return to government, he pitched this as a temporary assignment, a ninety-day contract.

After thinking about it for a while, Elizebeth said she'd do it on one condition: that she be allowed to work from home.

The coast guard agreed. Root gave her a metal badge that said SPECIAL AGENT, U.S. TREASURY in gold letters. Once every week or so, she traveled to Root's office in the main Treasury building, flashed the badge, picked up an envelope of unsolved puzzles, took the envelope home, solved the puzzles, returned the answers to Treasury, and picked up a fresh envelope.

A block of pale granite across from the White House, the Treasury headquarters was guarded on its south side by a bronze statue of Alexander Hamilton, the Founding Father who designed America's system of money and gave birth to the forerunner of the coast guard in 1790 by launching a fleet of ten small "revenue cutter" ships to catch smugglers and pirates. Fifteen thousand people in Washington now worked for the Treasury Department, and another forty-six thousand in field offices across the country, doing all kinds of tasks: minting coins and paper bills; collecting taxes and customs duties; tracking the output of factories, the price of gasoline, the size of the annual wheat harvest.

Elizebeth didn't have anything to do with these bureaucratic and economic functions. She was involved with the side of Treasury that investigated crimes.

The department contained no fewer than six separate law enforcement agencies: the Prohibition Bureau, the Narcotics Bureau, customs, the coast guard, the IRS, and the Secret Service. The six agencies had broad authorities to probe financial fraud and most any product or person that moved illegally across a border—guns, liquor, drugs, migrants, counterfeit money. The Treasury detectives were known as "T-men" in the press, as opposed to the "G-men" of the FBI, part of the Justice Department. And although the G-men of the FBI tended to get the glory when famous gangsters went down, thanks to the publicity genius of J. Edgar Hoover, it was the T-men of the Treasury, more often than not, who made the cases. Treasury was the center of the fight against organized crime. It was T-men who eventually nailed Al Capone for tax fraud. It was T-men who caught the kidnappers of the Lindbergh baby.

The spiritual leader of the T-men was Elmer Irey, a soft-spoken father of two in glasses and a wool suit. Irey ran the IRS's Special Intelligence Unit. A bloodhound with a nose for money trails, he referred to the six agencies as "the six fingers of the Treasury fist." Al Capone wasn't much bothered by J. Edgar Hoover, but he feared Elmer Irey, who liked to relax in the evening by reading the first few pages of a mystery story and writing the solution in the margin. Reportedly he was never wrong.

Before long, the T-men would learn to see Elizebeth Friedman, this petite mother of two in low heels, as one of their most potent weapons.

The rum kings of 1925 were not gentlemen. The gentlemen had already been driven out of the game. During the first few years of Prohibition, it had been possible for independent sea captains to make a living by racing crates of booze from the Bahamas to the Florida coast, for the cash and the thrill of it, but those days were over. The men in charge of the liquor rackets now were mobsters, killers, associates of killers, and shadowy corporations with intentionally understated names, such as the Consolidated Exporters Corporation of Vancouver, whose rum fleet would have been the envy of many small nations: sixty to seventy boats of various sizes, from enormous "mother ships" the size of baseball fields to small speedboats. The mother ships functioned as huge floating warehouses, anchored as far as sixty miles offshore, and capable of holding up to 100,000 crates of liquor. Consolidated's ships spanned the U.S. East Coast and the West Coast and stretched into ports in the Caribbean and South America. The rum business was hemisphere-wide: "The whole half of the world," Elizebeth said, "was interested in thwarting the prohibition law."

It was a daunting thing for her to face. She got to work.

During her first three months with Treasury in 1925, Elizebeth solved two years' worth of backlogged messages. Captain Root was so appreciative, he asked his bosses for money to hire Elizebeth for good: "Mrs. Friedman is the only person available with the required skill and experience." She accepted a full-time job breaking messages for all six Treasury law agencies and continued to work from home, delivering envelopes of solved messages to the office, carrying envelopes of unsolved messages back to 3932 Military Road.

"Mrs. Friedman," went a handwritten note from the Office of the Chief Prohibition Investigator. "Please see what you can do with this and return it to us." Several cryptograms were attached, including this one, sent from Halifax, Nova Scotia, to an unknown ship on the East Coast:

AWJTSSK JQS GBQKWSK LYMSE EJBCG SPEC QPFYEYQD
MYHGC PRPYC JWKSWE CWI PQTGJW EPFS VBSM
AWAJASTCE HJJS.
     BLACKCAM.

She saw that it was a relatively simple substitution cipher. Elizebeth penciled in the plaintext letters without much effort, revealing a note telling the rum captain to anchor near a New Jersey lighthouse.

PROCEED ONE HUNDRED MILES SOUTH EAST NAVISINK
LIGHT AWAIT ORDERS TRY ANCHOR SAVE FUEL
PROSPECTS GOOD

At first all she did was solve these messages, unraveling the codes of the rumrunners one by one. But most of the messages weren't this easy, and with each passing month, the task grew more difficult. The rum codes of the bigger operators were more secure than any military codes she had encountered during the Great War, demonstrating "a complexity never even attempted by any government for its most secret communications," particularly the codes of the Consolidated Exporters Corporation. The syndicate used different code systems for different sets of ships. Some code blocks were three letters long, some five letters, some four, and the smugglers changed the codes every few weeks, meaning that a broken code didn't stay broken. Consolidated relied heavily on a multistep processes of enciphered code, the messages like little onions, each layer created with a different technique that Elizebeth had to peel back: Starting with the ciphertext block MJFAX, she might have to decrypt it into another five-letter block, BARHY, which corresponded to 08033 in a widely available book of commercial codes used by legitimate businessmen to save on telegraph charges, from which she subtracted 1,000 to get 07033, which matched up with the English word ANCHORED in a second code book. "If I may capture a goodly number of your messages," she wrote, "even though I have never seen your code book, I may read your thoughts."

As good as she was at solving the individual messages, Elizebeth's ambition didn't stop there. She wanted to build a broader, more comprehensive system for extracting intelligence from wireless messages, a system that didn't exist at Treasury or anywhere else in the U.S. government because radio was still such a young technology.

Radio intelligence requires cooperation and sharing of information, and because Elizebeth was the only senior cryptanalyst in Treasury, the only one who knew how to break codes, she became a link between people in law enforcement who didn't usually talk to each other. She communicated with the T-men at the listening posts; depending on atmospheric conditions, the listening posts could hear all the way to Germany, a portent of things to come. She kept in constant touch with T-men in the field who gave her lists of ships as they came and went in U.S. ports. She sketched maps of the naval routes and maps of the radio traffic. She helped train T-men to use direction-finding machines that they loaded in the back of pickup trucks, driving up and down the coasts to find hidden pirate radio stations used by the rumrunners. The smugglers essentially acted like foreign spies in wartime, trading intelligence by radio from hostile territories, and to read their thoughts, Elizebeth had to act like a spy hunter, a counterespionage or counterintelligence agent—skills she would ply against the Nazis in the early 1940s.

"I sort of floated around," she said. "It seemed I went here, there, and everywhere I was needed." It was a manic and exhausting period in her life. She traveled to the West Coast and consulted with customs agents in San Francisco about gangsters running rum boats off the coast. She exposed a ship called the *Holmwood,* which was sailing up the Hudson River with 20,000 cases of liquor, disguised as an oil tanker called the *Texas Ranger;* the ship was halfway to Albany, New York, its destination, when the coast guard and customs seized it on the strength of Elizebeth's decryptions.

The names of ruthless men appeared in her solved puzzles: mafiosi and underworld bosses, racketeers and semi-legitimate CEOs, some of them hidden behind veils of complicated corporate structures and webs of interlocking investments. There was a fleet of rum ships on the West Coast controlled by two Canadian brothers named Hobbs. A California gangster named Tony "The Hat" Cornero owned some of the Hobbs boats. Their rival, for a time, was the Consolidated Exporters Corporation, the unstoppable Vancouver syndicate. Consolidated seemed to have an unlimited stream of money, pumped in by the Reifel family, patriarch Henry Reifel and sons George and Harry, hotel magnates in Vancouver. The Reifels also had an American investor, a colorful Bostonian named Joseph P. Kennedy—the father of John F. Kennedy, future president of the United States. Elizebeth noted in one report that in 1928 the Hobbs brothers appeared to sell their interests "to Joseph Kennedy, Ltd., of Vancouver, large holders of stock in the Consolidated Exporters Corporation."

Elizebeth wasn't afraid of these men, and she didn't hate them, either. She was doing a job, and the job was frequently exciting, each case a detective story. Elizebeth flicked invisibly at the world, setting great chains of events into motion. She went to Houston, Texas, at the request of a federal prosecutor and solved more than 650 messages in 24 different code systems used by rum syndicates in the Gulf Coast. Many of Elizebeth's solutions became evidence in a Texas rum trial involving a one-legged cabdriver named Louis "Frenchy" Armatou. Several of the messages seemed to involve a rum ship not at issue in the trial, the *I'm Alone.* Elizebeth's solutions triggered an international manhunt for the ship's captains. One fugitive, Marvin Clark, went on the run and was gunned down by an unknown assailant in New Orleans, and the other,

Dan Hogan, was arrested and developed a vendetta against Elizebeth—"he was in a very mean mood"—that required her to travel to a federal hearing surrounded by a security detail. And all the time she wrote to her children, to Barbara and John Ramsay, sharing news of her travels. "Only woman on plane," she wrote on a pad of yellow paper during a cross-country flight, at a time when flight in America was still new and wondrous. "Co-pilot as attentive as if I had been a young and pretty girl. The country, so stunning with its rivers and streams, green wooded mountains, and even flat farm country laid out in such patterns that it seemed as if a directing mind from above had planned it." When she got home, she took the kids out for afternoon tea and did puzzle games with them as they sipped their cups.

When the stock market crashed in 1929, demand for liquor only increased, along with Elizebeth's workload. She worried about the family's finances in letters to William. "Did the bank call the loan?" she asked in one letter, then added, as if it were no big deal, "Broke into another system this morning, messages unearthed from a safe."

The volume of intercepts pouring into Elizebeth's coast guard office grew and grew until she felt "almost buried under the press of duties." The grind of it was staggering: two thousand messages per month demanded her attention, and some twenty-five thousand per year. Not all of these contained information relevant to law enforcement, but she had to analyze all of them to know which ones to transmit and which ones to discard. She begged Treasury for help but all they allowed her was a single clerk-typist, a woman. Despite these meager resources, Elizebeth reported to her superiors in 1930 that she had solved twelve thousand rum messages in the previous three years, "which covered activities touching upon the Pacific Coast from Vancouver to Ensenada; from Belize along the Gulf Coast to Tampa; from Key West to Savannah, including Havana and the Bahamas; and from New Jersey to Maine."

Until 1930, almost all codebreaking for the U.S. government's planetary war against smuggling was handled by these two tired and perpetually overworked women, Elizebeth and her clerk, but that year Elizebeth decided she'd had enough and wrote a seven-page memo to coast guard commanders, proposing that they create a "central unit" for codebreaking. It wouldn't do to stumble along with two people anymore; there needed to be a proper team. At a minimum, she said, the unit should have seven employees, two cryptanalysts and five cryptographic clerks, earning combined salaries of $14,600 per year. She argued that her accomplishments of the past several years "can be increased a hundredfold by a sufficiently staffed and thoroughly organized unit" striking at the heart of the smuggling trade from "the most promising point of attack": the interception and solution of encrypted radio messages.

While she waited and hoped for help to arrive, Elizebeth did her best with the meager resources at her disposal. To build an archive of the smugglers' words and stay on top of any shifts in the codes, she made a carbon copy of every solved message, and when the stack of copies grew to be an inch thick, her clerk-typist bound the plaintexts into a volume, a book. By the start of 1931, Elizebeth had generated thirty of these books, a *Britannica* of criminal clockwork. She was compiling a secret history of her age, using the very words of its hidden kings.

Elizebeth and William still had a chance to work together in these years. The collaboration flowed in one direction only, from her to him. William's army work was too secret now. She got the sense that her husband found the smuggling puzzles to be a refuge from problems facing him at work; for him the rum war was a lark. They passed messages and worksheets back and forth, drawing cipher letters in the boxes of sheets of grid paper, snippets of potential plaintext: "Position," "Landing boat," "Is there any indication there will be trouble?"

One evening Elizebeth came home from a League of Women Voters meeting to find William examining a rum message. He looked up and smiled. "Andrew says send glass eye," he said. Elizebeth often uncovered personal notes from the underpaid and exploited sailors on the rum boats, funny or sweet messages to their wives or children, and this was one that she had decrypted and William had apparently seen. The actual text read, "Andrew says advise wife to send reserve glass eye." In a different message to shore, this same Andrew had requested "a pair of shoes, size 15." He must have been a very large man. Elizebeth and William shared a laugh as they tried to picture it: a luckless giant on a rum boat, holes in his shoes, missing a glass eye, pining for his wife.

Elizebeth finally got some relief from her workload in July 1931, when Treasury cleared her to recruit a codebreaking team of her own, at long last, according to the specifications she had laid out in her earlier proposal. Officially the unit would live within the coast guard, but it would break codes for all six Treasury agencies. The department gave Elizebeth funds for three junior codebreakers and two stenographers, along with her first dedicated office space, a small block of rooms inside the Treasury Annex on Pennsylvania Avenue, a building that resembled a classical Greek temple and housed the coast guard, customs, and the Bureau of Narcotics. The new space was a large open workroom with desks and small offices on the periphery that fit two to a room. As part of the deal she got a pay raise, too, from $2,400 to $3,800 per year, and a fine new title: Cryptanalyst-in-Charge, U.S. Coast Guard.

It was the first unit of its kind in Treasury history, and the only codebreaking unit in America ever to be run by a woman—another pioneering moment for Elizebeth.

The first thing she did was hire and train a staff. There was no available pool of cryptologists in the commercial sector; you had to train your own. Scouring civil service lists of applicants who had taken math, physics, or chemistry, Elizebeth was unable to find names of women, so she hired young men for the junior codebreaker jobs, setting them to work on practice problems she had designed, a beginner's course in codebreaking.

She had never been the boss of men before and she worried that her new employees might not accept her authority as a woman, but this concern proved unwarranted, with one exception, the all-time top scorer on the math exam, a young mathematics Ph.D. from Columbia University who was supposed to be a prodigy. He refused to complete Elizebeth's exercises and babbled nonsensically about an "indecipherable cipher"; she decided "he did not comprehend the English language" and replaced him with a candidate who showed a more practical cast of mind. Elizebeth's first two successful hires were Hyman Hurwitz, a twenty-one-year-old electrical engineer from Dorchester, Massachusetts, who liked to tinker with radios, and thirty-one-year-old Vernon Cooley, from Kalamazoo, Michigan, who had taught schoolchildren there and also worked for a time in the factory of the Kalamazoo Paper Company. Both were given the title of Assistant Cryptographic Clerk. A third young codebreaker-in-training soon joined them, Robert Gordon, who was twenty-three and hailed from Waco, Texas.

The three men, "able, agreeable, and cooperative," treated Elizebeth with respect, and soon they all settled into a productive routine, learning each other's quirks and strengths, dividing up tasks as a team. Asking if Elizebeth experienced sexism was like asking if Marie Curie did. Cryptology was a young field. It hadn't yet sorted itself into rigid roles by gender. And Elizebeth was exceptional. She deployed overwhelming mental firepower against the pull of gravity. She was Cryptanalyst-in-Charge.

By the end of 1932, when Americans elected Franklin Delano Roosevelt to his first term as president, Elizebeth's team at the coast guard was pound for pound the best radio intelligence organization in America. They knew how to extract information from clandestine radio networks, map the hidden structure of the transmitters, and hunt the people using them. This set of skills would later make Elizebeth an important figure in the quest to destroy the clandestine networks of Nazi spies. Her fight against smugglers was like target practice for the coming fight against fascism.

Americans weren't yet afraid of Nazis. They were too worried about scraping up money for food. By 1933 the Great Depression had put 15 million people out of work. Fabyan told William in a letter that the mayor of Aurora, Illinois, had shut down the town for five days to stop runs on the bank. "The world is a mess, and everything is going topsy-turvy," Fabyan said.

Elizebeth took her daughter to hear FDR's inaugural speech on March 4, 1933, a viciously cold morning. They walked to the Capitol from their house. "The only thing we have to fear is fear itself," FDR said, focusing on his plans to restart the economy. He did not mention Adolf Hitler or the Nazi Party, who had taken power that January, deploying mobs of men in brown uniforms and swastika armbands to crush dissent. The international press covered him like a normal leader. Many Germans did not think he would really do the things he had said he would do.

After FDR's speech people stayed outside in the freezing wind, cheeks pink as raw beef, waiting to see the parade. Elizebeth and Barbara handed out League of Women Voters fliers to the spectators. Elizebeth was pleased to see her daughter playing the role of "ardent worker."

Eighteen days later, in Germany, in a vacant gunpowder factory northwest of Munich, the Nazis opened the first concentration camp, Dachau. The occasion was announced by Heinrich Himmler, *Reichsführer* of the SS, at a press conference.

One month after that in Washington, at the end of April 1933, Elizebeth prepared to leave the city on her latest assignment for the T-men. "I pack my bag," she wrote, "and hug my children a good-by which is to last for a week or a month or longer, I know not, and board a train with a prayer that the new fields will be not impossible of conquest." She kissed William. He asked her to be safe and not take unnecessary risks. If she had known what awaited her in New Orleans, she might never have gone.

"**Please state your name**," America's top Prohibition official said. He had solemn eyes and spoke in a slow drawl.

"Elizebeth Smith Friedman."

"What is your occupation?"

"I am a cryptanalyst."

"And what are the duties of a cryptanalyst?"

"A cryptanalyst is a person who analyzes and reads secret communications without the knowledge of the system used."

It was May 2, 1933, and Elizebeth was speaking in a witness box in a federal courtroom, about to play her part in a giant conspiracy case against twenty-three suspected agents of the syndicate she had been tracking for years, the Consolidated Exporters Corporation. They had recently expanded into the Gulf of Mexico, directing a fleet of eight rum-running ships from a pirate radio station in New Orleans. T-men intercepted at least thirty-two coded radio messages and mailed them to Elizebeth, and her solutions exposed the ribs of the scheme: the names of the rum ships (*Concord, Corozal, Fisher Lassie, Rosita, Mavis Barbara*); their system of sneaking crates of liquor into lonely bayou towns on small boats called luggers and then unloading the crates onto freight trains, covered in sawdust. The government considered it "the greatest rum-running conspiracy since Prohibition," and now Elizebeth had been summoned to this federal courtroom in New Orleans to explain her methods to a judge and jury.

She wore a pink dress and a hat with a flower pinned to its brim. Directly in front of her was a stack of yellow papers with her message solutions. She looked out at the courtroom, its wooden pews crammed with spectators and journalists. The defendants sat together like a sports team on a sideline, in suits; the previous day in court they had been quietly switching seats to make it more difficult for witnesses to identify any one of them. The accused ringleader was Albert Morrison, a rawboned man in his sixties with white hair. He went by the aliases Charles Cosgrove, M. Ryder, A. A. Brown, Harry Hale, J. J. Jones, B. M. McGregor, and "Mr. Burk." The government believed that at least three of the other defendants—Nathan Goldberg, Al Hartman, and Harry Doe—were Chicago associates of Al Capone.

The government had spent $500,000 and more than two years on the investigation, marking this as a case they could not afford to lose, which is why the lead prosecutor today was the chief of the Prohibition Bureau himself, Amos Walter Wright Woodcock, a methodical former army colonel. He had argued for years that the only effective way to enforce Prohibition was to bring "a steady attack" against major crime syndicates and leave the small-time moonshiners alone. Big fish, not little fish. Consolidated was the whale.

Woodcock looked at Elizebeth in the witness box. "Have you a message which we will identify as 6:07 P.M., April 8, 1931?"

She shuffled her papers and located the sheet containing this message, which began, in code, QUIDS, ABGAH, FLASH, SLATE, FABLE, SHOOT, BOWSKY. She was about to read the solution aloud when a defense attorney objected, claiming that her testimony was "incompetent, irrelevant, and immaterial." Eight other defense attorneys leapt to their feet, joining the objection.

One was Edwin Grace of New Orleans, who handled Capone's appeals in federal courts. Grace was joined at the defense table by Walter J. Gex Sr., the patriarch of an influential family from the nearby Gulf city of Bay St. Louis, Mississippi.

"I believe I am asked my opinion of the reading of this message?" Elizebeth said, turning to the judge. "This is not a matter of opinion. There are very few people in the United States, not many it is true, who understand the principle of this science. Any other experts in the United States would find, after proper study, the exact readings I have given these."

"I ask all that be excluded," Gex said. "I think it is very improper."

Woodcock continued to walk her through the messages, and Elizebeth read her plaintext translations to the court, one at a time, interrupted by additional objections from the defense, accusations that she must have gotten her information from federal agents and not from codebreaking ("That certainly is some information somebody gave to that lady"). The syndicate had used a system of enciphered code—words that stood for letters in a cipher that stood for letters in a code. To solve it, Elizebeth had to rewind each step. A message that looked like

GD (HX) gm ga HX (GD) R gm OB BT HR CK 25 BT BERGS
SUB SMOKE CAN CLUB BETEL BGIRASS CULEX CORA STOP
MORAL SIBYL SEDGE SASH (?) CONCOR WITTY FLECK SLING
SMART SMOKE FLEET SMALL SMACK SLOPE SLOPE BT SA
back to the word SLDGE its SEDGE instead of SLDGE HW

became in plaintext

SUBSTITUTE FIFTY CANADIAN CLUB BALANCE BLUE GRASS
FOR COROZAL STOP REPEAT TUESDAY WIRE CONCORD GO
TO LATITUDE 29.50 LONGITUDE 87.44

When it was time for cross-examination, Gex rose from the defense table.

"How shall I address you," he said. "Madam or Miss?"

"I am Mrs. Friedman."

"Before you could properly translate these symbols somebody

had to tell you it was symbols in reference to the liquor transportation?"

"Oh no," Elizebeth replied, innocently. "I might receive symbols related to murder or narcotics."

"The same symbols these gentlemen used to mean what you say, whiskey, beer, position, could not have been made up by people in code for transportation of women from Europe?"

No, Elizebeth responded, "not with the meaning given them here."

"I move that all of the testimony of this lady be stricken out," said another lawyer, Maxwell Slade. The judge overruled him.

When there were no further questions from the defense, Elizebeth reached for her handbag, stepped out of the witness box, exited the courthouse into the damp spring air, and returned to Washington. Four days after her testimony, the jury convicted five of the syndicate's ringleaders, including Albert Morrison, who received two years in prison. Woodcock gave credit to Elizebeth, telling her superiors that she "made an unusual impression on the jury."

And not just the jury. Reporters covering the trial were taken with Elizebeth, describing her in stories as "a pretty government scrypt-analyst or 'code-reader,'" "a pretty middle aged woman," "a pretty young woman with a filly pink dress," and "a pretty little woman who protects the United States." This was her first sustained encounter with the press and it left a sour taste. She was still young enough to resent being called a middle-aged woman and thought the kinder phrases were badly written. The newspapers wrote about her again the following year when she returned to New Orleans to testify in the convicts' appeal. Facing off against Edwin Grace, Capone's attorney, Elizebeth grew impatient with his attacks on the validity of her science and told the judge she could settle the issue quickly if she had a blackboard. A bailiff found a blackboard in storage and wheeled it into the court. Elizebeth stood with a piece of chalk and diagrammed the rum ring's codes on the board until the jurors were nodding their heads and Grace was muttering that this was highly irregular. "CLASS IN CRYPTOLOGY," one newspaper blared the next day. The headline made her queasy. She had not signed up to be a public figure. She hoped the attention would die down. To her horror, it was only just beginning.

William went dark as his wife ignited and lit up the sky. The army kept his projects locked behind thick metal doors. He wasn't allowed to talk about his work, and she knew enough not to ask. "He never put into words and never asked me to put into words," Elizebeth said. The two cryptologists were so connected, so attuned to the slightest crease of an eyebrow or curl of a lip, that their faces had a way of becoming mirrors. A grim look on his face was instantly reflected in hers. Because neither wanted to cause pain in the other, the most painless course of action was often to make their faces like masks and not speak about forbidden things, though despite these efforts at self-control Elizebeth could tell when something was troubling her husband: "Many times there was a certain grim look that came around his mouth."

What she didn't know at the time, and didn't learn until after the war, is that the army had asked William to break a series of ciphers used by Japanese officials to encrypt their diplomatic communications, an enormous undertaking that would come to define his career and consume the next decade of his life.

He wasn't doing it alone. Like Elizebeth he was starting to build his own codebreaking unit, hiring junior cryptanalysts and molding them to his needs and his vision. The army had never agreed to such an expansion before, but the plates of government codebreaking had shifted due to a fateful decision made by the newly appointed secretary of state, Henry Stimson. A former artillery officer in the Great War, Stimson thought the idea of reading other nation's messages in peacetime was immoral, and upon learning that the State Department was paying codebreakers in New York to read the mail of foreign diplomats—Yardley and his American black chamber—the secretary was appalled. "Gentlemen do not read each other's mail," he was supposed to have said. Whatever the reason, Stimson decided to pull Yardley's funding in 1929, and the chamber was forced to shut down.

Because Yardley had developed some expertise with Japanese ciphers, the closing of his bureau left America with no ability to break new ciphers developed by Japan, a nation with a growing military and ambitions of empire. Worried about being left in the dark, the army turned to William, and in 1930 he launched a new army codebreaking unit that would later become the nucleus of the National Security Agency. William called his new organization the Signal Intelligence Service, or SIS.

The first three people he hired for the new unit were male mathematicians in their early twenties: Abraham Sinkov, Frank Rowlett, and Solomon Kullback. He gave them adjacent desks on the third floor of the Munitions Building, in a vault behind a thick steel door, and began to train them with dusty books and sample problems. William handed Rowlett a book by a German cryptographer named Kasiski. How is your German? he asked Rowlett. Rowlett said it was not good. William suggested he study the Kasiski text anyway. Now here was a book by an Austrian military cryptographer named Figl, and here were some of Friedman's own works, the "Riverbank Publications," written with Mrs. Friedman during the war. "Do you speak French?" Friedman asked Rowlett, who was sorry to say that he did not.

One morning, when William felt the three young men were ready, he popped his head in the vault and said they should come with him immediately. They followed the boss down a staircase and through a long corridor, where he took a left turn into an area that seemed deserted except for a steel door with a combination lock. Friedman reached into his coat pocket, removed a card, examined a series of numbers on it, and manipulated the lock until the bolt swung free. The door opened; behind it was a second steel door with a keyhole. Friedman fished a key from his jacket and jiggled it in the lock. The inner door opened into a completely dark room. The men waited outside as the boss disappeared into the dark and struck a match. A smell of smoke drifted out of the vault. Friedman found the dangling cord of a lightbulb and pulled, revealing a windowless room full of filing cabinets. He switched on three wall-mounted fans, turned to his young employees, and said, with gravity, "Welcome, gentlemen, to the secret archives of the American Black Chamber."

Yardley's files. The cabinets from his bolted bureau. Friedman had obtained them on the theory the files would prove useful to future army codebreaking projects, particularly regarding Japan.

Over the next fifteen years, William and his three young colleagues, joined by others, would drive themselves to the brink of collapse while working as codemakers and codebreakers alike, struggling to solve Japanese cipher machines of unknown design and applying what they learned about the weaknesses of the foreign machines to design new kinds of secure machines for America. At heart it was all the same business, the business of dominating secrets.

William first taught his deputies about cipher machines by challenging them to solve different machines that he himself had mastered years before (the Kryha, the Hebern), offering gentle hints when the young codebreakers got stuck. Then he set them to work on their first mysterious and yet-unconquered Japanese machine, which began transmitting in 1930: *Angooki Taipu A,* meaning "type A cipher machine" in *romanji,* the romanized form of Japanese that was used for transmitting messages. The name *Angooki Taipu A* was the codebreakers' sole piece of information about the machine. No one in the Western Hemisphere had ever laid eyes on it. There wasn't so much as a drawing of one, much less a physical copy. Internally, William and his colleagues referred to it by a nickname, "Red." Later, in 1938, the Japanese replaced Red with a more sophisticated machine, a significant upgrade: *Angooki Taipu B*. This one the SIS codebreakers called "Purple."

Red and Purple were used by Japanese diplomats in Nazi Germany to communicate with the Japanese government back home, which meant that solving the machines would give the SIS a most intimate view of strategic thinking in both Japan and Germany. To attack each system—first Red, then Purple—the American codebreakers needed to build their own bootleg versions of the Japanese machines, reverse-engineering them based on nothing but educated guesses from analyzing the garbled messages they produced. It was a task akin to building a watch if you have never seen a watch before, simply by listening to an audio recording of the ticking and clicking of its gears.

Even as he tried to infer the shape of foreign machines, William was building an American machine of his own, to protect American communications: the Converter M-134, a multirotor device of innovative design. He filed a patent in hopes of selling it one day on the commercial market. The patent was granted but held secret for many years, essentially nullfiying his rights. This would become an enduring frustration for William. As hard as he tried to earn money on the side, security concerns always got in the way.

Thanks to powerful brainstorms from Frank Rowlett, the M-134 eventually evolved into the SIGABA, an Egyptian Sphinx of a cipher machine with fifteen rotors and an ingenious mechanism that Rowlett called a "stepping maze." Up to four rotors might turn at the same time, with a single key press, and the rotors could be inserted in reverse direction. The army and navy distributed 10,060 SIGABA machines across every theater of the Second World War. President Roosevelt used SIGABAs to communicate from his Hyde Park home and when he traveled on the presidential train. The SIGABA was like an American Enigma machine or Purple machine, only inviolate. No enemy codebreaker, whether German, Italian, or Japanese, would ever manage to break it, despite strenuous efforts; the Nazis ultimately stopped intercepting SIGABA messages altogether, since they could not be read. The machine ensured "the absolute security of army and navy high command and high echelon communications," William later wrote with pride, and "contributed materially to the successful outcome of the war."

"Never said a word to me," Elizebeth swore.

All through the 1930s and the first half of the 1940s, the only thing he revealed to his wife about his work is that it involved Japan. That was all. He didn't talk about Red or Purple, or his own designs.

People wondered later how this could have been true, how a husband and wife, lying in bed together at night, both of them cryptologists, could possibly resist sharing the dramatic details of their work. No: "My husband never never opened his mouth about anything." She had to guess at the mind of her husband by watching changes in his behavior. He came home in the evening and said little. He opened a silver snuff box and inhaled the black tobacco dust. He was almost never cross, only withdrawn. The kids picked up on it. They noticed there were times when friends and neighbors were welcome to visit, when William cooked steaks on a backyard grill and laughed a lot and "was hugging me warmly," as Barbara later recalled, and other times he seemed trapped in a fugue state and was unable to come to the door. In the morning he paced through the house. At night he couldn't sleep. The bed would shake at 3 A.M. and Elizebeth would see him out of a corner of her eye, heading downstairs to make himself a sandwich.

He was not as sick as he would get in later years. She was not yet seeing him unable to get out of bed, lethargic, shaking, acutely depressed, talking casually of suicide, carrying a length of rope in the backseat of his car. She still thought of William's affliction as "nothing more or less than exhaustion" and refused to call it mental illness, a rational choice given the wide stigma against the mentally ill and the poor treatment options (no antidepressants). The chief psychiatrist at the city's preeminent hospital, Dr. Walter Freeman of George Washington University, was an early adopter of electroshock therapy and the inventor of the ice pick lobotomy, a cruel and unwarranted procedure that involved jamming a sharp metal stick through the back of the patient's eyeball while the patient was awake and wiggling it around until a sufficient amount of brain matter was turned to goo. There was a fair chance that if William sought mental treatment in Washington, he would end up under the care of Dr. Freeman or one of his disciples (and sure enough, eventually, William did).

So Elizebeth did her best in the 1930s to cover for her husband in public and shore him up in private, lending her strength to a person who seemed unwell without letting on to friends that he was unwell, out of loyalty to her lover and also a simpler kind of self-interest. The Friedmans had always refused to acknowledge clear distinctions between his career and her career, his adventures and her adventures, because it all sprang from those same fertile years at Riverbank when they worked so closely together and ultimately forged a sacred alliance to escape, and that bond had survived through the decades: "Any story of my experiences is quite inseparable from that of my husband, whose wizardries in cryptanalysis are of international note," Elizebeth once wrote. Their relationship was progressive in this sense, a joint bank account of the mind. It didn't seem like sacrifice for Elizabeth to help her husband when he was down and weak. It was like treating a wound on one of her own limbs.

The closest she ever came to explaining the politics of their marriage was in a letter years later to Barbara, recommending she read a novel called *Immortal Wife,* about a real American couple from the nineteenth century, Jessie and John Frémont. John was an army colonel who mapped the wilds of California, Jessie the feisty daughter of a senator, and they collaborated all their lives, Jessie believing that "to be a good wife a woman must stand shoulder to shoulder and brain to brain alongside her husband." As a husband and wife they attracted fame and controversy. Jessie took the fragmentary journals from John's expeditions and turned them into rich narratives that enthralled Americans. The Frémonts made money and lost it all; the U.S. Army court-martialed John on a bogus charge of insubordination; he suffered a nervous breakdown, deteriorating into a "gray-haired and sunken-cheeked person . . . his face at war with itself." Some of the novel's more vivid passages describe the efforts of Jessie to rebuild John's shattered confidence:

> Everything that happened to John must of necessity happen to her; when two people marry they cease to be purely themselves but step into a new and expanded character, the character of their marriage. . . . She used every art and guile known to the heart of woman to nurse him to health. As they rode . . . spirited horses through the forests she challenged him to race with her, complimented him on how beautifully he sat the horse. Sitting before the warmth and bright red flames of the fireplace she played up the hours and episodes he had enjoyed most in their years together, in which he had appeared to the best advantage, filled him with her pride in his accomplishments. . . . She played the temptress, wearing her loveliest gowns, using her most delicate perfumes, shamelessly arousing his sexual love for her, the love that always had been such a strong and potent force between them. . . .

William also told his son to read this book if he wanted to understand his own parents: "In many ways it parallels my life with your mother," he wrote, calling Elizebeth "a remarkable woman" whose "own indominable spirit helped me climb up out a psychological morass that was pretty deep and distressing." The Friedmans connected with the tale of the Frémonts, this pair of American explorers menaced by the whims of the army and the nuances of neurochemistry. It made them feel, as all good books do, less alone.

Their lives had become so isolated in Washington. When they first met and fell in love, at Riverbank, secrecy was about exploration and connection, a joyous hunt for a hidden order. Now it was a force of loneliness.

They rebelled.

Not openly. Not by breaking laws or leaking secrets. Instead they figured out how to use cryptology to reach out to people, to resist the isolation that their cryptologic careers enforced.

They began with their own children, teaching them simple ciphers when they were seven or eight: A=B, B=C, C=D. Barbara wrote letters in this cipher from sleepaway camp. "EFBS NPUIFS BOE EFBS EBEEZ . . . Dear Mother and Dear Daddy. We went on a canoe trip. We went about 12 or 14 miles. We did it in 1 day too. I paddled 1/2 the way. Love, Barbara." William and Elizebeth replied, "XF BSF QMFBTFE . . . We are pleased to know that you can handle a canoe. It is lots of fun. You will be surprised to learn that Pinklepurr had her kittens on August 2."

The Friedmans also shared cipher letters with their friends. Each December they sent a holiday card in the form of a puzzle. In 1928, it was a "turning grille," a square of red paper perforated with circular holes, its four sides numbered 1 through 4, with a single left-facing arrow that said "TURN." When this square was laid atop a separate sheet of paper containing a 9-by-11 grid of letters, certain letters showed through the holes: FOR CHRISTMAS GREETINGS IN 28. Turning the overlay 90 degrees clockwise, new letters appeared in the holes (WE USE A MEANS QUITE UP TO DATE), then again with a third turn (A CRYPTO TELEPHOTOGRAM HERE), and a fourth

(BRINGS YOU WORD OF XMAS CHEER). Another year William drew a picture of a tree with happy faces and sad faces hanging from its branches like fruit. "Friedman's Wishing Tree," the caption said. "Individual fruits biformed, differentiated by upcurving or downcurving crevice." It was a message written in Bacon's biliteral cipher: a nod to the Friedmans' own past, to their creation story at Riverbank, the Garden of Eden where they fell in love and were expelled for the sin of learning to distinguish reality from delusion. The happy face was the *a*-form in the cipher, the sad face the *b*-form. The plaintext read, "Season's Greetings from the Friedmans."

They took these games further by organizing live puzzle-solving events that were famous in their social group throughout the 1930s. Some of these "cipher parties" were scavenger hunts that sent guests winging through the city. Elizebeth handed you a small white envelope. You tore it open to find a cryptogram. The solution was the address of a restaurant. When you arrived, you ate the salad course, then solved a second cryptogram to discover the location of the entrée. Other parties were hosted at the Friedmans' home with food cooked by Elizebeth. A shy army wife arrived at 3932 Military Road one evening with her husband and panicked when the Friedmans handed her a menu in code. "The first item was a series of dots done with a blue pen," she later recalled. "The 'brains' at the party worked over the number of dots in a group when it occurred to me it had to be 'blue points'—oysters—and it was! I had done my bit, and from then on I was quiet."

On a different occasion, Elizebeth designed a menu that listed one of the courses as "An Indecipherable Cipher." A guest wondered if this meant "hash," a cryptographic term for a string of text that gets scrambled once and never unscrambled, like a door that locks forever behind you (today hashes are used to protect Internet passwords). The guest was delighted when Elizebeth arrived from the kitchen carrying a steaming plate of meat-and-potato hash.

The Friedmans received so much praise for these parties that William thought he saw a chance to make money: If the army wouldn't let him sell his cipher machines because they were state secrets, couldn't he bake some of the same ideas into a mass-marketed board game? In a heat of inspiration he tried to design a Monopoly of codes and ciphers. The Crypto-Set Headquarters Army Game was a folding piece of cardboard with a red spinning wheel; players had to solve puzzles to advance tokens from the start line to the finish. A second prototype, Kriptor, featured two ivory-colored rotating discs printed with hieroglyph-like symbols. Players had to play "codemaker" and "codebreaker," trading secret messages.

He thought Kriptor had commercial potential and sent it to Milton Bradley, makers of Battleship and the Game of Life.

In the hands of the Friedmans, cryptograms were like poems or songs, a way of telling friends and family they were part of something wonderful, a shared language, and that they were loved. Given the immense secrecy of their profession, reaching out to people in these elaborate and whimsical ways, widening the circle instead of narrowing it, was a kind of defiance, although, ultimately, their most defiant act during these interwar years may have been the simplest.

They built a library.

Soon after moving to Washington, Elizabeth and William began to collect books and papers about all things related to secret writing, including documents that touched on their own lives in cryptology and that were not classified or restricted to government vaults. They stored objects on bookshelves in the den of their house. The library was as broad and curious as its creators. It contained books the Friedmans loved and books they hated; books they felt committed sins against cryptologic accuracy, good prose, or both; books that were centuries old but still contained relevant cryptologic knowledge; books they enjoyed but did not understand (William was beguiled and frustrated by the fiction of James Joyce and Gertrude Stein, who wrote sentences so garbled they might as well have been in code). The most valuable book dated to the sixteenth century, *De Furtivis Literarum Notis,* by Giambattista della Porta, an Italian cryptologist and scientific rival of Galileo. The Porta book, with a burgundy binding and text in Latin—an extremely rare 1591 forgery of the 1563 first printing, one of only three such copies in the world—

had arrived in the mail one day from Riverbank, an unexpected gift from George Fabyan. Elizebeth was moved by this gesture, because she had never known Fabyan to give anything away without demanding something in return.

All the rest of their books were collected cheaply. Wherever they went in the world, the Friedmans rummaged through used bookshops for ten-cent treasures. They salvaged books that other people discarded. Once, at the Munitions Building, William fell into conversation with a white-haired Civil War veteran. "He came up to me one morning as I was coming into work," William recalled later, "and he whispered, 'They're burning things today.' And I said, 'Such as?' And he showed me these books." They were a complete set of Union army cipher books, precious items that William rescued from destruction.

Some of the most important items in the library were books of unsolved puzzles and historical mysteries that the Friedmans thought they stood a chance of solving, if they ever got the time. In this sense the library was an archive of their dreams, a set of escape hatches into other, lighter-hearted kinds of lives that seemed possible within the world of secret writing, lives of academic exploration and scholarship. Elizebeth had begun taking graduate classes at American University in 1929, working toward a master's degree in archaeology. She was drawn to several richly illustrated books of Mayan pictographs that had never been decoded; she sometimes fantasized about quitting the coast guard and flying to Mexico, climbing through Mayan ruins in a straw hat, making sketches. William pondered his copy of the Voynich Manuscript, an illuminated book of uncertain lineage, written in a delicate looping script that corresponded to no known language, illustrated with pictures of flowers. He studied a book about the Beale Treasure, supposedly a stash of buried gold and silver ingots in Beale County, Virginia, its location revealed by an unsolved cryptogram. William once joked that he "worked on the cryptogram, on and off, but only in my leisurely hours at home, nights, Saturdays, Sundays, and holidays."

They also collected their own writings in the library. The archives of the Riverbank code section had remained at Fabyan's estate, so the Friedmans didn't own their earliest worksheets, but in the years since they had kept or copied every other nonclassified scrap of paper that had passed through their hands.

The Friedmans weren't necessarily doing this to document the history of American intelligence and tell a renegade story about its birth, although this would be the ultimate, magnificent result. Rather, the Friedmans built an archive because that's what the best intelligence professionals do. They become librarians. It's no accident that J. Edgar Hoover got his start in government as an eighteen-year-old Library of Congress clerk, a job that gave him "an excellent foundation for my work in the FBI," he later said, "where it has been necessary to collate information and evidence." The FBI was a library of human fingerprints and human deeds. Elizebeth's thirty bound volumes of rum messages were a library of the outlaw seas. The files of Herbert Yardley's American black chamber, now controlled by William, were a library of diplomatic intrigues from many nations, and William's SIS unit at the army was fast becoming a library of as-yet-unsolved Japanese communications, an archive pitched against a rising fascist power.

For all the harmless innocence conjured by the word "library," the Friedmans knew the truth: a library, properly maintained, could save the world—or burn it down.

They took their home library seriously enough to follow the practices of professional librarians. William made his own children sign a checkout slip if they wanted to carry a book from one floor of the house to another, and whenever the Friedmans acquired a book, they pasted a custom bookplate inside the cover, a rectangle of card stock designed by a professor friend who studied Mayan writing. The illustration on the bookplate showed a crimson warrior swinging an axe down upon the skull of a human. Pictographs spelled out a warning to book thieves:

Lay ca-huunil kubenbil tech same
This our book we entrusted you a while-ago.
Ti manaan apaclam-tz'a lo toon
It not-being you-return-give it us
Epahal ca-baat tumen ah-men
Is-being-sharpened our-axe by the expert.

On the wall of the library, to reinforce the message, William and Elizebeth hung an axe with a wooden handle and a black blade.

They were not trying to be mean or intimidating. They were showing their reverence for knowledge. Knowledge is power, as Francis Bacon once said and as George Fabyan and Mrs. Gallup taught them years ago. The Friedmans had taken this precept to an extreme, structuring their whole lives around it—an attitude of sharp curiosity and ruthless self-honesty that defined them. They had carried that little kernel of Bacon's philosophy along with them when they escaped Riverbank, and afterward they stayed true to Bacon's idea in ways their mentors at Riverbank never did, because to really live a life in search of knowledge, you must admit when you are wrong.

Mrs. Gallup never acknowledged that her method was flawed. She never moved on to a new project, never left the estate at all, and Elizebeth and William did not write to her about the Bacon ciphers, feeling it would be unkind to keep restating their skepticism. As the Friedmans climbed onward to brilliant careers in Washington, Mrs. Gallup remained in a cottage at Riverbank, peering through a looking glass at old books, expenses paid by Fabyan, until her death in 1934. She died believing she was correct, that Bacon was Shakespeare, that she had discovered proof, that history would vindicate her.

As for Fabyan, he never quite said the Bacon cipher project was hopeless, that the messages were not really there, only that the argument was not winnable, that he and Mrs. Gallup had failed to make the case. "I have no facilities or knowledge by which to prove the authorship of any volume of Shakespeare," Fabyan told the *Chicago Evening American,* which mocked him: "E'en Colonel Fabyan, the seer of Riverbank, now stands agnostic on the rock of doubt, chewing the food of sweet and bitter fancy, and says he does not know. . . ."

During the final years of Fabyan's life, William's attitude toward his old tormentor softened considerably. They wrote to each other like old friends. They gossiped about Herbert Yardley. Fabyan called Yardley an "ass." William loved that. He noticed that Fabyan seemed glum and tired. His health was failing.

He complained of a hernia and a prostate gland that had to be "reamed out" by a surgeon. He saw another world war coming and hoped it wasn't true. "With love to Elizebeth and the family, and trusting we are not going to get into another war," Fabyan wrote at the end of one letter, signing off, "Always the same old, GF."

George Fabyan died two years after Mrs. Gallup, in 1936, at the age of sixty-nine. The Friedmans heard the news in a letter from an old Riverbank colleague who still lived there. "The Colonel's interest in life had slipped a good deal in the last year," the colleague wrote. A bout of laryngitis had worsened into pleurisy of the lungs and a cascade of medical complications. Fabyan suffered greatly in the last ten days of his life, feverish on the bed in the Villa that hung from chains, the bed swinging to and fro with the violence of his coughing. He said that when he died he wanted his employees to shut off the light in the lighthouse, the one that flashed two white lights and three red lights in a continual pattern every night, the coded message saying *keep out, keep out, keep out*. It went dark according to his wish.

Fabyan left behind far less money than anyone would have guessed. The Great Depression, along with his lavish expenditures on the laboratories and their science experiments, had nearly wiped him out. In his will he gave $175,000 to his widow, Nelle, and provided for a monthly stipend of $150 to his loyal Scottish secretary, Belle Cumming, for as long as she might want to live on the grounds as caretaker. Nelle died of cancer three years later, in 1939, and in 1946, Cumming was killed in a gruesome accident along with two other women when an oncoming train struck their car as it crossed the Geneva tracks. County officials purchased the estate for $70,500 and added it to the local forest preserve. Riverbank was public property, the once-mighty kingdom now an empty set of buildings and fields.

All along in his letters to Fabyan, William had been coaxing the old man to give his papers to a library: his own personal papers and also the voluminous files of the Riverbank Department of Ciphers. In Fabyan's last letter to William before he died, he had mentioned destroying "a lot of old correspondence on ciphers because I did not know whose hands it might fall into." Now

William wrote to his widow, Nelle. What was to be done with the files of the Department of Ciphers? These records were part of the Friedmans' own history and America's history too, and the Friedmans wanted to make sure they were preserved.

The reply came that Fabyan, as one of his last wishes, had ordered many records to be burned.

William and Elizebeth guessed that he did it to prevent embarrassing revelations about his many deceptions and schemes. Fabyan had chosen to go out in a flame of self-immolation, a Viking funeral of documents.

It wasn't as bad as they thought. While Fabyan had destroyed some documents, others survived and would eventually make their way to the New York Public Library. But the Friedmans assumed the worst. To them it was a tragedy: loss of history, loss of knowledge. No one else in America particularly cared. A dying man had burned some papers about cryptology. Cryptology is a profession of secrets. Secrets staying secret is the norm. Officials only get riled up when the opposite happens—when secrets are leaked, published, disclosed. And by now, the Friedmans and their colleagues were struggling to manage the consequences of the biggest leak of cryptologic secrets in the country's history.

"A lie!"

At home, William turned the pages of a book with a black cover, facial muscles tightening, writing annotations in the margins:

"This is a patchwork of misstatement, exaggeration, and falsehood."

"Lies, lies, lies."

Six crimson lines divided the book's cover into seven black rectangles. Inside one of the rectangles, crimson letters read,

THE AMERICAN BLACK CHAMBER

HERBERT O. YARDLEY

Yardley.

The poker player and codebreaker had not reacted well to the closing of his State Department bureau by Henry Stimson, the secretary who believed that gentlemen do not read each other's mail. When the government put Yardley out of work, he had been unable to find a new job in the Depression. Down to his last couple of dollars, he decided to spill his secrets, for money and revenge. He published a book, *The American Black Chamber*, in the summer of 1931, which became a bestseller and international sensation. It also hit the U.S. government like an exploding volcano, hurling rock and lava into the stunned surroundings and forcing the Friedmans to navigate the smoking landscape it left behind.

The book claimed to be the true story of the black chamber, "a glimpse behind the heavy curtains that enshroud the background of secret diplomacy." He framed it as an act of patriotism, and he had a case. Yardley believed it was dangerous and naive to stop breaking the codes of foreign governments. It made America weaker, more vulnerable. He believed there should be a debate, but it was impossible to have a debate about the kinds of things the black chamber did if the public didn't even know that it had existed. He argued that because the bureau was now closed, there was no harm in revealing its activities.

There was enough truth in the book to scandalize U.S. officials. Yardley disclosed, for instance, how his country had been reading the diplomatic traffic of Japanese ambassadors, information that Japanese readers were surprised to learn; the book sold 33,119 copies there. He also tantalized with anecdotes of sex and deception. "A lovely girl dances with the Secretary of an Embassy," Yardley wrote in the book's introduction, promising more details within. "She flatters him. They become confidential. He is indiscreet." He devoted a full chapter to a bewitching female spy named Madame de Victorica, "the beautiful blonde woman of Antwerp," who worked for the British and French during the Great War and employed ciphers and invisible ink. Yardley claimed that an unknown enemy once enlisted a blonde spy to seduce him at a New York speakeasy: "It did seem to me that she showed a bit too much of her legs as she nestled in the deep cushions." They got into a taxi and went to her apartment, where she promptly fell asleep on the couch; Yardley, suspecting foul play, searched her desk and found a note that said, "See mutual friend at first opportunity. Important you get us information at once." The next night, thieves broke into his office and rifled through the cabinets: "I took it for granted that they had photographed the important documents which they required."

This probably never happened. Yardley's colleague later said that the story of the blue-eyed blonde was a "damned lie," and that the only things taken from the office were a couple of bottles of booze. "It was my booze and I think [Yardley] took it himself." As for Madame de Victorica, she did exist, but Yardley embellished her biography. He admitted to friends that he fictionalized parts of the book. He compressed time, invented dialogue, added "bunk" and "hooey," and made no apologies: "To write saleable stuff one must dramatise."

William Friedman found this loose attitude intolerable. Truth was truth and anything else was fuckery. He also disagreed with Yardley's assertion that there was no harm in telling old stories. William thought that some of Yardley's revelations might startle adversaries into boosting the security of their communications, particularly Japan, thereby making the jobs of American code-breakers more difficult.

In the end, though, William resented that Yardley was telling a story about cryptology that William wanted to tell himself. William had always been a stifled writer, unable to publish what he would have liked. At Riverbank the obstacle was the ego of a rich man, George Fabyan, who insisted on taking credit for William's work, and now the reason was national security. He couldn't tell stories of his army feats because, unlike Yardley, he wasn't willing to violate his promise to keep America's secrets.

Yardley had actually sent William a copy of *The American Black Chamber* as a courtesy, signing his name on the inside cover. Beneath the signature, William wrote, "OMNIS HOMO MENDAX," which means, in Latin, "Every man is a liar." Then he began to write his own alternate history of the events and concepts Yardley had described, inside Yardley's book, in the margins: "A lie! Which can be so proved to be. See papers attached. Exhibit 1." *He attached exhibits to another man's book.* He underlined sentences, bracketed paragraphs, tagged words with asterisks, spangled pages with exclamation marks. Revenge by

annotation. Not content with his own annotations, he then circulated the copy of Yardley's book among four of his colleagues
in the army and MID, and they added their own annotations in
their own handwriting, a chorus of jeers and boos. William created a numbered key in the front of the book so that future readers could keep track of the different voices.

The fallout from the publication of Yardley's book was lasting
and broad. Intelligence bosses and lawmakers grew newly anxious
about the security of cryptologic information and moved to crack
down on future disclosures. Yardley became persona non grata in
U.S. intelligence circles for the rest of his life; exiled from his old
haunts, he made friends in Los Angeles and wrote screenplays for
Hollywood movies. In 1933, Congress passed a law specifically to
prevent Yardley from publishing a book of codebreaking yarns
focused on Japan; the new law, called "An Act For the Protection of Government Records" and derided by Yardley as the "Secrets Act," declared it a crime to reveal secrets about cryptologic
information. But stories about codes and ciphers only increased
after Yardley opened the gates. He had proven that there was a
market for them, particularly yarns about Yardleyesque women
who dealt in secrets, and when editors looked around for such a
woman, they did not have to look far.

Elizebeth Friedman was attractive. She was a mother. She was
American. She was testifying in open court against the criminal masterminds of her age. And unlike Yardley's women, she
was verifiably real in every detail. "Widespread interest in the
romantic stories of beautiful female spies, secret codes and ciphers which Yardley had told caused editors from this time on
to become keenly aware of the news value of such stories," went
a confidential memo later circulated by the U.S. Navy to warn
about the dangers of cryptologic publicity. "Consequently, when
in 1934 magazine and newspaper accounts broke concerning Mrs.
Elizabeth [*sic*] Smith Friedman, a Coast Guard Cryptanalyst, a
number of similar incidents followed."

"I'll confess, Mrs. Friedman, I was thunderstruck the other day
when I met you for the first time. I simply wasn't prepared to find
a petite, vivacious young matron bearing the formidable title of

Cryptanalyst for the United States Coast Guard. How did you ever get interested in the highly technical science of codes and crypts?"

"I never thought of my job as terribly unusual until the newspapers stumbled upon what I do for the government, Miss Santry."

Margaret Santry was a radio reporter for NBC. In 1934 she launched a series of interviews with "First Ladies of the Capitol," mostly socialites and the wives of congressmen, and in May she asked Elizebeth to speak about her career on a national NBC broadcast. Elizebeth brought her children to the NBC radio studio in Washington; commercial radio was less than a decade old, still slightly wondrous, and she thought the kids might enjoy seeing the technology. Barbara was ten, John Ramsay was seven. The NBC staff let the kids watch their mother from the glassed-in control room adjacent to the sound booth, analog indicator needles twitching back and forth as Santry peppered Elizebeth with questions.

"Does the habit of thinking in code ever creep into your family life?"

"I guess it's bound to—it's so much a part of our life," Elizebeth said.

"And are your children experts in code, too?"

Elizebeth replied that Barbara was "quite an expert for her age" and had sent her parents messages in cipher two years earlier, when Elizebeth and William sailed to Spain for an international conference about radio transmissions.

Santry asked, "And I suppose she wants to be a cryptanalyst like her mother when she grows up?"

"No, she wants to be a professional dancer," Elizebeth said. As for her son, John Ramsay "states he wants to be a code expert when he grows up," but "at present he is in the boats and guns stage."

"How do you solve the business of running a home and a family—and an important job—all at the same time, Mrs. Friedman?"

"Oh, it solves itself rather nicely; especially when I have such a grand housekeeper to look after things at present. I never really made any definite plans for a career, Miss Santry—it just happened."

This NBC interview was one of the rare media appearances that Elizebeth enjoyed; she preferred speaking with female reporters, and NBC treated her children with kindness. Most of the time she hated dealing with the press. Depending on the reporter, she found the experience irritating to unbearable, and altogether the articles and radio shows represented a real threat to her livelihood.

Codebreaking is a secret profession. Its practitioners aren't generally supposed to talk about how they do what they do, as demonstrated by the shock at Herbert Yardley's disclosures. Elizebeth wasn't anything like Yardley; he was motivated by money, and she was motivated by doing the job that the government was asking her to do. Elizebeth only discussed active cases in public when a prosecutor called her to testify in court, and at all other times she limited her comments to closed cases, and then only with the full authorization of the Treasury publicity office. She testified for a reason: to put bad guys in prison.

Yet Elizebeth was so talented at this task, and the trials so spectacular, that the resulting waves of publicity threatened to wash away her career, in an era when government was growing concerned about the security of secret information.

The publicity coincided with a rise in the drama and stakes of her cases. When Prohibition was repealed in 1934, destroying the market for bootleg liquor, several of the rum rings made a nimble transition to smuggling drugs, mainly opium and the drugs derived from it, heroin and morphine, which were refined by pharmacists and criminal gangs in the port cities of China. Elizebeth adapted. Bags of heroin were smaller than crates of booze, easier to conceal, which only made codebreaking more essential—the only reliable way to discover the drugs was to swipe the details of their location from the smuggler's own lips.

The drug networks were global in scope, spanning more lands and languages than the rum syndicates. Elizebeth coordinated the investigations with T-men in offices all across America, with members of the U.S. State Department, and with police inspectors in foreign countries. She worked with translators to read encrypted notes in Portuguese, Spanish, French, Italian, German, and Mandarin. She was able to take messages written in blocks of letters or numbers and trace these figures back to specific Mandarin words in commercial code books designed for Chinese merchants. Reporters and readers seemed amazed that she could break a code in a language she did not speak, but that was the power of having a system, a science. "The whole deciphering science is based on what we call the mechanics of language," she explained on NBC radio, making it seem easy. "There are certain fixed ways in which language operates, so to speak; and by studying the known elements and making certain assumptions, one can arrive at a result that usually does the trick."

Global heroin rings went up at the stroke of her pencil. Smuggling ships were skimmed off the ocean like fat from a simmering pot of soup stock. Each new feat startled loose a flock of news articles that sang about Elizebeth's previous feats and added one more verse to the ballad of her growing legend. She decrypted a stack of intercepted letters and telegrams exchanged between a member of Shanghai's fearsome Green Gang of criminal warlords and two brothers in San Francisco, Isaac and Judah Ezra, the dissolute twin sons of an upstanding Shanghai pharmacist. They had been smuggling opiates from Yokohama, Japan, to San Francisco on the passenger liner *Asama Maru,* the drugs hidden in barrels of tree-nut oil. The messages spoke of "wyset," "wysiv," and "wyssa" in various quantities; Elizebeth determined that "wyset" meant cocaine, "wysiv" was heroin, and "wyssa" was morphine. She alerted the T-men in San Francisco, who searched the *Asama Maru* when it arrived and found 520 tins of smoking opium, 70 ounces of cocaine, 70 ounces of morphine, and 40 ounces of heroin. Informed that their code had been broken, the Ezras quickly confessed and were each sentenced to twelve years in prison.

EZRA GANG FALLS IN TRAP OF WOMAN EXPERT AT PUZZLES, went a headline in the *San Francisco Chronicle.* SOLVED BY WOMAN. The press had a way of praising Elizebeth and condescending to her at the same time, professing amazement at the capabilities of the female brain. She had "supplied the Federals with enough dynamite to break up the ring," the paper said, and now "the Ezra boys have twelve years to figure out another code she can't break."

When the reporter asked her how she broke the Ezras' code, Elizebeth declined to say: "We have to keep our ideas secret so that we do not give other smugglers any new ideas." Worried that she was attracting too much attention, that T-men would feel slighted by her fame and that Treasury chiefs would fret about smugglers getting wise to her methods, she tried to get the press to stop writing about her. She begged reporters to credit the coast guard as a whole instead of one woman for solving the cases. She wrote apologies to colleagues and bosses, complaining that reporters would not leave her alone: "The mystery-lure of the words code and cipher, coupled with a woman's name, invariably inflames the news reporters and they start on the trail of a story."

Nothing worked. She had chosen a profession that continually immersed her in lurid realities. "She is entrusted with more secrets of the crime world and of federal detection activities than any woman in history," reported *Reader's Digest* in 1937, in a five-page feature that declared Elizebeth "Key Woman of the T-Men" and was mailed to more than a million subscribers. "When one of the Treasury Department's enforcement agencies gets the scent of a new international enterprise in smuggling, dope running . . . there is one unofficial order that sticks in every agent's mind: 'Get some of the gang's correspondence and send for Mrs. Friedman.' "

There were people in the U.S. military who didn't like that Elizebeth was talking about codebreaking in criminal trials and didn't much care that prosecutors were forcing her to do it. Her testimonies showed that she knew how to break many different kinds of codes, including codes in foreign languages, which "could lead to only one conclusion on the part of espionage agents—decryption of other nation's codes was in progress behind the scenes," according to the confidential 1943 memo. After the publication of "The Key Woman of the T-Men," the army's inspector general stormed into William's office and demanded to know why William had been mentioned in the article, as Elizebeth's husband and as the army's top codebreaker. Officials at both Treasury and the War Department had reviewed the article before publication and confirmed that it contained no classified information, but the inspector had not been consulted. William had no idea what to say. "I lost my tongue completely," he wrote, "and failed to ask for permission to sleep in the same room and/or bed with my wife." Journalists mentioned him in articles about Elizebeth because the two of them were married and shared a profession. If it was a problem for two cryptologists to be married, what was the solution?

The troublesome publicity came to a peak in the first months of 1938, when the buzz from the *Reader's Digest* profile dovetailed with a new round of press from a flashy drug trial in Canada. At the request of the Royal Canadian Mounted Police, Elizebeth flew to Vancouver to analyze messages discovered in a raid of a Chinese merchant's home and business. Police seized thirteen Luger pistols, four hundred Mauser clips, almost one hundred machine gun drums, numerous cans of opium, and two dozen coded cablegrams in a safe. Working with RCMP translators, Elizebeth solved the messages, exposing a smuggling ring that traded Canadian weapons to Hong Kong in exchange for drugs. The cablegrams were written in English ciphertext, four letters per block, which Elizebeth turned into Arabic numbers, that corresponded to Mandarin characters in a Chinese commercial dictionary, that stood for Mandarin words, that were translated into English words:

UUOO AMAS ANAG USOG UKUU IUUI AEIY thus became "Cable three thousand select fully Wat list," Wat Sang Co. being the name of the drug dealer's Canadian company. She found that the smugglers referred to opium as "ginseng" and "groceries," and guns and ammunition as "hams," "presses," and "tails."

Her testimony at the Vancouver criminal trial helped convict all five defendants, and a fresh wave of articles and headlines ap-

peared. CANADA SMASHES OPIUM RING WITH U.S. WOMAN'S AID.
WOMAN TRANSLATES CODE JARGON. The February 15, 1938, issue
of *Look* magazine included Elizebeth in a feature on "outstand-
ing" women "in careers unusual for their sex," along with a female deep-sea diver, a female conductor of a symphony orchestra,
and a silversmith. She had not provided *Look* with a photo or an
interview, but the magazine somehow got its hands on a gauzy
black-and-white photo of Elizebeth in a white robe. *Detective
Fiction Weekly* printed fourteen pages about Elizebeth's career
written by a newspaper editor who moonlighted in gangster fiction. "Lady Manhunter," he titled the piece, "A True Story." A
telegram arrived in Elizebeth's office from a magazine writer in
New York. "PLEASE COOPERATE BY ANSWERING QUESTIONS BELOW
BY RETURN SPECIAL DELIVERY TO ME," the man wrote. "FOR WHAT
DEPARTMENTS DO YOU DECIPHER MESSAGES? HOW MANY HAVE YOU
DONE? WHAT TYPES? SENT BY WHOM? HOW DO THEY FALL INTO YOUR
HANDS? . . . LIKES, DISLIKES, SUPERSTITIONS HAVE YOU, ANY FURTHER
ANECDOTES OF HUMAN INTEREST, HUMOR, OR UNUSUAL EXPERIENCES
WOULD BE APPRECIATED." Across the bottom of the telegram, in
large letters, Elizebeth scrawled in disgust, "Ad Absurdum!"

She was now the most famous codebreaker in the world, more
famous even than Herbert Yardley, the impresario of the American Black Chamber. And she was more famous than her husband,
too—a reversal from the longstanding pattern.

All their lives, William had been the celebrated one, the master, the genius, and she the dutiful wife, supporting him from the
shadows. But Elizebeth was now said to be the true genius in the
family, the driving force of the duo. A rumor spread that she had
taught her husband everything he knew about cryptology. People
started to approach William at military functions and regale him
with stories about Elizebeth's adventures, as if he were unaware of
them. He thought the inversion of the narrative was hilarious, "a
scream"; it had been silly before when everyone thought he was the
only genius in the couple, and it was silly now. The Friedmans had
always considered each other equals. He didn't mind the attention
swinging to Elizebeth. "When people introduce me and then say
that my wife is also etc & is really better at it, I invariably assent,
with a real smile," he told her in a letter. He was proud of her.

Elizabeth, of course, didn't think she was a genius; actual geniuses never do. Codebreaking to her was about teams, systems, cooperation.

She told people she never wanted to see her name in print again. In the final months of 1938, she got her wish. The U.S. government gave her a new assignment that was every bit as clandestine as her previous missions had been public. Soon the mark of her pencil, once celebrated on front pages, would become one of the most closely held secrets in America.

William left Washington for Hawaii in October 1938 on a secret mission to the Pacific. The army wanted to deploy the cipher machine he had invented, Converter M-134, and start using it to protect "highly secret communications" between its headquarters in Washington and army installations in San Francisco, Hawaii, and the Philippines. William's job was to carry two M-134s to each base, install and test them, and train army staff in their operation. He boarded a troopship, the *Republic,* in New York City, with a trunk containing six bulky cipher machines and some books of essays and poetry, and the ship steamed south to the Panama Canal and passed from there into the Pacific.

He was at sea during the atrocity known as Kristallnacht, the Night of the Broken Glass, when Nazi mobs in a thousand German towns murdered at least ninety-one Jews and set fire to synagogues. These were pogroms like the ones in Russia that had terrorized William's kin. The mobs attacked with stones, bricks, wood poles, and automobiles driven through plate-glass windows, destroying Jewish stores, Jewish hospitals, Jewish nursing homes, Jewish kindergartens, Jewish cemeteries, Jewish Scrolls of Law. On a busy Berlin street a cheering crowd hacked apart a grand piano with hatchets.

In Washington a few weeks later, Elizabeth fell ill. A creeping fatigue glued her to the bed for long periods. Not wanting William to worry, she concealed the details from him, only telling her sister, Edna, who assumed that her sister was simply overworked. Edna didn't know about Elizabeth's new mission, and William didn't, either. He was on the ship, cut off. His airmail letters arrived in Washington within a few days, but they were expensive.

Regular-mail letters took weeks; telegrams had to be short. The *Republic* had a radio transmitter for military use only. William was sometimes able to sneak on and send Elizebeth messages at her Treasury office.

From a few words in her letters and telegrams, he got the sense his wife was struggling, that something was wrong, but she didn't spell it out. He was sick of these distances and wished they could go back to the time when they were together all day long, solving puzzles side by side. On the ship he daydreamed about Riverbank. "I can almost forsee the time when you and I will be working together again in an office," he wrote to Elizebeth. "It would be a joy to have you alongside me again—to recapture those days when we worked side by side and stole kisses in the vault at Engledew—such passionate kisses on my part, and sometimes on yours."

Arriving in Hawaii in late November, William began tests of the cipher machine and bought gifts for his family: hula skirts for Elizebeth and the kids, and a glass vial of Shalimar perfume for Elizebeth. Then he sailed to San Francisco for more tests at a different army facility. He was back on the ship, heading homeward, by Christmas Eve. The crew of the *Republic* arranged a Christmas dance for passengers. He sat in the smoking room reading a book while others danced. After a colonel entered and shot him a dirty look for being a prude, William put the book down and danced a waltz with the daughter of the Cuban minister to Japan, a young beauty with a slight accent, "as brunette as can be."

He was still thousands of miles from Elizebeth when the new year dawned, and skimming along at fourteen and a half knots. During these first days of January 1939, Japanese pilots dropped bombs on civilians in the Chinese city of Chongqing. The cardinal of Munich praised Hitler's "simple personal habits." The Nazis announced the immediate deployment of hundreds of new U-boats, a smaller and swifter generation of submarine. A physics professor in a basement laboratory at Columbia University wrote in his diary, "Believe we have observed new phenomenon of far-reaching consequences"—nuclear fission, the splitting of uranium atoms into lighter elements, achieved with an atom smasher.

The motion of the ship disturbed William and he slept irregularly, conking out for hours in deck chairs at midday, sun cooking his pale skin, then twitching with insomnia at night, wandering the ship alone. He watched Hollywood movies in the evening, above deck, where the ship kept a film projector and passengers gathered in the open air. During Cecil B. DeMille's *Cleopatra,* the moon rose fat and white between the mountains near Monterey and the cryptologist's eyes wandered from the movie screen to the sky. He played shuffleboard with two congressmen. He fretted about the state of his patent applications and the low tide of his bank account. He wondered what was happening with the board game he sent to Milton Bradley. No one had gotten back to him. He learned later the game had stumped the firm's designers. It was one thing to design a fun code game for a party with friends and quite another to deliver a game in a cardboard box to a stranger.

On the ship he began a letter to Elizebeth that ultimately ran to thirty pages. He wrote on transparent tracing paper because it was thin and he could fit more pages in an envelope, saving on postage. Each day he added to the letter. His cursive handwriting smeared on the delicate clear sheets and the writing on the reverse showed through to the front. "I hope these sheets don't drive you wild to read," he wrote. "I wanted to get on as much as possible and use this paper to reduce the weight." He was trying to express a feeling that seemed too large for a single page or a moment's thought. He copied love poems by Tennyson and Thackeray from one of the books in his trunk. Thackeray wished to be a violet, plucked by his belle to live for an exalted hour "shelter'd here upon a breast / so gentle and so pure." He watched a romantic comedy starring Helen Hayes, *What Every Woman Knows,* about the shrewd wife of a British politician who helps his career behind the scenes, inspiring William to tell Elizebeth, "Well, that wasn't exactly news to me, my Darling. For I've known for a long time that you are the one in back of me and responsible for what little I've done. Had it not been for you I'd have been sunk long ago by unsolved infernal conflicts, by windy storms of emotion, by failure to keep up the fight when things seemed not worthwhile. . . . I know how much I owe to you—for love, for wisdom, for courage, and common sense."

Floating above the blue-black depths soon to be lethalized by German U-boats, he often stood at the top deck's rail looking out at the miles of ocean. Billions of droplets exchanging secret histories every fraction of a second. Trillions. Actual unbreakable code. One night when it was hot and he couldn't sleep he drank a glass of cold tomato juice and ate a few crackers and went up to the top deck. The place was deserted except for two or three men in chairs along the rail, smoking. He sat and pulled out the diary, flipping to the next empty page. He was a man on the verge of mental breakdown, writing a letter to his wife, who was also close to collapse, at a moment when the world was rearranging itself, and they would both have to give more of themselves than they had ever given before.

"The ocean," he said, "is as calm as a bowl of warm milk sitting on a table."

PART III

THE INVISIBLE WAR

1939–1945

CHAPTER I

Grandmother Died

Elizebeth Friedman, U.S. Coast Guard
Cryptanalyst-in-Charge, and a junior cryptanalyst,
Robert Gordon, puzzling out a problem together, 1940.

Lights out 'cause I can see in the dark . . .
—FUGAZI

The Second World War did not begin with a gunshot or a bomb. It began with a feat of deception involving elements long familiar to Elizebeth Friedman—a code phrase, a radio station, and a murder. The men responsible were Nazis, and they belonged to the same part of the Nazi state that would soon attract Elizebeth's deep attention.

At 4 P.M. on August 31, 1939, in a hotel room in a small Polish town four miles from the German border, a Nazi officer named Alfred Naujocks dialed a number in Berlin. Someone in Berlin picked up. A high-pitched voice said, "Grossmutter gestorben." "Grandmother died." Naujocks hung up. He went to gather his team, the six operatives he had brought across the border. "Grandmother died" was the signal to execute a preplanned mission at 8 P.M.

The mission was to provide Germany with an excuse to start a war. Hitler had already decided to attack Poland, to seize his neighbor to the east, but he did not want to appear as the aggressor, so a pretext was needed, a simulated attack on German forces that would allow Hitler to claim he was acting in self-defense and create confusion about where the truth really lay.

This is where Naujocks and his colleagues entered the picture. They would invent the proof of Polish aggression.

They belonged to the SS, the chief instrument of Nazi terror. They were the men in black, the storm troopers, numbering 250,000 by 1939. They wore the death's-head symbol on their uniforms, the skull and crossbones. As individuals they were like any other large group of humans, containing multitudes: opportunists, idealists, fanatics, scholars, mediocrities, petty crooks. But as a collective they became "the guillotine used by a gang of psychopaths obsessed with racial purity," in the words of the historian Heinz Höhne. It was SS men who built the concentration camps and managed the ghettoes and trains that herded and transported Jews and other minorities into the camps to be enslaved, tortured, and killed. They were the guards at Dachau and Auschwitz, the murderers of millions. They were the mobile killing units, the Einsatzgruppen, that swept in behind the advancing German military, shooting resisters and Jews. They were the Gestapo, the ruthless Nazi police. They were the Nazi intelligence men, the ones who spied on their fellow Germans to keep them in line, and they were the spies who worked undercover in other nations, extending the reach of the regime until it encircled the world. And they were not meant to be confronted or understood. One SS leader bragged that the organization was "enveloped in the mysterious aura of the political detective story." Elizabeth would spend much of the war trying to penetrate this veil.

After the SS men in Poland received the code phrase over the phone on August 31, they waited until just before dusk, then drove two cars through the pine forest toward their objective, a Nazi radio station that transmitted propaganda broadcasts. The plan was to pose as Polish insurgents, take over the station, and broadcast a message denouncing the Führer. They stopped near the station and met a Gestapo captain to pick up what they had been calling "Canned Goods": the unconscious body of an SS prisoner, a forty-three-year-old Catholic farmer named Franz Honiok. He had been shot, sedated, and dressed in a Polish uniform. His face was smeared with blood. Naujocks carried him to the steps of the station and left him there slipping away from his fatal gunshot wounds, then stormed into the broadcast area with his team, aiming a revolver at the staff: "Hands up!" One of the SS men spoke Polish. He grabbed the emergency microphone used for storm warnings. "Attention, this is Gliwice," he shouted in Polish, pretending to be an insurgent. "The radio station is in Polish hands." He called for an uprising. The men fired bullets into the ceiling to simulate an armed struggle.

Thousands of radio listeners heard the gunfire and the burst of Polish. Two hours later stations in Berlin were spreading news of the "Polish attack." The BBC in London reported that "Poles forced their way into the studio." And while diplomats around the world tried to get a fix on the truth, the Nazis were massing at the border. At dawn on September 1, 1939, the morning after the incident in Gliwice, the Wehrmacht sliced into Poland, forty-two divisions all at once, with one and a half million men. It was the middle of the night on the East Coast. President Roosevelt woke to a ringing phone at 2:50 A.M. He picked up the receiver at his bedside and heard the voice of William Bullitt, the U.S. ambassador to France. Bullitt was calling with the news from Warsaw: Germany had invaded Poland. The president sat up. "Well, Bill, it has come at last. God help us all." He lit a cigarette and started making calls, waking up the cabinet secretaries.

Later that morning Roosevelt called a quick press conference. The first question was, can America stay out of the war? Roosevelt said, "I not only sincerely hope so, but I believe we can."

He was speaking for most of the country. In the days that followed, as Britain and France declared war on Germany, the U.S. public met the news with relief. Europe would defeat fascism on its own. The fight was across the ocean, far from U.S. shores. Nazis seemed a safe distance away.

In private, though, Roosevelt and his advisers were planning for the worst.

For years now, they had been thinking about the possibility of direct Nazi attacks on the United States. It was obvious that no effective invasion could be launched from Germany itself; ships were too slow, and airplanes couldn't carry enough fuel to cross the ocean, drop bombs, and return home.

But there was a catch in this argument, an unnerving loophole: *What if the Nazis got control of South America?*

South America. It was neutral ground—for now. No government there had declared a position in the war. Brazil, Argentina, Chile, Paraguay, Bolivia, and Ecuador were content, for the moment, to sell beef and raw metals to the combatants and to measure the political winds.

But this would certainly change as the war evolved and politicians cut deals. Hitler had already shown an ability to destabilize foreign governments. That's how Austria had fallen to the Nazis, and Czechoslovakia, too. And President Roosevelt was convinced that if Nazism took root in South America, even in just a few places, it would pose a clear and present danger to U.S. cities like New York.

South America was very big: the land mass of Brazil alone was slightly larger than the entire continental United States. South America was also very close: as Roosevelt put it in a 1940 speech to Congress, "Para, Brazil, near the mouth of the Amazon River, is but four flying-hours to Caracas, Venezuela; and Venezuela is but two and one-half hours to Cuba and the Canal Zone; and Cuba and the Canal Zone are two and one-quarter hours to Tampico, Mexico; and Tampico is two and one-quarter hours to St. Louis, Kansas City and Omaha."

If Britain fell to Hitler, the thinking went, Nazi ships could move west and set up bases in South America, seizing its rich resources, the metals to make war machines and the food to sustain armies, and then U.S. coastal cities would be within reach of Nazi bombing raids. Some officials dissented from this view, namely at the State Department, which considered a Nazi invasion from South America to be unlikely, but the Reich's rapid military victories and the erratic behaviors of the Führer had caused many to revise their sense of what was possible.

There was another sound reason to worry about German influence in South America. Millions of Germans were already living there as colonists. They had emigrated in waves since the late nineteenth century, seeking land and work, 140,000 arriving between 1919 and 1933 alone, many fleeing the same desperate economy that fueled the rise of the Nazis.

The Germans had left a cold country of worthless currency and sailed into the crystalline seas hugging a warm and open continent. A "bewildering abundance" met them ashore. "Everything is violent—the sun, the light, the colours," the Austrian exile Stefan Zweig wrote of Rio de Janeiro, Brazil's glinting capital. "The glare of the sun is stronger here; the greens are deep and full; the earth tight-packed and red . . . Rather than encouraged, growth has to be fought, so as to prevent its wild power from overwhelming the efforts of mankind." In Rio the green leaves of palm trees burned white in the midday sun as if radioactive. Men wore suits of white linen that became soaked by sudden downpours and thunderstorms. Women strolled topless on Copacabana Beach. A cream-colored luxury hotel, the Copacabana Palace, faced the beach and the crystalline blue waters of the Guanabara Bay; Fred Astaire and Ginger Rogers danced in the hotel's ballroom in the 1933 Hollywood musical *Flying Down to Rio*.

Some of the arriving Germans decided to stay in Rio; others filtered into the surrounding Brazilian provinces, sparse rural lands of forest and cattle, and still others scuttled south along the coastline to the fast-growing industrial city of São Paulo, which Zweig likened to Houston, Texas, because of its abrupt rise from nothing: "There are times when one has the sensation of not being in a city, but on some gigantic building site." Farther south, another great city of the continent beckoned to the Germans: Buenos Aires, Argentina, a polyglot metropolis of four million, a chaos of automobiles and bookshops and neon lights and cobblestone streets where tango music descended from the open windows of brothels. The nation had grown rich from cattle and wheat raised in the Pampas, the flatlands to the west and south of Buenos Aires, where tens of thousands of Germans lived alongside the *gauchos* and their horses.

Wherever Germans settled in South America, they built German schools (two hundred in Argentina alone), German businesses, German radio stations, German newspapers, and transportation links back to the homeland. Zeppelins floated people and cargo from Berlin to Rio, and two airlines, Condor and LATI, connected South America and Europe. Condor was owned by Germans, LATI by Italians. A visiting U.S. consul reported "a fair sale for German Bibles" across three Brazilian states and that 20 percent of all residents spoke only German; parts of southern Brazil became known as Greater Germany. "The German spirit is ineradicably grounded in the hearts of these colonists," wrote a German physician, "and it will undoubtedly bear fruit, perhaps a rich harvest, which will not only prove a blessing to the colonies, but to the Fatherland." A German visitor to Brazil reported with pride, "Surely to us belongs this part of the world," and the Nazi ambassador in Buenos Aires, Baron Edmund von Thermann, believed that German Argentines must show "complete subservience" to "the ambitions and desires of the home country. Germans naturally count on these prosperous nuclei to assist eventually in the rebuilding of a new Germany."

A small percentage of German immigrants brought fascist politics to South America, starting local Nazi clubs and chanting Nazi songs, but these groups were small and disconnected, stagnant ponds of fascist fervor. The bigger rivers of fascist sympathy in South America coursed through the local populations. It was a time of protests, marches, fantasies of revolution. Right-wing parties and radicals on the continent found inspiration in Nazism. Followers of a Brazilian movement called Integralism raised their hands in Nazi-style salutes, wore uniforms of green (the men were "Green Shirts," the women "Green Blouses"), and goose-stepped through the streets of Rio. In 1938, a throng of Argentine youths marched into the Jewish quarter of Buenos Aires, chanting anti-Semitic slogans, "stripped to the waist like Mussolini, mustachioed like Hitler," writes one historian. "When enraged Jews attacked them, police arrested the Jews."

Similar movements were gaining followers in Chile, Bolivia, and Paraguay, and powerful local officials poured fuel on the fires. Across the continent, men who dreamed of leading their own regimes had risen to the top of police and military hierarchies; many had gotten their training from German officers. A group of Paraguayan officers formed a secret lodge, the *Frente de Guerra,* to organize an ultra right-wing revolution; their motto was "Discipline, Hierarchy, Order." The chief of the Paraguayan national police, wishing to honor the dictators of Germany and Japan, decided to name his son Adolfo Hirohito. In Argentina, a young military instructor with a Cheshire-cat smile, Juan Perón, was studying the leadership styles of Mussolini and Hitler and found much to admire. One Argentine general, Juan Bautista Molina, displayed so much zeal for National Socialism that even Thermann, the Nazi ambassador, found it "embarrassing."

Hitler appreciated this wellspring of sympathy in South America. His strongest affinity was for Argentina, which had protected German interests in the First World War while ostensibly remaining neutral. In June 1939, three months before he invaded Poland, Hitler met with Argentina's ambassador to Germany. Writes the historian Richard McGaha, "Knowing that war was going to break out soon," the Führer "cryptically stated that he hoped Argentina would stay neutral and that neutrality could be the basis of a closer relationship." Then Hitler launched into a tirade about America and England, saying that "the U.S. was the worst-governed country in the world," that Roosevelt wanted war "at the instigation of the Jews, who controlled industry and the press," and that England was "a paper tiger with its little fleet and meager air force."

In his mind the Führer had already added South America to the Nazi column. If he decided not to invade at this time, he would simply annex the continent after defeating Europe. As Baron von Thermann put it, "Once the war were decided in Germany's favor, her domination of Latin America would follow without much effort." This was the Nazi attitude, and it meant that the war in South America could not be a hot war, a war of soldiers and sailors in recognizable uniforms, a war of battleships and mortars and planes and bombs. Instead it promised to be a war of languages and secrets, codes and conspiracies, masks and seductions, wireless transmitters and cipher machines—the type of war where everything depended on the invisible flashes of energy radiating from a radio coil hidden on a farm or beneath the floorboards of an unremarkable house.

The term of art for an intelligence operation that must remain entirely concealed is "clandestine." If a clandestine job is successful, no one ever knows it happened. It is invisible. The war in South America would be the Invisible War.

There was a school in Hamburg where SS intelligence officers trained combatants for this war. Male party members were selected to receive a basic course in espionage tradecraft. They were taught to write letters in secret inks. The SS had developed a disappearing ink that actually looked like ink, bluish in color and carried in a regular ink bottle; a message written with this ink would turn invisible after a few minutes and could only be unmasked with a certain reagent. They learned how to operate a German-invented "microdot" camera that shrunk documents to the size of the dot above an *i*, allowing espionage reports to be concealed in otherwise innocuous letters, and they were shown different methods of writing messages in cipher, including an ingenious system for exploiting a popular novel, any novel, to generate garbled text.

For the purposes of spying, this hand technique was often preferable to cipher machines like Enigmas, which were bulky, harder to transport, and more incriminating if discovered. A novel aroused no suspicion. One in common use was *All This and Heaven Too,* a period potboiler about a French governess falsely accused of murder. *Would the unlucky Henriette Desportes manage to clear her name? Or would the conniving Parisian judge dispatch her to the dungeon?* German men abroad pressed their noses to the book, eyes wide, turning the pages quickly, underlining words—no, these were not Nazi spies, these were simply readers under the spell of a story, needing to know what happened next.

The SS instructors taught students how to transmit text in Morse code and how to operate shortwave radio transmitters and receivers. Radio technology had made giant leaps since the heyday of rum-running. A shortwave transmitter of moderate power could now fit in a suitcase. The transmitter was a small metal box with vacuum tubes on the inside and dials affixed to the cover, and the antenna was a long wire looped into a tight coil.

Portable transmitters in hand, the novice agents were dispatched to begin their espionage careers for the Führer. U-boats delivered some of the spies onto alien shores, and others parachuted from planes or sailed on neutral ships under phony names, sometimes getting caught by customs inspectors or police along the way, their radios confiscated. The SS issued all foreign spies two kinds of suicide drugs to ingest in case of arrest. The first was a tablet that caused death by heart failure within ten minutes, and the second was a powder that resulted in "a slow process of general collapse over a two-week period" when rubbed on the body.

If a spy managed to arrive at his destination with the radio intact, he unfurled the wire antenna and established contact with the fatherland, tapping out an encrypted message in the dots and dashes of Morse, the signal aimed at a receiving station in Hamburg or Berlin. Sometimes it worked, and the spy could be heard in Germany—there were no atmospheric disturbances, and the signal squeezed through the crowded frequencies—but storms and interference often fuzzed out the radio pings, making it necessary to build more powerful stations, which required a higher level of expertise. A *Funkmeister* was needed: a technical leader, a radio wizard, able to piece together clandestine radio transmitters in foreign lands. And this is why, in 1941, the Nazi SS dispatched its most capable *Funkmeister,* Gustav Utzinger, a twenty-six-year-old man with short brown hair and a chemistry Ph.D., to South America.

Elizebeth Friedman's next mission for America became the biggest secret of her life. She would never speak in detail about what she did between 1940 and 1945, even as an old woman, and the records of her work, the documents that now make it possible to tell the story, were classified after the war and locked away for a generation, unsealed only after her death. In the 1950s and 1960s, when she gave speeches or interviews about her career, she freely shared anecdotes about various colorful adversaries of the past—the millionaire George Fabyan, the rumrunners, the drug smugglers— but she skipped the Second World War entirely. These were the years when she disappeared into "a vast dome of silence from which I can never return," she said.

The one time she seems to have alluded to her wartime mission, briefly and vaguely, was in 1975, during an interview with her husband's biographer. She uttered a few words and then the transcript cut off.

"The spy stuff," she said. "That's what I did."

It wasn't Elizebeth's conscious decision to spend the war chasing Nazi spies. It was yet another "pure accident" in her career. This is how it always happened: She put her ear to the ground here and there to learn how the pieces of the world fit together. She figured out how to hear a new sound. Then men in uniform showed up at her side, asking questions, wanting to listen over her shoulder. This had been true at Riverbank two decades earlier, when "the world began to pop and things began to happen," as she put it once; it was true in the 1920s and '30s when she shone a floodlight on the American criminal underworld; and it was proving true again now, in early 1940, when she and her team identified a new and sinister set of voices in the intercepts furnished by the listening stations.

The basic rhythm of her typical weekdays had not changed since the early '30s. She was still working in her coast guard office at the Treasury Annex building near the White House, serving as chief of the Cryptanalytic Unit that she had founded in 1931 and nurtured ever since. Her three junior codebreakers, Robert Gordon, Vernon Cooley, and Hyman Hurwitz, the ones she had originally recruited and trained, were still with her, and a handful of women clerk-typists had also joined the team as support staff. Elizebeth, Gordon, Cooley, and Hurwitz often worked together at a long table in the office, analyzing the ever-replenishing piles of cryptograms that arrived from the coast guard listening stations, chewing the ends of their pencils, maps of the world pinned to the wall behind them, the clack of the clerks' typewriters filling the room.

Outside the door, they could hear the muffled noise of T-men going this way and that, customs men, narcotics men, IRS men, coast guard men. They pressed their foreheads to the intercepts, Elizebeth perhaps wearing a simple white high-collared dress, Gordon smoking a pipe in a suit and vest, chomping on the pipe and frowning at a page. Sometimes Elizebeth would stand up and disturb Gordon's cloud of smoke as she walked to a shelf to look at a piece of cryptologic literature or to examine one of the cipher machines she kept there in case she should encounter a message that had been generated by one. She had an Enigma machine on the shelf, an old version that had been freely available in the 1920s. She also had a Kryha there, the semicircular German device that William had once mastered.

Elizebeth reported to the chief of the coast guard communications section, a salty vice admiral named John Farley, and Farley reported to the secretary of the Treasury, Henry Morgenthau Jr., an old friend of President Roosevelt from a prominent Jewish family. Morgenthau was the kind of person Elizebeth tended to get along with—polite, educated, pragmatic—although Elizebeth came to dread phone calls from his devoted personal secretary, Henrietta Klotz, who had a habit of calling Elizebeth's office at 4:28 or 4:29 P.M., one or two minutes before the 4:30 close of the day, and making what Elizebeth called "rapid-fire dictator-sort of requests," demanding that Elizebeth and her team solve some difficult problem in an impossibly small amount of time. Morgenthau would usually phone Elizebeth the next day and reverse Klotz's order with bashful apologies.

Morgenthau needed Elizebeth to be happy. He now depended on her to perform one of the department's wartime functions. Smuggling wasn't what it used to be—the war had disrupted the drug networks and made business perilous—so Elizebeth's Cryptanalytic Unit had shifted its attention to British and German ships. The Treasury was responsible for enforcing U.S. neutrality laws, and foreign ships along the East Coast needed to be monitored for any violations that might cause diplomatic controversies. At Morgenthau's request, in 1938, the unit began to analyze the wireless messages of British cruisers and German merchant vessels. Elizebeth broke the codes of Nazi captains as they tested the limits of U.S. neutrality and provoked tense confrontations. In December 1939, a German freighter flying the swastika flag pulled suspiciously close to Florida shores and was chased by U.S. Army planes and a nearby British cruiser, *Orion*. Elizebeth decrypted the German captain's panicked messages home. It was the first gunfight of the war in American waters:

AM TRYING TO RUN INTO AMERICAN HARBOR PORT
EVERGLADES OR MIAMI CODE DESTROYED

THE CRUISER HAS TRAINED HIS GUNS AGAIN HE IS RUNNING
SLOWLY FORWARD

CRUISER NAMED ORION

THREE AMERICAN ARMY PLANES HOVERING OVER US

"Exciting, round-the-clock adventures," she said later about these episodes. But an even more intense mission was yet to come.

While monitoring these radio signals for her Treasury bosses and solving the puzzles that were given to her, Elizebeth started to detect a new import to the messages. In January 1940, with Hitler preparing to invade Scandinavia, dozens of mysterious encrypted texts piled up in Elizebeth's office all at once, apparently transmitted by several different unregistered radio stations and intercepted by U.S. listening stations.

At first, the messages looked similar to the thousands of smuggling messages she had solved before. They used the same kinds of call signs and similar frequencies. But after a brief period of confusion, Elizebeth realized that the messages hadn't been sent by smugglers at all. The plaintexts were in German. They contained sensitive information about the routes of U.S. and British ships and the capacities of U.S. factories. And according to the bearing fixes, the signals originated from unknown radio stations in Mexico, South America, and the United States.

It soon became clear that the stations had been built by Nazi spies to share sensitive information with their bosses in Germany, transmitting and receiving dots and dashes of encrypted text at the speed of light. A pair of stations exchanging wireless signals formed a "circuit," and each circuit was protected by a different code or cipher that had to be broken before the messages could be read.

These were clandestine circuits, meant to stay invisible, and it became Elizebeth's goal to pry them out of the dark while remaining invisible herself—an essential part of the job. She knew that if the spies discovered that she was breaking their codes and reading their messages, they would switch to more secure codes, and she wouldn't know what the spies were saying until she could break the new codes, which might take weeks or months. A spy who speaks in a broken code is "the goose that lays the golden eggs," as William put it once. If you want to keep gathering the eggs, you must not frighten the goose.

For this reason, Elizebeth's Cryptanalytic Unit "was probably even more secret than other [codebreaking] organizations," the NSA concluded after the war, "because it dealt with counterespionage." Counterespionage, counterintelligence—these are the formal terms for what Elizebeth was beginning to do. She was counterspying on foreign spies, serving as America's eyes and ears in the invisible world of fascist espionage. Today there are large sections at CIA and FBI that perform foreign counterintelligence, teams of American professionals who spend their days trying to monitor the activities of Russian and Chinese spies, but in 1940 there was almost nothing, and Elizebeth had to act with extreme caution every day. It was essential that her Nazi targets never learn that she existed.

The first few batches of eggs fell smoothly into her basket. As soon as Elizebeth began to analyze the clandestine circuits in 1940, she realized that the spies were relying on different kinds of hand ciphers, variations of tried-and-true methods. Some were familiar systems from the rum days, adulterations of commercial codes like the ABC code and the ACME code. These were solved in a snap. The key for one circuit was found to be 3141592, the first seven digits of the mathematical constant pi. Elizebeth called this circuit "the pie circuit." Sometimes the Germans sent the key at the start of the message and in groups of three or four letters instead of five, indicating that there was something special about these letters and giving away that they were a key.

When an unfamiliar system was encountered, and nothing was

known about the speakers "to provide an entering wedge," Elizebeth and her teammates tried to start with something small and simple. For instance, if they determined by a routine sort of check that they were dealing with a transposition system, with the letters mixed up instead of swapped out, they would look for common German words in the messages, like *zwo,* "two," which is a useful word to a codebreaker because it contains two low-frequency letters, *z* and *w,* which makes it stick out more. (The names of numbers were often spelled out in messages to eliminate potential confusion from dropped letters due to radio interference.) Another technique that often helped was to take multiple messages and stack them on top of one another, creating a "depth" of text that made it easier to identify patterns as opposed to analyzing one message at a time:

1	E	A	W	I	Z	T	Z	N	X	O
2	I	E	U	R	Y	R	X	F	E	H
3	U	I	U	H	Z	F	E	N	N	X

Here, Elizebeth was able to look at row 1 and anagram the letters, Scrabble-like, to make the word *zwo:*

1	Z	W	O
2	X	U	H
3	E	U	X

Now the columns were in a different order, and this new order gave a clue to the structure of the underlying cipher that allowed her to break it.

Essentially, Elizebeth's goal was to look at these daunting mountains of nonsense and chart a route up the slope in small discrete steps, each of which was like a little game—not quite child's play but not totally unlike child's play, either. And the games grew more intricate as the months went on and the coast guard codebreakers followed the intercepts.

Several sets of Nazi spies were using book ciphers similar to the ones that Elizebeth and William had long studied but with new twists. For instance, on January 1, 1940, she received her first intercept from a wireless circuit that linked Mexico with a radio tower in Nauen, Germany. The messages contained only eleven letters of the alphabet: *N, R, H, A, D, K, U, C, W, E,* and *L.* One message began

UHHNR LNDAL NURND WCNCK NRHLN DNRAN CHNDR UNDEN

Relying on intuition and experience, Elizebeth made a few quick assumptions. *N* was the most frequent letter. She guessed it was being used as a "word separator"—a space bar. She also guessed that because there were only eleven letters in the messages, one of which was a space, the letters must stand for the numbers 0 through 9. But which letters stood for which numbers? If she was correct, the spies might have used a key word to determine that. Elizebeth and her colleagues tried to find the key word by anagramming the eleven letters:

WACKELND RUH
WAHL DRUCKEN
ACH RUND WELK
DA LUNCH WERK
DURCHWALKEN

There it was: *Durchwalken,* a colloquial German word meaning "to give a good beating." This was probably the key:

D	U	R	C	H	W	A	L	K	E	N
1	2	3	4	5	6	7	8	9	0	-

Now Elizebeth was able to turn the letters of each message into numbers, using *N* as the separator:

UHHNR LNDAL NURND WCNCK NRHLN DNRAN CHNDR UNDEN
255-3 8-178 -23-1 64-49 -358- 1-37- 45-13 2-10-

Cleaning up the numbers, the line became

255-38 178-23 164-49 358-1 37-45 132-10

This looked like a book cipher to Elizebeth; the numbers probably corresponded to locations in some unknown book owned

by the spies. After translating the letters of several messages into
numbers, she saw that some number combinations appeared more
frequently than others: 1-1, 132-10, 343-2, and 65-12. The coast
guard codebreakers underlined these frequent combinations, and
"after a little experimenting the following was produced":

65-12	132-10	373-2	301-21	285-25	343-2
B	E	R	L	I	N

65-12	375-2	132-10	321-2	132-10	343-2
B	R	E	M	E	N

BERLIN and BREMEN, two German cities. (In some cases, a
letter like *R* was linked to a few different number combinations.)
These frequent letters gave her a start, and when able to solve the
code in full, Elizebeth identified the names of two known Nazi
agents in Mexico, MAX and GLENN, who would appear in
other messages in the future, linked to agents in the United States
and South America. The two Nazi spies were reporting to Berlin
on the movements of U.S. and British ships, making those ships
vulnerable to U-boat attacks.

Elizebeth solved their book cipher without needing to see the
book and did the same with messages that used other books: *The
Story of San Michele,* the memoirs of a Swedish physician; *Soñar
la vida,* a spy story by a female Mexican fascist; *O servo de Deus,*
a Portuguese novel. One Nazi spy proposed using the 1936 novel
Vom Winde Verweht—in English, *Blown Away by Wind,* i.e.,
Gone With the Wind—and asked Berlin to locate a copy. Berlin
replied that *Blown Away by Wind* was unavailable in Germany
and another book would need to be chosen.

Several Nazi agents, Elizebeth discovered, were using a copy of
the romantic novel *All This and Heaven Too* and a sophisticated
process that generated messages full of garbled letters instead
of numbers. Each spy had been assigned a unique identification
number, such as 7. To encrypt a message, the spy would take that
day's date, add the number of the day and the month to his identification number (for a January 10 message he would add 1 + 10 +
7 = 18) and turn to the resulting page in the novel (page 18). The

first words of the first line became part of that day's key—the key for transforming plaintext words into blocks of nonsense according to a Scrabble-like method that jumbled the letters by stacking them into columns. The rest of the key was taken from the first letters of unindented lines going down the page.

To solve the messages, Elizebeth first had to deduce that *All This and Heaven Too* was the novel these particular spies had chosen. To do this she went through the same process of reverse engineering that she and William applied in 1917 to solve the Hindu messages. Then she bought her own copy of *All This and Heaven Too* and kept it on her coast guard desk, allowing her to easily ungarble any new message sent with that system, flipping through the novel and underlining or circling the pieces of the daily keys in red pencil. Here is how she marked up page 15, where the novel's fictional heroine is deciding whether to become the governess for a hot-tempered Parisian family and move into their home:

Yet she did not dread the thought of entering it. The difficul-

ties (i)t presented would at least be stimulating. One would not

(p)erish of boredom in a place where charges of gunpowder might

(l)urk in unexpected corners to explode without warning. She felt

(o)ddly exhilarated—almost, she thought, as if she were about to

(s)tep upon a lighted stage filled with unknown players, to act a

(r)ole she had had no chance to rehearse beforehand. She must find

(t)he cues for herself and rely on her own resourcefulness to speak

(t)he right lines. Henriette Desportes's heart under the plain gray

Elizebeth wrote the letters of the key horizontally on a piece of graph paper and used it to fill in the German plaintext.

Her basic puzzle-solving style hadn't changed from the smuggling days, and it remained effective: a process of trial and error with pencil and paper, deduction and experimentation, granules of eraser dust swiped away with a flick of the palm. Her scrap papers still looked like the scrap papers of a person doing the newspaper puzzle page over Sunday-morning tea; she wrote no equations, only numbers and letters grouped and stacked in rows, columns, squares, rectangles, and more exotic shapes. This approach worked for her because over the previous twenty-five years, encountering tens of thousands of messages, Elizebeth had solved so many different kinds of puzzles that she knew how to find shortcuts, to identify patterns in fields of text that were like signatures telling her what to do next. She was a kind of human computer in this sense. Today, if you want a computer to recognize certain patterns, you can train it through a process of "machine learning." How do you get a computer to recognize a picture of a cloud, for instance? You feed it a lot of pictures and say, essentially, *This here is a cloud,* and *This here is not a cloud.* After the computer gains enough "training data," it's able to look at a new image, do some math, and say, *This is almost certainly a cloud.* By 1940, Elizebeth's brain had probably accumulated more training data about codes and ciphers than any other brain on the planet. She had just seen so many damn clouds. It's why she was able to make inspired guesses about puzzles. She may not have been writing equations, but she was thinking mathematically.

This is also why, in 1940, when Elizebeth encountered her first Enigma messages from a German Enigma machine, she didn't feel overly intimidated.

Enigma was a straightforward idea expressed in a diabolical device. In the simplest sense, it was a box that cranked out poly-alphabetic ciphers. Remember the secret messages that eight-year-old Barbara Friedman sent her parents from summer camp? A=B, B=C, C=D. That's a MASC, a mono-alphabetic substitution cipher. One cipher alphabet encrypts the whole message. Enigma was *poly* instead of *mono,* using multiple cipher alphabets per message.

Poly-alphabetic ciphers date to the sixteenth century and can be written by hand with the aid of pre-printed grids of letters or sliding strips of paper. Instead, Enigma did the job with three or more rotating alphabet wheels connected to electrical wires. The wheels lived inside a box with a typewriter keyboard on the outside, the keys arranged in a familiar order, starting with

Q W E R T Z U I O. Above the keyboard was a "lampboard" of the same twenty-six letters in the same order. When a writer pressed a key, such as Q, a different letter, perhaps Z, would illuminate on the lampboard—the cipher letter, lit by a small battery-powered bulb. Later, the recipient of the message, operating his own identically configured Enigma, would type Z, and Q would light up, decrypting the message letter by letter.

With each key press, an electrical circuit was completed, and Enigma stepped the right-hand wheel, shifting it one letter forward. Once the wheel stepped through all letters, it stepped the middle wheel by one letter, then the left-hand wheel. The motion was similar to a car odometer—after you drive 9 miles, the right-hand number flips to 0, and the next number to the left flips to 1—and it generated a seemingly random, nonrepeating sequence of 16,900 cipher alphabets before the three wheels returned to their starting positions.

Crucially, no letter could be enciphered as itself. If you pressed *j* a million times, you would never see *j* light up on the lampboard.

Although this was a known limitation of the machine, it seemed to pale in comparison with Enigma's flexibility. The wheels could be arranged in different orders (1-3-2, or 2-3-1), the alphabet rings on the wheels could be set at different starting positions on the wheels, and the starting letter of each wheel, as seen through a small window on the box, was another variable. The choice of variables comprised the machine's key—the starting configuration used to encrypt all messages on a particular day, week, or month, depending on how often the key was changed.

How many possible keys existed? Depending on the model of Enigma, the number of keys might be as large as 753,506,019, 827,465,601,628,054,269,182,006,024,455,361,232,867,996, 259,038,139,284,671,620,842,209,198,855,035,390,656,499, 576,744,406,240,169,347,894,791,372,800,000,000,000,000.

Each one of these keys produced a unique set of 16,900 alphabets before repeating.

All of this seemed to make the job of a codebreaker impossible. There were too many possibilities to comprehend, and then there were possibilities about those possibilities, and possibilities about those possibilities about those possibilities. Clearly, shortcuts had to be discovered, and by the late 1930s, finding these shortcuts—and conquering Enigma—was the biggest problem facing Allied intelligence. After Polish mathematicians made some early breaks into the device, the Germans kept changing its design and how it was used, so the battle over Enigma was ongoing, a cryptologic arms race. The machine had been clunky at first, weighing as much as one hundred pounds, but subsequent versions grew lighter and more compact. The German navy, the Kriegsmarine, first adopted them in 1926 and installed Enigmas in ships and U-boats, followed by other branches of the military, embassies, and intelligence services. In 1936, the Nazis banned all commercial sales of Enigma and began to improve the machine in secret, adding additional components and subtleties intended to make Enigma codes absolutely unbreakable. Different Nazi organizations developed their own variants. Germany withheld knowledge of these alterations from the enemy, as if Enigma were a submarine or a bomb.

To extract useful intelligence from an Enigma system, Elizebeth Friedman (or anyone else) needed to accomplish two separate and immensely difficult things. First, the machine itself had to be "solved," its inner workings deduced and mapped—the motions of its wheels and the maze of wires controlling them. This required some leap of human ingenuity, some feat of mathematical deduction or inspired guessing. Then, once the wiring was solved—the part of the system that generally didn't change—the keys had to be recovered, which changed at different intervals (month, week, day) depending on the practices of different Nazi services. If you found an Enigma key in the morning, you might go to bed at night and get locked out again in your sleep, and the next day you had to find the key again if you wanted to read the new day's messages.

There were too many Germans using too many Enigmas with too many shifting keys to ever recover the keys by hand, so codebreakers needed to build machines of their own to assault the enemy's machines, giant electro-mechanical contraptions and some of the first digital computers, too. Automation. Polish codebreakers were the first to solve Enigmas and automate the process of recovering keys. They built "bombes" that mirrored the Enigma rotors, ticking through possible alphabets until they found ones that might fit. Later, the British mathematician Alan

Turing discovered how to make bombes dramatically more powerful, based on mathematical principles and previously solved bits
of text known as "cribs"—a crib might be the name of a Nazi
officer, the time of day, or "Heil Hitler." His solutions were essentially search algorithms, ancestors of the Internet search algorithms of today. Turing's biographer calls these "search engines
for the keys to the Reich." It was anti-Nazi Google.

The British codebreakers worked at Bletchley Park, a mansion in the countryside outside of London. Bletchley grew from a
handful of people in 1938 to thousands by 1945, the bulk of them
women, recruits from the Women's Royal Navy Service who operated the bombes, among other jobs, and were billeted in large
country houses.

The Enigma codebreaking program would come to be known
as ULTRA; Enigma decrypts were stamped with the imposing
phrase TOP SECRET ULTRA as a reminder to handle them with the
utmost care. Later, America would join forces with the British,
assembling its own ULTRA factories in Washington and sharing
the burden. But early in the war, when Elizebeth and her coast
guard unit analyzed their first Enigma machine, ULTRA was a
strictly British franchise. There was no one to tell the Americans
what to do. They had to invent their own method.

At first, Elizebeth didn't know that she was dealing with an
Enigma at all. Enigma cryptograms look like lots of others, ge-
neric blocks of nonsense letters. In January 1940, coast guard radio monitors began intercepting one to five messages per day with
the call signs MAN V NDR and RDA V MAN. Elizebeth wasn't
able to make heads or tails of the first twenty or thirty messages
that were intercepted on this circuit. However, after accumulat-
ing a greater "depth" of messages, sixty or seventy, she was able
to write them one on top of another on a worksheet and see the
letters in a new way by gazing *down the columns.*

Enigma is poly-alphabetic. It creates a new cipher alphabet
with each key press. That's the beauty of the machine. But if an
Enigma user types a number of messages using the same starting position of the rotors, the first letter of each message will
use the same alphabet—and the second letter of each message
will use the same alphabet, and the third letter. In other words,

any individual message is full of alphabets, but if a codebreaker lines up the messages in a tower, *each column in the tower is mono-alphabetic*—one alphabet:

```
1  2  3  4  5  6  7
D  X  J  X  L  H  N...
L  W  S  X  I  Y  F...
M  H  O  S  S  L  C...
```

The letters in the first column here, D L M, all use the same alphabet. And the letters in the second column, and so on.

With only three messages, there isn't enough information to help the codebreaker. The "depth" is too low. There need to be more floors in the tower. At greater depths, closer to twenty messages and beyond, letter frequencies become visible:

```
1  2  3  4  5  6  7
D  X  J  X  L  H  N...
L  W  S  X  I  Y  F...
M  H  O  S  S  L  C...
M  A  P  A  C  T  Y...
F  P  W  S  G  S  C...
Y  Q  A  S  A  C  W...
N  S  H  W  U  F  C...
F  U  W  X  G  S  P...
M  B  D  W  X  U  O...
O  P  O  D  Y  X  L...
A  J  Y  S  X  F  D...
M  W  S  X  E  C  C...
```

M appears four times going down column 1. In this column, *M* might be equivalent to the letter *E,* the most frequent letter in German as well as English. In other columns, different letters might be equal to *E.* And now the codebreaker can use tried-and-true methods to fill in plaintext letters and piece together the adversary's words.

In this way, the technique of "solving in depth" can take a hard problem and turn it into a simpler problem. The trick is often to get the messages aligned in depth in the first place. If the Enigma user changes the starting position of the rotors from message to message, the floors of the tower have to be staggered to track with the shift in the starting position.

Figuring out how to align messages in depth is a subtle art. It can be done with clever guesswork and trial-and-error, and it can also be done by applying the principle of the Index of Coincidence, William Friedman's fundamental insight about the relationships between letters that sit in towers of text. Elizebeth tended to use both approaches in her work, but luckily, in this case, she didn't need to align the messages, because the senders had made a mistake by using the same starting position for all the messages. The messages were *already* in depth. Before long, then, the coast guard codebreakers were able to identify frequent letters in the columns and use those letters to piece together the plaintexts for most of the first batch of messages.

The words seemed to be in German.

Elizebeth and her colleagues still didn't know what type of cipher they were dealing with, so now they decided to write down the alphabets for many of the messages they had solved in depth, the ciphertext equivalents of the plain letters, to see if a pattern popped out. They quickly noticed that no letter was ever enciphered as itself: an *A* never meant *A*, a *B* never meant *B*. This suggested an Enigma.

They went to the shelf in their coast guard office and picked up their old commercial Enigma machine.

The codebreakers had already solved most of the messages, but now they wondered if they could solve the machine itself—the wiring. Knowing the wiring makes it easier to solve new messages. Without the wiring, they would have to repeat the laborious process of solving in depth every time the key changed. Their challenge now was to use the text they had recovered, the plain letters and the cipher letters, to work backward toward the unknown machine, almost like a police detective analyzes the spatter pattern of blood at a murder scene, starting with the red evidence and rewinding back to the moment of the crime, deducing from the crusts of blood the speed and angle of the knife.

Unbeknownst to the coast guard, groups of British and Polish codebreakers working on the Enigma problem had already discovered methods for working backward from the text to the machine. The Poles had done it with an algebraic approach, the mathematics of permutations, and one of the brilliant Bletchley codebreakers, a linguist and scholar of classical literature named Dilly Knox, had relied more on pattern recognition and a kind of alphabetic grid called a "rod square." But the coast guard didn't know about these approaches, and so, working in isolation, the codebreakers had to grope toward their own method. They poked and prodded and turned the wheels; they wrote alphabets on sliding strips of paper and moved the strips against one another, thinking.

It seemed to Elizabeth that there must be a fixed relationship between the alphabets she had already discovered by solving in depth—the plaintext letters and their cipher equivalents—and the motions of the Enigma's wheels. To test this hypothesis, she drew a number of diagrams that visualized the relationships between letters at each position of the machine. She wrote new kinds of towers of letters on the worksheets that were more like X-rays than photographs, probing more deeply into the identities at the heart of the Enigma, and immediately she saw clear patterns, hints of order and regularity.

Certain letters repeated vertically on the page, like LL and HH, and also pairs of letters, like SJ and EM. Elizabeth and her colleagues realized that these letter groups were telling them something about the spacing between pairs of wiring contacts on the Enigma's rotors. The maps were whispering secrets about the physical intricacies of the machine. Building upon these "remarkable results" over the following days, filling more worksheets to the brim with letters, drawing more towers and analyzing the patterns that appeared, the codebreakers managed to solve the wiring for all three wheels of the unknown Enigma. Then they were able to reveal the full plaintexts of all unsolved messages from the radio circuit.

The codebreakers now realized two slightly disappointing facts: The plaintexts seemed to contain no Nazi secrets; later the codebreakers learned that the messages had been sent by the neutral Swiss army, which sometimes used Enigmas to communicate in German. Then the coast guard shared the wiring diagram of the Enigma with William's codebreaking team at the army, in case it might be useful to them, and the army reported back that the diagram corresponded exactly to the wiring of a commercial version of Enigma.

Elizebeth had hoped that she was mastering a new kind of Enigma entirely. Still, it was a significant achievement. "This recovery of wiring assumed to be unknown was achieved without prior knowledge of any solution or technique and is believed to be the first instance of Enigma wiring recovery in the United States," her team wrote in a secret technical memo after the war. As far as Elizebeth and her codebreakers could tell, and they were hardly prone to bragging, they were the first Americans to solve an unknown Enigma.

Until this moment, cipher machines had always been William's territory, not Elizebeth's, but her solution of the commercial Enigma showed that she had a similar aptitude for solving machines, and this initial headfirst dive into the pool of Enigma codes would lead her to deeper waters later in the war. Of course, she didn't know this in early 1940. Demolishing that first Enigma was just work. She was confident enough in her abilities that solving an Enigma seemed like a reasonable and normal thing that she might accomplish with her team on a given week. She didn't brag or make a big deal. Anyway, there was no time. New puzzles were arriving at the coast guard all the time, new codes to break, along with increasing demands for assistance from outside agencies.

All along her plaintexts had been circulating through other parts of government. Each time her unit solved a message, a clerk typed the English solution on a fresh sheet of paper, a decrypt, and gave it to the coast guard chief of communications, Vice Admiral Farley, for dissemination. Depending on the content of the decrypt, the vice admiral might send a copy to navy intelligence (OP-20-G), army intelligence (G-2), the State Department, British intelligence, or the FBI. The decrypts were like blood cells in the veins of government, delivering the vital oxygen of raw intelligence, and as different intelligence agencies realized that Elizebeth had tapped into a trove of information about Nazi spies, they inevitably asked the coast guard for more decrypts. In the 1920s, she had complained about government men "appearing on my doorstep," wanting her to solve puzzles. They were still appearing on her doorstep, but now, instead of relatively anonymous T-men, they were some of the most powerful spymasters in the world.

J. Edgar Hoover liked to eat dinner at Harvey's restaurant on Connecticut Avenue, next to the Mayflower Hotel, a five-minute walk from his suite of offices in the Department of Justice headquarters, a gargantuan gray edifice near the National Mall. Harvey's had separate dining rooms for men and women. The ladies' dining room was on the second floor, accessible by a separate entrance at street level. The first floor was the gentlemen's restaurant and bar, with waxed floors and rich leather banquettes. It was one of those places in Washington where men of influence slurped oysters and let their guard down for an hour or two.

The FBI director's face was beginning to acquire some of the first creases and pouches that would characterize the eventual marble busts of him. He was forty-five years old, one of the few immutable objects in an ever-changing city. He wore white shirts, double-breasted Brooks Brothers suits, and a hat with a brim that could be turned up or down. His agents wore the same uniform. A pink expanse of forehead separated his bushy black eyebrows from his thinning hair. The large neat desk back at his office, a corner office on the fifth floor of Justice, contained a radio, usually a vase of fresh flowers, and a framed copy of "Penalty of Leadership," the text of a Cadillac advertisement from 1915. It read in part, "When a man's work becomes a standard for the whole world, it also becomes a target for the shafts of the envious few."

At Harvey's he usually ordered steak or roast beef and a Caesar salad. He ate at the same table every time, the most secure in the room, almost invisible from the door under a stairway. A reporter once watched Hoover sign twenty autographs during a single dinner at Harvey's. He liked to sit there with his chief deputy, armed bodyguard, and longtime companion Clyde Tolson. It was a table for four with just two chairs. There was always a bottle of wine waiting for Hoover at his table when he arrived—part of a ritual that he performed here.

William Friedman dined at Harvey's on occasion. There were times at the restaurant when the cryptologist sensed motion in his peripheral vision, when a shadow darkened the white cloth. He turned his head and saw Hoover standing there with the bottle of wine. Without saying a word, the director nodded and poured wine into the cryptologist's glass. He had respected William for years and appreciated his periodic assistance with FBI cases, with the little encrypted notes written by criminal suspects that William would solve in his free time and send back to the bureau.

Hoover was almost certainly aware of Elizebeth Friedman. But he would not yet have had many chances to cross her path. She wasn't allowed to eat in the gentlemen's dining room at Harvey's. There were a lot of male enclaves like this in the city, inaccessible to her. And Hoover was a chauvinist of the old school. When he first took charge of the bureau in 1922, there had been three female agents. He got rid of them. The next two female agents wouldn't join the bureau until after his death in 1972. He argued that women weren't agent material because they couldn't be taught to shoot guns. Female clerks and secretaries at the bureau had to wear skirts and weren't allowed to smoke at their desks as the men could. One of his least favorite people was Eleanor Roosevelt, who wrote him a mildly indignant letter after the FBI conducted an intrusive background check on one of her friends. Hoover compiled a secret dossier alleging she was a communist. "When a woman turns professional criminal," he wrote once, "she is a hundred times more vicious and dangerous than a man." Women at Hoover's bureau were only deemed fit for "boring clerical functions," according to the memoir of one longtime agent. "It was perfectly all right to bullshit 'em and ball 'em: Just don't tell 'em any secrets."

But by 1940 Hoover had gotten himself into a jam serious enough to require the technical assistance of a woman.

It had long been the FBI's job to disrupt espionage rings within U.S. borders. Any Nazi spies operating in America were Hoover's quarry. However, he didn't seem to be very good at catching them. He had built the bureau's name on its flashy investigations of jazz-age gangsters, men who enjoyed attention and went out in public with entourages. Counterespionage was another discipline entirely, a matter that required a certain finesse, and the bureau's first sizable Nazi spy case, in 1938, had ended in a public-relations disaster.

That year in New York, the FBI arrested a Chicago man of Austrian parentage, Guenther Rumrich, along with two associates suspected of spying for Nazi Germany. Then an FBI agent named Leon Turrou made the mistake of tipping off Rumrich's collaborators that an indictment was coming. They panicked and fled the country.

Newspapers mocked the FBI for letting Nazis slip through its fingertips, and U.S. intelligence agencies that had long resented the FBI found new reason for their scorn. Over the years, Hoover's insatiable hunger for publicity had caused a lot of bad blood; in the press he repeatedly claimed sole credit for investigations to which other agencies had contributed but were not free to discuss. The head of army G-2, George Strong, one of William Friedman's superiors, despised Hoover, and the navy OP-20-G chiefs couldn't stand him, either. Henry Morgenthau at the Treasury hesitated to even speak the director's name in meetings, and Hoover thought of him as "that Jew in the Treasury." When British intelligence officers started to arrive in Washington in 1940, hoping to forge links with U.S. agencies, they were shocked by this toxic atmosphere of mistrust and quickly traced the cause to Hoover. "J. Edgar Hoover is a man of great singleness of purpose, and his purpose is the welfare of the Federal Bureau of Investigation," a group of British operatives later wrote. "It was once remarked of a well-known Oxford scholar that, while he had no enemies, he was hated by all his friends. Something of the same kind would express the feelings towards the FBI of its fellow U.S. agencies."

For a man as vain as Hoover, and as publicity-obsessed, and as intensely disliked by rivals in his own government, the bungled Rumrich case represented both a personal black eye and a threat to the FBI's future authority. Somehow he needed to salvage the bureau's reputation, to prove that it was capable of catching fascist spies, and in 1939, he proposed a bold plan to do just that.

Hoover knew that the concept of "hemisphere defense" had become a fixation with Roosevelt and military chiefs: Guarding the United States meant guarding *the entire Western Hemisphere* from Nazi encroachment. In other words, it wasn't enough to fortify U.S. defenses. South America must be protected as well. Roosevelt talked about hemisphere defense in speeches, arguing that "no attack is so unlikely or impossible that it may be ignored," and Secretary of the Navy Frank Knox raised the specter of Nazi planes taking off from South American airfields in the night and dropping bombs on "our own women and children in our teeming seaboard cities." Seeing an opening, J. Edgar Hoover pressed Roosevelt to dramatically expand the FBI's jurisdiction. For the sake of "the common defense of the Western Hemisphere," Hoover argued, the FBI must be allowed to operate beyond U.S. borders. He demanded the authority to send men into South America, "to seek out and identify agents of the Axis operating in all the Americas, to ensure the ultimate safety of the United States."

Hoover got his wish in June 1940, with a presidential directive that represented a historic expansion of the FBI's power. For the first time, the bureau was free to dispatch agents into other countries. He created a new division called the Special Intelligence Service (SIS) and began recruiting agents for duty in South America.

Their mission would be to find and monitor the secret mail drops and radio stations used by the spies; to map the structure of their organizations and communications networks; to determine the true identities of the enemy agents; and to cooperate with local State Department officials and police in arresting the spies, seizing the radio stations, and destroying the rings.

A tall order. The first five SIS agents were dispatched to the continent in September 1940, one each to Peru, Uruguay, Argentina, Brazil, and Venezuela. Pale and corn-fed, they stepped off their planes into the lacerating sun of another continent. They wore snap-brim hats and looked like detectives that South Americans had seen in newspapers and movies. The agents knew little about codes, ciphers, or radio, these crucial tools of their adversaries, and didn't speak the local languages. The SIS man sent to Brazil had been given a crash course in Spanish. When he arrived, he realized, to his frustration, that the language of Brazil was actually Portuguese.

Hoover's men in South America were so unprepared that they

had almost no chance of catching the spies through old-school gumshoe tactics: interviewing associates, recruiting confidential informants, developing leads. They needed to know what the spies were saying to one another in private. They needed codebreaking. And this was exactly the problem.

To break codes, you need intercepts and you need codebreakers to solve the intercepts. The FBI had neither. It had no intercepts because it had no listening stations; when the bureau wanted intercepts it was forced to obtain them from the coast guard and the Federal Communications Commission (FCC). And when the FBI got these intercepts, it couldn't read them, because the FBI had no codebreaking unit. What it had instead was a Technical Research Laboratory, essentially a crime lab, a place where bureau technicians analyzed bullets, fingerprints, threads of fabric, and blood samples.

All of this spelled trouble for J. Edgar Hoover. At the very moment he was launching a hemisphere-wide hunt for spies who communicated in code, his bureau had no ability to discover what they were saying.

Around this time, Elizebeth received an unusual order from her Treasury bosses: They asked her to visit FBI headquarters. She was to teach codebreaking to an agent named W. G. B. Blackburn, an employee in the Technical Lab. Elizebeth proceeded to train Blackburn in codes and ciphers, much as she had trained her own junior colleagues, and Blackburn established a small Cryptographic Branch at the FBI, which would grow to the size of a handful of employees over the next several years, all of them codebreaking novices.

This still wouldn't do for Hoover's purposes. The Invisible War demanded a level of technical firepower and prowess that his Technical Laboratory simply did not command. What he required was the full assistance of a mature codebreaking organization, whether they wished to help him or not. He needed Elizebeth and the coast guard.

She read pacifist poetry. It resonated. She thought of her kids. Bar-bara was in her last year of high school and planned to attend college at Radcliffe, and John Ramsay was a fourteen-year-old freshman at Mercersburg Academy, an elite boys' school in rural Pennsylvania. He wasn't young enough to be safe from a military draft. War would scatter her family. She also worried about the fate of her team at the coast guard. She had built this little organization and it was good and she wanted to protect it from disruption. Codebreaking is delicate work. You have to look at the page and get all the letters aligned just right, then you have to look at your team and get all the people aligned just right, so that the flow of intercepts and records and ideas and solutions becomes as efficient as possible.

Elizebeth escaped Washington for a week in June 1940, traveling to Mexico on a quick vacation with daughter Barbara and sister, Edna. It was the last time in the next five years she would get a break, a chance to pause and look around and spend time with the women closest to her. They drove a beat-up rental car through the farmlands of Oaxaca and the mountain ranges of Puebla Cordoba, descending into canyons on the backs of burros. Elizebeth wrote to William, "All Mexico is so full of resounding cockcrows, pig-grunts, burro-brays, and church bells that all sleep is intermittent, at best." The two sisters had a great time and woke up early each day; Barbara wanted to sleep in and complained that the altitude made her knees wobbly. She was a gorgeous girl of seventeen now, six inches taller than her mother, confident and voluptuous. One day, when they were all on a plane above Oaxaca, Elizebeth happened to fall asleep in her seat, and when she woke, she saw Barbara up in the cockpit, next to the pilot. *Wait, what kind of airline was this?* Are girls just allowed to ride in the cockpit without their mother's permission? Isn't that unsafe? She felt like a mom.

The news of the war got rapidly worse while she was in Mexico. She had to stop reading the papers in the morning because it was too depressing. Nazi tanks were said to be plowing through the French countryside on the way to Paris. The Mexican papers seemed to think America was bound to join the war. The peso was rising, eating into Elizebeth's meager trip budget of fifty dollars. She airmailed William a letter about the rising cost of goods. In his reply he begged her not to spend more money than was absolutely necessary, "or we shall never never climb out of this morass of debt."

She wasn't sure if William was okay. He sounded sad and mopey in his letters. He said it had been rainy in Washington, and in the evenings he had been sitting alone with a pencil and a pad, listening to the rain on the roof, writing a technical paper on cryptology. He told her, "There won't be anybody [to] read this thing, I imagine, at least not for some centuries," and added a lament about the shackles of secrecy: "I wish I could write about forbidden subjects. What a story could be told."

By the time she got back to Washington—to home, husband, and job—the Nazis had entered Paris, hanging the swastika flag from the Arc de Triomphe.

CHAPTER 2

Magic

One day in September 1940, inside the windowless vault of William Friedman's army codebreaking unit, one of the two female codebreakers on the team, Genevieve Grotjan, stood at her desk and let the men know that she might have found something.

The men called her Gene. She was twenty-eight and quiet and wore rimless glasses. She had a background in statistics; she would later teach math as a professor at George Mason University in Virginia.

Gene Grotjan had been looking at raw Purple intercepts for hours, weeks, months when she called the men over. The team had not been able to penetrate the garbled Japanese text on the intercept sheets that had been streaming into the Munitions Building. But now Grotjan thought she had noticed two patterns that others had missed—subtle cycles of repetition, loops of letters in the text, much like the ones discovered by the coast guard in Enigma messages. Frank Rowlett, one of William's deputies, came over and looked at Grotjan's worksheets. Then he looked at her. He got the sense that her eyes were beaming through her glasses, that she was struggling to contain her emotions. Others started to crowd around the desk. Rowlett started jumping up and down.

"That's it!" he shouted. "That's it! Gene has found what we've been looking for!" Another man busted into a funny little dance. He threw his arms up in a victory pose. "Whoopee!"

Grotjan was a modest person. "Maybe I was just lucky," she said later in an NSA oral history. "I perhaps had a little more patience" than some of the other workers. She didn't become as animated as Rowlett because "I regarded it more as just one step in a series of steps."

William Friedman heard the commotion and shuffled into the room from his nearby office: "What's all this noise?"

Rowlett showed Gene's worksheets to the boss, and William agreed within seconds that they revealed a loose thread in the code. There was still more to do, but it was clear that the thread, if pulled, would allow the team, with great grinding effort, to recover the daily keys and consistently read the Japanese messages.

The men were bouncing and laughing with the excitement of the discovery. Friedman seemed almost sad. "Suddenly he looked tired," Rowlett later recalled, "and placed his hands on the edge of the desk and leaned forward, resting his weight on them." Rowlett knew that Friedman had been under a lot of stress, working sixteen-hour days for weeks, months, years. The younger man pulled out a chair for his elder. Friedman sat quietly for a few moments. Everyone was looking at him, waiting for his reaction. He turned to the codebreakers. "The recovery of this machine will go down as a milestone in cryptologic history," he said in a formal, distant voice. Then he left the room.

In codebreaking, the larger the success, the more it must be suppressed. Any leak might reach the enemy and cause them to switch to a new code system, destroying the value of the break. Heroes celebrate briefly and in secret. Someone went and got Cokes.

A few minutes later, the group dissipated, and everyone returned to their desks to explore the new textual terrain they had unlocked—everyone except Rowlett, who went looking for Friedman. The young man was still shaking, still full of adrenaline, wanting some kind of catharsis. The boss's lack of enthusiasm confused him. Rowlett found Friedman in his office, "sitting at his desk, studying some notes he had made on a pad. When I entered the room, he sat quietly, merely looking questioningly at me."

Over the next hours and days, the team kept applying pressure to the hairline fracture in the code until it shattered and the first bits of plaintext revealed themselves. Five days after Grotjan's discovery, on September 25, 1940, the team produced their first full decrypted message. It was a big moment. William and the rest of the codebreakers had never been able to look at the Japanese machine, or touch it. They had never seen a drawing of it, or a patent illustration, or a photo. Yet they now understood how it worked and how to recover the daily key for a given set of messages. People had reverse-engineered cipher machines and devices before, but nothing at Purple's level of complexity. Today historians of cryptology believe that in terms of sheer, sweaty brilliance, the breaking of Purple is a feat on par with Alan Turing's epiphanies about how to organize successful attacks on German Enigma codes.

Once William and his colleagues fine-tuned their bootleg Purple machine, the one they were building to help them read Japanese messages, they demonstrated it for the unit's commanding officer, typing out a sample of ciphertext and then decrypting it as he watched. A sheet of fresh plaintext inched its way out of the machine, and after the man grabbed it and looked for a few seconds, a smile lit up his face, and he congratulated the codebreakers on their "magnificent achievement." Then he rushed off to get *his* commanding officer, Joe Mauborgne, William's old friend. When they returned together, the first officer pointed to the machine and said to Mauborgne, "Last night your magicians completed the reconstruction of the new Japanese cipher machine," and the codebreakers repeated the demonstration for Mauborgne. "By God, it really works beautifully!" Mauborgne said.

Your magicians. It really did seem like that. Like magic.

MAGIC became the top-secret moniker for these Japanese decryptions, for the astonishing fountain of secrets that would keep gushing up all through the war, secrets of Japanese strategy and Nazi tactics that flowed across Japan's encrypted circuits and were tapped by the U.S. Army (and later by the U.S. Navy too, which solved Japan's naval cipher), giving Allied planners the drop on their foes. The first handful of decrypted messages turned into 20, into 100, into 1,000, piped directly from William's unit through the corridors of power. MAGIC beguiled all who touched it. Men read the daily "MAGIC Summaries" with bulging eyes and could not quite believe they were reading the authentic words and orders of imperial Japan. It was almost too good to be true. The president read MAGIC, and Army Chief of Staff George C. Marshall, and Secretary of the Navy Frank Knox, and, eventually, Prime Minister Winston Churchill in Britain, who insisted on getting MAGIC raw, unsummarized by his generals.

MAGIC led directly to bombs falling on imperial ships at Midway and other decisive naval battles. It caused the deaths of hundreds of thousands of Japanese and saved the lives of unknown numbers of Allies. MAGIC changed the war. It was also one of the great secrets of the war, exactly like ULTRA, the Enigma codebreaking program. These were tremendous military advantages that could not be revealed to the enemy lest the enemy get wise and cut off the stream of intelligence. The advantage "would be wiped out almost in an instant if the least suspicion were aroused regarding it," George Marshall wrote later in a classified letter that captured the value of MAGIC in the Pacific War against Japan:

> The Battle of the Coral Sea was based on deciphered messages. And therefore our few ships were in the right place at the right time. Further, we were able to concentrate our limited forces to meet their advances on Midway, when they otherwise would certainly have been some 3,000 miles out of place. We had full information of the strength of their forces in that advance, and also of the forces directed against the Aleutians, which finally landed troops on Attu and Kiska. Operations in the Pacific are largely guided by the information we obtain of Japanese deployments. We know their strength in various garrisons, their rations, and other stores available to them. And what is of vast importance, we check their fleet movements and the movements of their convoys.

The triumph over Purple would turn out to be William's last hurrah as a hard-core codebreaker, his final death-defying climb. From now until the end of his life, he would serve America as an inventor of cipher machines and an architect of intelligence institutions (and ultimately a critic of them as well). He had reached his peak. Elizebeth, though, was still climbing, and she couldn't see him up there, across the gap between their two towers, starting his descent. She couldn't share this victory with him, because on the day he and his team broke Purple, a historic achievement that had required all of his battered brain, all he had learned in his unexpected life of exploration with the woman who meant everything, he said nothing about it to her when he came home. He did not seem different to her than he did on any other evening. He said hello and asked what was for dinner.

Heavy air attacks on London had begun that month. The Blitz. On September 7, 1940, shortly after 5 P.M., a thousand German planes appeared in the sky above London. It was a bright blue afternoon. The planes arranged themselves in vertical formations, fighters and bombers. The fighters had bright yellow noses and tails. The bombers were black and set their sights on industrial facilities that lined the Thames River. The bombs destroyed factories, shock waves and oily smoke rippling out. British Spitfires gave chase to the German planes. "The sky seemed full of them," one British pilot later said, "packed in layers thousands of feet deep. They came on steadily, wavering up and down along the horizon. 'Oh golly,' I thought, 'golly, golly. . . . ' "

The bombings of London continued for fifty-six straight days. Sirens and shelters, blackouts at night. The Axis was growing bolder in the final months of 1940. Japan invaded Vietnam, expanding its empire in East Asia. The Nazis confiscated the private radios and telephones of Jewish families and cordoned off the Warsaw Ghetto with barbed wire, trapping 400,000 adults and children, most of them Polish Jews.

America didn't want war. Both major political parties still supported neutrality. The aviation pioneer Charles Lindbergh argued in popular radio speeches that it would be foolish and hypocritical to fight Germany. He said America had no standing to accuse the Nazis of aggression and barbarism because America had sometimes been aggressive and barbaric itself. Later he argued that American Jews were a "danger to this country" on account of their "ownership and influence in our motion pictures, our press, our radio and our government." Lindbergh became the public face and champion of an antiwar group called the America First Committee. "America First," a campaign slogan of Woodrow Wilson, had been adopted by the Ku Klux Klan in the 1920s. Within a year the America First Committee was holding rallies at Madison Square Garden.

The worsening of the war in Europe, combined with U.S. reluctance to fight, was about to drag Elizebeth into the orbit of a highly motivated and capable group of British spies. The British were afraid. They knew they didn't have the money, the people, or the weaponry to sustain a long fight against the Nazis. They needed America to join the war. Their survival as a nation depended on it.

In the early summer of 1940, British officers began to arrive in America on a covert mission. Some went to Washington, making the rounds of embassy cocktails and dinner parties, looking for all the world like bright young chaps out for a good time, and others worked in the heart of New York, in a Fifth Avenue skyscraper, the thirty-fifth and thirty-sixth floors of Rockefeller Center. The group included Ian Fleming, a handsome lieutenant with blue eyes and a smart blue naval officer's jacket, and twenty-three-year-old Roald Dahl, a tall, elegant Royal Air Force fighter pilot who looked a bit like Gary Cooper. These guys would both become famous fiction writers after the war; Fleming invented the character of James Bond, and Dahl wrote children's books about chocolate factories, flying peaches the size of zeppelins, and foxes who outwit monstrous humans. For now, though, Fleming and Dahl were spies. Dahl was a particularly good spy. He seduced actresses and heiresses in Washington, gathering gossip in bed, and he charmed the president and the first lady, becoming a regular guest at their Hyde Park, New York, home, where they spoke so freely with the young pilot that he had difficulty maintaining his composure: "I would do my best to appear calm and chatty," he later wrote, "though actually I was trembling at the realization that the most powerful man in the world was telling me these mighty secrets."

They called their organization British Security Co-ordination, an intentionally boring name meant to deflect scrutiny. BSC was really one of the most fantastic associations of men and women ever created. It had one thousand members who worked toward a single goal: ending American isolationism and pushing America into the war, by any means necessary. One BSC recruit had been told, "All I can say is that if you join us, you mustn't be afraid of forgery, and you mustn't be afraid of murder." BSC planted anti-Nazi information in the American press, some of it false, through relationships with columnists like Walter Winchell. BSC staged protests at rallies for isolationist politicians and dug up dirt on their pasts. It used sex to steal information, sending gorgeous female spies to seduce enemy diplomats and swipe documents. And BSC also hoped to apply British radio expertise to catch enemy spies operating in the Western Hemisphere, which put BSC in direct conflict with the formidable American who had already claimed that ground and was not eager to give it up.

Early on, the British tried to strike a deal with J. Edgar Hoover. Ian Fleming and his superior officer went to FBI headquarters one day in June 1941 and met with the director in his corner office. The white dome of the Capitol was visible from the window, beyond a set of stone columns at the back of the National Archives, the central repository of government records. Fleming and his colleague explained that they wished to partner with the bureau and share intelligence on the Nazi threat. Hoover listened politely, "a chunky enigmatic man with slow eyes and a trap of a mouth," in Fleming's description. Then Hoover said he couldn't help; U.S. neutrality rules prohibited him from giving aid to any combatant nation.

This was true, but it was also an empty excuse. Hoover didn't want the British operating in America because he saw them as a rival to the FBI. The British didn't care one way or the other. They needed a friendly American spy agency as a partner, and if Hoover wasn't willing to be that agency, for whatever reason, they would find another one, even if they had to create it from scratch. And so they did. They planted the seed that eventually grew into the CIA. Behind the scenes, the British argued to U.S. officials that the FBI was ineffective. The FBI had "no conception of offensive intelligence as we know it," wrote Captain Eddie Hastings, a retired Royal Navy officer, now working for BSC in Washington; according to Hastings, America needed a new agency capable of "offensive" spy maneuvers in foreign countries. In July 1941, Roosevelt established the Office of the Coordinator of Information, a new civilian intelligence organization attached to the White House. The following year, the Office of the COI was renamed the Office of Strategic Services, which was the forerunner of the CIA.

So this is where the CIA began—with J. Edgar Hoover telling the British to go to hell, and the British not appreciating it.

This was also when the British began making friendly advances toward Elizebeth Friedman.

The British already had a mature radio intelligence agency, the Radio Security Service (RSS), that excelled at the art of wireless interception. But due to sheer geography, the British listening posts couldn't hear signals from some parts of the globe. The men of British Security Co-ordination wanted access to any intercepted and solved messages that America happened to have. And they realized that when it came to radio intelligence and hard-core codebreaking, the place to be in America was the coast guard. Unlike the FBI, Elizebeth's unit had access to intercepts from its own listening stations, and its cryptanalytic section "was incomparably better than that of the FBI," in the British view, because the coast guard's codebreakers had spent the last decade testing their skills on smugglers, whose networks happened to look a lot like Nazi spy networks. "The whole system" of rumrunning had the air of a German spy network in miniature," BSC historians later wrote. "Hence, on the outbreak of war, the Coast Guard was already experienced in the tricks of the illicit wireless operator."

BSC sent a few men to meet with Elizebeth in Washington and chat about the problem of Nazi spies in the Western Hemisphere, and they all hit it off right away. The men had considerable expertise and experience in radio intelligence, particularly a husky, apple-cheeked colonel named F. J. M. Stratton, who had taught astronomy before the war, specializing in studies of supernovas, distant exploding stars that registered as sudden and perplexing balls of light on Stratton's photographic plates. Before that he served in the radio corps of the British army in the First World War, developing a reputation as the happiest man in the trenches despite sleeping only four hours a night. His fellow soldiers called him "Chubby" on account of his bulk and his jollity. Elizebeth thought he looked like Santa Claus.

As they got to talking that first time, Stratton and Elizebeth, they realized that if they combined their resources and their knowledge, they'd have a better chance against the Nazi spies than if they were working alone. The British operated radio posts across Europe staffed by 1,500 secret listeners, many of them volunteer hobbyists, and the intercepts from those stations would fill gaps in the intercepts from the coast guard and the FCC, and vice versa. When the British couldn't hear something, the coast guard could hear it, and when the coast guard couldn't hear it, the British could.

Aside from that, Stratton enjoyed deep connections to Bletchley Park and the already massive codebreaking operation there, where some analysts had been focusing specifically on Nazi spy codes. It might make sense to share knowledge.

By now Elizebeth and her coast guard codebreakers had also begun working directly with the FBI at the request of J. Edgar Hoover. He wanted assistance with several different unknown code systems. Elizebeth obliged. She found that some of the spies who interested the FBI were using book ciphers, and others relied on "turning grilles" much like the grilles that the Friedmans drew one year in their family Christmas card. The spies wrote letters in holes punched through a piece of paper of certain dimensions according to certain rules, and Elizebeth had to make five or six separate deductive leaps to figure out those rules to determine the exact shape of the piece of paper using only clues derived from the messages themselves.

Not only did Elizebeth break the codes for the FBI, she made special devices and tools for the G-men so that they could easily solve future intercepts on their own. For instance, when she solved a book cipher, she gave the FBI the name and description of the book, and when she solved a grille system, she made grilles for the FBI Technical Laboratory. In other words, when the FBI was able to solve its own messages, this was only because Elizebeth had given the Technical Laboratory the means of solution—the laboratory run by the G-man she herself had trained in 1940.

While Elizebeth solved these individual puzzles as fast as she could, immersing herself in the gritty details, the larger goal behind it all, preventing a fascist takeover of South America, remained an obsession at the highest levels of U.S. government. On December 29, 1940, in a "Fireside Chat" radio speech from the Diplomatic Room of the White House, FDR argued that it was time for America to rethink its role in the world. It was futile to hope that fascism would leave America alone if America returned the favor. Instead, the nation must become an "arsenal of democracy," a force to defend and spread freedom abroad. During the 36 minutes and 56 seconds of the speech, he mentioned South America twice and used the word "hemisphere" 10 times. He said, "Any South American country, in Nazi hands, would always constitute a jumping-off place for German attack on any one of the other republics of this hemisphere." Without getting into specifics, Roosevelt referred to "secret emissaries" of the Axis, fascist spies like the ones Elizebeth was tracking. "The evil forces which have crushed and undermined and corrupted so many others are already within our own gates. Your Government knows much about them and every day is ferreting them out."

Hitler responded that England would soon be destroyed along with all other "democratic war criminals" and promised a rapid Nazi victory within the first few months of 1941. On New Year's Eve, Londoners climbed into the blacked-out streets, over the charred remains of buildings, and sang "Auld Lang Syne."

Elizebeth heard the news four days later, on January 4, 1941. She rushed to Walter Reed General Hospital in northern Washington.

The main building was majestic, meant to be a comforting sight to wounded warriors: three stories of red brick, with soaring white columns in front. Staff directed Elizebeth to the Neuropsychiatric Section, a separate structure connected to the hospital by an underground tunnel. She found William there, confined to a large, noisy room. It was three and a half months after his team's breakthrough on the Purple code. Elizebeth counted between sixteen and twenty other psychiatric patients in the room, all men, including some who appeared very disturbed. She was scared and could see that William was scared, too.

Walter Reed was the nation's flagship military hospital, and for soldiers or officers suffering from physical injuries or infectious diseases, it was about as good as could be. During the Great War, men returning from the trenches, many with amputated limbs, would sit on the wide porch in wheelchairs, covered in blankets, looking out at the manicured grounds and the fountain whose bowl was ringed by four stone penguins standing atop concrete pedestals. Psychiatry, however, had never been a priority at Walter Reed or in the army as a whole, and in 1940 and early 1941, the energies of army psychiatrists were almost entirely geared toward keeping the mentally ill *out* of the army, not treating them once they got in.

Walter Reed's chief psychiatrist, Colonel William C. Porter, saw the job of the hospital's Neuropsychiatric Section as one of evaluation and processing rather than healing. The section did offer a range of treatments standard for the time, including chemical sedatives like Amytal, group therapy, and electroshock therapy, but it was too small to provide long-term care, so it functioned instead as a way station, a purgatory. When the section admitted a new patient, doctors and nurses examined him, studied his military records, and observed him for a period of weeks or months before deciding whether he should be discharged from military service. Depending on the decision, the patient was sent back to the army, or home to his family, or in many instances, to a mental asylum.

Sometimes, instead of discharging a patient from the army altogether, the section's doctors recommended he be transferred to a desk job, presumed to be less stressful. The idea that desk work itself might be a cause of debilitating stress—that the army now employed puzzle solvers, cryptologists, who bashed their brains against the stone of codes and bore the heavy burden of secrets— never occurred to the doctors of Walter Reed.

They didn't know what to do with William Friedman when he presented himself. William told the doctors he had collapsed several days earlier and believed he was having a nervous break-down. A psychiatrist asked a battery of questions about his job, his family, and his career. Without mentioning the Purple project, William said his work had been demanding lately. He felt a constant tension that interfered with his ability to function, and sleep provided little relief when he could manage to sleep at all.

The doctors assigned the cryptologist to one of the section's five mental wards. There were three wards for men and two for women, with a maximum capacity of 104 patients. Security guards patrolled the wards. William spent the next two and a half months here, inside the redbrick building, unable to leave until the staff completed their evaluation.

Elizebeth came to visit most days, taking the train to the 116-acre hospital campus and walking briskly past the main building with its cupola and fountain on her way to the Neuro-psychiatric Section. She always wanted to talk to her husband in private during these visits, to see how he was doing, to kiss him and say she loved him, but the setup made it nearly impossible, because the patients were forced to spend their days in the group room, and they all had to share the same psychiatrist, who consulted with each patient within earshot of the others. "In other words, the patient was isolated except for his fellow-patients," Elizebeth later told William's biographer, "who could discuss and consult with each other if they felt inclined to do so."

The patient. Seeing him there was horrible for Elizebeth, and the hospitalization represented such an obvious threat to his live-lihood in the army that she had to find ways to distance them both from what was happening. She refused to admit that her husband might have a serious mental illness. She thought the word "depression" was "too strong a term" and preferred "mood swings" or "downswings." At home she answered his personal mail, explaining that William was ill and would get back to people when he could.

Meanwhile, at the Munitions Building, his team of cryptol-ogists continued to harvest the fruit of Purple and plant seeds in new places. Having already built one replica of the Japanese machine, the SIS workers built several more, and in January, two of William's deputies, Abe Sinkov and Leo Rosen, sailed across the Atlantic with two Purple machines, delivering them to grateful British codebreakers at Bletchley Park. Now the British could make their own translations of Japanese messages. It was an important exchange of cryptologic knowledge between America and Britain, one of the first of many during the Second World War, although the British didn't return the favor yet—they weren't ready to share what they knew about German Enigma systems.

In March 1941, the staff of Walter Reed finally made their decision. William Friedman, they believed, should return to army duty. His nervous collapse was an "anxiety reaction" sparked by "prolonged overwork on a top secret project." The hospital discharged him on March 22 into Elizebeth's arms. He went back to work at the army on April 1.

He wasn't quite the same, and never would be. The breakdown and the hospitalization had changed his universe in ways it would take years to measure and understand. For one thing, the ordeal had planted doubt in the military bureaucracy that William Friedman was fit for service. It created a trail of medical documents that would chase him for years, popping up and causing havoc at the oddest times. Three weeks after he left Walter Reed, William received a letter from the army notifying him that he had been honorably discharged "by reason of physical disqualification"—no hearing, no chance at a defense. William made a vigorous protest, pointing out that the hospital had pronounced him fit, but the army forced him to retire and continue as a civilian; eventually he would need to sue to get his old rank and pay reinstated. Later, in 1946, checking his personnel file, William discovered that the government had him classified as a *temporary employee*. It was probably a paperwork snafu, but it struck him as a bizarre indignity—his twenty-five straight years of service to America had hardly been temporary—and his friends were so horrified on his behalf that they threw him a big surprise party at an officers' club and staged a mock court-martial as a send-up of the ridiculousness. The judges recorded their votes on a cipher machine, pronounced him guilty, and presented William with an aluminum medallion that read, "To Wm. F. Friedman for making the intelligible unintelligible and vice versa 1921–1946. Presented by those he has led astray."

William's illness also disrupted the balance of the Friedmans' marriage. William and Elizebeth had always acted as fierce equals. Modesties and flatteries aside, they lived as if neither was smarter than the other, or stronger, which was the truth. From here on, though, Elizebeth often had to be the stronger one, out of pure necessity. She had to care for William during his depressions and keep her job. They needed two incomes to pay the mortgage and their kids' private-school tuition. The Friedmans, like so many middle-class Americans who hurt for money and pinch pennies, were determined that their children receive the same educational opportunities as the "sons of capitalists," Elizebeth once wrote. And throughout the spring and summer of 1941, as William recovered, Elizebeth's job was only getting harder. The Invisible War was intensifying. The documents produced by her team now bore its mark. On the coast guard decrypts, in the lower left corner, beneath the letters of the plaintext, the same two words appeared, over and over, on page after page.

"German Clandestine."

CHAPTER 3

The Hauptsturmführer *and* the Funkmeister

Johannes Siegfried Becker, the most prolific and effective
Nazi spy in the Western Hemisphere during the Second World War.

No code is ever completely solved, you know.
—ELIZEBETH S. FRIEDMAN

The fact that Johannes Siegfried Becker is an obscure figure today, a man without a Wikipedia page, his name producing a few stray Google hits, is a testament to his skill as a spy and also the skill of the woman who became his nemesis, Elizebeth Smith Friedman. They were two cloaked particles meeting across a void at the speed of light and partially annihilating each other, leaving jets of alphabets, a spray of letters falling to the ground.

According to the FBI, which was slower than Elizebeth to

understand his significance, Johannes Siegfried Becker was "one of the most active as well as the most capable of German agents operating in this hemisphere during this war," a spy of rare vision and resourcefulness, directing endless funds and resources with a "deft Teutonic hand." He spoke German, Spanish, Portuguese, and English. He held the rank of SS-*Hauptsturmführer* in the elite Nazi security service, equivalent to a captain, and wore a gold ring carved with the SS's death's head symbol, "a sign of our loyalty to the Führer," Heinrich Himmler wrote in a letter of praise to Becker, and "a warning to be ready at any time to sacrifice our lives as individuals for the life of the whole." He had some forty-seven aliases and several false passports and moved freely across South America, recruiting spies in seven nations, organizing political plots and military coups with Nazi sympathizers, and building clandestine radio stations. In mid-1944 the FBI concluded that the activities of 250 Nazi agents in South America and twenty-nine radio stations could be traced back to Becker by direct or indirect steps.

Yet Becker did not appear on the FBI's radar during 1938, 1939, 1940, 1941, 1942, or 1943, and when the FBI finally began to hunt him, they were too late. He was the Invisible War's invisible shadow. He escaped every trap, slipped every net—except, at last, the one set by Elizebeth. And even Elizebeth would be astonished by Becker's ability to vanish.

For all his talents, Becker started out as something of a screwup, his first missions in South America marked by mediocre results and sexual improprieties. Between 1936 and 1939, while spying for the SS in Brazil and Argentina before the war broke out, he left a trail of irritated German expatriates, men who "disliked his manner" and found him vain. There appeared to be nothing exceptional about Becker. His Nazi Party number was 359,966, marking him as a relatively early convert to the cause of National Socialism but hardly one of its pioneers. He stood five foot ten, with wavy blond hair and a slight paunch, and his face was not overly handsome, leaving his acquaintances wondering how it was that Becker always seemed to have a girlfriend. For a time he worked for an Argentine firm as an importer of German children's toys and doll eyes, claiming to be a woodworking expert, when in truth he spent most of his days watching British ships come and go in the harbor and his nights prowling the bars and dance floors of the city, writing the phone numbers of prostitutes and fascist sympathizers in a personal address book. He caused a scandal in Rio de Janeiro by impregnating the wife of a Brazilian cabinet minister. The Nazi ambassador complained to Berlin that Becker was risking an international incident. No one in South America had anything good to say about Becker's personal habits, and all who met him were struck by his grotesquely long fingernails, which curled down like the talons of a predatory bird.

Still, Becker had one essential quality that set him apart from almost everyone else in his corner of the Nazi universe: he was *adaptable.*

Becker worked for a wide-reaching SS office that placed spies all around the world and communicated with them from a four-story building in Berlin that had once been a Jewish retirement home. Called AMT VI, the SS office employed five hundred people in Berlin and managed another five hundred spies in foreign countries. A minority of the spies were actual SS officers like Becker and the rest were considered "V-men" (*vertrauensmann* is German for "informer"), usually German expatriates and local fascists who wanted to help the cause. There was also a separate German agency, the Abwehr, that sent spies to foreign countries, but the Abwehr predated the Nazi movement, and SS leaders thought the Abwehr was insufficiently ruthless and possibly disloyal. They promoted their own AMT VI as the true Nazi foreign intelligence service.

Not just any Nazi could be selected as an SS intelligence officer, according to an SS handbook. He had to be the purest of Nazis, a man of "absolute loyalty and obedience to the Führer . . . Like the knights of the Holy Grail intelligence officers have the most noble task to protect the most valuable possession and its future realization: the blood of the Germanic race, the National Socialist ideology."

In practice, however, this meant that Becker's organization was riddled with amateurs promoted for their zeal instead of their knowledge. The leader of the South America section of AMT VI, Theodor Paeffgen, was a thirty-one-year-old bureaucrat with "no qualifications whatever for intelligence work," an American interrogator would later conclude. Paeffgen's previous job with the SS had involved "combating partisans" in Russia, a euphemism for killing Jews. Paeffgen's deputy was a former Gestapo thug named Kurt Gross, who badgered his spies in South America to send him packages of cognac, coffee, and silk stockings, and often made lewd comments to the buoyant, brown-haired young woman who managed the section's files, Hedwig Sommer, who had been forced into working for the SS against her will. (After the war, Sommer gladly told U.S. interrogators everything she knew about the section.)

These men cared mainly about ideology, not competence, and even when they made a rare exception, they were overruled by other fanatical organs of the Nazi state. One of the section's most talented spies was a Jewish man from Holland named Weinheimer who was working for the SS in the hopes of saving his family from the concentration camps. He had smuggled himself into Chile, posing as an immigrant, and according to Sommer he sent back a number of "highly regarded" and "very accurate" reports about political and economic trends in the Western Hemisphere. Then Weinheimer learned that the Gestapo had shipped his mother-in-law to the Bergen-Belsen camp. Kurt Gross asked the Gestapo to make an exception for the spy's kin, but Gross was unsuccessful, and the spy stopped sending reports. The Nazis lost one of their best agents because they wouldn't spare his loved ones from the death camps.

When it came to the nuts and bolts of intelligence, the SS bosses in Berlin didn't really know what they were doing—and neither did Siegfried Becker at first. But unlike his superiors, he was flexible enough to learn from his mistakes. He was a loyal Nazi but didn't concern himself with the intricacies of Nazi dogma. Hedwig Sommer liked him. "He was an intelligent person," she said. "He was sincerely desirous of doing a good job. Added to these attributes was the fact that he was something of an adventurer."

After the Nazi invasion of Poland in September 1939, Becker had left South America and sailed to Berlin, guessing that his bosses would want to modify his mission. He was correct. In meetings at the home office, the SS leaders told Becker that he was now their top agent in South America, and he needed to go back to the continent and recruit a team of spies.

Berlin gave Becker a trunkful of explosives for blowing up British ships in the harbor. Becker arrived in Buenos Aires with the trunk in December 1940 and was intercepted at the German embassy, where the ambassador opened the trunk, saw the bombs, imagined the diplomatic headaches they would cause, and ordered Becker to dump the bombs in the river. At this point, he abandoned the sabotage mission and began building his new spy network in earnest, traveling across the continent, from Argentina to Brazil to Bolivia to Paraguay, trying to convince German colonists to spy for the Führer.

Many of the would-be V-men in South America proved hopelessly ineffectual—one was a petty crook who seemed to do nothing but wander along the waterfronts carrying a revolver and scaring passersby—but in Rio de Janeiro, Becker soon met and cultivated a formidable spy named Albrecht Engels, a broad-shouldered German businessman with a thick mustache. Engels was already spying for the Abwehr, which meant that he and Becker were not supposed to work together, but neither man minded. Becker thought Engels was a perfect collaborator: married to a Brazilian woman, owner of a thriving firm in Rio, well liked by all Germans in the community.

And Engels, who went by the code name "Alfredo," was impressed by Becker. Ever since Engels started working with the Abwehr, he felt he had been dealing with imbeciles. His Abwehr colleague in São Paulo was a jittery mechanical engineer of Polish ancestry, Josef Starziczny, who went by the code name "Lucas." Starziczny was an elfin man with large ears who lived with his Brazilian mistress and talked too much. He observed the harbor, radioed reports to Germany with his own transmitter, and didn't listen to advice. He made Engels nervous. Becker was different, another caliber of spy: "the only real professional" in South America, Engels later told an FBI interrogator.

Engels's arrangement with Becker was strictly *improvisado*—making things work. Until this point, Engels's duties for the Abwehr had consisted mainly of scouring English newspapers and magazines (*Time, Collier's, Reader's Digest*) for information about U.S. politics. Becker single-handedly turned this press-clipping service into an actual spy network. He built a courier system to exchange information with Germany, convincing employees of the Condor and LATI airlines to carry spy messages in pouches on their flights to Germany and deposit the pouches at a firm owned by an SS man. He taught Engels how to use book ciphers and codes based on pencil-and-paper grids and turning grilles.

And in the spirit of *improvisado,* when Allied pressure shut down Condor and LATI flights to Germany in summer 1941, destroying Becker's courier service, he found ways to communicate wirelessly with Berlin. At first Becker paid a V-man to set up a small shortwave transmitter on the patio of a German expatriate's home. When the signal proved too weak, Becker coaxed the captain of a Swiss ship docked in the Rio harbor, the SS *Windhuk,* to allow the spies to borrow the ship's radio.

Becker signed his wireless messages with one of several code names. The main alias was "Sargo." Engels sent messages under his "Alfredo" alias.

It was difficult to get a reliable signal, and Becker, for all his ability, lacked the technical expertise. He asked the SS to send him a *Funkmeister,* a radio operator, and in September 1941 the SS dispatched Gustav Utzinger to Rio.

Utzinger was the opposite of Becker in many ways: a man of education, a trained chemist. He went by the code name "Luna." Clean-cut and athletic, with brown eyes and close-cropped brown hair, he had served in the 1930s as a *Funkmeister* in the German navy before joining the SS. Later, speaking to an American interrogator, Utzinger claimed that he acted out of "natural patriotic efforts for my Fatherland" and not "the most detestable tendencies of Nazi ideology." The interrogator didn't buy this. Still, after speaking with Utzinger for hours, the interrogator concluded that he was essentially an honest and even somewhat idealistic person: "an extremely able and personable young man who was the product of his era and who acted according to his own lights."

Becker arranged to meet Utzinger in a café outside of Rio to discuss the urgent need for reliable clandestine radio stations. Utzinger's first impression of Becker was unfavorable. He saw Becker as a man "with very little education and few moral scruples in pursuit of his ends." But Utzinger would come to respect Becker over the next several years as the two men worked to spread fascism across South America. In their separate realms of expertise— Becker in espionage, Utzinger in radio—they approached their jobs with the pride of craftsmen. Becker had the contacts and the vision. Utzinger had the technical skill. Soon the *Hauptsturmführer* and the *Funkmeister* would prove to be the most dangerous Nazis in the West.

In the beginning Elizebeth knew the *Hauptsturmführer* and the *Funkmeister* only by their aliases, "Sargo" and "Luna."

She first encountered these names in the late spring of 1941, at the tail end of William's hospitalization in the mental ward at Walter Reed. This is when the listening stations of the coast guard and the FCC provided Elizebeth with the first of thousands of intercepts from clandestine transmitters in South America, and she started doing what she had always done: smash the codes, recover the plaintexts, translate them into English, type the translations on fresh sheets of paper (decrypts) ready for study and dissemination, sift the decrypts for clues about the secret identities of the spies, keep immaculate records; build an archive, a library of enemy words.

The original messages had been written in German, Spanish, and Portuguese, and to recover the plaintexts and translate them, Elizebeth worked closely with her lead coast guard linguist, thirty-two-year-old Vladimir Bezdek, a handsome Czechoslovak army veteran with black hair and high cheekbones. Born in Czechoslovakia, Bezdek had escaped to America when the war broke out by sneaking onto a ship. He spoke eight languages fluently: Czech, German, English, French, Polish, Latin, Italian, Russian. He read dictionaries in his free time, for fun, so of course he and Elizebeth got along, checking in with each other throughout the day, puzzling out bits of language together.

It appeared that the Nazis had at least three separate clandestine radio stations up and running in South America. Two were in Brazil, on the eastern coast of the continent, and one was in Chile, on the western coast. The Brazilian stations were in Rio de

Janeiro and a suburb of São Paulo, about two hundred miles south of Rio. All three stations exchanged wireless messages with either Berlin or Hamburg.

Elizabeth gave each radio circuit an alphanumeric label to keep them straight, like 2-B or 3-A. The label was typed on the top of every decrypt from that circuit, beneath the word "S E C R E T," along with the date and time the message was sent, the original language (German, Portuguese), the radio frequency in kilocycles, and sometimes the first few groups of the message's raw ciphertext. Below the header came the plaintext message itself, in English, followed by three lines at the bottom identifying the decrypt as a coast guard product: "CG Decryption," "CG Translation," "CG Typed," the date it was typed, a serial number unique to the message, and the phrase "German Clandestine."

The code names of the suspected Nazi agents were always typed in capital letters, to make them stand out and help everyone on the team get familiar with this strange cast of characters scurrying across the continent next door. You had to get to know your adversary, to see into men's hearts and predict their behavior from a running conversation of potentially enormous stakes that no one else in the world was watching except you. If Elizabeth picked up an inch-thick stack of decrypts and flipped through them quickly with her thumb, as if shuffling a deck of cards, she could see the names of the Nazi agents flick past, a blur of SARGO SARGO SARGO LUNA UTZ ALFREDO LORENZ LUNA ALFREDO LUCAS ALFREDO LUCAS SARGO SARGO SARGO.

The number of times a certain name appeared in the messages was a rough indicator of that person's importance. SARGO appeared again and again on the decrypts. He also seemed to call himself SARGENTO, or JOSE, or JUAN. Elizabeth guessed that he was a Nazi spy chief of some kind. The individual known as LUNA tended to speak about technical issues, the details of radio transmitters; Elizabeth pegged him as a radio expert. He went by UTZ in addition to LUNA.

There was a third man in the messages, ALFREDO, who often mentioned his dealings with the other two—ALFREDO, a trusted colleague of SARGO and LUNA—as well as some other names, like HUMBERTO. For Elizebeth, seeing a name like HUMBERTO was a piece of luck, because it was longer and contained some less-frequent letters, like *M* and *B,* and it repeated across multiple messages as a predictable signature. It was a "crib," a piece of repeating text that gives the codebreaker a foothold. A British colleague of hers once said, "When you get a man with a nice long name with about twelve syllables, it can be of the greatest help to us." If Elizebeth could solve for HUMBERTO, she was well on her way to breaking the rest of the code.

At first the spies in South America were using book ciphers. Elizebeth solved them. She watched these men talk and plot and share information: reports of Allied ships in the Rio harbor, political developments in the United States, information about shipments of ores and weapons and beef, the health of crops, the number of planes being built in American factories. In September 1941, the agents switched to a grille-like cipher, and Elizebeth penetrated that, too. After she solved a message and the clerks typed the decrypt, Elizebeth and the other codebreakers and translators would perform a preliminary level of intelligence analysis, lightly marking up the decrypt with colored pencils, calling attention to proper names and places with check marks and sometimes stapling a handwritten note explaining who the speakers were, what function they served in the network, and what they seemed to be discussing. Then the decrypts had to be transmitted to other agencies: army intelligence (G-2), navy intelligence (OP-20-G), the State Department, the British. Another line was added at the bottom of the decrypt, sometimes in pencil, indicating its destination.

Regardless of a message's content, the coast guard provided copies of every solution from the South American circuits to FBI headquarters, at the request of J. Edgar Hoover. The bureau's newly created Special Intelligence Service then circulated the coast guard decrypts throughout the hemisphere, sending them to SIS agents on the ground in South America, giving the agents a leg up on their quarry.

Throughout 1940 and the first half of 1941, the coast guard was pumping solved puzzles to the FBI on a steady basis, dozens per week, hundreds of messages on each clandestine network and ultimately thousands taken all together. Yet this relationship between the coast guard and the FBI only went in one direction. SIS agents in South America never sent useful information or evidence to the coast guard codebreakers. Worse, the FBI systematically obscured all traces of the coast guard's deep involvement in the spy hunt. When Elizebeth sent them a decrypt, the FBI placed it in their own SIS filing system, with a new four-digit identifying number, and the FBI invented new names for the radio networks that Elizebeth had already named.

This is how the history of the Invisible War would become distorted; these are the small decisions that erased Elizebeth from the record and later allowed J. Edgar Hoover to take credit for her achievements. "A considerable amount of the investigation conducted relative to these espionage groups was based on information obtained from the messages transmitted to and received by the clandestine stations," the FBI wrote after the war in a three-volume history of the SIS. "The technical facilities of the Bureau were used to monitor the several German transmitters, and by analysis and coordination of information obtained from the decodes of the messages, furnished by the Technical Laboratory, and the intensive investigation by SIS representatives, the persons referred to in the messages were identified, their cover names ascertained, and their associates were established."

This is highly misleading. The decrypts were indeed "furnished" to agents in the field by the FBI Technical Laboratory, *after the coast guard had furnished the solutions to the Technical Laboratory*. The evidence is on the original documents themselves. Before the coast guard sent the FBI a decrypt, the coast guard clerks typed "SIS Dupe" at the bottom of the sheet, beneath the line that said "CG Translation" and "CG Decryption." These once-secret files, located in the National Archives and finally declassified in 2000, prove that the coast guard, not the FBI, solved these Nazi radio circuits.

Hoover's stinginess on these South American matters was difficult for the coast guard to understand, especially since the coast guard was simultaneously assisting the FBI with a massive spy investigation inside the United States. It centered on a South African man living in New York City, Frederick Joubert Duquesne, a big-game hunter with dark, floppy hair and a grudge against the British dating back to the Great War, when he was arrested carrying a file of newspaper clippings about bomb explosions on ships.

Bureau personnel called the case "the Ducase," riffing off the pronunciation of Duquesne. Several times a week during the spring of 1941, Duquesne went to an office on Ninety-second Street in Manhattan and met with another German spy, William Sebold, exchanging sensitive information about U.S. military capabilities and discussing the activities of more than thirty confederates who had been recruited as spies by the two men. Sebold used a clandestine radio station in Long Island to transmit the information to Hamburg, the messages encrypted with the book cipher based on *All This and Heaven Too,* which Elizebeth had already broken. What Duquesne didn't know was that Sebold was secretly working for the FBI as a double agent. Surveillance cameras in the walls of the office were capturing him on film, and the radio transmitter in Long Island was controlled by the FBI, which altered the information before sending it to Hamburg.

The bureau reached out to Elizebeth and the coast guard when the FBI radioman in Long Island received an unexpected request from Hamburg: Could he use the Long Island station to relay messages from Nazi spies in Mexico? The clandestine station there wasn't powerful enough to transmit all the way to Hamburg. The FBI agreed, but when the messages for relay started to come, they were in an unknown code.

Elizebeth broke it. The spies in Mexico turned out to be MAX and GLENN, the same agents she had tracked a year earlier.

She gave these plaintexts to the FBI and kept solving new messages sent from Long Island to Hamburg. By the summer of 1941 her team had decrypted hundreds of notes exchanged by Duquesne and the other members of the ring. These messages not only provided hard evidence against the spies that could be used in court; they also revealed links between the spies in New York and Nazi agents in South America and Mexico, pointing the FBI to suspects they hadn't known about before. The coast guard's patient codebreaking, combined with the FBI's surveillance footage and the cooperation of the double agent William Sebold, led to what J. Edgar Hoover called "the greatest spy roundup" in U.S.

history, a series of raids in June 1941 conducted by ninety-three FBI agents and sweeping up Duquesne and thirty-two members of his ring. Nineteen pleaded guilty to espionage charges and the remaining fourteen, including Duquesne, were put on trial three months later in Brooklyn. President Roosevelt followed the trial closely; if the time came for America to declare war, he needed to know there wasn't an enemy spy network on U.S. soil, able to perform sabotage. After six weeks of sensational testimony by FBI agents and Duquesne himself, all defendants were convicted, and the thirty-three spies were sentenced to three hundred years collectively.

The wild success of the "Ducase" had two large and lasting effects on America. The first was that it discouraged future Nazi attempts at spying within the borders of the United States. The second was that it made J. Edgar Hoover a legend. Hollywood later filmed a movie about the Duquesne spies, *The House on 92nd Street,* in close cooperation with Hoover himself. The Ducase "gave birth to the popular cultural belief that the Bureau was the nation's first line of defense against foreign and domestic espionage," writes the former FBI counterintelligence agent Raymond J. Batvinis. "It launched the popular myth of Hoover as the guardian of 'the American way of life.'"

Elizebeth, who received no credit for her contributions to the Ducase, wasn't nearly as impressed with the FBI's performance. It bothered her that FBI agents had described the spies' cryptographic practices in detail at the trial: "The FBI exposed the secret messages and methods without as much as asking a by-your-leave from the Treasury Department where the solutions and systems were achieved," she wrote later.

It seemed obvious to her that the FBI was too cavalier about publicity, and just as obvious that she couldn't do anything about it. The bureau was more powerful than the coast guard. What Hoover wanted, Hoover got, and that fall, as the Nazis marched on Moscow and the U.S. government shifted toward a war footing, Hoover continued to demand the coast guard's decryptions of spy messages, and Elizebeth's team continued to provide them.

During the final weeks before Pearl Harbor, October and November 1941, she could feel herself losing control of her codebreaking team. The military was starting to take over civilian functions. On November 1, a day after a Nazi U-boat destroyed an American ship off the coast of Ireland, killing more than one hundred sailors, Roosevelt signed an executive order declaring that the coast guard was no longer a Treasury agency. Instead, effective immediately, the coast guard was part of the U.S. Navy, and all coast guard personnel were subject to the authority of Navy Secretary Frank Knox. Basically, with a stroke of the presidential pen, Elizebeth and all her colleagues had been drafted into the navy.

She had no objection to working for the navy per se, but she was convinced that moving the team out of Treasury would disrupt their work and harm their effectiveness. Elizebeth complained to a Treasury undersecretary, Herbert Gaston, and Gaston relayed her objection at a Treasury staff meeting on November 5, 1941, in Secretary Henry Morgenthau's office.

The group that day consisted of thirteen men and Morgenthau's secretary, Henrietta Klotz. They gathered around his desk. The Oval Office in the White House was visible through a nearby window. Morgenthau could sometimes see the silhouettes of FDR and visitors moving around, and flashbulbs popping.

At 10:45 A.M., the men began to talk about income tax rates, agriculture legislation, the price of automobile tires, and the recent federal conviction of Nucky Johnson, the criminal boss of Atlantic City, of tax fraud. When it seemed like every issue had been addressed, every nugget of gossip shared, Morgenthau said, "Anything else?"

"One matter that doesn't much belong to me," Gaston said, "but since I took it up—I mentioned it to you on the phone, so I will mention it again. That is the matter of Mrs. Friedman."

Gaston, a birdlike man with wire-rimmed spectacles and a fine suit, explained to the group that Elizebeth wanted to stay at Treasury.

"She is very discontented about the prospect of having to work for the navy. She is very rebellious and gloomy."

No one was sure what to do. Should Morgenthau call up the secretary of the navy, Frank Knox, and ask for Elizebeth back? Should Treasury fight to keep her, or let the navy have her?

Gaston said, "She has a good organization. Perhaps it is as

good as there is in the government, as you know, on cryptanalysis." He added that Elizebeth "would have some usefulness" if she
stayed in Treasury and left the wartime spy-catching work to the
army, navy, and FBI. She could help the T-men investigate bank
accounts controlled by the Axis powers, for instance.

Morgenthau said that whatever happened, he did not want
to answer angry questions from a very gloomy and rebellious
woman. "I just don't want it to appear that I am taking the initia-
tive."

Then Harry Dexter White spoke up. Forty-nine years old
and balding, dressed in a three-piece suit and owlish spectacles,
White looked every bit the economics professor he used to be,
a man of charts and formulae. This impression may have been
carefully cultivated. Documents unearthed in the 1950s would
link White to Russia's top intelligence agent in Washington.
Some historians believe that during the 1930s and '40s, White
was spying for the Kremlin from within Treasury's innermost
sanctum.

In the meeting, White gave his opinion on the matter of Mrs.
Friedman. "I don't like to butt into this, Mr. Secretary, but I understood that she is one of the best in the country, is that correct?"

Gaston replied, "She says her husband is very much better than
she is, but I think she is very good."

Morgenthau ended the meeting with an intention to talk to
Elizebeth about the situation. He never did, though, because a
month later, on December 7, 1941, the Japanese attacked the U.S.
naval base at Pearl Harbor.

When news of the bombing reached the Friedman home, William
started pacing and stammering under his breath that he did not
understand. Elizebeth heard him say, "But they knew, they knew,
they knew," over and over.

He left immediately for the army's cryptologic bunker inside
the Munitions Building. Personnel started to stream in and mill
around. The colonels wore a variety of expressions, some red-
eyed and worn, others strenuously poker-faced. They heard that
two thousand Americans had been killed, maybe more, including

1,177 crewmen aboard the battleship USS *Arizona*, incinerated by an armor-piercing bomb that had burrowed its way into the forward ammunition hold. Twenty-one ships sunk, almost two hundred planes destroyed. A good portion of the Pacific Fleet lay at the bottom of the ocean.

Over the next few days, more than one codebreaker wrote his will. They witnessed the documents for each other.

The codebreakers had known for days, if not weeks, that a large Japanese attack was coming. William and the rest of his team had seen the MAGIC intercepts. It was obvious from MAGIC that Japan had been poised to strike; the only mystery was where. What surprised William on December 7 was not the attack itself but the location. He thought it would happen in Manila, not Pearl Harbor.

In the years that followed, William would become obsessed with the question of what went wrong. He analyzed thousands of pages of Pearl Harbor documents and wrote a three-volume report that boiled down to this: MAGIC had strongly indicated an attack on December 7, but the decrypts had gotten bottled up through a series of farcical missteps in the dissemination stage of the process, and U.S. leaders weren't alerted to the danger in time to take action. It was nuanced: The crucial MAGIC decrypts had been slow to arrive in Pearl Harbor partly because the military hadn't given the Pearl Harbor commanders a Purple machine of their own, a direct tap into the MAGIC fire hose. This decision had been made out of a reasonable desire to limit the distribution of Purple machines in order to minimize the chances of the Japanese learning about the MAGIC secret.

It was a prime example of the brutal choices that codebreakers must live with. Do you take risks to keep a secret that may save hundreds of thousands of future lives, or do you expose the secret to save a small number of lives right now? William once referred to this broad dilemma as "cryptologic schizophrenia," adding, "What to do? Thus far, no real psychiatric or psychoanalytic cure has been found for the illness."

Cryptologic schizophrenia may have explained an unusual personal interaction that the Friedmans had on the day of the attack, December 7. That evening, after work, Elizebeth and

William were at home when they heard the clack-ack-ack of their elephant door knocker. They found a red-faced British man on their doorstep: Captain Eddie Hastings, the BSC officer who had pushed America to create a new spy agency, the Office of the Coordinator of Information. Elizabeth had gotten to know Hastings, and so had William.

According to a declassified NSA report of a postwar interview with Elizabeth, what Captain Hastings did next would become "one of the most vivid recollections of her life." Hastings wobbled into the Friedman home and sat down. He mentioned Pearl Harbor. Then he started to laugh. The attack had just been announced a few hours earlier on the radio. Elizabeth looked at him, baffled, as he kept laughing. "Mrs. Friedman was shocked and offended," reads the NSA report. "Apparently Hastings found the surprise element of the attack amusing. Nevertheless their friendship continued."

Maybe Hastings was giddy from the stress of the day. Maybe he really did think it was darkly funny that MAGIC, for all its power, couldn't save the American sailors and pilots at Pearl Harbor, that the Americans could almost literally read the minds of Japanese leaders and yet fail to prevent a huge Japanese attack. Elizabeth would never understand the British man's laughter. It was one of those mysteries in the intelligence profession that leaves you to dangle, that you think about years afterward, that comes back to you in calm moments on a plane or in your bed at night, making you realize that as much as intelligence seems to be about knowing things, about gaining power through knowledge, it is just as much about not knowing them, or getting them wrong, or seeing other people get them wrong, and having to go on living with the uncertainty, with the not knowing, and thinking about what might have been.

"**Yesterday, December 7th, 1941—a date which will live in infamy—** the United States of America was suddenly and deliberately attacked by naval and air forces of the Empire of Japan."

Less than twenty-four hours after the bombings, President Franklin Delano Roosevelt stood at a podium in Congress, speaking into a bouquet of microphones, asking for a declaration of war. His son, James, stood next to him in his marine uniform. Earlier in the day, James and his father's aides had fitted his paralytic legs with the three metal braces that were required to support the president when he could not be seen in his wheelchair. Sixty-two million Americans listened to the speech by radio. It lasted seven minutes. Within an hour, Congress had authorized war with Japan. Three days later, Germany declared war on America.

Before a single regiment of U.S. soldiers set foot on European soil, the war changed American culture. It was a stomach that ingested a large diverse nation and started breaking it down into widgets. Hollywood movies and Disney cartoons were about the war now. Business was about the war. Work was about the war, and school was about the war. It was the only time before or since when Americans became emotionally invested in the idea of self-deprivation and frugality. Third graders roamed their neighborhoods in packs, gathering scrap materials, tires, and paper and cooking fat and old sneakers whose soles could be sacrificed for the rubber. The Big Three automakers stopped making cars and started making planes. Factory workers took secrecy oaths. Everybody had a secret now. The government issued ration stamps for eggs, milk, bread, gasoline, contained in ration books, manila-colored pamphlets. Elizebeth's ration book listed her height as five foot three and her weight as 120 pounds.

After Pearl Harbor all government matters were urgent and the military didn't want civilians in charge of sensitive functions. The coast guard decided to appoint a new chief of the Cryptanalytic Unit, Leonard T. Jones, a young lieutenant who had taken an army training course in cryptanalysis. Just like that, Elizebeth was demoted from Cryptanalyst-in-Charge to mere Cryptanalyst. She was no longer the leader of the team she had invented, staffed, trained, and nurtured.

This upset her, but she remained the unit's civilian commander and thought Jones showed promise as a codebreaker, so she didn't complain, and anyway, it wasn't like she had a choice. Men told her what to do, and her services were in high demand. Every few days someone was calling up Henry Morgenthau, wanting to bor-row Mrs. Friedman for various cryptologic tasks.

The battle for her attention rose to the highest levels of Wash-ington. She got to know James Roosevelt, FDR's son, and Wil-liam Donovan, a tall, irascible former army colonel with a manic personality, whose soldiers used to call him "Wild Bill." FDR had asked his son to help Donovan launch the Office of the Coordina-tor of Information, the spy organization that would become the OSS and later the CIA.

Donovan was starting from zero, in borrowed office space. One of the first things a spy agency needs is a way to commu-nicate securely with its people in the field. It needs codes and ciphers, and mechanical aids to generate them, and clerks to write them, and training for the clerks. Donovan didn't have any of this, and didn't know the first thing about codes or ciphers, so James Roosevelt approached Elizebeth to lend her expertise, and Donovan reinforced the demand by sending a letter directly to Henry Morgenthau that requested Elizebeth by name, citing an "urgent need for her services pending the establishment of our permanent code section" (Morgenthau grumbled at a staff meeting, "He wants Mrs. Friedman"). This became her first mis-sion after Pearl Harbor. Detailed to Donovan's office on a tem-porary basis, she spent three and a half weeks creating the first permanent cryptographic section for the proto-OSS and proto-proto-CIA.

She built it from scratch, making alphabet strips and other aids to generate ciphers, obtaining hard-to-find cipher devices through navy channels, installing the machines, and customizing them according to the new agency's needs. She interviewed potential cryptographic staffers and made recommendations to Donovan, which he ignored, treating her the whole time like a servant and failing to appreciate basic principles of communications security. When the job was complete, Elizebeth wrote him a seethingly polite letter that conveyed her feelings between the lines, her hor-ror that an important national function was going to be directed by a man who struck her as foolish and cavalier (Donovan's OSS would be defined by recklessness). She sent the letter via James

Roosevelt to make sure Roosevelt was aware of Donovan's short-comings as a guardian of information:

> *My experience and observations during my temporary duty*
> *with your organization, lead me to make the following*
> *recommendations:*
> *—That the representatives going to the field in every case be*
> *required to spend sufficient time to become thoroughly drilled*
> *in the systems of communication provided for them. This drill*
> *and resulting mastery cannot be accomplished in a few hours.*
> *It should extend for a few hours daily over a minimum of five*
> *days, and with certain types of mind a longer time will be*
> *required.*
> *—That a general indoctrination in and discussion of and*
> *handling of classified information be undertaken throughout*
> *your organization. . . . This matter of indoctrination is a long*
> *and difficult process. . . .*

She signed the letter "Dr. Elizebeth Smith Friedman" to under-score her credentials. (Her alma mater, Hillsdale College, had awarded her an honorary L.L.D. in 1938.) Then she returned to her own office, to her trusty desk and her fine coast guard col-leagues, relieved to be back "home."

Fresh piles of intercepts from South America awaited her there, and once again she dug in, eavesdropping on the latest activities of the spies. "Sargo" and "Alfredo" still appeared to be in charge of the network there, synthesizing information from their agents across Brazil and Chile and sending reports to Germany over the radio, but there was a new strain of malevolence in their messages. After Pearl Harbor, Brazil had declared solidarity with America, and the Nazis responded by going after Brazil, firing torpedoes at Brazilian ships for the first time. The positions of the ships were provided by "Sargo," "Alfredo," and their men. Outraged Brazilian authorities moved against German businesses. "Measures against members of the Axis are assuming drastic form," one spy in Brazil radioed to Germany. "Bank deposits already blocked. We are destroying all compromising documents, maintaining radio operation as long as possible. Heil Hitler." In January 1942 Hitler launched Operation Drumbeat, a coordinated U-boat assault on American and British merchant ships carrying war supplies, and "Sargo" and "Alfredo" helped with this effort, too. In three months the ruthless U-boats sent one million tons of material to the bottom of the sea and by summer 1942 the U-boat captains had murdered five thousand Allied seamen. "All along the Atlantic coast," writes the historian John Bryden, "Americans could look out and see plumes of smoke by day and red fires by night." The messages Elizebeth solved were dense with detail about Allied vessels coming and going in South American waters:

> MARCH 14, 1942 AT 0038
> Departed Montevideo: 4th (American SS) F.Q. *Barstow* to Curacao and (American SS) *Western Sword* to USA. Departed Rio de Janeiro: 11th (American SS) *Ruth* to Baltimore; 12th (American SS) *Lammot du Pont* to Buenos Aires. Arrived Rio de Janeiro: 12th (American SS) *Delmar* from New Orleans; and 13th (British MS) *Devis* from Glasgow.

Elizebeth passed these decrypts along the chain as quickly as she could, knowing that Nazi U-boats might already be hunting any of these U.S. or British ships and hoping that the Allied captains could be warned.

In the first weeks of March, Elizebeth also solved a sinister series of intercepts given to her by an FCC listening station on the coast of Rhode Island. The messages suggested the Nazis were preparing to destroy a troopship, the RMS *Queen Mary*, that was carrying 8,398 American servicemen:

> MARCH 7, 1942
> On board Queen Mary, Indians, Americans, Englishmen, tanks, disassembled airplanes. Came from Dutch Indies via South America.

> MARCH 8
> Queen Mary departed on March 8 1800 local time.

MARCH 12
The Queen Mary on the 11th at 1800 MEZ was reported by the ship Campeiro on the seas (near) Recife.

MARCH 13
The Queen Mary on the 12th at 1500 MEZ was reported near the coast at Ceara in the direction of Belem through Piratiny.

MARCH 14
Concerning Queen Mary, the troops of young people of white race number seven to eight thousand men.

As it turned out, Hitler had placed a bounty on the *Queen Mary:* any U-boat captain that destroyed her would win the Iron Cross with Oak Leaves and one million Reichsmarks. Elizebeth's decryptions (and similar decryptions provided by other Allied codebreaking units) were quickly shared with the *Queen Mary* captain, who was able to take evasive maneuvers, sneaking past a U-boat that was lurking in wait and saving the lives of more than eight thousand U.S. troops and his crew. It was a good example of why these clandestine radio circuits were important: as long as Elizebeth kept solving the messages, she could see danger coming, and America had an edge.

Which is why she grew increasingly confused in February and March when the spies started talking about being chased by police. Spies in Chile reported that their "hiding place has been searched three times." The men in São Paulo told Berlin that the temperature was 31 degrees Celsius and "getting worse"—in other words, they were feeling heat from police—then their transmitter went eerily quiet. In Rio, the Abwehr spy chief Albrecht Engels radioed, "Throughout country sharp police action against Germans." On March 17, 1942, he told Berlin that the docked Swiss ship whose radio they sometimes borrowed, the SS *Windhuk,* had been raided by Brazilian police, the second officer drowned in a struggle and the rest of the crew imprisoned.

Elizebeth, noticing this sharp uptick of panic in the messages she solved, guessed that authorities in Brazil and possibly Chile were conducting some kind of spy roundup, arresting Nazi agents and seizing their radio equipment. She couldn't tell if the FBI was leading the effort, local police, or a combination.

Whatever the case, it wasn't good. She and everyone else at the coast guard felt strongly that now was not the right time to move in and make arrests. The codebreakers were learning so much about the larger structure of the Nazi networks, more and more each day, and if the spies figured out that they were being chased because their codes had been broken, they would surely switch to new codes, perhaps stronger ones. And until Elizebeth managed to break the new codes, which could take weeks or months, depending on the level of difficulty, the Allies would be blind to Nazi activities across the continent. All of South America would suddenly go dark. If the spies started targeting another U.S. troopship like the *Queen Mary*, officials might not be able to warn the ship before a torpedo ripped into its hull.

A veteran codebreaker like Elizebeth understood these things. But the FBI, new to this line of work, did not, and before Elizebeth could figure out what was going on in South America with the police action and stop it from happening, FBI agents there grabbed the golden goose and cut off its head.

"Don't! You'll blow the house up!"

Josef Starziczny, a.k.a. "Lucas," the Abwehr agent in São Paulo, told the Brazilian detective to drop the suitcase. The detective set it down gently.

It was March 15, 1942. Elpido Reali had come to this house in the suburbs of São Paulo armed with a search warrant, intending to arrest a man he had been told was a Nazi spy. He knocked, entered, and immediately saw a spy camera, telephoto lenses, a darkroom, and a radio receiver. Starziczny's mistress was here, wearing a dressing gown, a confused look on her face.

"That suitcase," Reali said. "You said it will blow the house up."

Starziczny shook his head. No, there was no bomb inside.

Reali popped the latch and saw a portable radio transmitter.

A Kriegsmarine code book tumbled out. Starziczny had been using the codes to send the coordinates of Allied ships to German U-boats. Seeing the code book, the spy reached for a revolver on a nearby shelf, apparently planning to kill himself—"The Gestapo will never forgive me"—then thought better of it and allowed Reali to take him to the police station.

Soon after the arrest, in Rio, Albrecht Engels, a.k.a. "Alfredo," phoned Starziczny's house in São Paulo. Someone picked up the phone. Engels didn't recognize the person's voice. He hung up.

Engels assumed that Starziczny had been arrested and that he would not hold up under police interrogation (he was correct on both counts), and now Engels activated his emergency plan to protect the spy network he had spent months building under the guidance of Johannes Siegfried Becker, the brilliant SS captain. He arranged to move the radio transmitter in Rio to a new location, gave his code book and $89,000 in cash to a confederate, and fired off a string of wireless messages to Berlin warning them that the network was in danger, the last of these on March 18:

> *MEYER CLASEN in Porto Alegre arrested denounced LEO and ARNOLD thereupon ARNOLD arrested in Sao Paulo and transferred to Porto Alegre. I fear that MEYER (has) also given away (denounced) the radio procedure therefore I shall lie low until further notice.*

That day Engels was arrested by the Delegacia de Ordem Politica e Social (DOPS), the Brazilian federal police. They took him to one of their prisons, threw him into a dark cell with no toilet, told him to confess, and kept him there for weeks, punishing him with frequent interruptions to his sleep.

The DOPS took almost ninety members of the spy ring into custody over the next two months, at the insistence of the FBI's Jack West, the bureau's top man in Brazil, the head of its SIS operations across the country, and its legal attaché in Rio. West believed that the intensifying torpedo attacks on Allied merchant ships in the Atlantic meant it was time to take action against the spies who had been providing coordinates to the U-boat captains, and the bolder the action, the better. While the DOPS rounded up the spies, a young FCC employee named Robert Linx drove around Rio in an automobile full of direction-finding equipment, telling police where to find the clandestine radio transmitters, which were then seized and impounded.

According to *The Shadow War,* a 1986 account by the historians Leslie Rout and John Bratzel, "Jack West's conclusion was that piecemeal action was useless; a hard, sweeping blow" was the only way to take the Nazi radio stations off the air at once, to put the known spies in prison "and keep them off the airwaves."

But while the FBI's motive was sound, its tactics were questionable. The spies resisted their Brazilian interrogators. A month passed with few confessions. West grew impatient, suspecting that right-wing elements within the Brazilian police were working against him. And that's when he made the fateful decision to go above their heads. He took copies of the messages that Elizebeth and the coast guard had solved—verbatim copies of the spies' intercepted and decrypted radio messages—and showed *hundreds* of these messages to the president of Brazil, the foreign minister, and the air force minister. J. Edgar Hoover later confirmed that Allied agents "delivered the complete information" about the radios and the messages to the Brazilian government.

The gambit had the desired effect: Brazilian police started to get tougher on the prisoners. The men were stripped naked and questioned. Some police officers showed them pages of decrypts (Elizebeth's decrypts) and demanded that the prisoners fill in a handful of missing words. At least two men were beaten until unconscious. One had his fingers dislocated. One was repeatedly kneed in the scrotum while naked, and burned with cigarettes. One was interrogated in a six-by-three-foot cell with no bed. Police in São Paulo questioned one naked suspect for two days straight, pouring cold water on his skin and blasting a high-speed fan in his face until he lost control of his mental faculties. A few prisoners resisted. The Abwehr spy Friedrich Kempter went on a hunger strike after being given a plate of food full of rocks; the FBI grew concerned that Kempter would become too weak to talk and finally arranged a meal of steak and french fries. This was a rare occasion when the FBI intervened to stop brutality or torture; the rest of the time they either participated or looked the other way, in a grim foreshadowing of the bureau's future misadventures in Latin America during the 1970s and '80s.

The FBI's plan didn't work. The roundup failed. To deliver a deathblow to the Nazi network, to keep the spies off the airwaves, the FBI needed to get all the spies at once. But they didn't. Becker, the SS captain, remained at large—the FBI knew little about him anyway. Gustav Utzinger, the radio expert, also got away. He air-mailed one of the radio transmitters to Paraguay and boarded a Brazilian ship on a phony passport, picking up the transmitter in the Paraguayan capital of Asunción.

Even worse for the Allies, Albrecht Engels, the Rio businessman-turned-spy, was able to smuggle three long letters out of prison by passing them to the visiting wife of a colleague. In these letters he described the brutality of the police and alerted Berlin that the spies' codes had been broken and must be changed.

Berlin sounded the alarm across the clandestine network, telling all stations to interrupt communications with Brazil. "Warning," they radioed the spies in Chile. "Alfredo arrested. Take all precautionary measures, above all, separate."

Engels, trapped in his Rio prison cell, found comfort in the thought that the ones who escaped could now rebuild the network and make it more secure than ever, with new codes. He knew that Utzinger was still out there, somewhere. He wasn't sure about Becker. Engels asked a fellow prisoner who was about to be freed to go looking for Becker and send Engels a pack of cigarettes if Becker was safe. Soon afterward, a pack of cigarettes arrived at the prison for Engels. He was glad. As long as Becker and Utzinger, the *Hauptsturmführer* and the *Funkmeister,* were free, there was hope.

Circuit 3-N

A cipher message from Circuit 3-N, the Nazi clandestine radio link
between Argentina and Berlin, solved by Elizebeth's coast guard unit.

Villiam Friedman's depression returned in December 1941,
the month of Pearl Harbor. He had trouble sleeping and was
besieged by doubts and morbid thoughts. "Flight, fight, or neurosis," he wrote on a loose sheet of paper years later during a similar

period of depression, trying to describe the feeling. " 'Floating anxiety' which attaches itself to anything and everything. Fear that E. despises me for being such a weakling." It was scary for a man who prided himself on precision and rationality to feel like he was not in control of his mind or his body. He sometimes referred to this unpleasant condition as the "heebeegeebees," which he abbreviated as "hbgbs" in private notes to himself.

He did not seek help this time, did not go to a psychiatrist or check himself into a mental hospital—after his experience at the understaffed and punitive mental ward of Walter Reed in January 1941 he was not about to repeat this mistake unless completely desperate—and so, as always, the Friedmans concealed the seriousness of his condition to friends and family, and they continued to work, except for three consecutive days in the spring of 1942 when they stayed home to celebrate their twenty-fifth wedding anniversary. That was the celebration, sleeping in. It was amazing. They were so tired. Elizebeth went to the store and bought a whole chicken and some strawberries and figured she would cook their usual anniversary dinner, a simple feast of roast chicken and strawberry shortcake.

They hadn't told their friends about the twenty-five-year milestone but somehow the secret leaked, and that evening, to their delight, colleagues and friends knocked on their door, offering silver-anniversary gifts. Fred and Claire Barkley brought a sterling silver round sandwich tray; Jean Chase Ramsay wore a stunning silver dinner gown; Stub and Enid Perkins appeared with an array of flowers in a glass bowl, yellow and blue and white irises, blue delphinium, flame-colored columbine, white gypsophila. To these Elizebeth added pink and yellow roses she thought to pluck from her own rosebushes, and some white and yellow honeysuckle, too, and by the time the next-door neighbor brought two huge armfuls of his own scarlet roses, the house was dizzy with fragrance.

All day long, telegrams of congratulations arrived from friends near and far. Two of the telegrams were jokes written by William, notifying Elizebeth that she had been awarded an honorary A.B. degree from the Sorbonne, "Artiste de Boudoir," and also a D.S.M. from Harvard, "Doctor of Successful Marriage." In the second telegram he made light of his mental struggles and acknowledged his wife's patience and kindness during his periods of illness, though of course he did not use those words:

> *WHEREAS ELIZEBETH SMITH FRIEDMAN*
> *HAS CONDUCTED IMPORTANT SPECIALIZED*
> *RESEARCH EXTENDING OVER A PERIOD OF*
> *TWENTY FIVE YEARS IN THE VAGARIES AND*
> *IDIOSYNCRASIES OF ERRANT HUSBANDS; AND*
> *WHEREAS DURING THE CONDUCT OF SUCH*
> *RESEARCH SHE HAS BEEN SUBJECTED TO MANY*
> *HAZARDS INVOLVING CONSIDERABLE MENTAL*
> *ANGUISH, PERSONAL CHAGRIN, DAYS OF*
> *ANXIETY, AND NIGHTS OF SLEEPLESSNESS; AND*
> *WHEREAS SAID RESEARCH HAS RESULTED IN*
> *THE DEVELOPMENT OF ADEQUATE METHODS*
> *AND INSTRUMENTALITIES FOR THE CONTROL*
> *OF ONE HUSBAND, TO WIT, WILLIAM FREDERICK*
> *FRIEDMAN, AND HAS MADE HIM LIVABLE*
> *WITH . . .*

The children weren't there to celebrate with their parents. John Ramsay was finishing his sophomore year at prep school in central Pennsylvania, and upon graduation he planned to join the Army Air Corps and head straight to flight school. Barbara was between semesters of college and living in New York City, in an apartment on West Fifty-sixth Street, getting involved in leftist political causes and dating an activist named Hank. "Hank is beautiful," she wrote to William, "but we're so utterly different. He lived in the slums and led a gang (because he was the tallest and the biggest) and hated cops and swam in the East River. . . . And now we go to bars and stand at the rail with the workmen and talk about Leninism."

William had no interest in Leninism but told his daughter she had a good heart. "I hope you will let nothing interfere with your enthusiasm for helping where help is needed, but don't let the slow, snail's-pace progress upward and onward get you down," he wrote. "Remember always that the dawn of man's conscience is only 3 or 3½ thousand years behind us."

He had always found this a comforting thought, that the age of barbarism was not long past, that if humans failed to be kind it was because they were still children, historically speaking, and the idea rang true to him as he read and disseminated MAGIC intercepts through the spring of 1942, learning secrets about Japanese war strategy in the Pacific and helping to guide the American response. In June 1942, with the two opposing navies speeding toward a fatal clash at the Battle of Midway, William and his codebreakers moved from the Munitions Building to a new location, Arlington Hall, a former private school for girls located on the outskirts of the city. The army had taken over the hundred-acre campus to provide room for an expanded codebreaking operation. Meanwhile, the navy started transferring intelligence personnel to a similar facility, on Nebraska Avenue, also a former private girls' school, anchored by a five-story building dubbed the Naval Communications Annex.

These two campuses soon evolved into an American version of Bletchley Park, deeply secret compounds where workers solved puzzles behind barbed wire and never spoke about what they did. Many were women. It's where the machine era of cryptology began, the era of brute force, women operating machines the size of rooms, American bombes and some of the first IBM punch-card computers.

The women of Arlington Hall and the Naval Annex were mostly WACs and WAVES, members of the army and navy auxiliary programs designed to patch the wartime shortage of male labor. They lived together in barracks and apartments. Hundreds had been trained in secret cryptology courses offered at the Seven Sisters colleges, the likes of Bryn Mawr and Vassar and Mount Holyoke, the professors relying on exercises and concepts first pioneered by William and Elizebeth Friedman. Inside the high-security buildings encircled by barbed wire and guarded by U.S. Marines some of the women sat at long rows of desks, smoking and drinking coffee, and identifying cribs to feed into the bombes, while others operated the bombes that ticked and whirred as they explored the keyspaces of distant Enigma machines. The buildings were hot and unventilated. An Arlington Hall codebreaker named Martha Waller recalled that in the summer, it was often 90

degrees indoors at 8 A.M., and because of the wartime nylon shortage the women couldn't wear nylon stockings, so "we rejoiced in
going bare.... Sitting quietly at a desk, one could feel drops of
sweat rolling down one's legs."

Less than a year from now the navy would force Elizebeth and
the Coast Guard Cryptanalytic Unit to move to the Naval Annex,
but even then she would never work inside the large, hot rooms
with the rows of young women at desks. The unit remained separate, a small elite team doing its own thing. From time to time
she and her colleagues would take advantage of the Naval Annex's
technology to make progress on clandestine circuits, relying on
IBM punch-card machines to perform statistical analyses that
saved time in solving certain ciphers, but the initial assaults on the
puzzles emerged from their brains alone. Elizebeth was among the
last of the paper-and-pencil heroes. And in the summer of 1942, as
the U.S. and Japanese navies clashed in the Pacific and the Nazis
ordered French Jews to wear the Star of David, she was taking on
the most fiendish challenge of her career, for the biggest stakes.

Exactly as she had feared, the Nazi spies had changed their
codes after the March police raids in Brazil. As a direct result of
the FBI's roundup, "Germany was unmistakably informed that
the systems had been solved," wrote the uniformed commander
of her unit, Lieutenant Jones, and "the inevitable consequence was
that systems on all clandestine circuits were almost immediately
thereafter completely changed." Within two or three weeks the
Nazis were back online in multiple locations across South America, and from there the radio network expanded, adding nodes in
new cities and countries.

Elizebeth, Lieutenant Jones, and their coast guard teammates
watched in frustration as new circuits lit up throughout the summer and fall of 1942—two, then five, then fifteen—each using a
different and yet-unbroken code. It was as if the FBI had tried to
destroy an approaching asteroid with a single huge bomb but instead just blasted the rock into dozens of sentient fragments able
to regenerate and spread wreckage over a wider swath of earth.

The codebreakers were hardly the only Americans troubled by
the FBI's actions in South America. Intelligence chiefs at the army
and navy couldn't believe it either. "Unfortunately, the matter got

out of hand, and it became public knowledge that the ciphers used by the espionage agents in that territory were being read by our government," wrote Joseph Wenger, head of the navy's OP-20-G, in an internal memo. "It might be much more valuable to the military services to obtain the information flowing through clandestine stations than to close them up." The British were also taken aback—they had never trusted J. Edgar Hoover in the first place—and as more newspaper stories appeared in Brazil about the court cases of the arrested Nazi spies, local British officials had the stories translated and exchanged secret telegrams about the unfortunateness of it all. "You may care to read the attached rough translation of the story as it appeared in the Brazilian Press," one diplomat in Brazil wrote to an official at MI5, Britain's counterintelligence agency. "Rather shattering, I'm afraid."

The whole fiasco triggered four months of jurisdictional squabbles between the different intelligence services, conducted by memo and conference. The army, navy, and British complained about the FBI; the FBI pushed back. These fights resulted in a series of awkward compromises. The army and navy forced the State Department to promise that no further clandestine stations would be seized without their approval, and all parties recognized that the coast guard had the authority and the expertise to monitor clandestine circuits in the Western Hemisphere. But no one could stop the FBI from doing counterespionage in South America. Hoover had clear authority from the president.

And so, unable to remove his power on paper, the other agencies simply started to freeze him out in secret, routing information so that it flowed *around* the FBI as much as possible. British relations with the coast guard "grew steadily closer and more informal," according to the BSC history. "It was understood by both sides that no information received from the coast guard was to be divulged to the FBI." People literally whispered secrets to one another whenever an FBI representative was in the room; the British observed that "valuable items of intelligence were imparted, hurriedly and sotto voce, either before the meetings began or after they had adjourned."

In April, the navy ordered the coast guard to stop disseminating clandestine decrypts itself and provide the decrypts to

OP-20-G instead, for tighter control. That month, representatives of the army, navy, coast guard, British Security Co-ordination, and Canadian intelligence met at a weeklong conference in Washington to talk shop about radio and spies. It was chaired by Commander Wenger of OP-20-G. The FBI was not invited. Elizebeth was. On the day she explained the coast guard's approach, April 8, there were seven male naval officers in the room, three male army officers, four male Canadians, five male British, three male coast guard personnel, and her—listed at the top of her cohort in the meeting minutes:

U.S. Coast Guard
Mrs. Friedman
Lt. Comdr. Polio
Lt. Comdr. Peterson
Mr. Bishop

Her Cryptanalytic Unit had grown since Pearl Harbor, but it was still fairly small, with fewer than twenty cryptanalysts, translators, and clerks. They now worked as a team to fix the mess the FBI had caused: to break the new codes on the multiplying circuits and wrangle the chaos into some kind of order.

Until the arrests in March 1942, most spies had been using book ciphers or single transposition systems (a common, Scrabble-type cipher) that were similar in style and relatively easy to break. After March, the spies switched to weirder, harder stuff: running-key systems, double-transposition systems, poly-alphabetic substitution with columnar transposition. They started hopping on the radio at different times of the day in an attempt to avoid interception. They mixed and matched cipher methods in unpredictable ways. One of the new procedures relied on "rail-fencing"—a more sophisticated application of the same principle that Elizebeth and William once used for writing love notes, only now, instead of the plaintext being JE T'ADORE MON MARI or I LOVE YOU VERY MUCH, it was a snippet of German from a Nazi spy in France: HERZLICHE WEIHNACHTSGRÜßE UND WÜNSCHE ZUM NEUE JAHR: "Warm Christmas greetings and wishes for the new year . . ."

As soon as Elizabeth and Lieutenant Jones could get a handle on one circuit and break the codes, a new one came online. It was a cryptanalyst's nightmare. Here is a partial list of the circuits the unit was monitoring by the end of 1942:

3-G	Hamburg—Valparaíso (Chile)
3-J	Hamburg—South America
4-C	Lisbon (Portugal)—Lourenco Marques (Mozambique)
4-D	Madrid—West Africa
4-F	Hamburg—Lisbon (Portugal)
4-G	Stuttgart—Libya
4-H	Hamburg—Unknown
4-I	Hamburg—Bordeaux (France)
4-L	Hamburg—Gijon (northwest Spain)
4-M	Hamburg—Spain
4-N	Hamburg—Unknown
4-O	Berlin—Madrid
4-P	Hamburg—Madrid
4-Q	Hamburg—Tangier (Morocco)
4-R	Hamburg—Vigo (northwest Spain)
4-S	Berlin—Tetuan (Morocco)
4-T	Berlin—Teheran (Iran)
5-D	Hamburg—The Crimea (USSR)

This was a planet's worth of radio signals, a one-of-a-kind view of the earth as it convulsed with war, borders blurring, power shifting hands. The Nazis had invaded Crimea, in the Soviet Union. Iran was partially occupied by British and Soviet troops. Morocco was controlled by the Nazi-collaborationist Vichy government of France. Portugal was neutral and fiercely contested by both sides. Libya had been conquered, for the time being, by Italian and German troops. The Nazis had spies and saboteurs in all these places, sending back information over clandestine radio transmitters, and Nazi diplomats and even military officers sometimes borrowed the transmitters to speak with Hamburg and Berlin in times of transition and stress. The clandestine network, though built for espionage, was really just another communications channel, a way for Nazis of all sorts to share information in a fluid situation. Nikola Tesla predicted in 1926 that "when wireless is perfectly applied the whole earth will be converted into a huge brain." The clandestine network was the Nazi brain, fragmentary but already encircling the earth, and adding new synapses at a fearsome clip.

The information flickering through the brain wasn't necessarily accurate—some circuits seemed to contain little but "a miscellany of military information partly factual and partly distorted," wrote Lieutenant Jones in a 1944 memo, "evidently pieced together from barroom conversations with merchant seamen"—but even the garbage circuits were useful to monitor. "In all cases," Jones went on, "the reading of these circuits provided assurance that if any serious leak of vital information did occur we would learn of it almost before it became known in Germany and would be in a position to provide immediate safeguards." In other words, the Nazi brain served as a kind of early-warning system for the Allies. Beyond that, the solved messages provided a wealth of information about the Nazi grasp of American military capabilities, because when Germany asked the spies for information about U.S. ballistics or antiaircraft guns, they were revealing that this was information they lacked.

Elizebeth raced to stay on top of the shifting codes, the proliferating patterns. Her worksheets grew weird, beautiful. She filled the grid squares with letters and numbers that made different geometric shapes when you stepped back and looked at the worksheet from a distance. Some of the shapes were parallelograms, some looked like stairs, others like labyrinths. She pulled mischievous letters from the sky and sorted them on the page. The invisible world was all out of whack, misaligned, and she had this set of tricks to knock it back into order.

Once Elizebeth broke the code on a circuit, she solved every message that she came across, no matter how trivial or personal. Sinister messages, personal messages, sober messages that spoke of bombs and guns and ships and submarines, messages that just seemed bizarre. (Berlin to South America: "Can you procure details about the process of making explosives from cacao?" Cacao is the main ingredient of chocolate.) She learned intimate details of the spies' lives. Berlin informed a spy in Iceland that his wife, Erika, had given birth to a healthy baby girl, Jutta. Several times a family member of a spy was allowed to transmit a personal message from Berlin. One of the wives shared Elizebeth's first name, although she spelled it in the traditional way. "My dear JOHNY: My heartiest congratulations on your birthday and thousands of loving greetings and kisses from your Elizabeth."

Elizebeth Friedman solved and disseminated these messages like all the others. Such were her weapons against fascism: pencils, puzzles, circuits, names, dates, places, check marks, handwritten notes affixed to typed pages with staplers, stacks of solved messages rising with the hours and days and weeks.

And she stepped lightly as she worked. Her name did not appear on the documents of the Cryptanalytic Unit. She wasn't the top commander anymore, so she didn't write official memos to other parts of the intelligence community (Lieutenant Jones did that), and she didn't meet directly with FBI officials (Jones did that, too). Jones's name was the name typed on memos about clandestine radio traffic that circulated within the navy's OP-20-G. And although Elizebeth was not shy behind closed doors, sometimes quarreling with Jones about the direction of their work, disagreeing about which puzzles were more urgent or less urgent to tackle (she found his judgment clouded sometimes by careerism, a hunger for promotion that was irritating), she didn't mind being anonymous on the page. Her experiences as a cryptologic celebrity in the 1930s had convinced her that in this secret world, attention was a kind of poison. She said after the war that she considered herself "one of the workers" in the Cryptanalytic Unit—she was certainly being paid like a worker, not a leader, earning $4,200 a year, equal to $63,000 today, as a P-5 civil servant, a middle classification of government employee—and this combination of Elizebeth's tendency to minimize her own contributions and Jones's role as commander is one reason that her role in the war would go undiscovered for so many years.

Still, unavoidably, she left fingerprints in the records, little traces of her labor. Her initials, ESF, were typed at the bottom of some coast guard decrypts. Her handwriting appeared on many notes stapled to the decrypts, preceded by a code name, "GI-A," that meant "analyst" and was shared by others in the unit, and sometimes she wrote directly on the decrypts themselves, a characteristic burst of red or blue colored pencil next to a particularly crucial name or phrase, scribbled in the distinctive slant of Elizebeth's hand that had not changed since she was a college student writing in her diary.

At the end of the day, the unit was hers, not Jones's. She was the one who first proposed its creation in 1930. She staffed it. She trained the cryptanalysts. She guided its work over the years, building it into a powerhouse. This gave her an informal authority. People who had been paying attention knew that she was the beating heart of coast guard codebreaking. In December 1942, two British intelligence liasons met with Elizebeth in Washington to talk about closer coordination between Bletchley Park and the Coast Guard Cryptanalytic Unit. One of the liasons, Major G. G. Stevens, described the meeting in a "Most Secret" cable to his superiors in Britain and suggested that they arrange for a coast guard representative to visit Bletchley. "For this one would like to see Mrs. Friedman go," the British official wrote, "but probably at this end"—i.e., the Washington end—"it would be considered more suitable to send Lieutenant-Commander Jones, the official head of the section."

Given the tension of her job, the pressures of extreme secrecy, and the bleary afternoons stacking alphabets, Elizebeth was glad for her friendship with F. J. M. "Chubby" Stratton, the British astronomer and radio expert who looked like Santa Claus. He was a calming presence. He had a habit of appearing at Elizebeth's desk at unexpected times. She would be at her desk, nose buried in some problem, and look up to see him there, smiling. Without ever revealing anything about himself, he had a way of making you feel like you had known him forever, and that everything would be OK. And, of course, he was brilliant—a bit of a tinkerer. He loved the challenge of locating hidden radio stations. He had invented a device he called a "snifter," a little piece of electronics that fit in the pocket and allowed a person on foot to covertly pinpoint the exact location of a pirate radio transmitter within a building, accurate to the floor. He managed a staff of direction-finding experts who cooperated with the FCC to narrow down the location of pirate signals.

Through terrific effort, teamwork, and long hours, the coast guard codebreakers and their partners finally regained mastery over the clandestine circuits after an unpleasant period of blindness. By winter 1942, Elizebeth's eyes had adjusted. In the dark, lights out, she was watching them now, one letter of the alphabet at a time. She had the Nazi brain in a jar on her desk, alive and glistening, electrodes running out to her pencil. The unit had conquered every new circuit, every new code that came online that year.

Except one.

The circuit exchanged its first messages on October 10, 1942. The messages seemed to resist solution. She wondered if it might be an Enigma circuit, the messages encrypted by an Enigma machine of some kind.

She called it Circuit 3-N.

Presumably the messages on Circuit 3-N were sensitive enough to require a stronger-than-usual code, so she guessed that the messages must be important. She was more correct than she knew: in the end, the fate of the Invisible War would turn on the sinister frequencies of Circuit 3-N.

For now the anxiety of an unsolved puzzle was more than enough to motivate her. Here were some messages she couldn't read and she wanted to read them. The code simply had to be smashed.

At weekly radio intelligence meetings with her British and FCC colleagues, she talked about Circuit 3-N, and all the agencies chipped in with clues. The FCC determined that one of the stations was located in Europe and the other in South America, and the British confirmed the FCC's bearings.

New intercepts from Circuit 3-N arrived in Elizebeth's office each week. By December 1942 she had accumulated twenty-eight encrypted messages. A cursory analysis showed telltale signatures of an Enigma machine.

Elizebeth and the coast guard had already solved one Enigma, back in 1940, a commercial Enigma whose wiring scheme was already known. Now they were up against an Enigma that would turn out to be a machine of unknown wiring, never before solved. It was the real thing in the wild. She had to find a way in.

The Argentine sun felt good on his face and warmed his scalp through his thin layer of brown hair. Gustav Utzinger liked living in Buenos Aires. It was a beautiful old city—clean, musical, full of friendly people who did not ask too many questions. At a small shop in the heart of the city, 1511 Calle Donado, not far from the water, he built and repaired radios for legitimate paying clients.

He was twenty-eight. It did not seem crazy to imagine that he might survive the war, get married, and have children, either here or back in Germany. He had people in Berlin who adored him and worried about him. Utzinger's girlfriend worked at AMT VI, which gave her access to the radio transmitter, and she often sent him personal messages from her and from Utzinger's loved ones. She signed the messages "Blue Eye" and called him "Dark Eye." Sometimes "Blue Eye" transmitted a message written by Utzinger's grandmother in Berlin, who called herself "the Ahnfrau." He tried to take care of his grandmother from afar, sending packages of food from South America, tinned meat and coffee. She told him in radio messages that the packages were greatly appreciated and boosted her spirits. "I am sitting with . . . BLUE EYE and with a bottle of wine, celebrating your birthday," she wrote once. "On the one hand, I should like to have you here, on the other hand, I am proud of your accomplishments . . . received 2 packages this year. Was very pleased and thankful . . . I am well. Most cordial greetings. THE AHNFRAU." She asked him to remember to brush his teeth.

If it had been up to Gustav Utzinger, he may have spent the rest of the war running his little radio business there in Buenos Aires, making good money, placing advertisements in newspapers, doing everything on the up-and-up. But he was in the SS, the elite vanguard of the Nazi state, and the SS did not care about his personal dreams and ambitions, which is why, in the shop's basement, during the final months of 1942, he resumed his "natural patriotic efforts for my Fatherland," and made himself busy assembling and testing a new generation of clandestine radio sets, at the urgent request of his Nazi superiors.

It had been eight months since the FBI tried and failed to smash the Nazi network in Brazil. After the arrests, Utzinger had floated around for a while, trying to earn money to support himself. Escaping to Asunción, Paraguay, he got a job as a radio technician for the Paraguayan air force, which was commanded by an energetic fascist named Pablo Stagni. Under Stagni's wing, Utzinger built transmitters for the military and taught Paraguayan officers the fine points of radio. Sometimes he would see a few of his former Nazi colleagues kicking around Asunción, but all they did was talk—except for one day when they burned the reels of Charlie Chaplin's *The Great Dictator* in the public square.

It wasn't until Utzinger traveled to Buenos Aires that he got pulled back into the clandestine radio game.

He went there with Stagni, who was trying to convince the Nazi naval attaché, Dietrich Niebuhr, to sell him a gun sight for some ancient Krupp cannon that the Paraguayans had seized from the Bolivians during a previous war. Dietrich Niebuhr was far more interested in Utzinger than the cannon and started to bend the radio wizard's ear. Niebuhr said he was under extreme pressure from Berlin to build a clandestine radio transmitter here, and he needed Utzinger's help.

At this point in the war, Argentina was a far more amenable climate for clandestine radio than Brazil. The Brazilians had leaned toward the Allies as the war developed; Argentinians headed in the opposite direction. An American Jewish novelist named Waldo Frank toured Argentina in the summer of 1942 and noticed "a spawn of little nazi and nationalist papers," he wrote in an account of his journey. Frank traveled through cities and small towns delivering speeches about the value of democracy, and in every town a pro-Nazi newspaper attacked. He sensed that Argentina's conservative government, controlled by corrupt landowners, had "a very uncertain grip upon the country" and that fascists had infiltrated the police. By the end of Frank's tour, the government had declared him persona non grata, and before he could escape, five cops stormed into his hotel and beat him on the skull with truncheons.

Naturally the Nazis wanted to take advantage of this receptive atmosphere, especially since they were losing friends around the world at a rapid rate. Only two nations in the Western Hemisphere now maintained formal relations with the Reich—Chile and Argentina—and in early 1943, Chile would sever the link, making Argentina the lone holdout. This is why Berlin and Dietrich Niebuhr were desperate for a wireless link. Argentina was one of Germany's only listening posts in the hemisphere, one of the last places where it was still possible to obtain reliable intelligence about anything happening in the West, including the United States. They had to keep a line open at all costs.

Niebuhr gave Utzinger a powerful Seimens transmitter he had been keeping at the embassy, and in the fall of 1942, Utzinger took it to a small farm outside the city and began making tests. But as it turned out, Niebuhr wasn't the only Nazi in Argentina who needed radio assistance. Utzinger was also approached by an Abwehr spy named Hans Harnisch, code name "Boss," an employee of a German steel firm in Buenos Aires. He also wanted a wireless link to Berlin.

Then, in January 1943, the most mysterious spy of all suddenly appeared in Buenos Aires: *Hauptsturmführer* Siegfried Becker, a.k.a. "Sargo." He had stowed away on a Spanish ship, paying off the crew to smuggle him through customs.

And in his luggage, all the way from Germany, he had brought an Enigma cipher machine.

Becker looked the same as always—blond hair, nice clothes, a dirty mustache, bizarrely long fingernails—but Utzinger thought he detected a new gleam in Becker's eye. He was vibrating with ambition. Becker said that in the months ahead he would need Utzinger to transmit very sensitive information to Berlin. The "embassy crowd" must not be able to read the messages. Whatever Becker had in store for his next chapter, he intended to keep it close.

Utzinger now found himself in a nearly impossible position. Three men from three rival agencies had asked him for radio assistance: Niebuhr with the German embassy, Harnisch with the Abwehr, and Becker with the SS. In theory, he could create three separate radio stations, but this wasn't practical given the limitations of his own equipment. One powerful station was preferable to three weak ones. Instead of building three stations, then,

Utzinger decided to *trick Berlin into thinking that he had*. He designed a mythical radio organization that only existed on paper, a "Potemkin network," so that each agency in Germany would feel it controlled its own station in Argentina. He called this network "Bolivar" and told Berlin that it consisted of three sections, *Rot, Gruen,* and *Blau*—Red, Green, and Blue.

Red was Becker, Utzinger, and their SS collaborators.

Green was Hans Harnisch and the Abwehr.

Blue was the "embassy crowd."

The ruse gave Utzinger the freedom to run his operation as he saw fit without needing to explain his every technical choice to Berlin. He was determined to avoid the mistakes that had gotten the men arrested in Brazil, and with Becker's help he got to work. Together they recruited a new team of spies from Buenos Aires and surrounding towns, "42 loyal and seasoned collaborators," including German immigrants and working-class Argentines who believed in fascism. Utzinger selected new radiomen and drilled them in good security practices. Limit transmissions to short bursts; send decoy messages full of garbage text on prearranged frequencies; transmit at different times each week; never repeat the same message twice.

"I am teaching my boys tough wireless discipline," he radioed to Berlin. "The Yankees are copying every dot of our transmission."

In addition to the new radio link, the spies carved out another useful channel to Berlin, building a courier system that relied on Spanish sailors to smuggle packages and luggage on Spanish vessels. These men, paid in pesos and known as "wolves," allowed Becker and Utzinger to receive shipments of money, pharmaceuticals to sell for cash, and radio parts—items necessary to keep the network afloat. And through the wolf system and other sources, Utzinger also obtained crypto machines. He now had two Enigmas at his disposal, plus a pocket watch–size Kryha device called a Liliput, small and light and easily concealed.

By the end of February 1943, Utzinger and Becker had everything in place: a new radio system; crypto machines, including Enigmas, the best available, each message a perfect tiny fortress;

a team of collaborators. They were ready to transmit reports to Germany in volume. On February 28, Becker hopped on the radio and sent Berlin a cheerful progress update:

> *New organization established with LUNA. LUNA has*
> *assembled in splendid fashion a circle of co-workers. Spread*
> *and prepared so that I am able to start immediately with*
> *work according to plan. SARGENTO.*

Berlin replied with glee. "Old boy, now we are off," they radioed. "Test message. Cordial greetings to all. We are awaiting your blind traffic there on Monday, Wednesday and Saturday at 0200 and 0400."

The spies' SS leaders back home, the officers of AMT VI, could not have been happier to hear that a door was opening in Argentina. The winter months of late 1942 and 1943 had been grim ones in Germany. The Red Army was crushing the Wehrmacht in Stalingrad, and although Nazi censors blocked reports of losses, rumors leaked. In the snow-dusted cities of Germany, it was difficult to find toothbrushes, belts, bicycle tires, and toilet paper. Restaurants complained of patrons stealing glasses. On February 18, at a rally of twenty thousand in Berlin, Minister of Propaganda Joseph Goebbels finally admitted that the Battle of Stalingrad was lost and called for "a war more total and radical than anything that we can even imagine today," an apocalyptic death struggle against the Allies and against Jewry.

All in the homeland that season was darkness, danger, frost—and yet in Argentina, Becker and Utzinger reported, it was the height of summer, and they were making interesting friends.

Becker, for instance, was cultivating a relationship with Juan Domingo Perón, the future three-time president of Argentina, now just a young army colonel with a taste for moral larceny. (He lived with a fourteen-year-old girlfriend whom he called "The Piranha.") Perón belonged to a secret lodge of military officers, the United Officers' Group (GOU), that aimed to overthrow the Argentine president, and he was already thinking beyond his own nation. Inspired by Hitler's domination of Europe, Perón imagined himself at the head of a nationalist movement sweeping across all of South America.

His ambition exceeded his reach. He didn't yet have the contacts that a wider revolution would require. But Siegfried Becker did. During his decade with the SS, working in Germany, Brazil, and Argentina, Becker had gotten to know all kinds of influential South Americans: among them lieutenants, generals, diplomats, and police captains. He carried a small notebook with their phone numbers. These contacts—Becker's little black book of nationalists and fascists—made him an alluring figure to Perón and his friends, and soon the Argentines hashed out an informal deal with the Nazi spies.

Each side would get something it wanted. The Argentines would share secrets about the United States and protect the spies from the FBI and other Allied law agencies. In exchange, the spies would operate behind the scenes to extend Argentina's influence across South America, connecting revolutionaries in one country to like-minded men in another. They would plot coups, overthrow governments, and install fascist-friendly regimes.

The ultimate goal was to assemble a bloc of nations aligned against the United States. "Hitler's struggle in war and in peace will be our guide," Perón and his GOU plotters wrote in a covert manifesto. "With Argentina, Paraguay, Bolivia, and Chile, it will be easy to pressure Uruguay. Then the five united nations will easily draw in Brazil because of its type of government and its large nucleus of Germans."

With Brazil fallen, the "continent will be ours."

At her coast guard desk, Elizebeth reached for a fresh sheet of grid paper. Circuit 3-N. Argentina to Berlin. The unknown Enigma machine.

Twenty-eight unsolved messages from Circuit 3-N now sat in a pile on her desk. She wrote the twenty-eight ciphertexts on the worksheet in pencil, one on top of another, assembling a stack of text so she could solve the messages in depth, like she had done in 1940 to solve the commercial Enigma machine.

The twenty-eight messages all appeared to use the same key—a huge gift to the codebreakers from their Nazi adversaries. It made things easier and allowed Elizabeth to begin solving the individual messages.

She made a frequency count of the letters in the columns. The cipher letter *H* appeared seven times in column no. 2, four times in column no. 3, and so on. This was enough to start guessing at the plaintext letters. Each column was its own MASC, a monoalphabetic cipher, and she hopped around, penciling plain letters in the columns and guessing at German words across the rows, like *bericht* (report) and *wir hoeren* (we hear).

She was following the path she had blazed in 1940 with the commercial Enigma, using her experience with that old machine to discover "an entering wedge" with this new one, and then hammering the wedge until the damn thing split. The team was still relying on the geometry of patterns. Step one: Line up the messages one on top of another. Step two: Solve the plaintexts in depth. Step three: Use the resulting alphabets to deduce the wheel wiring. Step four: Exploit the wiring knowledge to reveal the new keys whenever the adversary changes them.

This time, the coast guard had some competition. Across the sea, at Bletchley Park, the secret British codebreaking campus, a unit called Intelligence Service Knox (ISK) was attacking the same Enigma, trying to solve it independently of the coast guard. Bletchley had been breaking spy ciphers since the start of the war. Now the two Allied teams each solved this new Enigma at about the same time, in December 1942, by different methods that ultimately got them to the same place.

The machine turned out to be a G-model Enigma designed for the Abwehr, similar to the commercial Enigma but with wheels that stepped less regularly. Within the high-security universe of Enigmas, it was a medium-security model, more puzzling than the commercial machine but less so than other Enigmas. Decades later, Elizabeth called the G-model a "less superior Enigma that was used by Germany and her confidential agents—her spies." At the moment, however, solving the machine felt like victory. "There was much celebration," an NSA historian reported after the war, in an interview with Elizabeth.

Now the coast guard could read the messages backward and forward—the old messages already intercepted and the new ones incoming. Looking at the plaintexts, the codebreakers confirmed that the South American end of the circuit was in Buenos Aires, Argentina.

Also, the names of three colors popped out in the messages: Green, Red, and Blue. The spies were tagging their notes by these color names, and each color seemed to represent a different spy leader using a different code.

It was as if multiple groups of Nazi agents had converged around Buenos Aires and were sharing this one circuit, pooling resources to make a kind of last stand.

Having broken into the Green messages, the coast guard codebreakers turned their gaze to the Red traffic, which they already knew was encrypted with a device that the Germans called Lily. "Following messages all enciphered with LILY," Berlin had radioed in February 1943. It seemed clear to Elizebeth and her coworkers that Lily was short for Liliput, as in a Kryha Liliput, a miniature version of the German cipher device.

The Liliput was somewhat more complex than earlier Kryha models, but as a rule, Kryhas were less secure than Enigmas. Elizebeth walked to her shelf, picked up a Kryha, and examined it with her colleagues. It was an older model, but the principle was the same. Two concentric alphabet wheels stepped against each other, the stepping regulated by a control wheel set to a certain starting position. After a month of work, the codebreakers recovered the key with the help of punch cards, the IBM crunching and tabulating the frequencies of certain juxtapositions of cipher letters that helped them understand how the letters were distributed and offered clues about possible keys that were then tested to see if they produced sensible plaintext.

Now the coast guard was able to read the bulk of the circuit's traffic, the Red messages as well as the Green. (They would never solve the Blue messages, but it didn't matter.)

And what two names popped out in the plaintexts?

"Sargo" and "Luna," Elizebeth's old friends.

She recognized the aliases of the spy and the radio wizard. She had first encountered them on the Brazil circuits in 1942, and now, apparently, the duo had reunited in Argentina.

This alone meant that Circuit 3-N was important—the presence of these two important individuals, "Sargo" and "Luna." But there was something else that concerned her. The men seemed to be building a completely new wireless organization, bigger than before. They were hiring radio operators and obsessively testing new radio equipment.

"We have antenna 100 meters long beamed on Berlin," the spies tapped out to Berlin one day. "Hope you like it."

It was clear to Elizebeth that the Nazi network in the West had shifted to Argentina, and that the men were preparing the ground for a significant espionage effort. Something big was about to happen.

Unfortunately for Elizebeth, right as she was starting to figure this out, the navy forced her to interrupt her work and pack up her office.

In March 1943, the coast guard codebreakers were ordered to move from their longtime home at the Treasury to the former girls' school now known as the Naval Communications Annex, the temporary wartime facility on Nebraska Avenue. Elizebeth had to pause her assault on Circuit 3-N while she got settled in the new location. She claimed an office on the second floor and spoke little to anyone outside of her own team. A coast guard employee who worked two doors down from her in the Annex later recalled, "Our work was compartmented. We went as a group for lunch, but the conversation was never about work, but concerned current events—the way the war was going, etc." Elizebeth worked a little with a few of the SPARS, members of the coast guard women's auxiliary, and she sometimes interacted with the young WAVES and WACs on a purely social basis, outside of the Naval Annex, inviting them to her home for tea and asking how they were getting along in Washington. One evening, Martha Waller, the codebreaker at Arlington Hall, returned home to her D.C. apartment to find Elizebeth and William Friedman sitting at dinner with five of her roommates. One of the roommates' father apparently knew William and had set up the dinner. Waller was dumbstruck. The Friedmans, she had heard, were legends. "I think I was mesmerized, and I know I said next to nothing,"

Waller recalls. Here is what she remembers about Elizebeth: "She looked, sounded, and behaved like the professor of English she might well have become."

The month after Elizebeth relocated to the Naval Annex, her husband left the country on his first big mission since his mental breakdown two years earlier. William Friedman traveled to Bletchley Park at the request of the army, which wanted him to negotiate an agreement to share information, expertise, and blueprints for building bombes. He succeeded, but his depression came back while he was overseas, manifesting as insomnia. He swallowed pink Amytal pills to knock himself out and in a personal diary of the trip he repeatedly mentioned sleep problems:

> TUESDAY APRIL 27 . . . *Bed at 10:30 but too tired for good sleeping.*
> SUNDAY MAY 2 . . . *Poor sleeping for some reason or other, maybe tetanus shot still working.*
> MONDAY, MAY 24TH . . . *Got up at about 1 a.m. and took two small pills from Washington cache but didn't do much good. Awoke early & not at all refreshed. Guess this work is very exhausting mentally & I hope to get through with it soon.*
> MONDAY, MAY 24TH . . . *I've noticed that on days when I am "tense" & have "heebeegeebees" I sleep well in night but when don't have them, sleep not so good. —Haven't had hbgbs for many days now. Wish I could solve this mystery of myself.*

Elizebeth didn't see these diary entries but imagined that William must be under great stress, and she worried about his mental state. Throughout April and May, she wrote him constantly while he was gone, sending at least fourteen letters to the military attaché at the American embassy in London for delivery to his secret location. In the letters she focused on small details of life in Washington: their lawn that was cracked and dried from the spring sun; a party with some family friends where they all drank too many old-fashioneds and stayed up until 2 A.M. singing songs around the piano. "I will create pictures for you of the scenes and people present and so seem like a momentary return for you." She mentioned a mutual friend, Colonel John McGrail of the signal corps, one of the intelligence professionals who had written critical annotations in William's copy of *The American Black Chamber*. Like William, McGrail was brilliant, prone to depression, and kind; he sent Elizebeth a corsage of violets on Easter, "the dear sweet thing," Elizebeth wrote to William, adding that McGrail "seemed more depressed than ever." She said that William's Telechron machine in the library, an electric clock, seemed to be ticking at faster-than-normal speed: "Even your telechron misses you. It is running crazily."

She mentioned that she had been skipping breakfast to stay slim and was smoking cigarettes in the evenings, a rare admission that she herself was stressed and overwhelmed by the mental strain of her job. One summer day, the temperature inside the Naval Annex reached 110 degrees, with no air conditioning. She was dripping sweat onto her worksheets. Everyone was. The navy commanders declined to send people home, saying that a war was on.

There was a real danger that if Elizebeth relaxed, even for a week, her Cryptanalytic Unit might fall behind and never catch up. Throughout the summer of 1943, the Nazis rolled out extensive reforms of their crypto systems to improve security. The orders were issued by a respected Nazi cryptographer, Fritz Menzer, who led a staff of twenty-five cipher experts in the High Command of the Wehrmacht. Spies who were still relying on hand ciphers discarded their old methods for *Oberinspektor* Menzer's new procedures, and the coast guard codebreakers had to adapt. They worked furiously to solve "Procedure 62," a double transposition system based on a thirty-one-letter key phrase. In this one, the letters of the message were scrambled once and then scrambled again, and only unscrambled by means of a key phrase that was itself scrambled based on the number of the month.

Clandestine circuits in Spanish-speaking countries began to use Procedure 62. Another new crypto system, "Procedure 40," adopted by spies in Madrid and elsewhere, was a substitution method combined with double transposition, both steps sharing the same key phrase. In the substitution step, the phrase was written in a 5-by-5 square of letters. One such key phrase that Elizebeth's team unearthed was the Spanish proverb *donde menos se piensa salta la liebra*. Written in the 5-by-5 square, skipping all repeated letters, it looked like this:

```
D O N E M
S P I A L
T B R C F
G H K Q U
V W X Y Z
```

Literally the proverb means "Where least expected, the hare jumps," though a better translation might be "Opportunity knocks where it is least expected."

And whatever else demanded her attention during the day, Elizebeth kept going back to Circuit 3-N, the link between Argentina and Berlin.

She was more convinced than ever that Circuit 3-N was the most important circuit of all. The volume of traffic abruptly rose in April and continued to expand every week. Some days they sent as many as fifteen different messages to Berlin and just as many replies traveled in the opposite direction—an explosion of new leads for Elizebeth and her team to chase down, a deepening abyss of Nazi text.

Germany seemed as hungry as ever for information about the United States, asking questions about U.S. weapons capabilities and political figures. ("Are there differences of opinion between Roosevelt and his Jewish advisers, above all Roseman, Morgenthau and Frankfurter?") Berlin often sent requests to the spies in long numbered lists:

1. The Fisher Co. in Detroit reportedly constructed a new anti-aircraft gun of about 12-centimeter caliber, which is fired by remote control. Urgent question: Construction, mode of action, performances.

2. What is manufactured in [Henry] Ford's shop in Iron Mountain? Size of the plant? Since when? Monthly production?

3. The USA armor bombs: What caliber? Kind of the material? Cross section plan, explosive charge, quantity of (5 letters garbled) and detonator. Is "Explosive D" used?

4. Details on new development and production of armor-piercing arms and in this, air bombs of USA and England in particular.

5. Details and particular on development and introduction of rocket weapons . . .

Such messages were concerning, of course, but they were also straightforward and familiar. Elizebeth had seen hundreds like them on other circuits. What was completely new and sinister about Circuit 3-N was the *political* intrigue shining through the plaintexts—a whole other level of conspiracy and malice. "Sargo" and "Luna" weren't just two guys transmitting the shipping news anymore. They had friends across the continent, poised to act in the name of revolution. They were building a secret army.

For one, it was clear to Elizebeth that the spies had forged an intimate working relationship with powerful Argentine figures. On June 4, 1943, a group of generals had occupied the *presidente*'s mansion in Buenos Aires, the Casa Rosada, deposing the old regime and installing a new *presidente,* Pedro Pablo Ramírez. The messages contained references to Ramírez (the Nazis called him "Godes") and other coup figures, including Juan Perón and Captain Eduardo Aumann, code name "Moreno," now a high official in the Argentine foreign ministry. Elizebeth noticed that a Nazi agent named "Boss" (Abwehr leader Hans Harnisch) was regularly meeting in secret with these men. "Another important conference with [*Presidente*] GODES revealed his willingness for energetic collaboration in the interests of the Axis powers," "Boss" radioed to Berlin on July 24, 1943, adding later that Aumann and others were "ready in every respect to promote mutual interests," and that "the USA is considered greatest enemy."

It wasn't shocking for Elizebeth to learn that Argentina, a supposedly neutral country, was cooperating with Germany behind closed doors. But the *scope* of the cooperation was surprisingly extensive. She was seeing glimmers of incredible clandestine missions. "Boss" wrote in one message, "Through our efforts Argentine Government has established close contact with nationalist groups in Chile, Bolivia and Paraguay; even with Brazil through V-men residing here." Argentina alone was not enough for the Nazis: They were conspiring to overthrow the governments of Bolivia, Chile, Paraguay, and Brazil. They were trying to run the table, to turn the continent fascist.

"Sargo" appeared to be the man behind the scenes, the invisible agent traveling from country to country, meeting with revolutionaries, bringing money and information, linking them to one another. He told Berlin he had secured the cooperation of the Paraguayan air force chief, Stagni, who "is completely in our camp" and was glad to fly him around the continent on Paraguayan planes. "Sargo" said he was optimistic about the prospects for a military coup in Bolivia, where he had formed a cell of conspirators with the Bolivian minister of mines and an attaché named Elias Belmonte. And as "Sargo" schemed and plotted in the dark, working on the gritty details of revolution, his colleague "Boss" continued to meet with the Argentines and talk about the big picture. In late August "Boss" attended "a secret session of high officers and officials" of the Argentine government and radioed the following to Berlin:

> Final objective is said to be formation of a bloc of South American countries, which would itself protect its interests, without tutelage of others who pretend to do this. Bolivia must not only be freed from USA influence, but also establish social justice. Argentine can carry out this together with or even in spite of the Bolivian government. Great progress was achieved in negotiations with Chile and further improvement is to be expected. The days of the Rios government with its ambiguous leftist policy are numbered. Chilean military circles had prepared everything in order to follow Argentina's example [i.e., to launch a coup] . . . haste is necessary . . . there is dissatisfaction with [Paraguayan president] Moringo. It was intimated that change in government there is not excluded . . .

Perhaps boldest of all, the spies seemed to be arranging a secret weapons deal between Argentina and Nazi Germany. They were trying to figure out a way to get guns and bombs from Berlin to Buenos Aires without the Allies knowing.

Elizebeth was able to follow every twist of the weapons deal in the plaintexts. The details, she discovered, would be negotiated in Berlin by an Argentine envoy, a local man who would sail from Buenos Aires under diplomatic cover. He had been promised meetings with Himmler and Hitler. "An agent will depart from Argentina to Germany," one of the spies informed Berlin in July. "Name, rank, mission to follow."

In a subsequent message, Elizebeth learned the man's name: Osmar Hellmuth.

Donde menos se piensa salta la liebra. She heard a knock.

Osmar Hellmuth had never felt so important before. He had never done anything quite this exciting. One minute you are at the German Club of Buenos Aires, having a nice conversation with some nice German fellows in riding boots, and the next minute you are back at your flat on the Calle Esmerelda with some different German fellows, discussing an international weapons deal, and you are introduced to a man with unusually long and curling fingernails and told that "this gentleman" will make all arrangements for you to sail across the world for a private audience with the Führer.

Hellmuth—forty, not too bright, heavyset, with a red mustache and red hair combed straight back—was "easy prey" for the Nazi spy ring, British officials would later conclude. A low-ranking naval officer, he handled minor diplomatic duties for the Argentine government. His portfolio was not enough to make him powerful. But it was enough to afford the kinds of diplomatic protections that brought him to the notice of Siegfried Becker.

It was the summer of 1943, and Becker's ambition was growing with his power. He now lived in an upscale neighborhood of Buenos Aires, not far from the palatial home of Juan Perón, whose star had been rising after the coup.

The two men spoke often, in secret, which is how Becker came to understand that the Argentines had a pressing desire for weapons. Fearing an invasion from Brazil, their enemy and main rival on the continent, they wanted Becker to help procure weapons from Berlin. This was a complicated request. At this stage of the war, weapons were hard to come by—Germany could not easily spare them—and Becker could not simply ask Berlin for weapons because the foreign ministry was controlled by a bureaucratic rival of SS foreign intelligence. The only way to get the weapons, then, was in secret, without "the embassy crowd" finding out.

This is when Becker approached Osmar Hellmuth, the former insurance salesman, and made him an offer.

Becker explained to the naive Argentine that the sale of weapons had to be negotiated in person. An envoy was needed, an intermediary. If Hellmuth agreed to accept the mission, Becker explained, he would board a ship in Buenos Aires, the *Cabo de Hornos*. The ship would sail to the port of Trinidad on the northern shore of the continent, where British officials searched all vessels bound for Europe, and then depart for Spain. Upon Hellmuth's arrival in Bilbao, Spain, he should check into the Hotel Carlton and wait for an SS agent to approach and speak the words "Greetings from Siegfried Becker." Hellmuth was to reply, "Ah! The *Hauptstürmfuhrer*!" The SS man in Bilbao would then arrange for Hellmuth to meet with Nazi leaders, including Himmler and Hitler, and after the deal was arranged, Hellmuth would be rewarded with a cushy consular job in Barcelona.

"I had a marvellous opportunity to go to Europe," Hellmuth explained later, "free of expense with a good salary, on an extremely interesting mission, and with good prospects." He didn't understand the risks. Becker did. Becker just assumed that if the mission failed, or if the Allies found out about it, no one would be able to trace its genesis back to him. In a brilliant espionage career, this may have been his one fatal mistake.

Becker wasn't alone in his miscalculations that summer. Gustav Utzinger, the ring's radio expert, was also failing to grasp the danger of his position. The Americans—Elizebeth Friedman and her team—had broken his cipher machines and were now reading his every transmission, but for Utzinger, the prospect of a Yankee breaking an Enigma machine was beyond his comprehension.

This wasn't to say that he slept soundly at night; like any good radio expert, Utzinger lived in a fog of professional paranoia. He simply assumed that any breaches of his network must be the fault of his incompetent counterparts in Berlin, a consequence of their "dilettantism and lack of imagination," in his words. They were always making stupid mistakes. They stayed on the airwaves for too long and repeated messages. One evening in July, the radio operator in Berlin transmitted the same message, "OK HELLO," for fifteen straight minutes over the same frequency. Utzinger scolded Berlin: "The enemy has such an easy time!"

Gradually, Utzinger's mood improved. By November 1943, the network seemed poised for a string of terrific successes. Becker assured him that it was almost time for the plotters in Bolivia and elsewhere to activate, the coups to be attempted. That month, Becker befriended a Chilean gunnery sergeant who had just returned from the United States after completing a one-year weapons course offered by the U.S. Navy. The Chilean's descriptions of U.S. naval capabilities and tactics, relayed to Berlin over the radio, amounted to a detailed American game plan against Nazi ships and U-boats:

In day fighting, the heavy artillery is said to use the following manner of ranging fire: salvos every 7 seconds. The first salvo 300 yards over the distance, measured by radar. The second salvo the radar distance. The fifth and sixth salvos 200 yards shorter or longer than the fourth salvo. In night fighting, the ranging is carried out by ladder and radar . . . Each shell contains aniline compound which produces intensive coloring of the water, so that the location of the fire can be observed . . . Depth charges [against U-boats] resemble English Vickers [depth charges]. Models with 300 and 600 lb. of TNT charge. The fuze contains 3.25 lbs. of granular TNT . . . exterior depth setting for 30, 50, 75, 100, 150, 200, 250, 300 feet . . .

The spies wrote this message and the others with their Enigma, ALCSA JYFMK JFNVH KYOIM, transmitting the letters in Morse, dot-dash (A), dot-dash-dot-dot (L), dash-dot-dash-dot (C), not suspecting that in Washington, an American woman was sitting at a loom, weaving these ugly loops of letters into a sensible fabric of plaintext.

Becker and Utzinger thought they had everything wrapped up tightly. As far as they could see, there was only one major loose thread in their system, one element beyond their control: Osmar Hellmuth, the former insurance salesman, about to become a diplomat abroad.

During the final days of September, Becker handed Hellmuth a letter detailing some precision radio instruments they wanted him to purchase and bring back. He also gave Hellmuth several trunks containing sixty kilograms of gifts for friends in Germany. He radioed to Berlin that Hellmuth was on his way. "HELLMUTH enjoys the absolute confidence of the Argentine Government," Becker wrote. "He is going to bring you lists of the government's wishes . . . The Argentine Government demands strictest secrecy about the mission." One last time, in Buenos Aires, the *Hauptsturmführer* wished his man luck, and on October 2, 1943, Osmar Hellmuth sailed into a faultless blue sea.

Elizebeth and her teammates solved the message about the Chilean sergeant who infiltrated the navy's gunnery school and escaped with its secrets. They watched in real time as the confidential agents of Germany and Argentina conspired to flip the chessboard of global politics together. Thanks to the Cryptanalytic Unit's successful assault on Circuit 3-N, Elizebeth understood the structure of Nazi espionage in South America. The money, the actors, the codes, the connections—she had the map now.

She forwarded her decrypts along the chain as fast as possible, often annotating them with brief written notes describing the named agents and explaining what they seemed to be doing.

Luna is probably Gustav Utzinger, the radio expert of the spy ring.

This is the latest in a series of messages dealing with the attempt of the German espionage ring, members of the Argentine general staff and the Brazilian integralists to form an anti-US bloc in South America.

These were ULTRA messages, stamped TOP SECRET ULTRA at the top.

By now, thanks to Elizebeth's ULTRA decrypts and those of British codebreakers, everyone in the English-speaking spy world knew that Argentina was conspiring with the Nazis. The ULTRA messages were clear.

But they were also forbidden fruit. The Allies would have liked to show the decrypts to the Argentine government and demand that they stop. But then, of course, the Argentines would know that the Allies had broken the Nazi codes, and then the Argentines would tell the Nazis, and the lifeblood of ULTRA would instantly stop flowing through the world of Allied intelligence—a catastrophe.

There seemed to be no way out of this bind until October 1943, when Elizebeth solved messages from Circuit 3-N describing plans to dispatch an envoy named Osmar Hellmuth to negotiate a weapons deal between Germany and Argentina. This seemed to provide the Allies with an unprecedented opportunity. A relatively obscure Argentine man, this Hellmuth, was working very closely with Nazis, and he was about to get on a ship and sail to Bilbao, Spain. Perhaps he could be intercepted en route and forced to divulge his Nazi contacts. The Allies could say they got the information from the confession, not the Enigma messages that Elizebeth had solved. Then British and U.S. officials could finally take steps to disrupt the Nazi network in Argentina without exposing the ULTRA secret.

This plan required a bold and possibly illegal act by the British. Hellmuth was a diplomat. He had protections. He could not simply be kidnapped. Could he?

The British yanked Osmar Hellmuth off his ship in the middle of the night. They ignored his diplomatic privileges. They did not listen to his protests and did not allow him to make a phone call to his embassy. The *Cabo de Hornos* had been docked in the port of Trinidad before departing for Spain. The British placed Hellmuth on a different ship, bound for England. It sailed east across the Atlantic and arrived in Portsmouth, England, on November 12, 1943. From there the confused and indignant Hellmuth was carried to a secret interrogation facility in a mansion in southwest London called Camp 020, part of a network of nine interrogation centers operated during the war by MI5, British counterintelligence.

Argentine officials made frantic calls to British diplomats and asked where their citizen was. The British diplomats claimed they didn't know.

He was placed in solitary confinement at Camp 020 for the first two weeks. The guards confiscated his seven trunks of gifts for German officials and the letter with the details of the precision instruments he was supposed to purchase. Then the prisoner was brought to see the commandant of the facility, Colonel Robin Stephens, a broad-shouldered man who wore the tan wool jacket of a British military commander, medals pinned above the left breast. A monocle affixed to his right eye made him look like a broken owl. His men called him "Tin Eye." His motto regarding the art of interrogation was "truth in the shortest possible time." He always said he did not believe in torture—he claimed he was so skilled at eliciting confessions that he did not need to use it— but after the war former Camp 020 prisoners told credible stories of being beaten by Stephens's men, whipped, subjected to mock executions, deprived of sleep for long periods, made to stand in excruciating "stress positions," and starved.

"I am speaking with authority," the commandant told Hellmuth, "and full authority of Great Britain in war. My observations do not invite any replies from you and I shall regard any interruption as an incipient indiscipline."

Stephens started to berate the prisoner. He said that the British had confiscated his letter asking him to buy precision instruments for the spies of the Reich, proving that he was a Nazi spy. He said Hellmuth was a Nazi stooge, in way over his head, playing a game he did not understand. Stephens made fun of the gifts in Hellmuth's trunks: "There are about seven trunks of gifts. Food for the pot-bellied masters of Germany. Silk stockings for their clod-hopping women. Chocolates for their sniveling children." Stephens said that Hellmuth's only chance for escaping this facility was to tell the full truth and reveal everything he knew about the Nazi spy network in South America. "When you lie, we shall know instantly."

Hellmuth stalled for a few days as the interrogators of Camp 020 demanded information. Initially the British found him to be "possessed, almost arrogant." Hellmuth claimed he did not know much about SS intelligence activities in Argentina; he was familiar with one or two Germans from social circles and that was all. The interrogators asked about the man they knew only as "Sargo"—the man they suspected was "the head of Himmler's faction in Buenos Aires" and "the prime mover of the secret mission." The power behind the scenes, the Nazi mastermind in South America. Who is "Sargo"? What is his real name? What is his rank? Hellmuth said he did not know. He began so many sentences with the word *probablemente*, "probably," that the interrogators became infuriated and banned him from saying it ever again.

Eventually, the interrogators said that if he didn't start talking, his treatment in the camp "must necessarily deteriorate," a hint he would be tortured.

Hellmuth softened. He told the British about how Argentina had worked with the SS to depose the Bolivian government and install a Nazi-friendly dictator. He told them about the clandestine radio stations and the radio technician who ran them, Gustav Utzinger, code name "Luna."

Most important of all, Osmar Hellmuth divulged the secret that his British interrogators had been burning to know. Hellmuth's confession, made possible by Elizebeth's decryptions, would soon spark a fantastic chain of global events, forever turning the tide of the Invisible War in the Allies' favor.

Who is "Sargo"?

Poor and unlucky Osmar Hellmuth, cold and alone, had been wondering for weeks if his personal sense of honor was worth this suffering, and now he decided it was not. "Sargo," he told his captors, was Siegfried Becker, an SS captain, age thirty-two, five foot ten with a strong build and blond hair.

CHAPTER 5

The Doll Lady

While the Nazi spies waited for news of Hellmuth's mission, not knowing that he had been kidnapped, they upgraded their security procedures in Argentina.

Since arriving here, Becker and Utzinger had encrypted hundreds of messages by typing them on Lily, their miniature Kryha. These were Red messages, meant for their SS superiors in Berlin. The springs and holes of the Kryha were now wearing out from overuse, and Utzinger asked Berlin to smuggle them a new cipher device through Becker's network of wolves.

Instead of a Kryha, Berlin sent a new Enigma machine.

"Enigma arrived via RED," Utzinger reported to Berlin on November 4, 1943. "Thank you very much." He typed this message on his older Enigma machine, the Green machine. He went on, "From our message 150 we shall encipher with the new Enigma . . . LUNA."

"It is a birthday surprise for LUNA," Berlin replied.

Utzinger now possessed three Enigmas. Throughout November and December 1943 he flashed Enigma messages to Berlin, newly confident in the security of his codes and increasingly upbeat about the prospects of his spy organization.

Momentum in South America seemed to be shifting in their favor. Right-wing movements continued to surge, and Becker and Perón were making progress in flipping governments. On December 20, 1943, a right-wing Bolivian general named Gualberto

Villarroel occupied the presidential palace in La Paz with his troops, assuming power in a successful coup. Becker and his Bolivian conspirators had set the coup into motion. The spy was elated. He arranged a meeting between Juan Perón and a Brazilian Integralist leader; Perón told the Brazilian that the "first fruit" of the continent-wide revolution had been plucked and would soon spread to "Chile, Paraguay, Peru and even Uruguay." The bloc of Hitler-inspired regimes, the Argentine revolution, the transplantation of fascist ideology to the soil of South America—these were no longer dreams but projects already begun, and any project begun has a chance of being completed.

Eight days after the coup in Bolivia, on December 28, 1943, Becker and Utzinger radioed warm greetings to Berlin, signing the message, as ever, with their code names. "We extend hearty wishes for a happy, successful New Year to all our loved ones and comrades in the war-torn homeland," wrote the SS spies. "Our thoughts are always with you and our Führer. SARGO, LUNA, and all collaborators."

Elizebeth never experienced a catharsis like soldiers do on a battle-field, a decisive moment when she got to stand over her fallen enemy with a sword and plunge a killing stroke into his heart. Rather, all through the war, she dissected fascists in the dark. If you were her adversary you never felt the blade go in. You bled slowly, painlessly, for months or for years, from tiny internal wounds, and then sometimes there was a terrible morning when you woke up groggy and confused, and your kidney was sitting in a bowl of ice on the counter.

She knew about the new Enigma machine sent to Argentina in November 1943—the Red Enigma—because the spies had discussed its delivery in Green messages and she had been reading those for a while.

The Red Enigma posed a new challenge for the codebreakers. It turned out that Berlin had forgotten to include keys in the shipment. This forced them to send a new key to Argentina over the radio. To protect that key, to keep it extra safe, Berlin decided to double the crypto. They sent Argentina a series of twenty-seven messages that were encrypted twice: first with a Kryha machine (using a new Kryha key), then with the new, Red Enigma. In other words, the plaintexts were typed on the Kryha, generating ciphertexts, and then those ciphertext letters were typed on the Enigma, generating a second set of ciphertext letters.

It wasn't possible for the coast guard to read the messages in depth because the crypto was doubled. The messages were like gnarly logs of wood covered with two distinct layers of bark. All the same, with hints from three sources—their own prior solutions, the navy's IBM machine, and the Germans themselves—the codebreakers found a way to strip off the bark and saw the wood into neat two-by-fours of plaintext. In an earlier message, Berlin had said that the new Kryha key incorporated part of the old key, which the coast guard already possessed. This reduced the number of possibilities for the new key, allowing the codebreakers to write a punch-card program that sorted the ciphertexts and aligned them in proper depth. Now the codebreakers could solve the plaintexts as usual and work backward toward the wiring.

During December 1943 and January 1944, as the coast guard worked toward a complete solution of the Red Enigma, wheel wiring and all, the team's prior solutions started to pay off on an international scale. Solving messages on Circuit 3-N had given Allied officials a priceless view of Nazi espionage in South America, and now, with Osmar Hellmuth's confession in hand, American and British diplomats were able to hammer Argentina for its cozy relations with Nazi spies. The pressure proved too great, and on January 26, 1944, the Argentine government announced that it was severing all relations with Germany and Japan. Argentina had been the Nazis' last friend in the West, the "last neutral bulwark," and now that bulwark was destroyed.

The following month, the coast guard solved the wiring of Red, their third Enigma of the war.

On February 19, 1944, Elizebeth's commander, Lieutenant Jones, sent a secret cable to Bletchley Park informing the British of the team's achievement:

CG have solved ... red. Details later.

Five days later, Jones transmitted the wiring details for all three wheels:

Following is wiring for new ... red machine ...
Outside wheel
P R Y B G A U T E V M K C Q D S J W L O F Z I X H N ...

The British codebreakers, always competitive, sent a cable back, notifying the coast guard that they had just solved Red themselves:

Many thanks. As this has just been solved here, details not required.

The SS in Berlin heard that Argentina had severed relations with Germany the same way everyone else did. They saw it in the news: a Reuters report of January 26, 1944. That day Berlin sent a panicked radio message to its spies in Argentina, begging for information on what happened. "We urgently need reports, whether it is true, and reports on the backgrounds and purpose."

The South America section of AMT VI had recently gained a new commander: Kurt Gross, the corrupt former Gestapo agent who nagged his agents to send him chocolates, and he now made a series of catastrophic assumptions. Unable to imagine that Elizebeth or any other adversary had been able to break their codes, Gross guessed that one of Becker's "wolves," the Spanish couriers, had betrayed him. "We cannot ward off the impression that there is a leak in your courier organization," he radioed to Argentina. "We suggest urgently once more that you scrutinize the men most critically."

Gross told the spies in Argentina to redouble their efforts. "Operation concerning USA and South America must now even more go on in full revolutions," Gross radioed. "Himmler's motto for 1944 is: 'We shall fight as long and no matter where, until the damned enemy gives up.'"

Throughout the early months of 1944, Allied planes bombed

Berlin. The headquarters of AMT VI suffered a direct hit. The spies' family members in Berlin used AMT VI's wireless to report that they were still alive. "Dear DARK EYE," began one message from Utzinger's girlfriend, "Blue Eye." "Air raids of Tommy cannot shake us . . . With the old zest and with 'Heil Hitler.' Your BLUE EYE." Utzinger's grandmother, "the Ahnfrau," wrote, "Here, life goes on, in spite of everything. With scornful and triumphant laughter, amid the ruins of homes, we take up our work immediately, with suppressed fury, in cardboard and wooden compartments."

Gross needed to hear good news from Argentina. His requests to the spies took on a newly apocalyptic tone. In one message, he asked Becker and Utzinger to investigate U.S. chemical weapons stocks and U.S. vulnerability to chemical warfare attacks:

> *Chemical warfare materials: What types of fluids or solid substances are on hand or in production? Their appearance? Effect on the body, eyes, respiratory system, clothes, metals. . . . What are the enemy's means of protection against gas?*

The same week that Argentina announced its break with Germany, in late January 1944, a sharp-eyed reporter for the London *Sunday Express* noticed U.S. warships massing off the wharves of Montevideo, Uruguay, almost within sight of the Buenos Aires waterfront. He made some phone calls and learned that something big and strange had just happened in the clandestine world of spies and counterspies. He wrote a story.

BRITAIN SMASHES SOUTH AMERICA SPY RING
Argentine H.Q. of Hitler's best agents

MONTEVIDEO (URUGUAY): The hulls of American men-o'-war were spotted in the sparkling waters of La Plata estuary. The warships were a sign that not much longer would secret German radio stations flash sailing dates and expected routes of Allied troopships and merchantmen to waiting U-boats. They were a sign that the Nazi dream of forming a solid block of Fascist dictatorships across South America is ended now that the Argentine has at last severed relations with Berlin and Tokyo . . . the coup that ended the Argentine's neutrality, fostered for years by Axis diplomacy and money, was the arrest of Osmar Alberto Hellmuth . . .

Within a few more weeks the "Hellmuth Affair," as it came to be known, was splashed across newspapers around the world. It had all the elements of a movie: Nazi masterminds, a pawn doing their dirty work, a kidnapping, a secret interrogation. It was already being spun into legend. Songs were being sung. Young Ziegfield, a popular calypso singer in Trinidad, performed a "Security Calypso" about the hapless Argentine named Osmar Hellmuth:

> He was on his way to cause a lot of trouble
> Osmar Hellmuth that Argentine Consul
> Caught with sufficient incriminating documents
> Offer no alibi to establish innocence . . .
> Just for the matter of a few paltry cents
> He sold his people and lost their confidence
> But in the end they will shoot the scamp
> Because he is safely locked up in one of Britain's Internment
>   Camps.

This publicity created immediate problems for two key institutions. One was the Argentine government. The Hellmuth Affair showed that Argentina was colluding at the highest levels with the Nazi state. This was very dangerous for Argentine politicians; they knew that if the Nazis lost the war, the Allies would surely punish them as collaborators. The public break in relations with the Reich wasn't going to be enough. Argentina needed to do something else, something bigger, more dramatic, to show that it wasn't a tool of Berlin.

The second institution that was unhappy about the Hellmuth Affair was the FBI. Here was the juiciest spy tale in the world, playing out within the bureau's jurisdiction, yet the bureau had been the last to know. The coast guard and the British had withheld their decrypts, fearing that the FBI would leak them. On December 16, 1943, an FBI assistant director sent a peeved memo to Hoover: "It is certain that the information was actually obtained from decodes of the extremely active clandestine radio traffic between Argentina and Europe, which decodes we have been endeavoring to obtain for a long period."

If there was anything J. Edgar Hoover hated, it was being out of the loop. But he was about to get the bureau back into the picture.

Argentina needed an "out." It needed a big public display of neutrality, a story to tell about its independence from Nazi influence, and the more spectacular, the better. Well, the FBI was good at telling stories. Perhaps the bureau could provide one.

The story, the "out" chosen by the FBI, was the story of Siegfried Becker.

Becker: a character out of a novel. A Nazi spy with long curling fingernails. A man who carried explosives in trunks and Enigma machines in his luggage. A seducer of the wives of Brazilian politicians. A stowaway on ocean-crossing ships. An SS-*Hauptsturmführer* who wore the ring of the death's head, a gift from Himmler. A friend of secret fascists in high places. A plotter of coups. A Johnny Appleseed of pirate radio, the human link between clandestine radio stations that spanned an entire continent.

The FBI had obtained some of this information from earlier coast guard decryptions and much of it from their own interrogations of the spies they arrested in Brazil in 1942. They reasoned that they could share the information with Argentina without exposing the ULTRA secret. The bare facts of Becker's career were already lurid, and in the FBI's hands, they could be arranged into a tale of a master spy, a kind of Nazi James Bond.

Francis Crosby, the FBI's legal attaché in Buenos Aires, made the case in a February 15, 1944, letter to Hoover, accompanied by a memorandum on Becker. "The memorandum was written with a view to furnishing the Argentine government with the material necessary for a spectacular spy story," Crosby wrote. "Becker would make excellent copy and perhaps provide the Argentines with the 'out' they are seeking in the Hellmuth case. The rational[e] which occurred to us is about as follows. Becker had an extremely successful career as an espionage agent in several other neighboring republics. . . . Details about the cases in which he figures would make excellent copy. However, this 'master spy' did not fall into the hands of the law until he ran afoul of the extremely astute police of the splendid Republic of Argentina, who immediately upon learning of his activities in Argentina, terminated his brilliant career." Then came Crosby's three-page memorandum for the ambassador, a pulpy, magazine-ready narrative:

> To the best of our knowledge, the deft, Teutonic hand of Siegfried Becker first appeared on this hemisphere when Becker organized clandestine radio station CEL in Rio de Janeiro. . . . It is possible to follow Becker into clandestine radio station CIT, also in Brazil . . . in Ecuador . . . clandestine radio station PYL in Chile, and a group in Mexico. . . . The hand of Becker is in some instances heavy and immediate, in others light and remote. However, it is possible to demonstrate the connection at all times. . . .

While the FBI circulated the legend of Siegfried Becker through diplomatic channels, they were going after him and his men on the ground in South America. The bureau now had two dozen SIS agents across the continent, covering all the major cities, finally acclimated to their local posts, cozy with local police, and able to get intelligence from confidential informants. Hoover ordered all of the FBI attachés across Latin America to search their files for any information on Becker, "one of the most important German agents in Latin America, who is presently a fugitive and believed to be hiding somewhere in Argentina." At the same time, SIS agents like Crosby sprang into action on the ground, scoping out bars and restaurants where Becker was said to hang out and interviewing prostitutes he had known.

The FBI also began to close in on the pirate radio stations themselves, the ones Elizebeth had been listening to all along, remotely, from her office in Washington. American agents went climbing over Chilean and Paraguayan mountains with handheld direction-finders, and FCC technicians drove through Buenos

Aires and the surrounding countryside with direction-finding automobiles.

Stations were seized, one by one, the radio equipment impounded. This was fun for the FBI agents, this classic police work. They were good at it. It was what they did best. They weren't codebreakers. They were investigators. They were cops. They built cases from physical evidence and in-person conversations, and they arrested people.

Utzinger wanted to go dark. That was his first instinct. In Argentina, when the SS radio expert heard that the Hellmuth mission had failed in the worst possible way—that the envoy had confessed to the British and the Argentine government was now breaking relations—Utzinger thought the spies should stop using the radio entirely. Something was not right. He needed some time to think.

He was overruled. Berlin insisted that the radio transmissions continue and even be increased, and Becker assured him that the break in relations was a "sham." According to Becker's military and political contacts, the Argentine government was just making some noise to appease the Americans. Everything would be fine once the commotion died down.

Having little choice in the matter, Utzinger continued to transmit. But his work with Becker only grew harder throughout the first months of 1944. The two men sensed they were being surveilled, followed. Utzinger took what security precautions he could, moving the transmitter to a friend's farm outside Buenos Aires and hiding it beneath a chicken coop. The authorities found it anyway. One day in February 1944, when Utzinger and Becker were somewhere else, Argentine cops raided the farm and arrested several of their collaborators, taking the V-men to a police station and applying painful shocks with the *picana eléctrica* (electric cattle prod) to extract confessions. A prisoner named Gaucho, who may have been Becker's bodyguard, had his eardrum destroyed. Another, Herbert Jurmann, a Hitler Youth leader, decided to commit suicide rather than confess, throwing himself out of a third-story window. "He fell on February 19," Utzinger radioed to Berlin, "faithful to his oath, for us a model and obligation."

Jurmann's death cast a pall over the work of the spies in Argentina. They were frightened and demoralized. The Reich seemed increasingly distant and German defeat increasingly plausible. No one was sure they wanted to die for National Socialism anymore. The Spanish wolves grew less willing to cooperate; two of the wolves were captured and hanged by the British. Utzinger's grandmother, "the Ahnfrau," sent him an ominous message over the radio, asking where he had placed their ancestral papers, the important documents of their family history. She feared that the papers would be destroyed by bombs.

Utzinger still could not shake the feeling that something was very wrong, that the Allies could read his words. But he assumed that the problem was the incompetence and poor security habits of his counterparts in Berlin. Over and over they had demonstrated themselves to be idiots; he had no reason to think that anything more subtle was amiss. So he tried his best to keep the radio network operational while maintaining a self-protective level of paranoid awareness. When looking at the raw intelligence gathered by his V-men, he paid special attention to news about Allied spy hunts, anything that might give him advance warning of a raid and save his skin.

He picked up an intriguing piece of news that he shared with the home office on March 22. According to the U.S. press, a "Mrs. VALERIE DICKINSON, N.Y." was being charged with treason. (Her first name was actually Velvalee.) She owned a shop in New York that sold dolls and doll clothing. She had sent suspicious letters to an address in Buenos Aires. The letters appeared to concern dolls, but the Americans believed Dickinson was a Japanese spy, communicating in code. The press was calling her the Doll Lady.

Velvalee Dickinson whirled around on the two FBI men and tried to scratch out their eyes. It was January 21, 1944. The agents had staked out the vault at the Bank of New York, waiting for Dickinson to walk in and open her safe-deposit box, and as soon as she did, unlocking a drawer that contained $15,900 in cash, the FBI agents said they had a warrant for her arrest. Dickinson shouted that she didn't know why. She was fifty years old, a widow, a frail-looking ninety-four pounds, with brunette hair. She made such a kicking commotion that the men had to pick her up by the armpits and carry her away.

The FBI arrested Dickinson because of five suspicious letters that had been previously intercepted by postal inspectors and forwarded to the bureau. The letters talked about dolls and the condition of dolls, some of which were damaged: "English dolls," "foreign dolls," a "doll hospital," and a "Siamese dancer" doll "tore in middle." The first of the five letters read in part, "You asked me to tell you about my collection. A month ago I had to give a talk to an art club, so I talked about my dolls and figurines. The only new dolls I have are these three lovely new Irish dolls. One of these three dolls is an old Fishermen with a Net over his back. Another is an old woman with wood on her back and the third is a little boy." The letter had been addressed to Señora Inéz Lopez Molinari in Buenos Aires. No such person existed; the letter was returned to the address listed on the envelope, the address of one of Dickinson's customers, a Mrs. Mary E. Wallace in Springfield, Ohio, who was confused to read the letter, as she had not written it.

Dickinson owned a doll shop on Madison Avenue in New York and had developed a reputation for her artistry—she sold dolls for as much as $750 apiece—yet the bureau discovered that she had fallen into debt after the death of her husband, that she was a member of the Japanese-American Society, and she had visited the West Coast in January 1942, immediately after Pearl Harbor. The FBI tested the shapes of ink on the letters against Dickinson's seized typewriter and confirmed a match; the bureau's investigation also revealed social ties between Dickinson and Japanese consular officials.

After the agents arrested Dickinson in January 1944, a federal prosecutor took up the case: Edward C. Wallace, the U.S. attorney for the Southern District of New York. Wallace had worked with Elizabeth in the smuggling days and hoped to get her opinion on the Doll Lady's letters, but first he called the supervisor of the FBI's New York office and asked if the bureau had any objection to showing Elizabeth the letters.

Within the FBI, the prosecutor's request provoked a remarkable exchange of at least eight phone calls, teletype messages, and memos that traveled up the chain from the FBI's New York office to Washington and ultimately to the desk of J. Edgar Hoover. The gist of these communications was that the prosecutor, Edward Wallace, wanted Elizebeth and spoke highly of her—"According to Mr. Wallace," an FBI agent in Washington wrote in a memo to Hoover's deputy, "Mrs. Friedman and her husband, who is a cryptographer for the Army, are recognized as the leading authorities in the country and have written numerous books on the subject"—but FBI agents worried that Elizebeth would siphon publicity from the bureau, stealing its spotlight. The agents seemed reluctant to speak of Elizebeth as an independent analyst separate from her husband. Although no one had ever discussed involving William in the case, the supervisor of the FBI's New York office fretted that the Friedmans, *plural*, "might, in the event of a successful espionage prosecution, attempt to lay claim for any work that they might have performed in this connection."

The New York office sent Hoover a teletype on March 18, 1944: "ADVISEASTOSUBMISSIONQUESTIONEDLETTERSTOELIZABETH FRIEDMAN FOR EXAMINATION." Hoover responded with a dismissive shrug of a memo: "Concerning the project to submit the documents to Mrs. Friedman. . . . There appears no point is to be gained by multiplying the number of examiners." But he posed no formal objection, so U.S. Attorney Wallace went ahead and sent Elizebeth the Doll Lady's letters, and Elizebeth analyzed them and crystallized her thoughts into a five-page letter before traveling to New York at the feds' expense to discuss the case with Wallace in person.

"My dear Mr. Wallace," Elizebeth began in her letter, "Within the last two days I have spent a few hours examining the Dickinson letters. I am setting forth here some queries and statements which may be accepted for what they are worth, mindful of your statement on the telephone that you hope to obtain 'leads,' and that you understand that the code in the letters is the 'intangible' type of method not susceptible to scientific proof."

After making it clear that this was not the usual sort of cryptanalysis that she did, that this was only her *opinion*, Elizebeth went on to discuss what the Doll Lady was really talking about when she talked about dolls.

The letters, she said, were a textbook example of "open code," a way of communicating secretly out in the open, without necessarily arousing suspicion. "Granddaughter's doll" in one letter might refer to a U.S. ship that had been damaged at Pearl Harbor and was being repaired. "Family" meant the Japanese fleet. "English dolls" meant three classes of English ships, such as a battleship, battle cruiser, or destroyer. Where Dickinson wrote, "One of these three dolls is an old Fishermen with a Net over his back" and "another is an old woman with wood on her back and the third is a little boy," she probably meant, "One of these three warships is a minesweeper, and another is a warship with superstructure, and the third is a small warship." ("Destroyer?" Elizebeth guessed. "Torpedo boat? Auxiliary warship?")

Elizebeth also pointed out that the street number of the address in the five letters—Señora Inéz, O'Higgins Street, Buenos Aires—was given as five different numbers (1414 O'Higgins, 2563 O'Higgins, etc.), suggesting that the messages were never meant to reach their destination and were intended to be intercepted en route, in an airline pouch or a censorship office, by a friendly Axis confederate.

Elizebeth's letter shows her analytical brilliance; it also shows her native cautiousness, her reluctance to say anything that couldn't absolutely be proven. Words in an open code can have multiple meanings. She didn't want to testify in court for this reason. Hoover saw it differently. To him, the vagueness of an open code was an advantage, not a disadvantage, enabling his agents "to give the more extended estimates and alternative possibilities" during cross-examination.

The FBI had gathered other damning evidence against Dickinson, including the unexplained cash and her relations with Japanese officials, and the government charged Dickinson with espionage, as a spy for the imperial Japanese government. The charge carried the death penalty. As far as anyone knew, Dickinson was the first woman to be accused of espionage on American soil since the war's start. "So far," wrote the *Washington Sunday Star*, "on this side of the water, Mrs. Dickinson is the woman spy of this war." During her first court appearance in New York in May 1944, Dickinson looked subdued, wore a black hat pinned with imitation white flowers, and twisted a handkerchief behind her back with black-gloved hands, glancing around the courtroom at the FBI agents and prosecutors and reporters: "Who are all these people?" she said. When the prosecutor spoke, "She even yawned, decorously behind a hand," the *Washington Times-Herald* reported.

Dickinson pleaded guilty. At her sentencing three months later she denied that she was a spy, breaking down in court and swearing that she didn't know a "battleship from any other ship except that it's larger." She ended up with ten years in prison and a $10,000 fine.

All through these proceedings, Elizebeth stayed out of the public eye. When the Doll Lady case was over, the conviction won, the FBI, as always, informed the press of its heroism, feeding the dramatic details of "the War's No. 1 Woman Spy" to reporters. "What made her become a Japanese spy?" asked the *Star*. "One FBI man who questioned her advanced the idea that she was an introvert, embittered by life, the frustration of childlessness." Elizebeth was not mentioned in the coverage. The articles said variously that the code had been cracked by "FBI cryptographers" or "a check with the Navy." Hoover himself wrote about the Doll Lady in *The American Magazine,* calling her "one of the cleverest woman operators I have encountered. Cultured, businesslike, cunning, and, despite her 45 years of age, most attractive, she presented one of the most difficult problems in detection the FBI has tackled in this war."

And while readers learned of the Doll Lady's treachery from Hoover, the woman who analyzed the Doll Lady's letters in her spare time, quietly, as a side project, returned to her primary task of hunting Nazi spies.

Through the rest of 1944, as Elizebeth and her coast guard team continued to decrypt Nazi radio messages, she noticed an uptick in paranoia in the plaintexts, a creeping sense of doom. All along, Elizebeth had been watching the Nazis build a spy network, and now, after invisibly undermining that network, she was watching it die—and stepping on its neck when need be.

She tracked the spies as they tried to escape or hide, decrypting their desperate wireless notes. Becker went into seclusion in

April 1944. "He is hidden in the center of Buenos Aires," Utzinger wrote. "He works at night only. Keep your fingers crossed for him. LUNA."

On August 11, 1944, Utzinger sent one of his final radio messages: "The enemy succeeded in locating two of our stations in 60 days."

Seven days later, on August 18, 1944, Gustav Utzinger, a.k.a. "Luna," was arrested by the Argentine federal police along with forty of his associates.

Utzinger later described the scene to an FBI interrogator. On a table at the jail, police arranged items they had taken from Nazi Party members years earlier, including swastika flags, pictures of Hitler, and hunting weapons, and put those next to a resistance coil from the spool of a film projector. Utzinger had managed to destroy his radio equipment before capture, so the police claimed the coil was a "bobina de tanque de un transmisor potente de los espías nazis"—*tank-spool of a powerful transmitter of the nazi spies*—and photographed the whole cornucopia to release to the press. It was for show; the Argentines were mainly concerned with covering up their own links to the Nazis. Five days after Utzinger's arrest, Juan Perón himself appeared at the jail and told police that "the investigation was to show merely the breaking up of a great German intelligence machine; and that the mention of contact with any political or military personalities or their foreign colleagues must be suppressed."

This wasn't the end of the spy hunt in South America. Siegfried Becker, the wily SS captain, remained at large, the target of a deepening manhunt by FBI agents and police in multiple countries, and Elizebeth would continue decrypting clandestine messages across dozens of Nazi radio circuits for the next thirteen months. But the arrest of Gustav Utzinger in August 1944 marked the beginning of the end of the Invisible War—"the final chapter to any effective espionage activity" carried out by Nazis in the Western Hemisphere, according to Utzinger's colleague, Hedwig Sommer. Nazi spies would never again pose a threat to America.

Elizebeth had not done it alone. It was a team effort, with strong work from Chubby Stratton and other British officers, the FCC, and the FBI. But in her own piece of the war, Elizebeth was a central figure. She had to smash the codes on the page before anyone else could smash the spy networks on the ground. "Technical advantages played a big role in the undercover struggles," Rout and Bratzel concluded in 1986's *The Shadow War*. "Technical brilliance in cryptography and radio interception plus hard work by field agents proved to be the unbeatable combination which made victory possible." The two historians credited the FBI for both the fieldwork and the technical brilliance (the coast guard's files were classified at the time), and authors of more recent books have also praised the bureau for destroying the Nazi networks in South America. But the FBI didn't intercept the messages. It didn't monitor the Nazi circuits. It didn't break the codes. It didn't solve any Enigma machines. The coast guard did this stuff—the little codebreaking team that Elizebeth created from nothing.

During the Second World War, an American woman figured out how to sweep the globe of undercover Nazis. The proof was on paper: four thousand typed decryptions of clandestine Nazi messages that her team shared with the global intelligence community. She had conquered at least forty-eight different clandestine radio circuits and three Enigma machines to get these plaintexts. The pages found their way to the navy and to the army. To FBI headquarters in Washington and bureaus around the world. To Britain. There was no mistaking their origin. Each sheet said "CG Decryption" at the bottom, in black ink. These pieces of paper saved lives. They almost certainly stopped coups. They put fascist spies in prison. They drove wedges between Germany and other nations that were trying to sustain and prolong Nazi terror. By any measure, Elizebeth was a great heroine of the Second World War.

The British knew it. The navy knew it. The FBI knew it. But the American public never did, because Elizebeth wasn't allowed to speak. She and every other codebreaker who worked on ULTRA material was bound by oath to keep the ULTRA secret. Even if she had been free to discuss her triumphs, explaining them to the public would have taken some time.

J. Edgar Hoover did not have these constraints. His power

allowed him to manipulate the press and disclose secrets without consequence. And because his agents were old-school detectives, not technical wizards like Elizebeth, Hoover was able to frame the Invisible War in terms of instantly familiar images: disappearing inks, saboteurs, hidden cameras, police raids on clandestine radio stations, gumshoes in snap-brim hats.

So this was the picture of the spy hunt that the public ended up receiving. They got Hoover's story, not Elizebeth's.

Hoover made sure of it. In the fall of 1944, with the Wehrmacht collapsing across Europe, and the Red Army moving toward Berlin, he launched a publicity blitz to claim credit for winning the Invisible War.

He published a seven-page story in *The American Magazine* titled "How the Nazi Spy Invasion Was Smashed." The sub-headline read, "One of the great undercover victories of the war—the defeat of a vast Axis plan to penetrate South America—is revealed here for the first time by the Director of the FBI." Hoover claimed that his bureau had disrupted seven thousand Axis operations, catching two hundred and fifty spies and seizing twenty-nine radio stations, and that these actions had "stopped Hitler in South America. He needed his men and radio stations to carry out his plans of conquest and sabotage. . . . Without his machine he was lost." Hoover did not mention Elizebeth or the coast guard, but he did thank the policemen of Brazil and Argentina, a significant number of whom were fascists, torturers, or both.

He also starred in a fifteen-minute film that was shown to U.S. troops abroad, *The Battle of the United States,* a highlight reel of the Invisible War.

The film, made with the help of beloved Hollywood director Frank Capra, opens with a blast of patriotic music and a waving American flag that fades to a shot of Hoover at his FBI desk, sitting in front of the Stars and Stripes. Hoover wears a pin-striped gray suit and a garish tie. His hair is neatly combed. His hands are clasped on the desk. "I want to talk to you fighting men and women about the Battle of the United States," he says, looking into the lens.

Cut to a shot of a wooden door marked FBI CONFERENCE ROOM. The door opens and the camera moves inside. Seven FBI agents are gathered around an oversize map of South America. Ominous music. The map expands to fill the entire screen. Animated planes fly above the map and drop bombs on the Panama Canal; radio towers writhe with cartoon electricity; a cartoon Nazi appears, dressed in fatigues, holding a bayonet. He stands on Argentina, facing the United States and growing taller until he lunges and stabs Wyoming.

The film goes on to describe roll-up of the radio networks and the destruction of the Duquesne spy ring as solo feats of FBI tenacity. In the final scene Hoover faces the camera and addresses the intended audience, the troops still fighting overseas in the last months of the war. "The attack against all German and Japanese agents in this country," he says, music swelling, "was as vigorous and as victorious as the attacks you have made against the enemy." He ends with a flourish: "We of the FBI feel that we are part of a team, to make America a great and decent place to live. We are on that team, all of us, together."

In December 1944, around the time U.S. troops were watching Hoover's film, Elizabeth sat down to write the Friedman family Christmas card. It wasn't a clever puzzle or a game like in years past. It didn't contain any pictures of Bill and Elizabeth and the kids, or any secret messages. It was just an old-fashioned letter in plain English. After four years of war, and tens of millions of people killed, writing a letter felt like a good and human thing to do.

She typed

*BULLETIN ** 1944 ** FRIEDMAN*

at the top of a fresh white sheet of paper.

"We keep wondering what has transpired in the lives of our friends, and their families," Elizabeth continued. "Perhaps they too are wondering about us."

The director of the FBI had been boasting about catching spies he did not really catch. Elizabeth, who did catch them, bragged about her family.

"Bill, Will, Billy," she typed, and summed up what was allowed to be said publicly about her husband's professional activities during the war:

> Having been retired from active duty in 1941 for physical disability, he has spent from nine to sixteen hours a day carrying on, doing a terrific war job. (Such things make liars of the Army Medical Service, and who doesn't know of like instances?) In March of 1944 he was the first man to be awarded the highest distinction given by the War Department: the Exceptional Service Award with gold wreath, equivalent to the Distinguished Service Cross awarded to those persons who in the field of combat, "perform exceptional service over and beyond the call of duty."

"P.S.," William added, out of modesty. "I didn't write this.—Bill."

As for Elizebeth's own accomplishments in the year 1944, there was nothing to report. She wrote that she was "just carrying on a routine navy job, in an unglorious fashion, unlike her distinguished husband."

"P.S.," William wrote. "Elizebeth always was, is, and continues to be the most fascinatin' woman I've ever known."

Elizebeth spent the rest of the Christmas letter discussing her clever children. John Ramsay was now manager of the football team at his prep school, chairman of the Dance Committee, and president of the Senior Club. Barbara was learning Spanish at her job with the U.S. Office of Censorship in Panama. "There is no rationing in Panama," the mother reported, "cigarettes are plentiful and eight cents a pack, living expenses are about one third of their cost in the States, and her work, she says, is fascinating." Elizebeth signed off by wishing all of the Friedmans' friends and loved ones "a reunited family in 1945."

It was the spring of incendiary bombs. The Allies lit German cities on fire, one after another, in the first three months of 1945. The quantities of bombs were measured in thousands of tons. The RAF dropped more than thirty thousand tons of bombs on Germany in the month of January alone. On a single night in March, the RAF sent 223 planes above Würzburg, dropping bombs, lighting fires that tore through the beautiful old wooden buildings and sent people fleeing through the blazing streets to the river. A grandmother clutched her grandson to her chest, trying to protect him from the flames. Their bodies were found melted together.

The death marches from Auschwitz began in January 1945 as the Red Army closed in from the east. SS guards led the prisoners away from the camp on desolate roads and forced them to walk until they collapsed, then shot the ones who remained standing. The Allies liberated the Dachau concentration camp on April 29. Hitler killed himself on April 30.

On April 19, 1945, eleven days before Hitler's suicide, Siegfried Becker, "Sargo," the greatest Nazi spy in the West, was arrested in Buenos Aires by the Coordinación Federal, the Argentine state police. He had dyed his hair black and was living with a girlfriend named Teresa. There were twenty-six thousand pesos in his pocket, and police confiscated an address book containing the names of more than one hundred associates in Buenos Aires, Barcelona, Bilbao, Rio, São Paulo, Hamburg, and Berlin.

Police took him to Villa Devoto prison, located in a poor neighborhood in northwest Buenos Aires. Becker was outraged. He started to talk. He gave statements about his links with high Argentine officials all the way up to Perón, who was now preparing to run for president.

The statements were quickly doctored, but Becker had made his point: He could hurt Perón if he opened his mouth. Officials responded by treating him as gently as possible. Perón assigned his personal bodyguard, Major Menendez, to look after Becker's health and well-being, and Becker was allowed to spend much of his day in the warden's office instead of a cell. At Christmastime, Becker sent "a huge basket of delicacies and champagne" to officials of the Coordinación Federal and arranged for a friend to deliver seven stuffed turkeys to the prison.

Gustav Utzinger, "Luna," the radio expert of the spy ring, heard about the champagne and turkeys straight from Becker. They happened to be detained in the same prison, Villa Devoto, at the same time, although Utzinger hardly enjoyed Becker's privileges. Horrified by the corruption and brutality of the prison officials, Utzinger frequently defied them and was punished with solitary confinement and beatings.

In February 1946, Argentine voters elected Perón to his first term as president. He moved into the palace with his second wife, the alluring actress Eva Duarte. He no longer feared what the spies might reveal; he was powerful now, insulated from consequence. So he released them. Utzinger tried to start a radio business but was later re-arrested and deported to postwar Germany. Becker stayed in Buenos Aires and was never heard from again. According to the Argentine journalist Uki Goñi, the *Hauptsturmführer* used his connections to smuggle Nazi war criminals into Argentina, helping them escape prosecution. It is likely that he lived to be an old man in Buenos Aires and died a natural death there.

Some of Becker's friends across the continent did not enjoy his luck. On July 21, 1946, an angry mob invaded the presidential palace in La Paz and fatally shot Gualberto Villarroel, who had ruled Bolivia since Becker's coup. The president's corpse was thrown from a palace balcony and hung from a lamppost in the public square.

The Friedmans had been tired and stressed for years now. They had continued to function in spite of it, had kept going in to work every day, Elizebeth to the Naval Annex, William to Arlington Hall. But at the start of 1945, when it seemed like the enemy was on the run, and it finally felt permissible to relax ever so slightly, their bodies fell apart in unison. William got bronchitis and limped around the house, wheezing. Elizebeth took care of him and slept a lot on weekends. On George Washington's birthday, February 22, 1945, they were invited to a dance party with some army friends. Elizebeth realized that it had been years since she had put on nice clothes to go out and have fun, and she allowed herself to spend an extra-long time getting dressed, dabbing Shalimar perfume on her neck, choosing pearl earrings, finding just the right dress, before heading out the door with her hobbling husband, trailing a cloud of warm, luxuriant scent.

President Roosevelt died on April 12, of a brain hemorrhage. Elizebeth was crushed. She had never liked politicians, but Roosevelt was the exception, a man who seemed both decent and brilliant, who believed in democracy, science, equality, and international cooperation, values she held dear, and Elizebeth feared that in his absence, "evil influences" like the Ku Klux Klan would sweep the country: "Our country *will* go on. But who can say what catastrophic results will come from his going? Or worse, the results, hidden, subversive, that take place with no fanfare, no appearance on the surface, but so quietly, hiddenly evil."

The next month, a family friend died of a sudden illness: Colonel John McGrail, the army intelligence expert who had sent Elizebeth an Easter corsage of violets when William was away at Bletchley Park. The Friedmans felt that McGrail had worked himself to exhaustion in the war. During the colonel's burial at Arlington National Cemetery, Elizebeth stood at the side of his widow, Florence, and held her arm as six white horses carried McGrail's flag-draped casket to the grave.

Elizebeth didn't think there was any pattern to death. It was random and cruel. She only hoped it would not strike her own loved ones. They were still scattered across the planet, subject to the winds of the war. John Ramsay, in flight school in Alabama, told Elizebeth on the phone that he had been ordered to report to an unknown destination. He said, with trepidation, "I will let you know where I am, when I am." William kept murmuring in her ear that the U.S. Army was preparing a final mission for him, an assignment that would take him to Europe for weeks or even months. And Barbara was still in Panama, dating young navy officers about to report to the front and worrying they would be killed. Elizebeth tried to prepare her daughter for that possibility: "Remember, darling, my blessed fatalistic philosophy—when the number is up, he goes; whether he is on a davenport in his own home, or firing broadsides at the enemy." She signed off, "Your getting-to-be-an-old-woman, Mother."

Maybe it was her physical exhaustion, or her grief, or her ongoing fear for the safety of her family, but when Elizebeth heard on May 8 that the Nazis had formally surrendered to the Allied forces, that the war with Germany was over, it did not feel over. "We have difficulty believing it is really true," Elizebeth wrote to her daughter in Panama. "New York, we hear, celebrated. But work went on, we didn't even stop to hear the proclamation."

The picture from the news was fluid and confusing. Japan had not surrendered. Its diplomats spoke with a new humility but the generals vowed to fight on. Stalin's Red Army stood poised to occupy defeated territories in the East. Stalin was rumored to be building an atom bomb. The Americans were rumored to be building an atom bomb. Elizebeth wrote to Barbara, "It's absolutely terrifying."

The heat of the Washington summer grew brutal. In the master bedroom of 3932 Military Road she slept with the windows open in the vain hope that a draft might blow in. John Ramsay sent her a poem he had written in flight school, describing his desire to live the adventurous life, "to grip the rock-bound cliffs / that jealously guard the house of wisdom." His mother, delighted by her son's ambition and proud of his interest in poetry, re-copied the poem in a letter to her daughter.

In early July, Elizebeth learned that William's orders had finally come through. He was being sent to Europe for a ninety-day assignment with the Allied forces there. It would be his final mission of the war.

William told his wife he would be conducting research. He made it seem routine. It was not routine. He had been recruited for a mission called TICOM, a joint U.S. and British effort to seize intelligence secrets from former Nazi territories.

One historian has called TICOM, short for Target Intelligence Committee, "the last great secret of World War II." The aim was to preserve Western dominance in whatever the next war might be: perhaps, it seemed, against the Soviet Union. This meant preventing knowledge and technology from falling into the hands of Joseph Stalin. The Western Front needed to be scoured and picked clean of secrets—any Nazi cryptologic inventions snapped up, any information about MAGIC or ULTRA secured—for the United States and Britain to maintain a codebreaking edge in future battles. And in pursuit of this goal, William Friedman would soon journey into the deepest, most secretive chambers of the

Third Reich. He would find himself riding a tram into thin air to reach the Kehlsteinhaus, or Eagle's Nest: the private mountain lair of Hitler himself.

William and Elizebeth drove to the military air terminal together on July 14, 1945, a warm, clear blue morning in Washington. He wore his dress uniform of a khaki jacket, a khaki cap, khaki pants, and brown boots. She thought he looked handsome and wished the army allowed the men to sew stars on the shoulders. She kissed him goodbye and he boarded a Douglas C-54 transport plane along with twenty-three other passengers: army officers, several scientists, a WAC stewardess. Then Elizebeth went back to the family's car, which was parked on a little hill, and waited until the C-54 took off, watching it rise into the clouds, shrinking into a speck of silver.

CHAPTER 6

Hitler's Lair

William Friedman gripped the sides of a jeep as it shook and rumbled up the twisting incline to the laboratory of the Nazi scientist. The codebreaker was in a small town in Bavaria miles from any battlefield, ascending a mountain called Feuerstein. The jeep continued to climb. Looking backward William could see the intact homes and shops of the town below, growing smaller in the distance. Up ahead, the Laboratorium Feuerstein loomed into view. The impression was of reaching a castle on a mountaintop. The building was enormous. Red Cross signs were painted on the roof, a ruse to pass off the institute as a hospital and prevent RAF pilots from dropping bombs.

Dr. Oskar Vierling, an engineer from a poor family, had run this place. Before the war Vierling specialized in acoustics research, investigating the properties of sound and inventing new kinds of instruments; his "electrochord," an electrical organ, was a favorite of the Nazi minister of propaganda, Joseph Goebbels, who adapted the instrument for party rallies, using it to play forceful blasts of chords at predetermined spots in his speeches. In the late 1930s the Nazis insisted that Vierling focus his activities on war machines, and the Laboratorium Feuerstein, named after the mountain which it crowned, began to fill with scientists and assistants under the Doktor's direction, as many as two hundred people.

The Allies wanted to know what Vierling had invented during

the war, particularly any devices related to intelligence or cryptology, so they had dispatched William Friedman, along with a dozen colleagues from army intelligence, to make an inventory of the laboratory. Earlier the defeated Nazis had ordered Vierling to destroy all of his inventions, but he had hidden his favorites in a locked room in the basement and was now glad to show them to the Americans.

The interior of the lab was cavernous, Gothic. William felt like a character in a murder mystery, about to meet an intricate demise. He analyzed Vierling's prototypes. There was a machine said to encrypt the human voice the way that Enigma encrypted text, and a device for scrambling speech to make it unrecognizable to anyone listening on a wiretap. There was an "acoustic torpedo," a machine that shot bullets of sound; a coating for submarines that made them invisible to radar; and a "speech stretcher," an audio playback device that sped or slowed a recording without changing the pitch. It was hard for William to avoid drawing a comparison between Vierling and George Fabyan, the American robber baron who built a deviant temple to science—this place was like a Nazi Riverbank. Inside the Laboratorium Feuerstein, William ate a brief supper of hot dogs, potatoes, coffee, and crushed peaches, and he stayed late into the evening with his American colleagues, the topics of their conversations growing spookier, starting with Albert Einstein and the theory of relativity before veering off to more occult topics like the possibility of extrasensory perception.

Vierling's laboratory was only one target of TICOM, the mission to lock down the intelligence secrets of the war. The Allies deployed six TICOM teams to Europe starting in April 1945, each containing eight to fifteen intelligence personnel from both the United States and Britain. William's team began its work in late July. The mission carried him for hundreds of miles across southern Germany, France, and Czechoslovakia. He took notes. He had spent the war in office buildings. This was his first look at the landscape of physical battle, and it filled him with "a heavy feeling of sadness" difficult to describe. The German countryside was intact, the plots of wheat and rye and barley ripening, the green forests seemingly untouched, but the people were broken, and the machines were broken. The highways between cities were full of what the army called DPs: displaced persons. Mothers and fathers walked along the road with their children, carrying personal belongings on their backs or hauling small quantities of wood in handcarts, to burn for fuel. Wrecked trucks and tanks had been tipped into ditches, and women wept in the backs of clattering wagons.

He ate a C-ration for the first time, canned meat and beans. It didn't agree with his stomach. He ate Spam. He stayed at an army installation code-named BARN and thought that if it had been up to him he would have named it something less boring, like LEPIDOPTERA, the scientific name for a butterfly, but then he thought about it some more and realized that LEPIDOP-TERA would be more difficult for soldiers to remember and they might end up getting lost. He rode at slow speed through the fire-bombed cities of the Reich, through Frankfurt, Nuremberg, Aschaffenburg, and Würzburg, feeling like an ant crawling across the body of a dead child. "The destruction to be seen in cities such as these should be noted by anybody who believes in war," he wrote, "because it can tell more about what happens in modern warfare than reams of literature." More than once, while scouring an abandoned Nazi garrison, the men on William's team found copies of his own cryptologic publications from years past, translated into German or French by the Nazis. This was hardly a surprise, given the ubiquity of William's contributions to the science, and the men got a kick out of it, grinning as they showed him these discoveries—and the bibliophile in William could not resist taking these extremely rare documents as souvenirs and keeping them for his own library—but it was the strangest compliment. Imagine walking into the devil's library and seeing your book on his shelf.

William always believed the war was worth fighting. But he saw it as a grim duty, not a crusade, and his experience of fighting the war had permanently destroyed his faith in the way the world was put together. Earlier that year, his daughter asked him in a letter if he believed in Zionism, the project to create a Jewish homeland. He said no. "Zionism is only one of many virulent forms of a detestable disease known as 'nationalism,'" William wrote to Barbara. "The sooner we realize that we are all God's children regardless of color, race, creed, nationality, etc., the better for all nations and the world as a whole." He didn't believe in nations anymore, not even his own. This is what he had tried to tell his daughter. The world is very fragile, more fragile than it is healthy to believe if you want to get out of bed and make it through the day.

William did not sleep well the night before he visited Hitler's alpine lair: Kehlsteinhaus, the Eagle's Nest, a meeting center and getaway built by the party in 1939 and given to the Führer as a gift on his fiftieth birthday. An army friend of William's made steak sandwiches with raw onion at 1 A.M., and the cryptologist woke to the smell and downed the steak with Scotch. At 8 A.M. he rode in the army staff jeep at the front of a fifty-vehicle convoy to the town of Berchtesgaden, overlooking the Austrian Alps. The path to the lair went straight up a mountain for four and a half miles, no guardrails to prevent an errant car from plummeting several thousand feet to the valley. The driver kept the jeep in first gear all the way.

The convoy reached a plateau with a parking area at the base of the lair, and William entered a hundred-foot-long tunnel protected by two enormous, heavily ornamented bronze gates. The tunnel was wide enough for three automobiles and brightly lit with electric lamps. At the far end of the tunnel, an elevator shaft the height of a fifteen-story building rose the rest of the way, to the mountain's pinnacle. The elevator operator was the same German who had worked for Hitler and his deputies all during the war. William talked to him for a bit. "He gave us a little speech in his defense, saying that he was assigned to the job—the Nazis wouldn't let him get away, etc., etc."

The ride to the top took three minutes. Then William and his fellow officers were led through a passageway and into the first chamber of the lair, a small, wood-paneled dining room where Hitler held banquets with visiting world leaders. The Führer didn't visit often—the long automobile climb made him impatient, and the change in atmospheric pressure disagreed with his constitution—but Goering and Ribbentrop spent a lot of time at the Kehlsteinhaus, and Hitler's mistress, Eva Braun, loved to entertain friends here. She lugged her Scottish terriers up the mountain and let them romp in the thin air. She threw a wedding party for her sister and the groom, an SS captain later shot by Hitler for desertion.

Beyond the dining room and down a few stairs was an octagonal room with thick granite walls and a 360-degree view of the Alps. From this height the mountains appeared at eye level, jagged triangles of green and blue, like incisors rising from earth's jaw, "indescribably beautiful," William thought. A third room contained a fireplace of red marble, a gift from Mussolini. Most of the furniture was intact. The Americans milled around like tourists atop the Empire State Building, unsure what to do with themselves after the first minute or two. William took pictures and wished he had brought a movie camera.

The group descended in the elevator and drove a mile back down the mountain to a level area where the Nazi party had built Hitler a separate residence, a private house. The front of the house was a twenty-five-foot-wide plate-glass window with no glass in it. The building had been almost completely destroyed by an RAF blockbuster bomb and by subsequent visits from American soldiers, who wrecked what was left of the house before the army stopped them. William thought it was a shame. "I think it is too bad that this whole installation was not left absolutely intact to serve as an everlasting and terrible monument to the folly of a people led to perdition by a madman's lust for power."

The floor of Hitler's house was littered with chunks of rock and marble. William reached down and picked up a piece. When he got home, he decided, he would keep it on his desk, as a reminder. "I shall have it made into a paper-weight."

That day, back in the States, Elizebeth was in Michigan, visiting her sister. She stayed at a hotel in Ann Arbor that brought her a bowl of ice in the afternoon. She looked at this simple object, this bowl of smoking ice, the unthinkable luxury of it, in wonder and awe, and remembered that it was not normal for humans to spend their afternoons trapped inside a 100-degree building in Washington, sweating through their clothes, solving puzzles to save the free world. She remembered she was alive.

One evening she read an issue of *The New Yorker* straight through.

Germany was only the first leg of William's TICOM mission. The second leg took him to Bletchley Park, headquarters of British codebreaking, in late July. He arrived there on July 28, eight days before America dropped the first atomic bomb on Japan.

As he had done on his previous visit to Bletchley, William kept a detailed diary. One entry described a meeting with Alan Turing: "At 1535 a visit with Dr. Turing. He is leaving GC&CS, to my surprise. Says he's going into electronic calculating devices and may come to the U.S. for a visit soon. Invited him to visit us if he comes to Washington." This turned out to be the final encounter of William Friedman and Alan Turing. The two geniuses would never see each other again. In 1952, the British government stripped Turing's security clearance on grounds that he was a homosexual, and officials coerced him into taking estrogen injections. Turing's maid later found him dead of an apparent suicide, a half-eaten apple by the side of his bed, traces of cyanide in his blood. A government witch hunt had destroyed one of the war's greatest heroes.

William's goal in England in July 1945 was to learn how the war had looked to his Nazi adversaries, the code and cipher experts employed by the Reich. He read cryptologic materials seized by British intelligence and observed interrogations of Nazi prisoners. Twice he visited a manor in the country village of Beaconsfield, noting that he could "say no more." This was a POW camp where he encountered at least three high-value German POWs, including two of Nazi Germany's top cryptologists, Dr. Wilhelm Fricke and Erich Hüttenhain. William didn't conduct the interrogations but he did observe and suggest questions. After listening to the POWs and analyzing the documents, William concluded that Germany had never lost faith in the security of the Enigma machine. They thought Enigma was unbreakable all the way to the end. He was proud to learn that Nazi codebreakers had never managed to defeat America's best cipher machine, the SIGABA, which he had invented with Frank Rowlett.

He had a lot of downtime in England. The pace of things in the codebreaking offices had slowed. The buildings seemed to be emptying out. At night he took Amytals and crawled into bed. One evening a friend took him to a burlesque show in London, the famous *Les Folies-Bergère*. They sat in two plush seats in the front row. The women wore g-strings and glittery, spangly tops, and William admired the looks of intense concentration on their faces, their cool self-possession. "The girls devote their complete and absorbed attention to their work—not even a glance or a wink at any member of the audience."

Once or twice in the slack moments of the days, the British asked him to tell stories about crazy old George Fabyan and his merry band of conspiracy theorists. People seemed to love the Riverbank stories, and William loved to tell them. It was funny how he felt more and more generous toward Fabyan by the year. You get older and want to connect to the people who understand. You try to speak with the young and find that something is wrong with your ears. They use their own slang, their own code, and you start to feel nostalgic about your former enemies, who at least shared the same intense moment on earth and spoke words you could understand. Besides, if not for George Fabyan, William would not now be carrying a piece of Adolf Hitler's smashed marble floor in his pocket.

It had all begun in the most bizarre fashion.

William was in London on August 6, 1945, the first day of the nuclear age. He was asleep and dreaming in his room at the Hunt Hotel when the atomic bomb fell on Hiroshima and destroyed it in seconds, its people and its history, like a page torn from a book. A latch opened in the American B-29 at sixteen minutes after midnight, London time, which was 8:16 A.M. Hiroshima time, and the soldiers on the plane became the first humans to see, from a safe distance, what this technology could do to a living city. "Here was a whole damn town nearly as big as Dallas," the radio operator of the *Enola Gay* later recalled, "one minute all in good shape and the next minute disappeared and covered with fires and smoke." Above Japan, the plane peeled away from the mushroom cloud, while in London the cryptologist's eyeballs darted frantically behind closed eyelids. He was having a sex dream about Enid, the wife of his friend Stub. When he woke in the morning, he felt confused. He had never thought about Enid that way. He wrote in his diary that he would have to tell her. She'd think it was funny.

It took a while for the news to reach London, and William was busy with his work, so he didn't hear about Hiroshima until breakfast on the morning of August 7. He went to lunch that day with Eddie Hastings, the Royal Navy captain who had visited the Friedmans' house on the day of Pearl Harbor, and after drinking a few martinis in befuddled silence, William and Hastings spoke about the bomb. They were in agreement. They didn't understand why it was necessary to kill so many civilians merely to demonstrate the bomb's power. Hastings thought "it was [a] serious mistake to drop the first one on a big city—should have stated the case, given warning, dropped 1st one on a vacant area & then make renewed call to surrender," William recorded in his diary. "I think he is right. Early reports indicate over 350,000* people wiped out in Hiroshima—perfectly ghastly, no matter who the enemy may be." He added, "We all here agree that this new weapon represents the last call on man to give up war—or else!"

The second atomic bomb fell two days later, on Nagasaki.

At home in Washington, waiting for the war to end and for her family to come home, Elizebeth sat on the porch in the evening and wrote letters to the children and William, taking breaks to go upstairs and listen to radio bulletins. The weather turned cool and rainy. Each time she wrote to William she had to use a different address because he seemed to be moving all over the place. She worried he wasn't getting her letters. His letters reached her after a time lag of seven or eight days, which meant that their letters were crossing midstream.

---

* The true number was closer to 75,000 dead and another 75,000 injured.

On August 7, when the radio carried news of the atomic bombing of Hiroshima, the news didn't bother Elizebeth as much as it did William. She wrote to her husband that day, "Everyone is saying this will end the war P.D.Q."—pretty damn quick. "I wonder! Much too good to be true, say I."

The morning after the Nagasaki bombing, August 10, William got up in his London hotel, shaved, showered, ate a breakfast of bacon and egg, and took a bus to one of the city offices of his British colleagues. It was a bright day and the sun felt good on his face. After lunch he decided to get out and watch some tennis at a nearby public court, a mixed-doubles match, and he was enjoying the high quality of play when at 1 P.M. a U.S. Army lieutenant came running to tell him the Japanese had accepted surrender terms presented to the emperor. Unofficially, the war was over.

Word reached Elizebeth at the Naval Annex that day, and at 7:30 P.M. she started a new letter to her husband, writing at the top, "A day we will remember!" She told him she was getting tired of managing everything at home, fixing things around the house, sopping up after a small flood in the basement, getting the car ready for its annual inspection ("the fenders cost $30, the other work $13"), "all chores and no play with my Sweetheart. But maybe if V.J. comes true, we can both go play for a long vacation."

"By the time this reaches you," William wrote four days later, "the end of the war will be a fact." He said he was sorry she was tired and knew it was hard to be alone. He suggested sweeping the leaves from the sewer inlet on the side of the house to prevent rainwater from backing up and leaking into the basement. William also replied to a warm letter from John Ramsay, who had asked his father which books he should read to educate himself in spare hours at the Army Air Corps barracks. William recommended *So Little Time,* a war novel by J. P. Marquand ("so good"), and ended the ten-page letter by praising his son's vocabulary: "You've improved remarkably in penmanship and format. You do yourself proud, in fact. I found only one or two orthographic irregularities or aberrations (misspellings, to you!). They are of no consequence. Dad."

Japan surrendered unconditionally on August 14. President Truman declared a two-day holiday for federal workers. A crowd of 75,000 gathered at the White House, ringing bells, blowing horns. Elizebeth stayed in and tried to recover from a stomach bug, drinking clear consommé and ginger tea. "Bobbie, darling," she wrote to her daughter, who was scheduled to return from Panama in September, "I sure am counting days—only 23 more and you will be here! Oh, frabjous, frabjous day!"

She watched the lights come on in America. Gas stations resumed normal operations. The newspapers said nylon hose would be available soon. Shoes by Christmas. The military was starting to release large numbers of personnel who were no longer needed. Arlington Hall ordered a 50 percent staff cut by September 30 and another 25 percent cut by December 31. The Naval Annex in Washington had been emptying for weeks and was down to a skeleton force, quieter than Elizebeth had ever seen it. The temporary workers, including most of the WAVES, no longer necessary in peacetime, had been released without so much as a thank-you cake, and permanent employees were escaping the Annex for offices in more modern buildings.

Elizebeth knew she needed to make a decision about her future, about what to do with her postwar life. Her superiors at the coast guard said they wanted her to stay, to keep her codebreaking unit together in peacetime and return to smuggling investigations, but she couldn't see the point. There wasn't a lot of smuggling traffic anymore.

She noticed with detached amusement that American intelligence officials were scrambling to cast themselves and their agencies in a favorable light so that they might keep their jobs in peacetime. "The O.S.S. is starting a deluge of publicity," Elizebeth wrote to William, mentioning the wartime spy agency of Wild Bill Donovan, for whom she had worked in the early months of the war. "Fight against extinction, I suppose."

Elizebeth heard a radio interview with a New York man who taught cryptology classes. He was discussing the importance of codebreaking to the Allied victory. Elizebeth knew him to be a minor figure and wondered, in an offhand way, how he ever got the spotlight.

She turned fifty-three years old on August 26, a cold Sunday. Her coast guard colleague, Lieutenant Jones, came over with his wife, Gertrude, and Elizebeth Friedman, secret hunter of Nazis, cooked a dinner of minced clams in cheese sauce, a tossed French salad, and hot borscht that everyone agreed tasted wonderful in the cold evening. William spaced out his birthday presents to Elizebeth over four days, starting the previous Friday with a cable from Bletchley Park. ABSENCE ON YOUR BIRTHDAY DARLING SADDENSMEBUTHOPEFLORALAMBASSADOR . . . WILLVOUCHSAFEUNDYING LOVE. On Saturday a mutual friend hand-delivered a box of perfume to her door. Sunday morning, a dozen roses came, along with one of his business cards. The front said only, MR WILLIAM F. FRIEDMAN, and on the back he had written, "I love you! I love you! I love you! Bill." The next day, Monday, a letter arrived from William, written eight days earlier and timed to reach her right now. "I find it hard to tell you how much I miss you and love you—you're the most wonderful person to have for wife, helpmate, lover, and all," he wrote. "Save some special kisses for me when I get back. . . . I miss you."

She realized how tricky it must have been to coordinate all these gifts and messages across the distances of the war, the disrupted postal and cable lines, and land them to her at the exact moment of his choosing, around her special day. The timing alone was a performance of devotion.

"Dearest," Elizebeth wrote, "what a darling you are!"

He was ready to come home "very soon," he told her in a letter from London three days after her birthday. He said he was already thinking about what would come next for him and for the family. He wanted to make the kind of money that would give them the freedom to travel and pursue their dreams. National security concerns had always prevented William from profiting from the cipher machines he invented, but the war was over now. Surely he would be permitted to patent his ideas and commercialize them? When he was finished describing these thoughts to Elizebeth he rotated the sheet of paper and filled the side margin: "I LOVE YOU! I LOVE YOU! I LOVE YOU! *VERY MUCH* (I shall have that printed as a border on my special stationery to you.)"

Four days after that, on September 2, in Tokyo Bay, General

Douglas MacArthur accepted the Japanese surrender aboard the battleship USS *Missouri*. Elizebeth heard Truman say on the radio, "It is our responsibility—ours, the living—to see to it that this victory shall be a monument worthy of the dead who died to win it." She agreed that there was no point in having fought a war to preserve freedom if people used that freedom to start more wars. She wondered if William heard the same Truman broadcast. "You are the dearest and best husband any woman ever had!" she wrote on September 4. "Roses lasted until today. All, all my love, Elsbeth."

Around September 12, in Prestwick, Scotland, William finally boarded an army transport plane. He flew to the Azores, then Bermuda, then New York, each minute of the long flights an agony of anticipation. It was raining when he landed in New York and got on a train for Washington. The sky when he stepped out of Union Station was a slab of gray and the air was violent with fat drops that followed him to his office at Arlington Hall. Arlington Hall was like Bletchley Park had been, emptier than he ever remembered it, big empty rooms and echoing hallways and a handful of people carrying boxes around and packing up files. He could not concentrate on anything because he knew he would see Elizebeth soon. He waited out the day and it was still raining like crazy when he left the heavily guarded military facility and shoved his dripping luggage in a taxi and rode home, to the house at 3932 Military Road.

Elizebeth opened the door. She cried out in joy. His clothes were wet. His mustache was wet. She reached up and threw her arms around him and squeezed as hard as she could.

The months after V-J Day were a period of limbo for U.S. intelli-gence. All the agencies were thinking about how to extend the gains of the war and also justify their own existence in peacetime, when the government would surely contract. The future of cryptology was especially murky.

It was obvious to William and many others that there ought to be a centralized cryptologic function in America, one agency that gathered intelligence from wireless signals and broke the codes that must be broken. As an elder in the cryptologic community, a person who had not only invented many of its tools but also built a successful organization within the army to apply those tools, William was involved in these discussions at the highest levels—discussions that would give birth, in 1952, to the National Security Agency. In the meantime he entered a phase of furious personal documentation, writing technical descriptions of his cipher machines and applying for new patents in hopes of commercializing the inventions.

Elizebeth was documenting, too; not for commerce but rather for teaching and history. At the Naval Annex she sorted through the voluminous files of her coast guard unit, tens of thousands of intercepts, worksheets, memos, translations, and decrypts. Working with Lieutenant Jones and other colleagues, she produced a detailed technical account of their unit's work between 1940 and 1945, a 329-page book that detailed all forty-eight of the Nazi clandestine radio circuits and how the coast guard broke the codes. The book was secret, meant only for other intelligence agencies to use as a reference and perhaps also for historians of codebreaking in the far future. Five copies were printed, with dark green covers, and every page of every copy was stamped TOP SECRET ULTRA.

With the technical history complete, Elizebeth was told to mark a percentage of the unit's documents for preservation and destroy the rest. She decided to keep four thousand decrypts— the typed, solved messages from the forty-eight Nazi radio circuits. These she organized for transport to the classified areas of the National Archives in Washington. The phrase "government tombs" occurred to her. That's what it felt like. She was burying her experiences in Uncle Sam's mausoleum.

When the task was done, Elizebeth prepared to leave the Naval Annex for the last time. The navy forced her and all other departing workers to sign secrecy oaths that demanded their silence unto death. They could never tell anyone what they did in the war, under penalty of prosecution, for as long as they lived. They could not even tell their grandchildren.

At the end of her final workday, Elizebeth walked down the

stairs from the second floor to the first, went out past the turnstile where the first marine guard stood watch, then past the second marine guard, to the other side of the barbed-wire fences, until she was standing on the sidewalk on Nebraska Avenue. She crossed the street, paused for a few seconds, and looked back at the grubby, flat-roofed building where she had spent her war. She knew in that moment that she would never again return "to that particular form of endeavor"—breaking codes for the coast guard. "I was back in the world-at-large once more," she wrote later. "It was the end of a Period, an Era."

She was still a coast guard employee, and soon Elizebeth found herself back at her old desk in her old prewar office in the Treasury Annex, near the White House. But she had an exit plan. She was only going to stay long enough to complete a single job. At the Naval Annex she had sorted and filed the records of her clandestine war against the Nazis. Now, at Treasury, she needed to do the same for her smuggling cases of the 1920s and '30s. The smuggling records had been gathering dust during the war—"thrilling records in many respects, detective stories of high interest in many cases," Elizebeth recalled. "The past had been rich in accomplishments. I should see that everything was prepared for posterity to comprehend, if posterity should ever choose to examine the archives."

From the late fall of 1945 to summer 1946, Elizebeth conducted her last campaign for the United States: organizing and indexing the paper archive of her cat-and-mouse tussles with rum lords and drug gangs. Because the records were old and contained no national secrets, she was allowed to keep personal copies for her own library. Then, the task complete, she recommended to Treasury that the department abolish her coast guard unit, along with her job, on the grounds that it served no national purpose in peacetime. They obliged. On August 14, 1946, the coast guard notified her that, "In view of the curtailment of cryptanalytic activities previously performed by the U.S. Coast Guard, it has been necessary to effect a reduction in personnel," and she was hereby terminated at the close of September 12, 1946. Her salary at the time, the most she ever earned, was $5,390, or $67,000 in today's dollars.

J. Edgar Hoover used his influence to expand the FBI after the war. Elizebeth used it to get out of the game.

She had never really wanted to be a government employee anyway. It was only the constant requests from "people on my doorstep" that had gotten her into it in the first place. Now, with the war over, her thoughts turned to projects and desires she had put on hold to serve her country. She still wanted to finish her long-in-progress children's book about the history of the alphabet. She wanted to visit Barbara at Radcliffe and see how John Ramsay was living at the Army Air Corps base in Biloxi, Mississippi. And she wanted to reconnect with William and find a way to collaborate with him. The Friedmans had lived for years in an awkward and isolating silence, working in separate but adjacent government bunkers, afraid to speak freely even in their own home. No more! Goodbye to that! They wanted to work together on something again, and they had the perfect idea.

Elizebeth and William had never lost their fascination with the varieties of occult theories they first encountered in their youth at Riverbank; they never stopped wondering why people believed things that weren't true. The previous December, when the war was still on, they had attended a sold-out Washington show by the Amazing Dunninger, the foremost mentalist of the day. A New Yorker with a poof of brown hair and a tuxedo, Dunninger was both debunker and illusionist; he explained onstage how spirit mediums usually worked, showed that he was not using any of those tricks—and then read the minds of audience members anyway. William and twenty-five other intelligence men planted themselves here and there throughout the crowd at Constitution Hall in an attempt to learn his methods and "came away with *theories* as to how it's done, but no proof," Elizebeth wrote in a letter to her daughter. "The mere fact that Dunninger is still going strong is proof that human beings, the credulous dears, *want* to believe in the mysterious and supernatural."

It had not escaped Elizebeth and William that many people continued to believe the theory that the two of them had rejected in their earliest days at Riverbank, way back in 1917: that Francis Bacon placed cipher messages in Shakespeare's plays.

The community of Bacon obsessives was still around, alive and kicking, publishing new articles and arguments. After Mrs. Gallup died in 1934, followed by George Fabyan in 1936, the Baconians lost two of their most famous and energetic proponents, but others picked up the torch. In 1938 the son of Teddy Roosevelt, Theodore Jr., asked the Friedmans for an opinion on a cipher system devised by an economist named Dr. Walter McCook Cunningham. Roosevelt Jr. was vice president of the Doubleday publishing firm and Dr. Cunningham had submitted a manuscript about his cipher. The method was based on anagrams, and the Friedmans quickly recognized it as bunk. To demonstrate the cipher's folly, they applied Cunningham's method to a page from *Julius Caesar* to produce the following message, which they sent to Roosevelt Jr.:

> *Dear Reader: Theodore Roosevelt is the true author of this*
> *play but I, Bacon, stole it from him and have the credit.*
> *Friedman can prove that this is so by this cock-eyed cypher*
> *invented by Doctor C.*

The experience got them thinking that they should lay out their skeptical arguments in a book of their own, explaining once and for all why these ideas about secret messages in Shakespeare were only fantasies. The Friedmans obtained a pittance of a book deal from a British publisher (advance on royalties: 250 pounds) and went to work. For the sake of the project, they decided to sell their beloved house on Military Road and bought a spacious, high-ceilinged house on Capitol Hill within walking distance of two libraries where they needed to do research, the Folger Shakespeare Library and the Library of Congress. Many who live on Capitol Hill are lobbyists. The Friedmans moved there to be close to libraries.

They transported their own precious books and papers to the new house, reassembling their private library in the den of the second floor, and rehung the axe on the wall as a warning to potential book thieves. And together, researching and writing, they galloped back through the past, weighing the arguments of Baconians and cutting them to pieces. In their hands *The Shakespeare*

Ciphers Examined became a story about the drug of self-delusion
and the joy of truth. One section analyzed the cipher system of a
French general that had revealed the secret phrase IF HE SHALL
PUBLISH. The Friedmans showed that the cipher could just as
easily have produced the text IN HER DAMP PUBES. George
Fabyan received the full brunt of their scrutiny. The Friedmans
wrote that while Fabyan possessed "great natural gifts of energy
and dynamism," he was a salesman, not a scientist, and suppressed
facts he didn't like. As for Mrs. Gallup, "a sincere and honourable
woman, and no fraud," she "found in her texts what she wanted
to find" and "was therefore at the mercy of the promptings of her
expectant mind."

The Friedmans wrote with a ruthless honesty because that's
who they were as people. Still, working on the book made them
realize how much they owed the misguided mentors of their
youth. In the preface they thanked Mrs. Gallup, "whose work
on the question of Shakespearean authorship aroused our life-
long interest in the subject," and they thanked Fabyan, too—for
introducing them to Mrs. Gallup. They were genuinely grateful.
Elizebeth said she and her husband had decided to "give the devil
his due," and in later years Elizebeth even went as far as admitting, "Vile creature that he was in many ways, George Fabyan
really launched two or three things that were of vital importance
to this country," which was true. For all his malice and supersti-
tion, Fabyan threw enough money at actual scientists to accelerate the discovery of actual knowledge. He funded investigations
of Nature with a fortune that other tycoons would have spent on
yachts and jewels. He succeeded in creating the first real codebreaking institution in America, Riverbank Laboratories, an
idea factory christened by wartime realities. It not only forged a
new science of immense power; it also spawned a love affair that
spread the science and ultimately sharpened it into an antifascist
weapon. The modern-day universe of codes and ciphers began
in a cottage on the prairie, with a pair of young lovers smiling at
each other across a table and a rich man urging them to be spectacular.

Until she started researching the book in 1946, Elizebeth
always insisted that her life in secret writing was an accident, a

series of unpredictable chases, mazes, escapes, and detective capers. Now, viewing her life from a distance, she understands there might be order in it after all, a taut line stretching back through the decades and terminating at that mad place on the prairie.

To help herself write vividly about Riverbank, Elizebeth sits in the new house on Capitol Hill. She closes her eyes. She tries to imagine herself thirty years earlier, in the summer of 1916, a young woman at a rich man's estate, unmarried and free, her whole life in front of her.

A fragrance of overripe banana wafts up. William's fruit flies in the windmill.

The fire pit at night. The chemical reek of a mortar bursting near the ordnance lab. Fatty pork on her dinner plate from pigs slaughtered at Fabyan's word.

Silver blade of river, dome of prairie sky.

She remembers riding bicycles with her friend William Friedman, rushing past lawns and flowers thickened with summer rain, a blur of green and pink. She remembers the low Illinois sun streaming through the windows of the Lodge as she works there with Mrs. Gallup, struggling to see what the older woman saw, squinting through a magnifying glass at a page of Shakespeare, trying and failing to free the imprisoned ghost of Francis Bacon.

Mrs. Gallup and Fabyan keep telling her, try harder. The messages are there.

And there comes a day when Elizebeth just thinks: no.

There is nothing wrong with me. What's wrong is with other people.

This is the moment that hurls her out to the rest of her life. The savaging of Nazis, the birth of a science: It begins on the day when a twenty-three-year-old American woman decides to trust her doubt and dig with her own mind.

The room is dark but her pencil is sharp. An envelope of puzzles arrives from Washington, sent by men who have the largest of responsibilities and the tiniest of clues. With William she examines the puzzles. He is game, he looks at her with eyes like little bonfires, he is in love with her. She is not in love yet but she would not be ashamed to fall in love with such a bright and kind person. She stares at the odd blocks of text and starts to flip and stack and rearrange them on a scratch pad, a kindling of letters, a friction of alphabets hot to the touch, and then a flame catches and then catches again, until she understands that she can ignite whenever she wants, that a power is there for the taking, for her and for anyone, and nothing will ever be the same. The ribs of a pattern shine through. Something rises at the nib of her pencil and her heart whomps away. The skeletons of words leap out and make her jump.

Girl Cryptanalyst and All That

The Friedmans in their home library, 1957.

The government came for their books on an otherwise ordinary Tuesday in 1958. Scattered clouds, cool midwinter sun. William and Elizebeth were inside their home on Capitol Hill and heard a knock on the front door. They opened it and saw at least three men from the government. Behind them, on the street, was a rented truck, as if the men planned to remove something large from the house.

The Friedmans let them inside. One of the men, S. Wesley Reynolds, was the NSA's director of security. A second man worked for Reynolds, and a third worked for the U.S. attorney general.

The men asked to see the home library. The Friedmans brought them up to the second floor.

Elizebeth was sixty-six now, William sixty-seven. His health was precarious but the men didn't know that. They said they had orders to remove a list of books and documents that the NSA wished to reclassify according to a Defense Department order of July 8, 1957, Directive 5200.1, which declared that cryptologic documents previously marked "Restricted," a low level of classification, were now upgraded to "Confidential," a higher level. To the horror of the Friedmans, the men started to pull things off the shelves. They removed forty-eight items, including an entire personal safe full of William's documents, several manuals he had written about cryptology, envelopes of his lecture cards and notes, and his own articles from every phase of his career, including Riverbank, forty years ago.

According to a rumor that later spread through the agency, William "went berserk and he was throwing books around and saying, 'Take this, take that.'" The junior NSA employee who went to the house denied this but admitted that both Friedmans appeared "obviously upset by the action being taken." The NSA's Reynolds wrote in a memo three days later, "Mr. Friedman voiced no objections to my taking this material, however, it was quite obvious that he felt deeply hurt and that the material was being taken for reasons other than Security. He stated that this material deals with the history of cryptography and should belong to the American people."

William didn't understand why information about hand ciphers from the First World War needed to be seized. The ciphers were obsolete. Was it really necessary to seize papers from 1917 and 1918? To raid their home, their sanctuary, their archive of knowledge? He told a friend, "The NSA took away from me everything that some nitwit regarded as being of a classified nature."

As the men worked, carrying files out to the truck, Elizebeth looked on in silent rage, barely suppressing her tongue. She considered this a violation of their privacy and worried it was bad for William's health, which had corroded in the thirteen years since the war, darkening with the mood of a city where counterintelligence had become an obsession. Soviet spies had stolen nuclear secrets from the Manhattan Project, and the FBI and the House Un-American Activities Committee went hunting for communist agents. "The mad march of red fascism is a cause for concern in America," J. Edgar Hoover said to HUAC, promising that the bureau would attack and expose "the diabolic machinations of sinister figures." Senator Joe McCarthy destroyed people's careers with no evidence at all.

William's depression had returned in 1947. At first he complained to a doctor of "psychic giddiness" while walking and playing golf; the condition manifested itself as a tendency to walk to the left. The giddiness was followed by increasingly severe bouts of insomnia. Unable to sleep, on January 23, 1949, he checked himself into the psychiatric ward of the Veterans Administration hospital in Washington, where doctors placed William with a group of deeply psychotic patients. He hated it there. He went home and continued to deteriorate. By January 1950, William was unable to work or solve puzzles, his mind and muscles seeming to move at one-third or one-quarter speed, and suffering from acute despair. He had suicidal thoughts. His son found a rope and a noose at the house. A friend noticed a length of rope in the backseat of William's car and asked about it. William replied in a joking tone, "I'm looking for a tree to hang myself."

Desperate for a solution, he sought out a new psychiatrist in March 1950, Dr. Zigmond Lebensohn of George Washington University Hospital, who was an early proponent of electroshock therapy. William agreed to try it. The first course of shocks began on March 31, 1950. The legendary William Friedman was repeatedly electrocuted while awake, possibly without muscle relaxants (they were not widely used at the time), a heavily padded tongue depressor placed in his mouth to prevent him from breaking his jaw by grinding his teeth when the seizure hit. After six courses of shocks, five to fifteen shocks per course, William was sent home on April 11, 1950. Lebensohn observed that the patient "was almost elated when he was discharged and in a characteristically effusive way he kissed the nurses goodbye in a rather avuncular fashion. About a month or so later I saw him and his wife at a Toscanini concert at Constitution Hall."

William's illness took a toll on Elizebeth. Hair graying at the temples, perhaps shrunken by an inch (she considered herself five feet and two inches tall now instead of five three), she was 110 pounds and thinner than she'd been since she was a girl. In the polite phrasing of a girlfriend, "Anxiety kept her figure slim." Retired from government and earning a tiny pension, she spent increasing amounts of time taking care of William. On mornings when he was depressed, she helped him get dressed, drove him to work, walked in with him, placed a pen in his hand, and moved his hand to get the pen moving. She answered his professional mail when he was incapacitated in mental wards. Somehow she still made time for friends and hobbies. She surprised her friends by getting serious about cooking, hosting dinner parties themed around the dishes of India, Mexico, Italy: "I found it an outlet for some hidden creative instincts perhaps." She looked after her neighbors, once appearing on a sick neighbor's doorstep with a tray of roast lamb, roast potatoes, gravy, and a yellow rosebud in a vase. She stayed active in the League of Women Voters, researching the legal status of women, international relations, finance, and the urgent need for D.C. statehood. "At the drop of a hat," she wrote, "I will turn on a spigot labeled SUFFRAGE FOR THE DISTRICT OF COLUMBIA!"

It often seemed that she had forgotten her own career in codebreaking, that she was content to see her identity and history wash away. This wasn't the case. In 1951 she received an invitation to speak about her life in codebreaking to a women's social club in Chicago founded by the first female judge in Illinois. At first Elizebeth urged the group to reconsider: "That part of my life is over, my dear," she wrote the chairwoman. "You are asking a Has-Been to speak! Your audience will feel cheated, I am sure." But then she wrote a speech and traveled to Chicago with a suitcase full of lantern slides and at least fifteen mutilated sheets of paper, typed and cut with scissors and taped back together into a new order while she had agonized about what to say, and as soon as Elizebeth introduced herself to the women of the club, the beautiful hopeful postwar women of Chicago, they were hanging on her every word.

Speaking in a pink ballroom at the Blackstone Hotel, where the women had gathered for a dinner-dance, Elizebeth made it clear she wasn't free to talk about her life during the Second World War, but she was happy to share anything else, to answer any question at all. "Perhaps you may think that the expression 'code and cipher expert' describes a person who must live in a world apart," she said, then explained why this is a misconception. Your child's report card is a code. A is good, F is bad. It's not a world apart. It's just the world.

Elizebeth showed slides of code messages from her famous cases. The *I'm Alone*. The heroin network of the Ezra twins, "SOLVED BY WOMAN." The polite Canadian gangsters of the Consolidated Exporters Corporation. The women of Chicago kept her there, asking questions, transfixed, and afterward, Elizebeth received more speaking invitations, traveling to Detroit and giving her talk to a pair of neighborhood groups in private homes; one of the groups asked her questions for two and a half hours. They seemed to think that the story of Elizebeth Smith Friedman was one of the greatest they had ever heard.

Every once in a while, the urge struck Elizebeth to write it all down in one place. She wondered if history would remember her. One winter she and William traveled to England and attended a luncheon at Cambridge with two of their colleagues from the war, including Elizebeth's cheerful comrade, the astronomer Chubby Stratton. The men at the table got to talking about the war. "As befits a woman in the monastic traditions of Cambridge, I said little," Elizebeth recalled later, "but my own recollections began to boil up from the cauldron of memories."

After the luncheon she took out a sheet of lined yellow paper, wrote "FOREWORD" at the top, then described her feelings after V-J Day in 1945, when she "folded my tent to steal away" from the coast guard after six years of "exciting, round-the-clock adventures as we counter-spied into the minds and activities of the agents attempting to spy into those of the United States." She continued for seven pages, hinting at the dramas and capers of her war without going into specifics, the way an author does at the beginning of a book.

If Elizebeth intended this to be her memoir of the Invisible War, she never wrote the rest. The seven handwritten pages and a typed version of the same are all that exist. She later tucked the typescript into a manila folder marked "foreword to uncompleted work."

President Truman established the National Security Agency on November 4, 1952, at the peak of McCarthy's popularity and two and a half years after William's shock treatments. The NSA fused the signals intelligence units of the army and navy into one organization, including the unit that William founded and nurtured between the wars.

From the start the NSA was the most secret of agencies, basic facts of its existence concealed. William accepted a job there as a counselor and adviser, a role befitting a respected elder. But the agency had less and less use for him as it grew through the 1950s. It hired thousands of young linguists and cryptanalysts who were trained by the textbooks William wrote but who didn't necessarily listen when he spoke. It broke ground on a new campus in Fort Meade, Maryland, where today at least twenty thousand people work inside two large cubes of eavesdropping-resistant blue-black glass, and invested heavily in computers for breaking codes. William thought computers were "mostly nonsensical and completely nitwit gadgets for daily affairs," he wrote in a morose letter to the historian Roberta Wohlstetter. And as the NSA grew larger and stronger, it began to use that strength in ways that made William uncomfortable. It scooped up enormous quantities of signals seemingly because it could, towering haystacks of intelligence that would make it difficult to find the needles, and it continued to conceal and classify more and more kinds of documents that William thought should be publicly available. At other times in his life he had argued for greater secrecy, as when he objected to Herbert Yardley's book in the 1930s; now he muttered darkly to friends about a "secrecy virus" loose in government.

He suffered his first heart attack in April 1955, followed quickly by a second while in the hospital recovering from the first. That fall William retired from the NSA as a full-time employee. The agency gave him a nice ceremony and a consultant contract to keep him around; the director of the NSA at the time, Ralph Canine, admired William. Then a new director replaced Canine, a man with more inflexible views about secrecy and no personal fondness for the great codebreaker, and the agency raided the Friedmans' home library, and William became depressed again. He wanted to criticize the agency in public, to sound the alarm about the secrecy virus, but feared the NSA would withdraw his security clearance, severing him from his community and many of his own writings.

Whether or not the agency was specifically trying to humiliate him or just rigidly following regulations, William *felt* persecuted, and in his mentally delicate position, the ordeal was enough to push him to the edge. "Frightening to be alone [with] suicidal thoughts," he scribbled on a loose sheet of paper. "For fifty years have struggled with this off and on. . . . Repression by secrecy restrictions—fear of punishment chimerical but still there."

As his disillusionment with the NSA intensified into full-blown paranoia, he reconsidered his long intent to donate his papers to the Library of Congress. He couldn't bear to hand over the contents of his private library, his proudest possession, to the same government that had sent men to raid it. After some thought he decided instead to bestow his archive to the George C. Marshall Foundation, a private institution at the Virginia Military Institute in Lexington, Virginia. With Elizebeth's help he began organizing and indexing his vast trove of treasures in preparation for transfer to the Marshall Library: thousands of books, papers, memos, photographs, prototype board games, and other cryptologic curios. For a brief time the project seemed to revive him. "I now have a great desire to live," he wrote, "to bring the Marshall Foundation project to a completely satisfactory conclusion." His body did not cooperate. He suffered more heart attacks. His feet swelled so much he could not climb the stairs at the Folger Shakespeare Library when he went to hear lectures.

Elizebeth cared for him as always, taking notes on his condition in a daybook.

> MARCH 15, 1969: *Bill had fall in night. Confused and loss of memory momentarily.*
> JULY 20: *MAN ON THE MOON. ES & WFF watched on CBS until 3 a.m. when Neil Armstrong and 'Buzz' had finished moon walk and return to the module.*
> SEPTEMBER 24: WFF *birthday. Asked for spare ribs!*

A few minutes after midnight on November 2, 1969, he had his last heart attack and stopped breathing. Elizebeth called the doctor. William could not be revived. The doctor stayed at the house until after 2 A.M. to comfort her while William's body was taken away.

Overwhelmed, she picked up the daybook, out of habit.

> *My beloved died at 12:15.*

She started a brief letter to Barbara, who was traveling in Rome.

> *Dear heart be courageous. Your beloved father died. . . .*
> *Rejoice that he suffered only a very short time.*

More than 750 letters and cards of sympathy arrived at the house over the next weeks. Joseph Mauborgne called William "the greatest brain of the century," a man with an "ever shining place in history." The novelist Herman Wouk wrote to Elizebeth, "His effect on world history was incalculable, greater than that of kings & captains. Yet what a modest man!" Juanita Morris Moody, a codebreaker who got her start at William's Arlington Hall in 1943 and went on to supervise the NSA's Soviet desk, told Elizebeth that her husband was the last of his kind: "Our business now involves many more people and disciplines," Moody wrote. "It has become more abstract and impersonal. There are no more William Friedmans nor will there ever be."

Elizebeth received, from the Board of Management of the Cosmos Club, the men-only social club in Washington to which William had belonged, a "Woman's Privilege Card," granting access to the club's facilities for a period of two years.

She designed his tombstone.

WILLIAM F. FRIEDMAN
LIEUTENANT COLONEL
UNITED STATES ARMY
1891 ••• 1969
KNOWLEDGE IS POWER

Elizebeth decided to embed a secret message in the stone, in Bacon's cipher, in the letters of Bacon's quote. She specified that certain letters be carved with serifs and the rest without. The serifs were the *a*-form, sans-serifs the *b*-form:

KnOwl / edGeI / spOwE
(*a*- & *b*-forms shown as lower & upper case)
babaa / aabab / aabab
W / F / F

WFF: her husband's initials. It was a signature in cipher.

The army buried William with full honors at Arlington National Cemetery, the casket draped with a flag and carried by six black horses along the winding roads of the cemetery to the grave, accompanied by drummers. People from every branch of the military attended the funeral, and so did the antiwar U.S. senator from Minnesota, Eugene McCarthy. Elizebeth and the children were amazed to see him. The kids had worked on his 1968 presidential campaign. It turned out that McCarthy had worked as a codebreaker at Arlington Hall in 1944 under William's command. The family had had no idea.

After the funeral John Ramsay sent an emotional thank-you letter to McCarthy. "Your presence there seemed to make the idea of a military funeral a little more bearable for all of us. . . . I thought you might like to know that my father was a gentle and peaceful man who detested killing and war, secrecy, spying and all the things you and I hate. But he had a mad love affair with the world of secret writing to which he devoted his life and for which he felt many deep pangs of guilt. In spite of all his honors, he was not a happy man."

Elizebeth became William's avenger. Bitter about his treatment over the years by the army and the NSA, and worried that his contributions would be forgotten or erased, she set out to make sure that William received the credit he deserved. She took on this burden at the expense of curating her own legacy, which her grief and her anger now made a secondary concern.

Immediately after his funeral, in the now-empty house, she sat at William's own desk, the one with the 1918 KNOWLEDGE IS POWER photo under the glass, and worked to complete the annotated bibliography of his papers. The task occupied her for eight to ten hours a day. She mourned her husband while writing crisp descriptions of his articles and books on index cards. She did it out of a sense of duty to William, who would have wanted the project completed, and she also hoped that the collection, once open to the public, would entice a first-rate historian to write a biography of William, a book to cement his reputation.

The Marshall Library paid for a typist to help her one to two days a week and it still took months to finish the 3,002 cards for the 3,002 unique items in William's collection. Then she arranged to transport all of the material from Washington to the library, three hours south. Men came to the house one day in 1971 and loaded the boxes into trucks, along with William's desk. She told friends it felt like watching Bill die all over again. She followed the trucks on the highway in her beat-up, ten-year-old Plymouth, engine wheezing all the way to Lexington: "I guess I'm just a little old lady standing in the center of ruin and decay."

At the Marshall Library she worked six-hour days to manage the details of the transfer, making sure the papers were handled just so, out of love and respect for Bill. The archivists were thrilled to have her guidance (she "was entertained like a queen," she said) and got her on tape speaking about the donated materials, the Friedmans' life together, and Elizebeth's own career. And though she kept the focus on Bill, she also told stories about herself and donated thousands of her own personal papers to the

Marshall, separate from her husband's collection. Elizebeth's papers included documents she had preserved from the smuggling era of the 1920s and '30s, personal letters, her unfinished book manuscripts, diaries, and a lot more, but she had not indexed and annotated the collection like she did with William's. The archivists helped organize Elizebeth's files into twenty-two archival boxes, reverently stored behind the metal doors of the vault on the first floor.

In years that followed, researchers journeyed to the Marshall Library and used the Friedman files to write books that wouldn't have been possible before. The author James Bamford relied partly on William's collection to piece together his 1981 book, *The Puzzle Palace,* the first popular history of the NSA, whose publication the agency tried and failed to stop. The NSA sent representatives to the library twice, in 1979 and 1983, each time removing an unknown number of William's items, but the Friedmans had done such a careful job of indexing that a sharp-eyed professor at Virginia Military Institute, Rose Mary Sheldon, noticed that about 200 of the 3,002 index cards were missing. Sheldon submitted a series of Freedom of Information Act requests that eventually prodded the NSA to release 7,000 additional Friedman documents. In the last two decades the agency has gotten more comfortable telling its history—today it holds public cryptologic history conferences and operates a museum—but it took a while, and in the meantime, the Friedmans had created this alternate archive, beyond U.S. government control, where anyone could learn about U.S. codebreaking.

Even so, the attention of researchers fell lopsidedly on one Friedman and not the other. Elizebeth's papers at the library, unindexed and therefore mysterious, largely gathered dust while people explored William's. The world forgot about her and remembered him, which is what she had expected anyway. In 1975 the NSA informed Elizebeth that it planned to name the main auditorium at Fort Meade in William's honor and asked her to inspect and approve a bronze bust of his head. She attended the dedication ceremony. The NSA men's chorus sang "The Testament of Freedom." The following year a biography of William was published, *The Man Who Broke Purple,* which Elizebeth felt was a competent account of her husband's professional life but did not capture "the man I knew and loved."

She struggled in her final years as her savings dried up and her arteries hardened. She missed Bill so much. In her letters she sounded like a battle-hardened version of the girl who set Riverbank aflame, quick as ever but no longer joyful. "There is just one thing in this world I would now advise all unborn babies," Elizebeth typed one morning in a long letter addressed to no one ("I just had to blow off some steam"). She continued, "Either be born Rich or BE BORN POOR. It is we in between who PAY-PAY-Pay-y-y-y." She disliked the direction her field was taking, its increasing reliance on computers. She gave an interview to a *Houston Chronicle* reporter who found her "lounging in a turquoise silk robe from China, a gift from her husband in 1928." She told him computers are a curse. "The problem with machines is that nobody ever gets the thrill of seeing a message come out." She let her children know she wanted her body to be cremated when she died, with no funeral services. "In a few years there will be no place left on earth to bury any one, and before too long, I think, all cemeteries will have to be disposed of," she wrote. "Why add one jot or tittle to the mess already in existence?"

Elizebeth was eighty-eight when her arteries failed. She died on October 31, 1980, in a nursing home in Plainfield, New Jersey, four days before Americans elected Ronald Reagan to his first term as president.

The public response to her death was more muted than it had been for William's eleven years earlier. The *Washington Post* and *New York Times* printed respectful obituaries of Elizebeth. None of the obituarists mentioned her feats of codebreaking in World War II; almost certainly none of the writers were aware.

At Arlington National Cemetery her ashes were scattered atop William's grave and her name carved beneath his:

BELOVED WIFE
ELIZEBETH SMITH FRIEDMAN
1892 • • • 1980

For years, nothing much happened.

It took a while for people to rediscover Elizebeth. Bit by bit, people went looking. Mostly women. They suspected there was more to her story than had been told, and they were right. A historian at the Department of Justice, Barbara Osteika, located records of Elizebeth's old smuggling cases and came to see Elizebeth as a "beacon of hope" for women in federal law enforcement, a trailblazer. An FBI cryptanalyst, Jeanne Anderson, who solves the handwritten code and cipher notes of suspected criminals, found transcripts of Elizebeth's trials from the 1930s and studied them for guidance on speaking to juries. And although Elizebeth had never worked there, she also won fans at the NSA, where female cryptanalysts rose to distinction after the war, including Juanita Morris Moody, who briefed U.S. leaders during the Cuban Missile Crisis, and Ann Caracristi, who became the agency's number-two official.

In the 1990s the NSA renamed its auditorium. The William F. Friedman Memorial Auditorium is now the William F. Friedman and Elizebeth S. Friedman Memorial Auditorium. As of 2014 there is a second auditorium in the Washington area bearing her name, at a Justice Department building, thanks to a campaign launched by Barbara Osteika. Above the doors it reads, ELIZEBETH SMITH FRIEDMAN, PIONEER OF INTELLIGENCE-LED POLICING.

These things happened for two reasons: because women went looking for Elizebeth's ghost, and because her ghost was making noise in the archives. She was there inside the Marshall Library, rattling the doors of the vault, and she was in the "government tombs," the National Archives, where her records from the Invisible War were finally declassified. The ghost also cried out from unexpected places. Three of the index cards in William's collection contain brief, verifiably true comments about how J. Edgar Hoover and the FBI took credit for feats of spycatching actually performed by Elizebeth and the coast guard. These comments were obviously written by Elizebeth—William wasn't in a position to know. Each card is a knife slipped between the ribs of Hoover, Elizebeth's patient revenge.

She intended to use all of these archives to write her own story. She never got around to it. Maybe she lost hope. But the files are exactly where she left them, the fragments of an extraordinary life. The files have a weight to them, a texture. They can't be erased any more than Elizebeth's legacy can be erased, because her legacy is embedded in our lives today, in our smartphones and Web browsers, in the science that powers secure-messaging apps used by billions, in the clandestine procedures of corporations and intelligence agencies and in the mundane software loaded onto the iPhones in our pockets.

Secret communication is still a dance of codemakers and codebreakers, locks and lockpickers. The locks are different now, of course. With computation as an aid, everything has been massively sped up and mathematized beyond anything Elizebeth would have comfortably understood. But the game is still based in patterns. Someone designs a pattern that looks like mere clutter, and someone else tries to rearrange the clutter into a picture. Over and over again, gazing at what seemed random in the world, Elizebeth found a tiny spot of sense, and then she stood on that spot and invented a system to transform the rest of the landscape all the way out to the horizon, and this is still the process today. Codebreaking is work and patience and method and mind. And Elizebeth had more of these qualities than perhaps anyone else in her time.

She always remained a little sphinxy. Up to the end of her life she hesitated to blurt out all her secrets, to answer every question in movie detail, whether out of modesty, habit, fear of prosecution, or an appreciation for mystery.

"There are plenty of mysteries that you can leave dangling," she told the NSA's Virginia Valaki during their discussion in 1976. "Enough to allure a reader, I'm sure."

"I've been trying to put together the pieces," Valaki said. "We'll never make the whole picture . . . at least we'll get some of the perspective straightened out."

Valaki was one of Elizebeth's descendants, part of the next generation of women codebreakers who prospered after the war. She first joined the agency in 1954 as a linguist and now edited the NSA technical journal *Cryptolog*.

"Well, thanks again, Mrs. Friedman," Valaki said.

"Well, don't thank me," Elizebeth said. "It's been interesting."

"Sometime I myself would love to do a profile on you," Valaki added.

"Oh!" Elizebeth said.

"Girl cryptanalyst and all that. I would think it would be extremely interesting for people to read."

"What happened the other day?" Elizebeth said, asking the question to herself. She said she had been out in the city, walking on Capitol Hill, when she realized that a couple of young women nearby had seen her and were talking about her. Elizebeth recognized one of the women. They had crossed paths somewhere years earlier, in a professional capacity, and Elizebeth was tickled by the fact that these women considered her some kind of noteworthy figure. "Oh my!"

Valaki shut off the recorder. She and Elizebeth spoke for an unknown amount of time, possibly about mutual acquaintances at intelligence agencies. Then the recorder started again, and before too long, the conversation wound to a close.

They checked the time.

"You mean to say it's only five minutes after one?" Elizebeth said.

"My heavens!" Valaki said.

It had been so long since Elizebeth had talked about her life smashing codes that a simple conversation felt like an opera.

"I'll bet no two women ever said as many words in [so] short a time," Elizebeth said.

The transcript notes that the women laughed.

ACKNOWLEDGMENTS

I owe a lot to the people who shaped this book:

My editor, Julia Cheiffetz at Dey Street. Julia's passion for Elizebeth's story was always there, even when I didn't exactly know how to tell it, and our conversations enriched the book immeasurably. I'm grateful for her sharp eye, her instincts, and her belief. Thanks also to Sean Newcott, Lynn Grady, and the rest of the team at Dey Street: Tom Pitoniak, Kendra Newton, Heidi Richter, Dale Rohrbaugh, Paula Szafranski, and Owen Corrigan.

My agent, Larry Weissman. I am so glad to have the benefit of Larry's counsel and his sensibility for narrative nonfiction. I feel the same about his unflappable partner, Sascha Alper. I can't imagine writing books without their guidance and friendship.

Librarians and archivists: This book would not exist without the archivists who preserved, indexed, and annotated the Friedmans' files with such care. Paul Barron and Jeffrey Kozak at the George C. Marshall Foundation are wonderful humans, and their library is just one of the great American places. I was amazed by Rose Mary Sheldon, the Virginia Military Institute classics professor who spent years assembling her epic "The Friedman Collection: An Analytical Guide." She did it as a labor of love—didn't earn a cent—and was generous with her time and wisdom. The NSA historian Betsy Rohaly Smoot and NSA librarian Rene

Stein shared expertise and files with me. Hannah Walters at the Fabyan Villa Museum showed me around what remains of George Fabyan's Riverbank and answered numerous questions about Riverbank in its prime. Thanks also to Thomas Larson at the New York Public Library's Manuscripts and Archives Division; Jessica Strube at the Geneva History Museum; JoEllen Dickie at the Newberry Library; and the staff of the National Archives at College Park, Maryland.

Kari Walgran has been a friend and sounding board for years. Some of my favorite parts of the book grew from her questions and comments on drafts. Malcolm Burnley and Kirsten Hancock were capable research assistants who found and flagged important files. Phil Tomaselli turned up materials about the Nazi spy hunts in the UK National Archives in Kew. Eduardo Geraque in São Paulo sent documents from police archives there. Linda D. Ostman is a hero for discovering the transcript of the 1933 Consolidated Exporters case in a Texas court repository. I also appreciate research performed by Beth Robertson and Lisette Lacroix in Canada.

Thank you to the American women who spoke to me about their cryptologic experiences in World War II: Judy Parsons, Martha Waller, Pat Leopold, and Helen Nibouar.

I appreciate the historians, cryptologic obsessives, and technology enthusiasts who shared their time and wisdom. Philip Marks, the British expert in machine ciphers, was extremely patient in explaining Enigma systems and reviewing technical passages. Craig Bauer's engaging books about cryptology helped me navigate the subject, and conversations with Craig were always clarifying. The historian Richard McGaha helped me chart a path through the crazy waters of espionage and counterespionage in Argentina. The renegade Canadian author John Bryden pointed me toward the coast guard's clandestine decrypts in the National Archives. Jason Vanderhill in Vancouver knows everything there is to know about Canadian rum syndicates. James Somers is the kind of friend you want to have if you're writing about technology, a terrific writer who is also a programmer. I enjoyed meeting and talking with Barbara Osteika at ATF, a relentless researcher, and William Sherman, the Renaissance scholar who told me about the Riverbank cipher collection at the New York Public Library. Any cryptologic or historical errors in the text are mine.

Thank you to friends who provided advice, encouragement, leads, etc.: Carrie Frye, Sasha Issenberg, Eileen Clancy, Christi Bender, John Whittier-Ferguson, Nathalia Holt, Elonka Dunin, Josh Dean, Jason Leopold, Roy Kesey, Ann Daciuk, Sheila Liming, Puneet Batra, Chris McDougall, Stephen Rodrick, Steve Volk, Samantha Newell, Rob Morlino, Neel Master, Elon Green, and my excellent magazine colleagues—Greg Veis and Rachel Morris at the Huffington Post Highline, and Kristen Hinman and Michael Schaeffer at *Washingtonian*.

I'm indebted to the University of Michigan and the Knight-Wallace Fellowship program for inviting me and my family to Ann Arbor in 2014 and 2015. In a lot of ways, this book is a direct result of the rare alchemy of that program. Thank you so, so much to Charles and Julia Eisendrath for one of the best years of my life, Birgit Rieck and the fellow fellows, John DeCicco, and Carl Simon and the Center for the Study of Complex Systems. And I will always be grateful to Matthew Power for encouraging me to apply in the first place.

Thank you to Duchess Goldblatt for allowing me to borrow one of her lovely sentences.

Finally, thank you to my family: Frank, Sharyn, and Lauren Fagone; Gloria Jewell; Lynn and Rich Bauer; and the Howell clan. Most of all, thank you to the bright, adventurous women in my life, Dana Bauer, and our daughter, Mia Fagone. Dana and Mia inspired the book and kept telling me they wanted to read it. To the two of you:

```
O V M I D A D O O S S D A N E L I T
L E U A D N N H G H I O C Y B I E ?
I Y O A N A A M Y I T N E O U E V !
```

NOTES

ABBREVIATIONS

ESF	Elizebeth Smith Friedman
WFF	William Frederick Friedman
ESF COLLECTION	Elizebeth S. Friedman Collection, George C. Marshall Research Foundation (Lexington, Virginia)
WFF COLLECTION	William F. Friedman Collection, Marshall Foundation
NARA	U.S. National Archives and Records Administration (Washington)
NYPL	Bacon Cipher Collection, New York Public Library, Manuscripts and Archives Division (New York)
NSA	William F. Friedman Collection, U.S. National Security Agency, 2015 release (nsa.gov)
TNA	The National Archives of the UK (Kew, United Kingdom)

AUTHOR'S NOTE

xii *"the world's greatest"* David Kahn, *The Codebreakers: The Comprehensive History of Secret Communication from Ancient Times to the Internet,* rev. ed. (New York: Scribner, 1997), 21.

"Singlehandedly, he made" Ibid., 392.

"CRYPTOLOGIC PIONEER" Program for "Dedication Ceremony, William F. Friedman Memorial Auditorium," May 21, 1975, box 14, file 12, ESF Collection.

"She and her husband" Memorandum from Chief of Communications to Chief (redacted), November 8, 1949, box 12, file 15, ESF Collection.

"Mrs. Friedman and her husband" U.S. Department of Justice, Federal Bureau of Investigation, memorandum, *Subject: Velvalee Dickinson,* R. A. Newby to D. M. Ladd, March 14, 1944. Obtained under the Freedom of Information Act from FBI; received December 2015.

xiii *"We try to tell people"* Jeffrey Kozak (director of library and archives at the Marshall Foundation) in discussion with the author, January 2015.

xiv *an elite codebreaking unit* "History of USCG Unit #387," Record Group 38, Crane Material, Inactive Stations, box 57, 5750/2, NARA. This is a 329-page technical history of ESF's coast guard unit between 1940 and 1945, a thick bound volume written in 1945 or 1946. The unit had multiple names over its lifetime—the Coast Guard Cryptanalytic Unit, Coast Guard Unit #387, then OP-20-GU, and later OP-G-70, after the unit was absorbed by the navy in 1941—but it's all the same organization, founded by ESF in 1931 and evolving as it faced different challenges through the end of the war. Every page of the technical history is stamped TOP SECRET ULTRA, including the cover. No author is listed; it was probably written by ESF's coast guard commander, Lieutenant Leonard T. Jones, in collaboration with her and other codebreakers on the team. It wasn't declassified until 2000.

tracked and exposed them ESF and her coast guard team preserved the decrypts that they generated during the war—the typed sheets of solved messages. These are located in two places at NARA. An incomplete set of decrypts is in RG 38, Records of the Office of the Chief of Naval Operations, CNSG Library, boxes 77–81. The bulk of the decrypts are in RG 457, Messages of German Intelligence/Clandestine Agents, 1942–1945, subseries SRIC, boxes 1–5. More than any others, these are the records that made it possible to figure out what ESF really did in the war and why it mattered. It is not good etiquette to cry out in joy when you are researching in the National Archives, but I may have done that when I read the decrypts for the first time. I'm indebted to the Canadian historian John Bryden for flagging the importance of these documents in his excellent book *Best-Kept Secret: Canadian Secret Intelligence in the Second World War* (Toronto: Lester, 1993).

CHAPTER 1: FABYAN

3 *a female representative* Transcript of ESF interview with Virginia T. Valaki, November 11, 1976, transcribed January 10, 2012, NSA Center for Cryptologic History. Obtained under the Freedom of Information Act from NSA; received October 2015; originally requested by G. Stuart Smith. Valaki was a cryptolinguist for the NSA and retired in 1994 after a forty-year career; she died in 2015. See "Virginia T. Valaki," obituary, *New Haven Register,* June 7, 2015, http://www.legacy.com/obituaries/nhregister/obituary.aspx?pid=175022791.
   "Do you want a cigarette" Ibid., 1.
4 *eighty-four years old* ESF was born on August 26, 1892, in Huntington, Indiana. See Official Personnel Folder, box 7, folder 3.
   "Nobody would believe it" ESF interview with Valaki, November 11, 1976, transcribed February 16, 2012, 10.
   "I'd be grateful" ESF interview with Valaki, November 11, 1976, transcribed January 10, 2012, 1.
   six slightly different answers ESF interview with Valaki, November 11, 1976, transcribed February 16, 2012, 6–13.
5 *thought to tell the story* Ibid., 6–7.
   June 1916 Transcript of ESF interview with Forrest C. Pogue, May 16–17, 1973, box 16, folder 19, ESF Collection, 3.
   a chauffeured limousine Ibid., 2.
   five foot three ESF's ration book from the Second World War lists her as five foot three and 120 pounds, and she writes elsewhere that she was a bit smaller as a young woman.
   dark-brown curls and hazel eyes Though reporters sometimes called her eyes blue, and a 1930 oil painting of ESF shows her eyes to be a deep forest green, her children later insisted to a potential biographer of their mother that her eyes were really hazel. See Katie Letcher Lyle, "Divine Fire: Elizebeth Smith Friedman, Cryptanalyst," unpublished manuscript, July 4, 1991, ESF Collection, two PDF files, 175.
   crisp gray dress ESF interview with Pogue, 3.
6 *more than a foot* Richard Munson, *George Fabyan: The Tycoon Who Broke Ciphers, Ended Wars, Manipulated Sound, Built a*

Levitation Machine, and Organized the Modern Research Center (North Charleston, SC: Porter Books, 2013), 3. Fabyan was six foot four, Elizebeth five three at most.

6 *impression of a windmill*   ESF interview with Valaki, November 11, 1976, transcribed February 16, 2012, 8.

"Will you come to Riverbank"   ESF interview with Pogue, 2.

"Oh, sir"   Ibid.

"That's all right"   ESF interview with Valaki, November 11, 1976, transcribed February 16, 2012, 8.

lifting her by the arm   Ibid.

meek because she was small   ESF autobiography (unpublished manuscript), ESF Collection, PDF file, 2.

"odious name of Smith"   ESF diary, April 22, 1913, box 21, folder 1, ESF Collection. She also wrote in this entry that she hated the name Smith because it seemed terribly unfair for a lover of words to be saddled with a name so lexically vanilla: "Call it vanity if you will—but how should you like to have a name for which you couldn't have even the fun of looking up the etymology?"

"I feel like snipping"   Ibid.

7 *John Marion Smith*   "Geneaology from notes of ESF," July 23, 1981, box 11, folder 21, ESF Collection.

served in local government   "Addenda and Corrections to biographical data re Elizebeth Smith Friedman," box 11, folder 21, ESF Collection.

"My Indiana family"   Lyle, "Divine Fire," 166.

Sopha Strock   ESF Personal History Statement, box 11, folder 16, ESF Collection, 3.

grown up and scattered   Ibid., 13.

"We call a lot of things luck"   ESF diary, July 1, 1913.

"from her father"   Mary Goldman to Vanessa Friedman, February 15, 1981, box 12, folder 14, ESF Collection.

8 *seamstress for hire*   ESF diary, February 27, 1913.

underlining the pages   ESF's volume of Tennyson, box 22, ESF Collection.

Erasmus who "believed in one aristocracy"   ESF, "The Need for Erasmianism," box 12, folder 8, ESF Collection.

"I sit stunned"   ESF, "After Senior Philosophy Course," 1915, box 12, folder 9, ESF Collection.

9 *"passed away"*   ESF diary, March 20, 1913.

"I have marvelous abilities"   ESF diary, June 22, 1913.

"Very suggestive"   ESF, "The Need for Erasmianism."

"my musical heart was carried"   ESF diary, July 14, 1913.

10 *"it reveals the naked man-soul"*   Carleton Miller to ESF, July 22 [1915?], box 1, folder 44, ESF Collection.

10 *"mental question mark"*   ESF diary, January 29, 1916.
substitute principal   ESF interview with Valaki, November 11, 1976, transcribed February 16, 2012, 7.
a county high school   It was the public high school in Wabash, Indiana. See "Education and Experience," ESF Personnel Folder.
Almost 90 percent   Hans Joerg-Tiede, *University Reform: The Founding of the American Association of University Professors* (Baltimore: Johns Hopkins University Press, 2015), 14.
939 women   National Center for Education Statistics, *120 Years of American Education: A Statistical Portrait,* ed. Thomas D. Snyder (Washington, D.C.: U.S. Department of Education, Office of Educational Research and Improvement, 1993), 83.
62 women   Ibid.
"little, elusive, buried splinter"   ESF diary, October 10, 1914.
"I am never quite so gleeful"   Ibid., July 2, 1913.

11 *More than a thousand*   *Official Report of the Proceedings of the Sixteenth Republican National Convention* (New York: Tenny Press, 1916), 11–13.
rained most every day   Associated Press, "Republican Conclave Depressed by Weather; Shows Little Enthusiasm," *Chicago Daily Tribune,* June 9, 1916.
the political delegates   Ibid.
baseball parks   I. E. Sanborn, "Rain Stops Cubs; Double Bill Today with Herzog's Reds," *Chicago Daily Tribune,* June 21, 1916. See also James Crusinberry, "Sox Lose Chance to Rise by Rain in Mack Series," *Chicago Daily Tribune,* June 9, 1916.

12 *died on a steamship*   Paul Finkelman, "Class and Culture in Late Nineteenth-Century Chicago: The Founding of the Newberry Library," *American Studies* 16 (Spring 1975): 5–22.
had to be free to use   Ibid.
wealthy Chicago businessmen   Ibid.

13 *dreamlike White City*   "World's Columbian Exposition of 1893," Chicago Architecture Foundation, http://www.architecture.org /architecture-chicago/visual-dictionary/entry/worlds-columbian -exposition-of-1893/.
a day of demonstrations   "Under 10,000 Wheels," *Chicago Tribune,* August 27, 1893.
twice as large   The main building of the palace covered nine and a half acres and the U.S. Capitol building spreads across four acres. See *Encyclopaedia Brittanica,* New American Supplement to the New Werner Edition, s.v. "World's Fairs"; and Architect of the Capitol, "About the U.S. Capitol Building," https://www.aoc .gov/capitol-buildings/about-us-capitol-building.
one hundred thousand people   "Under 10,000 Wheels."

13 *builders completed construction*   "History of the Newberry
    Library," https://www.newberry.org/newberry-library-history
    -newberry-library.
    "a select affair"   *Chicago Times,* July 17, 1887, cited in
    Finkelman, "Class and Culture in Late Nineteenth-Century
    Chicago."
    five-story building   Finkelman, "Class and Culture in Late
    Nineteenth-Century Chicago."
    fill out a slip   Ibid.
14 *hundreds of incunabula*   Ibid.
    Arabic script   "Frequently Asked Questions about Audrey
    Niffenegger's The Time Traveler's Wife," Newberry, https://
    www.newberry.org/time-traveler-s-wife. See also Lawrence
    S. Thompson, "Tanned Human Skin," *Bulletin of the Medical
    Library Association* 34, no. 2 (1946): 93–102.
    six thousand books   Finkelman, "Class and Culture in Late
    Nineteenth-Century Chicago."
    a haul that included   "Chicago Gets a Prize: Librarian Poole's
    Report on the Probasco Collection," *Chicago Daily Tribune,*
    November 22, 1890.
    Romanesque lobby   Finkelman.
    mounting exhibitions   Jo Ellen Dickie (reference librarian,
    Newberry Library), e-mail message to the author, January 4, 2017.
    13 inches tall and 8 inches wide   The name of this particular
    Folio is "Winsor 17" and it now resides in the special collections
    department of the Bryn Mawr College library in Pennsylvania.
    Anthony James West, *The Shakespeare First Folio: The History of
    the Book,* vol. 2 (New York: Oxford University Press, 2003), 233.
    an engraving of a man   The Bodleian First Folio: digital
    facsimile of the First Folio of Shakespeare's plays, Bodleian Arch.
    G c.7, http://firstfolio.bodleian.ox.ac.uk/.
    The text said   Ibid.
15 *"that an archaeologist has"*   ESF autobiography, 1.
    One of the librarians   ESF interview with Valaki, November 11,
    1976, transcribed February 16, 2012, 7.
    Richmond, Indiana   Ibid.
    "something unusual"   Ibid.
    reminded her of Mr. Fabyan   Ibid.
    "young, personable"   ESF autobiography, 1.
    too startled   Ibid.
    "Shall I call him up?"   ESF interview with Pogue, 2.
    "Well, yes"   Ibid.
    be right over   ESF interview with Valaki, November 11, 1976,
    transcribed February 16, 2012, 8.

16 *any minute*   Ibid., 8. Elizabeth recalled that Fabyan arrived "before you could have hit a button."
"This is Bert"   Ibid. Bert is spelled "Burt" in the NSA transcript but his name was Bert Williams, according to John W. Kopec, *The Sabines at Riverbank: Their Role in the Science of Architectural Acoustics* (Woodbury, NY: Acoustical Society of America, 1997), 29.
Chicago & North Western   ESF interview with Valaki, November 11, 1976, transcribed February 16, 2012, 8.
"Where am I"   Ibid.
she remained still   Ibid.
She smiled at him   Ibid., 6.
17 *within inches*   Ibid.
"WHAT IN HELL DO YOU KNOW"   Ibid.
something stubborn   Ibid., 9.
turned her head away   Ibid., 6.
"That remains, sir"   ESF interview with Pogue, 3.
most immoral remark   ESF interview with Valaki, November 11, 1976, transcribed February 16, 2012, 9.
a great roaring laugh   Ibid.
he began to talk of Shakespeare   ESF autobiography, 2.
18 *he believed what he was saying*   ESF eventually came to believe that Fabyan was deceptive in how he promoted his ideas but he did seem to earnestly believe them.
He said that a brilliant female scholar   ESF autobiography, 2.
350-acre estate   Munson, *George Fabyan*, 3.
Teddy Roosevelt, his personal friend   Ibid., 13.
P. T. Barnum   Ibid.
Famous actresses   Ibid. The actresses included Mary Pickford, Billie Burke, and Lillie Langtry.
19 *a second limousine*   ESF interview with Valaki, November 11, 1976, transcribed January 12, 2012, 6.
came to a stop   ESF autobiography, 3; ESF interview with Pogue, 3.
a two-story farmhouse   Ibid., 4; author's visit to the Fabyan Villa Museum, Geneva, Illinois, March 19, 2015.

CHAPTER 2: UNBELIEVABLE, YET IT WAS THERE

21 *A naked woman*   John W. Kopec, *The Sabines at Riverbank: Their Role in the Science of Architectural Acoustics* (Woodbury, NY: Acoustical Society of America, 1997), 36–37.
sign that read Fabyan   Ibid.
22 *satisfy his lust*   Ibid.

22 *two white flashes*   Norman Klein, "Building Supermen at Fabyan's Colony," *Chicago Daily News*, April 22, 1921.
The electric trolley   Richard Munson, *George Fabyan: The Tycoon Who Broke Ciphers, Ended Wars, Manipulated Sound, Built a Levitation Machine, and Organized the Modern Research Center* (North Charleston, SC: Porter Books, 2013), 48.
bombs exploding   Kopec, *The Sabines at Riverbank*, 42.
warplanes buzzing   Ibid.
"A Garden of Eden"   Mme. X, "A Visit to a Garden of Eden on Fox River," *Chicago Daily Tribune*, October 2, 1921.
"Fabyan's colony"   Klein, "Building Supermen at Fabyan's Colony."
"a wonder-working laboratory"   "A Wonder Working Laboratory Near Chicago," *Garard Review*, November 1928, 1.
"one of the strangest"   Klein, "Building Supermen at Fabyan's Colony."

23 *"one of the greatest"*   "Varying the List of Clubs . . ." *Cincinnati Star*, December 21, 1923, Box 14, "The Ideal Scrap Book," NYPL.
"one who has achieved"   "Scientist Spends Millions in Experiments to Develop Flapper into Perfect Woman," *Evening Public Ledger* (Philadelphia), July 18, 1922.
"the man of a thousand interests"   "War on Debutante Slouch Is Started by Col. Fabyan," July 5, 1922, Box 14, "The Ideal Scrap Book," NYPL.
"the lord and master"   Klein, "Building Supermen at Fabyan's Colony."
"Chicago inventor"   "Flywheel Discs Cut Resistance," *Kansas City Journal*, March 13, 1923.
multi-millionaire country gentleman   "Fabyan Tries to Rear Perfect Flapper on Farm," *Chicago Herald Examiner*, July 6, 1922.
"the seer"   Leroy Hennessey, "Twas Bill! Nay, Bacon! But Now E'en Fabyan Knows Not Who Did Shakespeare," *Chicago Evening American*, January 1922, Box 14, "The Ideal Scrap Book," NYPL.
"the caliph"   "Col. George Fabyan Declares War on Profiteers," Box 14, "The Ideal Scrap Book," NYPL.
"Credible persons"   Cinderella, "Chicagoan Wins Name at Sculpture," *Chicago Daily Tribune*, June 1, 1915. This seems to be a legend; staff at the Fabyan Villa Museum told me that Fabyan only rode in a zebra-drawn chariot once, not twice a day every day.
donations . . . board meetings   Munson, *George Fabyan*, 4.
The black sheep   Ibid., 20.
$3 million fortune   Ibid., 10.

23 *striped seersucker cloth* Ibid., 22.

24 *"Ripplette"* Ripplette ad, *Farmer's Wife* (St. Paul, Minnesota), January 1, 1927.

"I ain't no angel" George Fabyan to WFF, June 10, 1926, WFF and George Fabyan Correspondence, Item 734, WFF Collection.

The steel magnates of Pittsburgh Andrew Carnegie and Henry Clay Frick, who hated and trolled each other. Frick built his mansion one mile from Carnegie's and vowed to make his rival's home look "like a miner's shack" in comparison. Christopher Gray, "Carnegie vs. Frick, Dueling Egoes on Fifth Avenue," *New York Times,* April 2, 2000.

a 165-room castle "Other Features Around Hearst Castle," California State Parks, http://hearstcastle.org/history-behind -hearst-castle/the-castle/.

"Some rich men" Klein, "Building Supermen at Fabyan's Colony."

Aspirin, vitamins Aspirin was discovered in 1897, vitamins in 1912, blood types in 1900; medical X-rays began in 1895.

Einstein's theory He published his theory of general relativity in 1915. American Institute of Physics, "2015: The Centennial of Einstein's General Theory of Relativity," https://www.aip.org /history-programs/einstein-centennial-2015.

swarm of bees "Col. Geo. Fabyan Soon to be a Miller De Luxe," *Chicago Herald,* July 12, 1915, reprinted in Kopec, *The Sabines at Riverbank,* 30–32.

25 *"Do you ever think"* Klein, "Building Supermen at Fabyan's Colony."

"community of thinkers" Ibid.

an ultraquiet test chamber Kopec, *The Sabines at Riverbank,* 59–73.

the buzz of a stray mosquito "A Wonder Working Laboratory."

a pencil writing on paper Ibid.

"racket ogre" "Fabyan May End Noises of City," *Aurora Beacon* (Aurora, Illinois), April 24, 1921.

"Look through this telescope thing" " 'Lord of Riverbank' Works in $100,000 Laboratory; Would Find Deafness Cure," box 14, manila folder of newspaper clippings, NYPL.

26 *one hundred or more* Klein, "Building Supermen at Fabyan's Colony."

"Over there in that hothouse" Ibid.

"bobbed blonde hair" "Scientist Spends Millions."

low-security juvenile prison L. Mara Dodge, " 'Her Life Has Been an Improper One': Women, Crime, and Prisons in Illinois, 1835 to 1933" (Ph.D. diss., Univeristy of Illinois at Chicago, 1998), 535–41, 718–19.

26 *cottage built with a donation*   Kopec, *The Sabines at Riverbank*, 37.
    required to undress   Ibid.
    "The results of our experiments"   "Scientist Spends Millions."
27 *"in his effort to impress"*   Ibid.
    told an Illinois historian   Munson, *George Fabyan*, 50.
    "the beams would creak"   Ibid.
    "recalled looking out the windows"   Ibid.
    "The staff in charge"   Austin C. Lescarboura, "A Small Private Laboratory," *Scientific American*, September 1923, 154.
28 *an X-ray screen*   Kopec, *The Sabines at Riverbank*, 36.
    $750,000 worth of radium   Munson, *George Fabyan*, 50.
    discovered in 1895   "This Month in Physics History: November 8, 1895: Roentgen's Discovery of X-Rays," *American Physical Society News* 10, no. 10 (November 2001), https://www.aps.org /publications/apsnews/200111/history.cfm.
    "Every so often the world"   Lescarboura, "A Small Private Laboratory."
    then disappeared   ESF interview with Valaki, November 11, 1976, transcribed February 16, 2012, 9.
    aristocratic appearance   ESF autobiography, 3; ESF interview with Valaki, November 11, 1976, transcribed January 12, 2012, 6.
29 *lived and worked here*   ESF interview with Valaki, November 11, 1976, transcribed February 16, 2012, 9.
    freshen up   ESF interview with Valaki, November 11, 1976, transcribed January 12, 2012, 1.
    striking new clothes   Ibid.
    sat on the bannister   ESF interview with Pogue, 6.
    a slim man   Ibid.
    a neat bow tie   Ibid.
    reminded Elizebeth of Beau Brummell   ESF interview with Pogue, 6.
    polished his boots with champagne   "Fashions of Hunting," *Baily's Magazine of Sport and Pastimes* 65, nos. 431–36 (1896): 163.
    Swedish and Danish servants   Transcript of ESF interview with Ronald Clark, handwritten note on page 7, March 25, 1975, box 16, file 22, ESF Collection.
30 *chickens, ducks, sheep, and turkeys*   Munson, *George Fabyan*, 63.
    prize-winning livestock   Kopec, *The Sabines at Riverbank*, 30.
    head of the table   ESF autobiography, 3.
    J. A. Powell   ESF interview with Clark, 5.
    "cause the University of Chicago"   "Here Are a Few Expert Suggestions for First Press Agent of U. of C.," *Chicago Tribune*, September 5, 1909.

30 *Bert Eisenhour* ESF interview with Clark, 5.

a country bumpkin ESF interview with Valaki, November 11, 1976, transcribed 16 February 2012, 5.

The dominant personality that night ESF autobiography, 3.

31 *"Mrs. Gallup had dwelt"* Ibid.

men's pajamas ESF interview with Valaki, November 11, 1976, transcribed January 12, 2012, 7.

a pitcher of ice water . . . an enormous bowl of fresh fruit Ibid.; ESF interview with Pogue, 5.

assigned an employee I'm inferring this from the fact that Elizebeth doesn't say in her autobiography or later recollections that Fabyan was her tour guide. I think if he had done it himself, she would have said that.

a new laboratory Kopec, *The Sabines at Riverbank*, 3–4.

Professor Wallace Sabine Ibid.

ordnance building Ibid., 42; ESF interview with Pogue, 3.

known as the Villa Munson, *George Fabyan*, 25.

32 *suspended from the ceiling on chains* ESF autobiography, 5.

Taxidermized animals Personal visit to the Fabyan Villa Museum, Geneva, Illinois, March 19, 2015.

a life-size marble statue F. Edwin Elwell, *Diana and the Lion* (sculpture, 1893), displayed in the Palace of Fine Arts in the White City, acquired by George Fabyan after 1917, according to a placard in the Fabyan Villa Museum.

A curving path Munson, *George Fabyan*, 59–60; Kopec, *The Sabines at Riverbank*, 27–28.

33 *Tom and Jerry* Ibid., 2.

flowing southward Wikipedia, s.v. "Fox River (Illiois River tributary)," last modified May 1, 2017, https://en.wikipedia.org/wiki/Fox_River_(Illinois_River_tributary).

two bridges "Fabyan Estate Viewed from the Southeast," map, in Kopec, *The Sabines at Riverbank;* Munson, *George Fabyan,* 5.

bought the windmill in Holland As is often the case with Fabyan, the truth here is actually weirder than the legend. Fabyan didn't buy the windmill in Holland; he bought it from a German craftsman in Lombard, Illinois, paying the modern equivalent of $2 million to take it apart, lug it across the prairie, and reconstruct it on the opposite bank of the Fox River. See "Fabyan Windmill," Kane County Forest Preserve District, http://www.kaneforest.com/historicsites/fabyanwindmill.aspx.

Elizebeth sat down with Mrs. Gallup ESF autobiography, 2.

two or three hours Ibid.

oversize sheets of paper Several of these large sheets are preserved in box 14, NYPL.

33 *rolled them out*   ESF interview with Valaki, November 11, 1976, transcribed January 12, 2012, 7.
    placed weights on the ends   This is my own inference, from having handled these scrolls myself at NYPL. They really are like window blinds; if you don't put the weights on them, they snap back into a scroll.

34 *would be twofold*   ESF autobiography, 5.
    popped in briefly   ESF interview with Valaki, November 11, 1976, transcribed January 12, 2012, 7.
    another bowl of fresh fruit   Ibid.
    "a mixture of astonishment"   WFF and ESF, *The Shakespearean Ciphers Examined* (London: Cambridge University Press, 1958), 210.
    five thousand women   George Morris, "Clothing Wet, Ardor Undampened, 5,000 Women March," *Chicago Daily Tribune,* June 8, 1916.

35 *Water poured*   Ibid.
    "right of each state"   Republican Party Platform, June 7, 1916, http://www.presidency.ucsb.edu/ws/?pid=29634.
    idolized the suffrage pioneers   ESF League of Women Voters report on International "Equal Rights," April 6, 1933, box 7, folder 5, ESF Collection. See also ESF to Miss Belle Sherwin, President, National League of Women Voters, April 14, 1933, box 7, folder 6, ESF Collection.
    "No woman's rights"   ESF diary, January 29, 1916.
    she reviewed her options   ESF interview with Valaki, November 11, 1976, transcribed January 12, 2012, 7.

CHAPTER 3: BACON'S GHOST

37 *a worksheet of white paper*   "Actors' Names—Shakespeare Folio 1623," box 15, folder "Elizebeth Smith," NYPL.

38 *"The Names of the Principall Actors"*   The Bodleian First Folio: digital facsimile of the First Folio of Shakespeare's plays, Bodleian Arch. G c.7, http://firstfolio.bodleian.ox.ac.uk/.
    devout Christian   WFF and ESF, *The Shakespearean Ciphers Examined* (London: Cambridge University Press, 1958), 189.
    "Surprise followed surprise"   Elizabeth Wells Gallup, "Concerning the Bi-literal Cypher of Francis Bacon: Pros and Cons of the Controversy" (1902; Internet Archive, 2008), 60, https://archive.org/details/concerningbilite00gall.
    "The sole question is"   Ibid., 65.

39 The New Atlantis   Francis Bacon, *The New Atlantis* (1627; Project Gutenberg, 2008), https://www.gutenberg.org/files/2434/2434-h/2434-h.htm.

39 *Mark Twain believed it* Mark Twain, *Is Shakespeare Dead?* (1909; Project Gutenberg, 2008), https://www.gutenberg.org /files/2431/2431-h/2431-h.htm.

So did Nathaniel Hawthorne Nina Baym, "Delia Bacon: Hawthorne's Last Heroine," *Nathaniel Hawthorne Review* 20, no. 2 (Fall 1994): 1–10, http://www.english.illinois.edu/-people -/emeritus/baym/essays/last_heroine.htm.

can be anagrammed WFF and ESF, *Shakespearean Ciphers,* 110.

2+1+3+14+13 Ibid., 179, 181.

40 *Orville Ward Owen* Ibid., 63.

most scientific and plausible yet WFF and ESF, *Shakespearean Ciphers*, 188.

The method had been demonstrated Francis Bacon, *De Augmentis Scientarium,* translated by Gilbert Wats (Oxford, 1640); pages relevant to ciphers in Wells Gallup, "Concerning the Bi-literal Cypher," 23–27.

anything by means of anything Ibid.

the new alphabet Ibid.

41 *don't have to be* a *and* b Ibid.

42 *scoured photo enlargements* "A CATALOGVE," box 13, folder 11, NYPL.

Then she drew charts "Alphabets for the Catalogue of the Plays," box 14, NYPL.

"Queene Elizabeth is my true mother" Elizabeth Wells Gallup, *The Biliteral Cypher of Sir Francis Bacon Discovered in His Works and Deciphered by Mrs. Elizabeth Wells Gallup,* 3rd ed. (1901; Internet Archive, 2008), 166, http://www.archive.org /details/biliteralcyphero00gallrich/.

"Francis of Verulam is author" Ibid.

"Francis St. Alban, descended" Ibid.

"You will either finde" Ibid., 165.

43 *her 1899 book* Wells Gallup, *The Biliteral Cypher,* 1st ed.

a secret king WFF and ESF, *Shakespearean Ciphers,* 192–94.

clandestine society of engineers Richard Munson, *George Fabyan: The Tycoon Who Broke Ciphers, Ended Wars, Manipulated Sound, Built a Levitation Machine, and Organized the Modern Research Center* (North Charleston, SC: Porter Books, 2013), 103.

"Here are 360 pages" "Bacon-Shakespeare: Mrs. Elizabeth Wells Gallup Throws New Light Upon the Mystifying Question—The Bi-Literal Cipher," newspaper article, Box 14, clipping file in wooden box marked "California glace fruits," NYPL.

Skeptics questioned the veracity WFF and ESF, *Shakespearean Ciphers*, 196–99.

43 *"impossible to those"* Ibid., 198.
traveled to Oxford, England Ibid., 202.
44 *Mrs. Gertrude Horsford Fiske* Ibid., 196.
Mr. Henry Seymour Ibid.
Mr. James Phinney Baxter Ibid., 224.
"acoustical levitation device" John W. Kopec, *The Sabines at Riverbank: Their Role in the Science of Architectural Acoustics* (Woodbury, NY: Acoustical Society of America, 1997), 4–6.
Eisenhour couldn't get it to work Ibid.
"The inheritance" Four-page typewritten draft beginning "The use and the commixture," box 13, blank folder between folders 14 and 15, NYPL. The text in this draft appears similar to Fabyan's published introduction in *The First of the Twelve Lessons in the Fundamental Principles of the Baconian Ciphers, and Application to the Books of the Sixteenth and Seventeenth Centuries* (Geneva, IL: Riverbank Laboratories, 1916).
a photo enlargement ESF describes the process she was taught by Gallup in *Shakespearean Ciphers,* 209, and ESF's worksheets from her earliest deciphering tests at Riverbank—eight sheets total—are in box 15, folder "Elizebeth Smith," NYPL.
45 *particularly fond of puzzles* ESF interview with Ed Meryl, March 1939, box 17, folder 14, ESF Collection.
She got stuck WFF and ESF, *Shakespearean Ciphers,* 210–11.
eight hours "Actors' Names," ESF wrote in pencil at the top of the worksheet, "8 hours' work," and marked the time and date of the solution: 10:30 A.M., June 5, 1916.
twenty-four-word plaintext translation Ibid.
46 *Elizebeth always asked Mrs. Gallup* WFF and ESF, *Shakespearean Ciphers,* 210–11.
Ragtime music "Col. Geo. Fabyan Soon to be a Miller De Luxe," *Chicago Herald,* July 12, 1915, reprinted in Kopec, *The Sabines at Riverbank,* 30–32.
a series of loudspeakers Ibid.
47 *sisters from Chicago* "An Investigation of the Newest Bacon-Shakespeare Cipher Theory," *St. Louis Post-Dispatch Sunday Magazine,* July 9, 1916, in box 14, "The Ideal Scrap Book," NYPL.
"Our experience at Riverbank" George Fabyan to chief of the MID, War Department, March 22, 1918, RG 165, Records of the Military Intelligence Division, Entry 65, box 2243.
looking glass WFF and ESF, *Shakespearean Ciphers,* 190.
attempting to complete Ibid., 208; Kate Wells to George Fabyan, n.d., box 14, NYPL.
resembled a piece of art Elizebeth Wells Gallup black notebook with red spine, box 14, NYPL.

47 *small wooden boxes*   Various boxes of news clippings, box 13 and 14, NYPL.

48 *"We lived hard and fast"*   ESF interview with Valaki, November 11, 1976, transcribed February 16, 2012, 10.
tiny salaries   ESF interview with Pogue, 5.
minor idle rich   Ibid.
invited Elizabeth to climb in   ESF interview with Valaki, November 11, 1976, transcribed February 16, 2012, 5.
"no billiard ball"   Mme. X, "A Visit to a Garden of Eden on Fox River," *Chicago Daily Tribune,* October 2, 1921.

49 *head would blow off*   ESF interview with Valaki, November 11, 1976, transcribed February 16, 2012, 5.
afraid he'd catch cold   Katie Letcher Lyle, "Divine Fire: Elizebeth Smith Friedman, Cryptanalyst," unpublished manuscript, July 4, 1991, ESF Collection, two PDF files, 44.
leisurely rides   ESF autobiography, 8–9; ESF interview with Pogue, 6.
sandhill cranes and red hawks   Gerald M. Haslam, "The Fox River Settlement Revisited: The Illinois Milieu of the First Norwegian Converts to Mormonism in the Early 1840s," *BYU Family Historian* 6 (2007): 59–82.
At twenty-five   WFF was born September 24, 1891, so he would have been just shy of twenty-five; they were a little more than one year apart in age.
old, creaky structure   Transcript of ESF interview with Marshall Research Library staff members Tony Crawford and Lynn Biribauer, Tape #5, June 6, 1974, 5.
teensy-weensy flies   Ibid.
quickly, then die   Ibid.

50 *one bottle of flies into another*   I am relying on my memory of performing this exact type of *Drosophila* experiment in high school. Thanks to my AP Biology teacher, Mr. Anderson.
150 workers   Norman Klein, "Building Supermen at Fabyan's Colony," *Chicago Daily News,* April 22, 1921.
Susumu Kobayashi   Kopec, *The Sabines at Riverbank,* 27.
Jack "the Sailor"   Ibid.
Belle Cumming   Ibid., 50.
Silvio Silvestri   Ibid., 4, 26.
"Achieve success!"   ESF interview with Valaki, November 11, 1976, transcribed January 12, 2012, 10.
handed out shiny dimes   Munson, *George Fabyan,* 6.
demonstrate how a snake   Ibid.
wearing red diapers   Kopec, *The Sabines at Riverbank,* 29.

51 *crops, genetics, and Francis Bacon*   Ibid., 13.

51 *so did Lillie Langtry*   ESF interview with Clark.
Billie Dove   Kopec, *The Sabines at Riverbank*, 23.
Richard Byrd   Munson, *George Fabyan*, 13.
the elegant Billie Burke   Ibid.
met and talked with Lillie Langtry   Ibid.
"star-complex and hero-worship"   ESF narrative of Gordon Lim case, 1937–38, box 6, manila folder of Lim case material, ESF Collection.
buy a new wardrobe   ESF autobiography, 6.
"so typically Fabyan"   ESF interview with Valaki, November 11, 1976, transcribed January 12, 2012, 14.
52 *bugler who played reveille*   Munson, *George Fabyan*, 58.
an honorary one   Kopec, *The Sabines at Riverbank*, 22–23.
the Fabyan Scouts   Ibid.
the Fox Valley Guards   Ibid.
screaming at the offender   ESF autobiography, 5.
stoking the coals   Ibid.
steel I-beams   Kopec, *The Sabines at Riverbank*, 52.
seventy-five plows   Klein, "Building Supermen at Fabyan's Colony."
Temple de Junk   Ibid.
He published a book   George Fabyan, *What I Know About the Future of Cotton and Domestic Goods*, 2nd ed. (Chicago, 1900).
one hundred blank pages   Munson, *George Fabyan*, 5.
53 *One day he walked past*   Ibid., 7.
"a very bright man"   ESF interview with Valaki, November 11, 1976, transcribed January 12, 2012, 1.
longer than a newspaper headline   ESF autobiography, 8.
repeat back verbatim   Ibid.
showed Elizebeth a prototype   ESF interview with Marshall staff, Tape #5, June 6, 1974, 5.
crawled on their stomachs   Klein, "Building Supermen at Fabyan's Colony."
54 *Professor So-and-So*   ESF interview with Valaki, November 11, 1976, transcribed January 12, 2012, 8.
"We'll get along fine"   Ibid.
all expenses paid   WFF and ESF, *Shakespearean Ciphers*, 205–6.
"useless Bacon-Shakespeare controversy"   Letter on Bliss Fabyan & Company letterhead, September 1916, box 13, blank folder between folders 14 and 15, NYPL.
getting to the bottom   "An Investigation of the Newest Bacon-Shakespeare Cipher Theory."
"hard, cold facts"   Letter on Bliss Fabyan letterhead, 1916.
disorient the guest   WFF and ESF, *Shakespearean Ciphers*, 206.

55 *convinced that the work was solid* "An Investigation of the Newest Bacon-Shakespeare Cipher Theory." In this article the *Post-Dispatch* reporter briefly describes meeting twenty-two-year-old Elizebeth: "Miss Smith told me that when she went to Riverside [*sic*] she did not believe there was anything to the bi-literal cipher theory. Now, she says, she hasn't the slightest doubt."

beginning to doubt WFF and ESF, *Shakespearean Ciphers*, 211.

John Matthews Manly Eric Powell, "A Brief History of the English Department at the University of Chicago," September 2014, https://english.uchicago.edu/about/history.

"wrassle" ESF interview with Valaki, November 11, 1976, transcribed January 12, 2012, 12.

Manly pushed her on the shoulder Ibid., 13.

"Oh, my!" Ibid.

strained credulity WFF and ESF, "Elizabethan Printing and Its Bearing on the Biliteral Cipher," in *Shakespearean Ciphers*, 216–29.

"See how she leans" William Shakespeare, *Romeo and Juliet*, ed. Brian Gibbons (New York: Bloomsbury, 1980), 2.2.23–25.

56 *never once suspected* WFF and ESF, *Shakespearean Ciphers*, 264.

"She could go through the texts" Ibid.

comparing herself to Galileo Elizabeth Wells Gallup, "Bacon's Lost Manuscripts, A Review of Reviews," box 14, clipping file in wooden box marked "California glace fruits," NYPL.

57 *business cards* "Francis Bacon," February 10, 1917, box 14, magenta-colored scrapbook, NYPL.

"Riverbank Laboratories are a group" Letter on Bliss Fabyan letterhead, 1916.

58 *a town in Russia called Kishinev* Ronald Clark, "Preparation," in *The Man Who Broke Purple: Life of Colonel William F. Friedman, Who Deciphered the Japanese Code in World War II* (Boston: Little, Brown, 1977), 7–26.

fluent in eight languages Ibid.

"research and ingenuity" Ibid.

an expert in heredity ESF interview with Pogue, 5.

"an agricultural expert" George Fabyan to WFF, June 14, 1915, Item 734, WFF Collection.

"I realize the value" WFF to George Fabyan (undated), Item 734, WFF Collection.

59 *vague, long-winded riff* George Fabyan to WFF, August 12, 1915, Item 734, WFF Collection.

"I want the father of wheat" Ibid.

"Jewish Invasion" Burton J. Hendrick, "The Jewish Invasion of America," *McClure's Magazine*, March 1913, 125.

59 *"nervous, restless ambition"*   Ibid., 127.
60 *made him miserable*   Clark, "Preparation," 16–17.
   "My idea of real love-making"   Lyle, "Divine Fire," 59.
   youthful fascination   WFF, "Edgar Allan Poe, Cryptographer,"
   in *On Poe*, ed. Louis J. Budd and Edward Harrison Cady
   (Durham, NC: Duke University Press, 1993), 40–54.
   The plot of the story   Edgar Allan Poe, "The Gold-Bug," *Dollar
   Newspaper* (Philadelphia), June 23, 1843, 1 and 4, https://www
   .eapoe.org/works/tales/goldbga2.htm.
61 *Americans associated codebreakers*   WFF, "Edgar Allan Poe,
   Cryptographer."
   a sketch of a long-stemmed plant   WFF, "Cipher Baconis
   Gallup," box 13, folder 5, NYPL.
   a gray duration of pitiless wind   Kopec, *The Sabines at
   Riverbank*, 3.
   sit on his lap   Lyle, "Divine Fire," 53, 60.
   wondering the same   ESF autobiography, 9: "As we were thrown
   together so much in our examination of the cipher proofs, we
   had many . . . talks ourselves. Even that first summer we began to
   wonder about the authenticity of Mrs. Gallup's 'solution.' "
   There are no hidden messages   The Friedmans, as careful
   scientists, never really said it this starkly, but I think this is
   what they believed. On the last page of *Shakespearean Ciphers*
   (288) they suggest that if the people looking for hidden
   messages in Shakespeare taught themselves the true science of
   cryptology and applied it to their efforts, the whole dispute
   "might cease altogether"—in other words, if the seekers really
   understood cryptology, they'd also understand that their quest
   is doomed.

CHAPTER 4: HE WHO FEARS IS HALF DEAD

63 *"and then begins step step leap"*   Anne Carson, *Float* (New
   York: Knopf, 2016), 138.
   burned its way from hand to hand   Barbara W. Tuchman, *The
   Zimmermann Telegram* (New York: Ballantine, 1958), 160–72.
   At 11 A.M. on February 27   Ibid., 172.
   "Good Lord!"   Ibid.
   Germany to Mexico on January 16   Ibid., 145.
   a series of number blocks   Ibid., 201.
   toiled for a month in a secret office   Ibid., "A Telegram Waylaid,"
   3–24.
64 *"We intend to begin"*   Ibid., 146.
   outrage against Germany   Ibid., 184–86.

64 *Her father was there*   ESF to WFF, January 31, 1917, ESF
Collection.
pinkish fluid   Ibid.
"My book-bag lies here unopened"   Ibid.
65 *"yours, Elsbeth"*   ESF to WFF, February 7, 1917, ESF Collection.
"one of the truest friends"   Ibid.
"rocking"   Katie Letcher Lyle, "Divine Fire: Elizabeth Smith
Friedman, Cryptanalyst," unpublished manuscript, July 4, 1991,
ESF Collection, two PDF files, 53, 60. Lyle notes that at some
point William also provided this "rocking" to Elizabeth's elder
sister, Edna.
"I love you / Elsbeth"   ESF to WFF, January 31, 1917.
seized by a new impatience   ESF to WFF, February 7, 1917.
seemed a little cruel   ESF interview with Marshall staff, Tape #2,
June 4, 1974, 14.
they buttonholed him   ESF autobiography, 10.
He shouted them down   Ibid.
66 *length of three miles*   Richard Munson, *George Fabyan: The
Tycoon Who Broke Ciphers, Ended Wars, Manipulated Sound,
Built a Levitation Machine, and Organized the Modern Research
Center* (North Charleston, SC: Porter Books, 2013), 8.
"Gentlemen," he wrote   George Fabyan to War Department
Intelligence Office, 15 March 1917, Item 734, WFF Collection.
67 *"There were possibly three"*   ESF autobiography, 11.
nine-year-old   "A Brief History: The Nation Calls, 1908–1923,"
FBI, https://www.fbi.gov/history/brief-history.
only three hundred agents . . . half a million dollars   Regin
Schmidt, *Red Scare: FBI and the Origins of Anticommunism
in the United States, 1919–1943* (Copenhagen: Museum
Tusculanum, 2000), 83.
April 6, 1917   "April 6, 2017: The 100th Anniversary of the
American Entry into World War I," American Battle Monuments
Commission, https://www.abmc.gov/news-events/news/april
-6-2017-100th-anniversary-american-entry-world-war-i.
seventeen officers   Joseph W. Bendersky, *The Jewish Threat:
Anti-Semitic Politics of the U.S. Army* (New York: Basic Books,
2000), 49.
codes and ciphers an "emergency"   Ralph Van Deman to Acting
Commandant, Army Service Schools, Fort Leavenworth, Kansas,
April 17, 1917, Item 734, WFF Collection.
a radio signal from a plane   Paul W. Clark and Laurence
A. Lyons, *George Owen Squier: U.S. Army Major General,
Inventor, Pioneer, Founder of Muzak* (Jefferson, NC:
McFarland, 2014), 187.

68 *stand eye to eye*   ESF interview with Valaki, transcribed
February 16, 2012, 2.
"the two greatest people"   Joseph Mauborgne to WFF and ESF,
January 8, 1956, box 1, folder 21, ESF Collection.
"to take immediate advantage"   Joseph Mauborgne to chief of the
War College Division, April 11, 1917, Item 734, WFF Collection.
"your exceedingly kind"   Ralph Van Deman to George Fabyan,
April 18, 1917, Item 734, WFF Collection.

69 *intercepted by covert means*   WFF, Lecture V, 107, in *The
Friedman Legacy, Sources on Cryptologic History*, no. 3 (Center
for Cryptologic History: 2006).
BGVKX   ESF and WFF, "Riverbank Problems in
Cryptanalysis," no. 1, Item 290, WFF Collection.

70 *403,291,461,126,605,635,584,000,000*   This is the number of
permutations of 26 letters, a quantity written as 26! and calculated
by multiplying 26 x 25 x 24 x 23 x 22 x 21 x 20 x 19 x 18 x 17 x 16
x 15 x 14 x 13 x 12 x 11 x 10 x 9 x 8 x 7 x 6 x 5 x 4 x 3 x 2 x 1. See
Wolfram Alpha, https://www.wolframalpha.com/input/?i=26!.
A thousand computers   Lambros D. Callimahos, "Summer
Institute for Mathematics and Linguistics," lecture, NSA, Fort
Meade, Maryland, 1966, NSA Reading Room, https://www.nsa
.gov/resources/everyone/foia/reading-room.
Monks, librarians, linguists   David Kahn, *The Codebreakers:
The Comprehensive History of Secret Communication from
Ancient Times to the Internet,* rev. ed. (New York: Scribner,
1997), is the definitive history of the field, with hundreds of pages
of this stuff.

71 *a Belgian countess named Alexandrine*   Nadine Akkerman, "The
Postmistress, the Diplomat, and a Black Chamber? Alexandrine of
Taxis, Sir Balthazar Gerbier and the Power of Postal Control," in
Robyn Adams and Rosanna Cox, *Diplomacy and Early Modern
Culture* (London: Palgrave Macmillan, 2011), 172–88.
an early example   Ibid. Akkerman argues that Alexandrine's
Chamber of the Thurn and Taxis might have been the very first
black chamber in Europe.
"What if this countess"   Ibid.
Parker Hitt and Genevieve Hitt   Betsy Rohaly Smoot, "Pioneers
of U.S. Military Cryptology: Colonel Parker Hitt and His Wife,
Genevieve Young Hitt," *Federal History* no. 4 (2012): 87–100.
"This is a man's size job"   Ibid.
"Good work, old girl"   Ibid.

72 *a serious book*   Parker Hitt, *Manual for the Solution of Military
Ciphers* (Fort Leavenworth, KS: Press of the Army Service
Schools, 1916).

72 *Aimed at Army units* Ibid., 1–3.

letter is E WFF and ESF, *An Introduction to Methods for the Solution of Ciphers,* Riverbank, no. 17 (Geneva, IL: Riverbank Laboratories, 1918), 2.

a unique signature Hitt, *Manual for the Solution of Military Ciphers,* 4–14.

It looks like this The frequency table here is one I made myself while playing around with the techniques described in ESF's codebreaking book for young adults.

74 *"certain internal relations"* WFF and ESF, *An Introduction to Methods for the Solution of Ciphers,* 6.

"There lives more faith" Alfred, Lord Tennyson, "In Memoriam A.H.H.," https://www.poetryfoundation.org/poems -and-poets/poems/detail/45349.

75 *TZYTV* ESF and WFF, "Riverbank Problems in Cryptanalysis," no. 5b, Item 290, WFF Collection.

"The thrill of your life" ESF codebreaking book (unpublished manuscript), box 9, file 12, ESF Collection.

Able for the letter A WFF and ESF, *An Introduction to Methods.*

A single miscopied letter Ibid., 3–4.

76 *The less you had to think about* Ibid.

never deviated Ibid.

black with white erasers One of the old pencils is on display under glass inside the Fabyan Villa, now maintained as a museum by the Forest Preserve District of Kane County, Illinois.

graph paper WFF and ESF, *An Introduction to Methods,* 3.

never threw anything out Ibid.

"a group of two operators" Ibid., 4–5.

Elizebeth filled the margins Hitt, *Manual for the Solution of Military Ciphers,* ESF's copy with annotations, Item 150, copy no. 3, WFF Collection.

77 *eight pamphlets* The first seven were written before William deployed to France in 1918 and the eighth, *The Index of Coincidence,* was published in 1920.

"rise up like a landmark" Kahn, *The Codebreakers,* 374.

Methods for the Reconstruction of Primary Alphabets WFF and ESF, *Methods for the Reconstruction of Primary Alphabets,* Riverbank No. 21 (Geneva, IL: Riverbank Laboratories,1918).

Methods for the Solution of Running Key Ciphers This is the Running Key paper, Riverbank No. 16. In ESF's interview with Valaki, Valaki asks her, "You participated in writing one of the manuals though, didn't you . . . one of the, ah, Riverbank books?" Elizebeth replies, "Yeah, the running-key cipher. Ah, admitted . . . Even in those days I was admitted to have been one of the authors." ESF interview with Valaki, November 11, 1976, transcribed January 12, 2012, 8.

78 *the drafts marked up*   "Chapter II: On the Flexibility of Mind Necessary for Cryptographic Analysis," box 14, folder 2, NYPL. This is a partial typed draft of the eventual Riverbank No. 17 with editing marks in both WFF's and ESF's handwriting.

the historical sections   "Appendix I, Historical and General," box 14, folder 2, NYPL. This is a typed draft of pages 7 and 8 from Riverbank No. 17. Whoever typed the draft didn't include the name of the author, but it's fascinating to note that at the top of the draft's first page, the words "By Elizabeth Smith Friedman" have been added in pencil—in WFF's handwriting.

"our pamphlets"   WFF to ESF, 15 July 1918, box 2, folder 14, ESF Collection.

"a piece of work"   ESF interview with Valaki, November 11, 1976, transcribed January 10, 2012, 4.

"Mrs. Friedman had a tendency"   Marshall Foundation to Vanessa Friedman, October 6, 1981, box 12, folder 15, ESF Collection.

"It may be egotism on my part"   George Fabyan to WFF, January 12, 1922, Item 734, WFF Collection.

79 *Seven of the eight*   Kahn, *The Codebreakers,* 374.

"That World War I leapt on"   ESF interview with Valaki, transcribed February 16, 2012, 2.

"Nothing was ever"   Ibid., 10.

"I don't think I remember"   ESF interview with Valaki, transcribed January 12, 2012, 12.

"I feel no confidence"   Ibid., 8.

80 *all the way from Scotland Yard*   ESF autobiography, 18–24.

"quite baffling"   WFF to Travis Hoke (reporter at *Popular Science Weekly*), January 21, 1920, box 6, folder 13, ESF Collection. This is a detailed narrative of the process used by WFF and ESF to solve the conspirators' cryptograms.

81 *Of 100,000 total words*   Ibid.

82 *"I challenge anybody"*   Thomas J. Tunney, *Throttled: The Detection of the German and Anarchist Bomb Plotters in the United States* (Boston: Small, Maynard & Co., 1919), 89. See also Rose Mary Sheldon, "The Friedman Collection: An Analytical Guide," rev. October 2013, Marshall Foundation, PDF file, 167, where the Friedmans point out that their solutions were included in this book without credit.

"someone had to stay behind"   ESF autobiography, 24.

83 *erupted in spectacle*   Tunney, *Throttled,* 103–4; ESF autobiography, 23.

83 all of the codebreaking   ESF autobiography, 13.

to be indecipherable   R. H. Van Deman, "Memorandum for Chief Signal Officer: Subject: Cipher with Running Key," March 16, 1918, Item 734, WFF Collection.

84 *"What Colonel Fabyan"*   WFF and ESF, *The Shakespearean Ciphers Examined* (London: Cambridge University Press, 1958), 287.

invented a new word   WFF, "On the Flexibility of Mind," box 14, folder 2, NYPL. This is a typewritten draft of a Riverbank Publication passage where you can actually see WFF cross out the word "decipherer" and write "cryptanalyst" above it.

85 *Callimahos took up snuffing*   Lambros D. Callimahos, "The Legendary William F. Friedman," *Cryptologic Spectrum* 4, no. 1 (Winter 1974): 9–17.

"cursed by luck"   Callimahos, "Summer Institute."

"Even if he computed odds"   Callimahos, "The Legendary William F. Friedman."

"Everything he touched"   Ibid.

"God-given"   Fred Friendly, remarks at ESF's funeral, November 5, 1980, in "Elizabeth Smith Friedman," *Cryptologic Spectrum* 10, no. 1 (Winter 1980): box 16, file 24, ESF Collection.

"I was never able to decide"   J. Rives Childs to Vanessa Friedman, September 28, 1981, box 12, folder 14, ESF Collection.

86 *There's a now-famous story*   ESF autobiography, 27–30.

87 *might have used key words*   Ibid.

"I was sitting across the room"   Ibid.

"springlike elasticity"   Ibid.

"it did not occur"   Ibid.

88 *"The female mind"*   WFF, "Second Period, Communications Security" (lecture), 50, NSA.

"I came to the end of my rope"   Ibid., 49.

"a wonderfully warm man"   ESF to Barbara Tuchman (undated), box 14, folder 11, ESF Collection.

"the smartest man who ever lived"   ESF codebreaking book, 65.

an article about ciphers   John Holt Schooling, "Secrets in Cipher IV: From the Time of George II to the Present Day," *Pall Mall Magazine* 8 (January–April 1896): 609–18.

This "Nihilist" cryptogram   Ibid., 618.

89 *"The meaning of the cipher"*   Ibid.

"met up with that message"   ESF codebreaking book, 65–66.

"courage"   Craig P. Bauer, *Unsolved! The History and Mystery of the World's Greatest Ciphers from Ancient Egypt to Online Secret Societies* (Princeton, NJ: Princeton University Press, 2017), 145–46.

89 *"Of course, when I learned"*   ESF codebreaking book, 65–66.
   wanted her all the time   WFF to ESF, December 21, 1938,
   large blue binder of letters donated by John Ramsay Friedman,
   Marshall Research Library, ESF Collection. See also WFF to
   ESF, September 9, 1918, box 2, folder 16, ESF Collection.
   removed the pin   WFF to ESF, September 9, 1918.
   imagined a life with her   Ibid.
   "The glacial undercurrents"   "Intermarriage of Jews Presents New
   Angle of Problem," *Jewish Criterion* (Pittsburgh), March 9, 1917.
90 *"A part cannot become merged"*   " 'Harper's Weekly' Weakness,"
   Jewish Criterion (Pittsburgh), February 24, 1905.
   "WILL THE JEWS COMMIT SUICIDE"   Charles Fleisher,
   "Will the Jews Commit Suicide Through Mixed Marriages?"
   Jewish Criterion (Pittsburgh), October 25, 1907.
   "your soul and spirit and heart"   Lyle, "Divine Fire," 85–86.
   "skinned to a frazzle . . . You're lots smarter"   WFF to ESF,
   August 7, 1917, box 2, folder 15, ESF Collection.
   "You can soar away"   Lyle, "Divine Fire," 85–86.
   "Oh Divine Fire Mine"   WFF to ESF, undated letter, box 2,
   folder 13, ESF Collection.
   whose fire is in Zion   Isaiah 31:9.
91 *realized that electronic circuits*   C. E. Shannon, "A Symbolic
   Analysis of Relay and Switching Circuits," *Transactions of the
   AIEE* 57, no. 12 (1938): 713–23.
   secret NSA projects   Transcript of Solomon Kullback oral
   history interview with NSA, August 26, 1982. Kullback discusses
   the NSA's interest in Shannon's research and says, "We had very
   close contacts with the Bell Laboratories. They were very, let's
   say, willing to work along with us."
   communicating through a noisy system   C. E. Shannon, "A
   Mathematical Theory of Communication," *Bell System Technical
   Journal* 27, no. 3 (July 1948), http://ieeexplore.ieee.org
   /document/6773024/.

CHAPTER 5: THE ESCAPE PLOT

93 "To be your North Star"   ESF to WFF, February 7, 1917, box 2,
   folder 1, ESF Collection.
94 *"I miss you infinitely"*   Ibid.
   "I shall work for you"   Ibid.
   I have dreamed about you   Ibid.
   "Anyway, Billy Boy"   Ibid.
   "Work hard on the letter tests"   ESF to WFF, January 31, 1917,
   box 2, folder 1, ESF Collection.

94 *"There was a time"* ESF diary, 46.
a careful, unemotional tone WFF to ESF, September 9, 1918,
box 2, folder 16, ESF Collection.
He confessed later Ibid.
95 *getting married would be silly* Ibid.
fetch a newspaper ESF autobiography, 14–15.
forced William to change Ibid.
"It just didn't go down" ESF interview with Clark, 11.
96 *"make a mark in something"* WFF to ESF, July 24, 1918, box 2,
folder 14, ESF Collection.
they went missing "William Friedman and Miss Elizabeth [*sic*]
Smith Were Married Monday," *Geneva Republican* (Geneva,
Illinois), May 23, 1917.
light-colored striped pants "Bride and Groom William F.
Friedman and Elizebeth S. Friedman," photograph, 1917, ESF
Collection.
A rabbi named Hersh Ibid.
The wedding announcement Ibid.
a story about a Selective Service bill "Sheriff Richardson Gets
Official Notice: The Sheriff Has Received Plans for Draft of
Eligibles," ibid.
"Mr. Friedman came to Riverbank" "William Friedman and
Miss Elizabeth [*sic*] Smith."
admitted this later ESF diary, 62.
"Splash!" Arthur Stringer, *The Prairie Wife* (Indianapolis:
Bobbs-Merrill, 1915), 3.
97 *"I am learning"* ESF diary, June 20, 1917.
a wire to Riverbank ESF to WFF, May 8, 1917, box 2, folder 1,
ESF Collection.
"I am cast into a whirl" Ibid.
"You would have thought" Ronald Clark, *The Man Who Broke
Purple: Life of Colonel William F. Friedman, Who Deciphered the
Japanese Code in World War II* (Boston: Little, Brown, 1977), 39.
She moved from Engledew Cottage into the windmill WFF
to ESF, June 1918, "Installment #3," box 2, folder 13, ESF
Collection.
98 *invited the Army to Riverbank* Jack Lait, "Recruit Rally Thrills
Throng," *Chicago Herald*, July 9, 1917.
A U.S. Army captain Ibid.
"Better to go and die" Ibid.
three hundred and fifty dollars Ibid.
He began to pester Fabyan ESF interview with Pogue, 65
Fabyan always waved him off Ibid.
99 *intercepting the Friedmans' mail* Ibid; ESF autobiography, 16.

99 *secret listening devices* ESF interview with Valaki, transcribed January 12, 2012, 4.

Mr. Powell ESF diary, August 13, 1917.

"My dearest" ESF diary. The slip of paper from WFF is inserted between pages 42 and 43.

"My heart sang" ESF diary, August 13, 1917.

throwing her arms WFF to ESF, December 21, 1938, ESF Collection.

"My Lover-Husband" ESF diary, August 13, 1917.

100 *started to dry up* ESF autobiography, 16, 26.

named Herbert O. Yardley David Kahn, *The Reader of Gentlemen's Mail: Herbert O. Yardley and the Birth of American Codebreaking* (New Haven, CT: Yale University Press, 2004).

"Why did America" Herbert O. Yardley, *The American Black Chamber* (Indianapolis: Bobbs-Merrill, 1931), 20.

a shark at poker Kahn, *The Reader,* 3.

Known officially as MI-8 Ibid., "Staffers, Shorthand, and Secret Ink," 28–35.

101 *"just off the street"* ESF interview with Valaki, transcribed January 12, 2012, 4.

"What was taught was taught" ESF interview with Valaki, transcribed February 16, 2012, 11.

booked the largest hotel ESF interview with Marshall staff, Tape #2, June 4, 1974, 11.

William and Elizebeth taught class ESF interview with Pogue, 64–66; ESF interview with Clark, 6.

stationed in paradise John W. Kopec, *The Sabines at Riverbank: Their Role in the Science of Architectural Acoustics* (Woodbury, NY: Acoustical Society of America, 1997), 41.

lavish military ball Ibid.

four of the officers' wives George Fabyan to chief of the MID, March 22, 1918.

102 *gathered outside the hotel* Clark, *The Man Who Broke Purple,* 47; Kopec, *The Sabines at Riverbank,* 47–48.

Each person stood for a letter Ibid.

the glass surface of his work desk WFF's desk is preserved at the Marshall Foundation, glass removed.

In May 1918 ESF diary, 43.

boarded a train to Chicago WFF to ESF, June 8, 1918, box 2, folder 13.

103 *"I, a mere woman"* ESF, "Pure Accident," *The ARROW,* box 12, folder 9, 401, ESF Collection.

"heartache of separation" ESF diary, July 1918 entry, 44.

"a calm Whole" Ibid.

103 *"The work is so hard"* WFF to ESF, July 24, 1918, box 2, folder 14, ESF Collection.
"out of the clear blue" WFF to ESF, August 26, 1918, box 2, folder 15, ESF Collection.
"On Saturday Col. M" WFF to ESF, November 10, 1918, box 2, folder 18, ESF Collection.
Frank Moorman WFF, "Six Lectures on Cryptology by William F. Friedman," Lecture V, 117, in *The Friedman Legacy, Sources on Cryptologic History,* no. 3 (Center for Cryptologic History: 2006).
"Love-girl" WFF to ESF, July 21, 1918, box 2, folder 14, ESF Collection.

104 *the .45 pistol* WFF to ESF, July 23, 1918, box 2, folder 14, ESF Collection.
French woman he called Madame WFF to ESF, October 6, 1918, box 2, folder 17, ESF Collection.
cigarettes as torches WFF to ESF, September 9, 1918, box 2, folder 16, ESF Collection.
based on six letters WFF, Lecture V, 109, in *The Friedman Legacy.*
"how much 'group work'" WFF to ESF, August 4, 1918, box 2, folder 15, ESF Collection.
didn't care for the taste WFF to ESF, July 6, 1918, box 2, folder 14, ESF Collection.
Lemonade Ibid.
nursing highballs WFF to ESF, October 15, 1918, box 2, folder 17, ESF Collection.
always regretted WFF to ESF, July 6, 1918, box 2, folder 14, ESF Collection. "Before I left home the Colonel's advice was that I forget poker completely."
spent time in France Kahn, *The Reader,* 45–49.
"I must confess" WFF to ESF, December 16, 1918, box 2, folder 19, ESF Collection.
each time he struck a match WFF describes this ritual in WFF to ESF, December 19, 1918, box 2, folder 19, ESF Collection.

105 *"Do you miss your Biwy Boy"* WFF to ESF, October 6, 1918, box 2, folder 17, ESF Collection.
"the many imperfections" Katie Letcher Lyle, "Divine Fire: Elizebeth Smith Friedman, Cryptanalyst," unpublished manuscript, July 4, 1991, ESF Collection, two PDF files, 86.
"good lover" Ibid.
"towered above me" Ibid., 87.
with the windows open WFF to ESF, October 6, 1918.
a recurring dream WFF to ESF, July 21, 1918, box 2, folder 14,

ESF Collection. "Most of my dreams of you have pictured me
as losing you, and I awoke trembling and with a deep fear in my
heart."

105 *"You didn't yike me"*   WFF to ESF, undated letter (December
1918), box 2, folder 19, ESF Collection.
"no money and a lot of debts"   WFF to ESF, August 4, 1918, box
2, folder 15, ESF Collection.
fixing rare grammatical mistakes   WFF to ESF, November 3,
1918, box 2, folder 18, ESF Collection. WFF circled the word
"vastly" and scrawled next to it, "Why Billy! Don't you know
better than to split an infinitive or something!"

106 *"This cable will read"*   WFF to ESF, September 20, 1918, box 2,
folder 16, ESF Collection.
a lock of her hair   WFF to ESF, December 26, 1918, box 2, folder
19, ESF Collection.
It's likely she destroyed them   This is Lyle's conclusion
in "Divine Fire," 84, and I tend to agree; Lyle was able to
interview Elizebeth when she was alive, and I think "Divine
Fire" is excellent on the personal issues Elizebeth faced in her
twenties.
Riverbank as R.   WFF to ESF, January 28, 1919, box 2, folder
20, ESF Collection.
G.F.   Ibid.
B.C.   Ibid.
Fabyan's "excesses"   Ibid.

107 *"You are perfectly right"*   WFF to ESF, November 10, 1918.
she revealed something   WFF to ESF, October 7, 1918, box 2,
folder 17, ESF Collection.
"Honey, I could have committed"   Ibid.
later confided to friends   Lyle, "Divine Fire," 96.
"Honey, don't be afraid"   WFF to ESF, August 30, 1918, box 2,
folder 15, ESF Collection.
German prisoners of war   Heber Blankenhorn, *Adventures
in Propaganda: Letters From an Intelligence Officer in France*
(Boston: Houghton Mifflin, 1919), 82. Blankenhorn was a MID
captain working at GHQ in Chaumont, same as William, and he
wrote beautifully about life there.
a group of American soldiers   Ibid., 135.
blew up bombs   Ibid., 136.
hung lanterns   Ibid.
he stayed indoors   WFF to ESF, November 10, 1918, box 2,
folder 18, ESF Collection.
"Home does not entail"   Ibid.

108 *"Elsbeth, my Dearest"*   Ibid.

108 *"The signing of the Armistice"*   WFF to ESF, December 16, 1918, box 2, folder 19, ESF Collection.
"What shall I say"   Ibid.
He had to stay in Chaumont   WFF to ESF, November 26, 1918, box 2, folder 18

109 *its Code and Signal Section*   Office of Naval Intelligence to ESF, n.d. [fall 1918?], Item 734, WFF Collection.
"of the greatest value"   John M. Manly to ESF, September 12, 1918, Item 734, WFF Collection.
how small an electron is   WFF to ESF, January 2, 1919, box 2, folder 20, ESF Collection.
"Can't two perfectly"   WFF to ESF, December 16, 1918, box 2, folder 19, ESF Collection.
"a long enough vacation"   George Fabyan to WFF, November 13, 1918, Item 734, WFF Collection.
"I refuse to have anything"   WFF to ESF, January 28, 1919.
"I don't want to flatter ourselves"   Ibid.
a love note in cipher   WFF to ESF, undated letter beginning "Good Morning, Flower Face Mine," box 2, folder 20, ESF Collection.

110 *"I am wondering how you are"*   George Fabyan to ESF, November 2, 1918, box 1, folder 42, ESF Collection.
blue colored pencil   George Fabyan to ESF, September 26, 1918, Item 734, WFF Collection.

111 *divide-and-conquer*   George Fabyan to ESF, January 6, 1919, box 1, folder 42, ESF Collection.
"Does he suppose"   WFF to ESF, January 28, 1919.
"see them in hell"   Fabyan to ESF, January 6, 1919.
"old man going down hill"   Ibid.
"I am inclined to agree"   ESF to George Fabyan, January 9, 1919, box 1, folder 43, ESF Collection.
"Won't our reunion"   WFF to ESF, February 5, 1919, box 2, folder 20, ESF Collection.
They stayed in the East   ESF autobiography, 33.
couldn't go back to Riverbank   Ibid.

112 *"The War will not make"*   WFF to ESF, October 6, 1918, box 2, folder 17, ESF Collection.
return to his first love, genetics   Ibid.
"extraordinary gift"   ESF autobiography, 34.
"Everybody said"   ESF interview with Clark.
"Come back to Riverbank"   ESF interview with Marshall staff, Tape #2, June 4, 1974, 12.
"He had us followed"   ESF interview with Pogue, 70.
in their own house   Ibid., 72.
raises never materialized   Ibid.

112 *shoved the report into a drawer* WFF and ESF, *The Shakespearean Ciphers Examined* (London: Cambridge University Press, 1958), 217–21.

113 *a crowning scientific achievement* David Kahn, *The Codebreakers: The Comprehensive History of Secret Communication from Ancient Times to the Internet,* rev. ed. (New York: Scribner, 1997), 376–85.
"The Index of Coincidence" WFF, *The Index of Coincidence and Its Applications in Cryptography* (Washington, D.C.: US Government Printing Office, 1925).
6.67 percent James R. Chiles, "Breaking Codes Was This Couple's Lifetime Career," *Smithsonian* (June 1987): 128–44.
statistics with cryptology Kahn, *The Codebreakers,* 376–85.
first in France WFF to Nelle Fabyan, October 30, 1937, Item 734, WFF Collection; ESF autobiography, 37–38.
"Fabyan's skullduggery" Herbert O. Yardley to WFF, August 14, 1919, Item 734, WFF Collection.
leapt at the chance Joseph Mauborgne to WFF, November 27, 1920, Item 734, WFF Collection.
"a great misfortune" Ibid.
"as powerful as he is ruthless" WFF to Joseph Mauborgne, November 29, 1920, Item 734, WFF Collection.
"I expect a lively row" Joseph Mauborgne to WFF, December 16, 1920, Item 734, WFF Collection.

114 *overly cruel* ESF interview with Clark, 11.
just as tricky ESF interview with Pogue, 72.
"our secret plot" Ibid., 73.
One morning Ibid.
the three o'clock train Ibid., 72.
an eerie calmness WFF to Joseph Mauborgne, December 16, 1920, Item 734, WFF Collection.
William assumed Ibid.
"after a very limited" WFF to ESF, January 28, 1919.
"Oh, you are some partner" WFF to ESF, January 28, 1919, "P.S." on a separate page, ESF Collection.
"By the end of the war" ESF, "Pure Accident."

PART II: TARGET PRACTICE

119 *"To work in this field"* Niels Ferguson, Bruce Schneier, and Tadayoshi Kohno, *Cryptography Engineering: Design Principles & Practical Applications* (Indianapolis: Wiley, 2010), 8.
"completely inadequate" WFF, "Second Period, Communications Security" (lecture), 45, NSA.

119 *"Military, naval, air"* Ibid., 45–46.
   two sides of the same coin WFF, "Communications Intelligence and Security Presentation Given to Staff and Students" (lecture, Breckinridge Hall, Marine Corps School, April 26, 1960), 5, NSA.

120 *"All the countries of the world"* ESF interview with Clark, 16.
   1,400 newly hired Prohibition agents Thomas V. DiBacco, "Prohibition's First 'Dry' New Year's Eve," *Washington Times,* December 30, 2015, http://www.washingtontimes.com/news/2015/dec/30/thomas-dibacco-prohibitions-first-dry-new-years-ev/.
   "to make the celebration" Ibid.
   the Munitions Building "Main Navy and Munitions Buildings," Histories of the National Mall, http://mallhistory.org/items/show/57.
   fourteen thousand army and navy workers Ibid.
   His starting salary "1921: William Friedman Joined War Department," National Cryptologic Museum Foundation, https://cryptologicfoundation.org/m/cch_calendar_mobile.html/event/2016/07/01/1467349200/1921-william-friedman-joined-war-department/74534.
   Elizebeth's was $2,200 ESF Personnel Folder, "Personal History Statement," July 1, 1930.

121 *a piano studio above a bakery* ESF interview with Pogue, 77.
   Mauborgne visited with his cello ESF interview with Clark.
   pedestrians stopping Ibid.
   with two dozen people "Morning in New York," in David Kahn, *The Reader of Gentlemen's Mail: Herbert O. Yardley and the Birth of American Codebreaking* (New Haven, CT: Yale University Press, 2004), 50–62.
   the messages of Japanese diplomats Ibid., 63–71.
   wasn't skilled enough to go further John Bryden, *Best-Kept Secret: Canadian Secret Intelligence in the Second World War* (Toronto: Lester, 1993), 88–89.
   apartment for his mistress Kahn, *The Reader,* 48.
   "an edge on her" ESF to WFF, "Friday, 2:30 P.M.," summer 1921, box 2, folder 2, ESF Collection.

122 *first scientifically constructed* ESF autobiography, 39–40.
   "all the difference" ESF interview with Clark, 16.
   designed to survive capture David Kahn, *Seizing the Enigma: The Race to Break the German U-boat Codes, 1939–1943* (Boston: Houghton Mifflin, 1991), 33.

123 *the application of electrical current* WFF, Lecture V, 156–57, in *The Friedman Legacy, Sources on Cryptologic History,* no. 3 (Center for Cryptologic History: 2006).

123 *William asked Hebern* WFF, "Second Period, Communications Security" (lecture), 18, NSA.

"discouraged to the point of blackout" ESF interview with Pogue, 43.

"As I was tying my black tie" WFF, "Communications Intelligence and Security," 34.

"P.S." WFF to George Fabyan, March 10, 1924, Item 734, WFF Collection.

"It's a striking paradox" WFF to George Fabyan, June 23, 1926, Item 734, WFF Collection.

124 *two discs of alphabets* WFF, "Six Lectures on Cryptology by William F. Friedman," Lecture VI, 149, in *The Friedman Legacy, Sources on Cryptologic History,* no. 3 (Center for Cryptologic History: 2006)

2.29×10^{82} Lambros D. Callimahos, "The Legendary William F. Friedman," *Cryptologic Spectrum* 4, no. 1 (Winter 1974): 9–17.

the number of atoms Thought to be 10^{80}. Wikipedia, s.v. "Observable universe," last modified April 22, 2017, https://en.wikipedia.org/wiki/Observable_universe.

"The number of permutations" WFF, "Communications Intelligence and Security," 30.

demonstrated his mastery Callimahos, "The Legendary William F. Friedman."

125 *"but it helps"* Rose Mary Sheldon, "William F. Friedman: A Very Private Cryptographer and His Collection," *Cryptologic Quarterly* 34, no. 1 (2015): 20.

Rochefort recalled later Captain Joseph J. Rochefort, interview by U.S. Naval Institute, Annapolis, Maryland, 1983, 45–47.

"Here is a bunch of messages" Ibid., 47.

had to stop breaking codes Ibid., 45.

"no one that could compare" Ibid., 40.

a young German engineer "The Man, the Machine, the Choice," in Kahn, *Seizing the Enigma*, 31–48.

"clever inventor" WFF, "Six Lectures on Cryptology by William F. Friedman," Lecture VI, 153, in *The Friedman Legacy.*

126 *"stay home and write some books"* ESF interview with Marshall staff, Tape #5, June 6, 1974, 8.

excited about her books WFF to George Fabyan, August 10, 1926, Item 734, WFF Collection.

"I am all alone" WFF to John M. Manly, February 4, 1922, box 13, folder 22, ESF Collection.

"I drove 36 miles" ESF to WFF, letter marked "Monday,

9:30 P.M." by ESF and "Washington era between 1921 and 1923"
by archivists, box 2, folder 2, ESF Collection.

126 *midwestern stranger*   Ibid.

"My dear"   ESF to WFF, "Monday, 9:30 P.M."

127 *move out of the city*   Katie Letcher Lyle, "Divine Fire: Elizabeth
Smith Friedman, Cryptanalyst," unpublished manuscript, July 4,
1991, ESF Collection, two PDF files, 126.

kryptos   Dictionary.com, s.v. "crypt," accessed May 10, 2017,
http://www.dictionary.com/browse/crypt.

128 *Pinklepurr*   Lyle, "Divine Fire," 218.

A. A. Milne poem   A. A. Milne, "Pinkle Purr," in *Now We Are
Six* (New York: Puffin Books, 1992), 89.

in the morning   Lyle, "Divine Fire," 128.

a former boxer   WFF, "Second Period, Communications
Security" (lecture), 20, NSA.

a pug nose   Ibid.

"Now," Elizabeth wrote   ESF codebreaking book.

"little book"   Ibid.

a children's history of the alphabet   ESF children's history
of the alphabet (unpublished manuscript), box 9, file 14, ESF
Collection.

129 *knock on her door*   ESF interview with Marshall staff, Tape #5,
June 6, 1974, 8.

A retired astronomy professor   Craig P. Bauer, *Unsolved!
The History and Mystery of the World's Greatest Ciphers
from Ancient Egypt to Online Secret Societies* (Princeton, NJ:
Princeton University Press, 2017), 500–3.

A man sent a bomb to Huey Long   WFF backfile part IV, "(FBI
J Edgar Hoover) Bank Robbery. Cases with Dept. of Justice,
Ohio State Penitentiary, and Post Office Inspection Service,"
Item 849, WFF Collection.

a plot   Ibid.

a congressional committee   ESF autobiography, 41–42.

caught the eye   Ibid.

130 *notes scrawled with code*   Ibid.

refused to pay him   ESF autobiography, 43.

she wore it around her neck   ". . . the one-of-a-kind Evalyn
Walsh McLean," PBS Treasures of the World, http://www.pbs
.org/treasuresoftheworld/hope/hlevel_1/h3_ewm.html.

an anti-Semitic weekly newspaper   "The International Jew:
The World's Problem," *Dearborn Independent* (Dearborn,
Michigan), May 22, 1920.

intelligence reports about Jewish activities   Joseph W. Bendersky,

The Jewish Threat: Anti-Semitic Politics of the U.S. Army (New York: Basic Books, 2000), xiii–xiv.

130 *"When they couldn't get him"*   ESF interview with Marshall staff, Tape #5, June 6, 1974, 8.

"second-hand"   ESF interview with Clark.

"Sad for me"   ESF interview with Marshall staff, Tape #5, June 6, 1974, 8.

131 *"I didn't want to work for the Navy"*   ESF interview with Pogue, 41–42.

She took Driscoll's place   Ibid.

left the navy after five months   Ibid.

suppressing her own desires   "Sopha was a shadowy figure, her life obviously deeply and exclusively involved in her anatomical destiny, that of bearing ten children over an eighteen year period." Lyle, "Divine Fire," 165.

"Often I feel"   WFF to ESF, July 31, 1918, box 2, file 14, ESF Collection.

"Sometimes I wish"   Ibid.

"a queer sensation"   Ibid.

"we could help her"   George Fabyan to WFF, February 24, 1924, Item 734, WFF Collection.

132 *she would be all right*   WFF to George Fabyan, February 25, 1924, Item 734, WFF Collection.

black woman named Cassie   ESF autobiography, 63.

constantly perplexed   ESF copy of George Fabyan's *What I Know About the Future of Cotton and Domestic Goods*, 2nd ed. (Chicago, 1900), containing diary entries about Barbara and John Ramsay, box 21, folder 1, ESF Collection.

shared her dolls with Krypto   Ibid.

his mirth excited   Ibid.

top volume on the Victrola   Ibid.

a doctrine of no doctrines   Ibid.

"let the rest take care of itself"   Ibid.

"She strings together consonants"   Fabyan, *What I Know About the Future*, Item 602, WFF Collection.

133 *3932 Military Road*   ESF Personnel Folder.

203 slow small patrol boats   David P. Mowry, "Listening to the Rum-Runners: Radio Intelligence During Prohibition," 2nd ed., Center for Cryptologic History, 2014, 16.

five thousand miles of coastline   Ellen NicKenzie Lawson, *Smugglers, Bootleggers, and Scofflaws: Prohibition and New York City* (Albany: State University of New York Press, 2013), 7.

in radio prowess   Commander J. F. Farley, "Radio in the Coast Guard," *Radio News* (January 1942): 43–48.

134 *a ninety-day contract* ESF, "Personal History," Personnel Folder, July 1, 1931.

work from home ESF interview with Clark, 15.

Fifteen thousand people U.S. Treasury Department, *Annual Report of the Secretary of the Treasury on the State of the Finances for the Fiscal Year Ended June 30, 1926* (Washington, D.C.: Government Printing Office, 1927).

six separate law enforcement agencies "SA Eliot Ness, a Legacy ATF Agent," Bureau of Alcohol, Tobacco, Firearms and Explosives, https://www.atf.gov/our-history/eliot-ness.

135 *a soft-spoken father of two* Robert G. Folsom, *The Money Trail: How Elmer Irey and His T-men Brought Down America's Criminal Elite* (Washington, D.C.: Potomac Books, 2010), 313.

"the Treasury fist" "T-men (1947) Quotes," IMDb.com, http://www.imdb.com/title/tt0039881/quotes.

solution in the margin "Elmer Irey Retires: Boss of Treasury T-men Was One of World's Greatest Detectives," *Life* (September 2, 1946).

independent sea captains Frederick Van de Water, *The Real McCoy* (Mystic, CT: Flat Hammock Press, 2007).

the envy of many small nations ESF, "History of Work in Cryptanalysis," April 27–June 1930," box 4, folder 17, ESF Collection; ESF, "West Coast," narrative of West Coast smuggling operations, box 4, folder 23, ESF Collection.

"The whole half of the world" ESF interview with Clark.

136 *"Mrs. Friedman is the only person"* Roy A. Haines, Acting Prohibition Commissioner, to Civil Service Commission, April 22, 1927, box 4, folder 16, ESF Collection.

"see what you can do" Office of Chief Prohibition Investigator to ESF, February 1, 1926, box 4, folder 10, ESF Collection.

sent from Halifax, Nova Scotia "1,000, Following for Your Information," January 29, 1926, box 4, folder 10, ESF Collection.

"most secret communications" ESF, "History of Work in Cryptanalysis," April 27–June 1930," box 4, file 17, ESF Collection.

used different code systems ESF, "History of Chief Smuggling Interests on the Pacific Coast," box 4, folder 23, ESF Collection.

137 *"If I may capture a goodly number"* ESF, "A Cryptanalyst," *Arrow* (February 1928), box 12, folder 9, 531–34, ESF Collection.

maps of the radio traffic ESF, "Chart Showing Operations of Liquor Smuggling Vessels as Directed by Short Wave Radio Through Secret Systems of Communication, Pacific Coast," August 1933, box 6, file 1, ESF Collection.

"I sort of floated around" ESF interview with Clark.

138 *She traveled to the West Coast* ESF autobiography, 52.

138 *sailing up the Hudson River*   Transcript of ESF interview with
Margaret Santry, NBC national radio broadcast, May 25, 1934,
box 19, folder 6, ESF Collection.
brothers named Hobbs   ESF, "History of Chief Smuggling
Interests."
"to Joseph Kennedy, Ltd."   Ibid. For more on Joseph Kennedy's
connections to Vancouver rum interests, see Stephen Schneider,
Iced: The Story of Organized Crime in Canada (Mississauga,
Ontario: Wiley, 2009), 207.
Elizebeth wasn't afraid   ESF interview with Ed Meryl. Asked if
she was ever in physical danger due to her work, ESF responded
dismissively—"Not that I know of"—despite the fact that she
needed bodyguard protection during the *I'm Alone* proceedings.
She went to Houston, Texas   ESF autobiography, 52, 92–95.
650 messages in 24 different code systems   ESF, "History of Work
in Cryptanalysis."
a one-legged cabdriver   "Bond for Armatou Lowered to $250,"
Galveston Daily News, March 26, 1930.
139 *"he was in a very mean mood"*   ESF interview with BBC for
"Codebreakers" TV special, box 15, folder 5, ESF Collection.
"Only woman on plane"   ESF Vancouver trip log, written for
her children while flying cross-country, October 16–17, 1937, box
16, folder 2, ESF Collection.
out for afternoon tea   ESF to WFF, 1932, box 2, folder 3, ESF
Collection.
"messages unearthed from a safe"   ESF to WFF, October 21,
1937, box 2, folder 4, ESF Collection.
"under the press of duties"   ESF to Josephine Coates, January 23,
1930, box 1, folder 2, ESF Collection.
two thousand messages per month   ESF, "History of Work in
Cryptanalysis."
a single clerk-typist   ESF, "Memorandum upon a Proposed
Central Organization at Coast Guard Headquarters for
Performing Cryptanalytic Work," box 5, file 6, ESF Collection.
twelve thousand rum messages   ESF, "History of Chief
Smuggling Interests."
a seven-page memo   Ibid.
140 *thirty of these books*   Ibid.
a refuge from problems   ESF interview with Clark, 15–16.
"Position," "Landing boat"   See intercepted rum messages and
worksheets in box 4, folder 14, ESF Collection.
Elizebeth came home   ESF interview with Clark, 15.
"pair of shoes, size 15"   Ibid.; ESF, "History of Chief Smuggling
Interests."

141 *a codebreaking team of her own* "History of USCG Unit #387," Foreword.

Cryptanalyst-in-Charge ESF, "Memorandum upon a Proposed Central Organization."; U.S. Treasury Department, Office of the Secretary to ESF, June 30, 1931, ESF Personnel Folder.

Scouring civil service lists ESF autobiography, 53.

worried that her new employees Ibid., 56.

the top scorer Ibid., 53.

"he did not comprehend" Ibid., 54.

Hyman Hurwitz Hyman Hurwitz, Official Personnel Folder, National Personnel Records Center, National Archives at St. Louis, requested September 2016.

142 *Vernon Cooley* Vernon E. Cooley, Official Personnel Folder, National Personnel Records Center, National Archives at St. Louis, requested September 2016.

Robert Gordon Robert E. Gordon, Official Personnel Folder, National Personnel Records Center, National Archives at St. Louis, requested September 2016.

"able, agreeable, and cooperative" ESF autobiography, 55.

a productive routine Ibid.

pound for pound the best This is my own conclusion. It was shared by the future officers of British Security Coordination, who arrived in the United States in 1940 and determined that Elizebeth's coast guard unit was the most effective in the country. See British Security Coordination, *The Secret History of British Intelligence in the Americas, 1940–1945* (New York: Fromm International, 1999), 469–70.

"The world is a mess" George Fabyan to WFF, January 23, 1932, Item 734, WFF Collection.

Elizebeth took her daughter ESF untitled two-page narrative about her work with the League of Women Voters, box 7, folder 6, ESF Collection.

"The only thing we have to fear" Franklin Delano Roosevelt, "Inaugural Address," March 4, 1933, http://www.presidency.ucsb.edu/ws/?pid=14473.

143 *outside in the freezing wind* James A. Hagerty, "Roosevelt Address Stirs Great Crowd," *New York Times,* March 5, 1933.

handed out League of Women Voters fliers ESF League of Women Voters narrative.

"ardent worker" Ibid.

Eighteen days later, in Germany "Dachau Opens," United States Holocaust Memorial Museum, https://newspapers.ushmm.org/events/dachau-opens.

at a press conference "Himmler sets up Dachau," The Nazi

Concentration Camps, Birbeck University of London, http://www.camps.bbk.ac.uk/documents/003-himmler-sets-up-dachau.html.

143 *"I pack my bag"* ESF, "Pure Accident."

"Please state your name" United States v. Albert M. Morrison et al. (E.D. La. 1933), No. 16,981, May 2, trial transcript vol. 1, 141.

twenty-three suspected agents "Wireless Station Operator Called in Rum Ring Trial," *Times-Picayune* (New Orleans), May 2, 1933. The *Times-Picayune* says there were twenty-four defendants; ESF says twenty-three, on page 80 of her autobiography. ESF never miscounts things, so I am going with her number. A few of the indicted men, including Isadore "Kid Cann" Blumenfeld, a gangster from Minneapolis, never appeared in court, creating some confusion.

a fleet . . . a pirate radio station United States v. Morrison, No. 6,981, May 1, trial transcript vol. 1, 47–62. This is Woodcock's opening statement, which gives an overview of the case.

at least thirty-two coded radio messages Ibid., Bill of Exceptions, Exhibit X 29.

mailed them Ibid., May 2, trial transcript vol. 1, 170.

144 *their system* Ibid., Bill of Exceptions.

"the greatest rum-running conspiracy" "Greatest Liquor Plot Case Trial Delayed 35 Days," *Times-Picayune* (New Orleans), April 15, 1932.

a hat with a flower "Code Expert Testifies at Trial," *Times-Picayune* (New Orleans), May 3, 1933.

a stack of yellow papers United States v. Morrison, No. 16,981, May 2, trial transcript vol. 1, 150–55.

quietly switching seats Ibid., May 1, trial transcript vol. 1, 64–76.

a rawboned man Ibid.

the aliases Ibid., May 1, trial transcript vol. 1, 47.

"Mr. Burk" Ibid., May 1, trial transcript vol. 1, 64–76. "Mr. Burk" was what Morrison called himself when he dealt with the shortwave radio operator in New Orleans, Charles Andres.

Nathan Goldberg, Al Hartman, and Harry Doe "Bond Reductions Ordered for Six Held in Rum Plot," *Times-Picayune* (New Orleans), April 15, 1931.

$500,000 and more than two years ESF autobiography, 80; the first radio messages from the ring were intercepted in March 1931.

methodical former Army colonel S. J. Woolf, "Col. Woodcock: Leader of the Dry Army," *New York Times*, November 2, 1930.

"a steady attack" Ibid.

"Have you a message" United States v. Morrison, Nó. 16,981, May 2, trial transcript vol. 1, 143.

144 *"QUIDS, ABGAH"* Ibid., Bill of Exceptions, Exhibit X 29.
"incompetent, irrelevant" Ibid., May 2, trial transcript vol. 1, 145.
Eight other defense attorneys Ibid., May 2, trial transcript vol. 1, 1.

145 *handled Capone's appeals* "Capone Renews Fight for Freedom,"
Town Talk (Alexandria, Louisiana), April 30, 1934.
Walter J. Gex Sr. "Leader of City and County Passes Away;
Last Rites Monday P.M.," *Sea Coast Echo* (Bay St. Louis,
Mississippi), February 1937.
"I believe I am asked" Ibid., 145–47.
"That certainly is some information" Ibid., 162.
"GD (HX) gm" *United States v. Morrison*, No. 16,981, Bill of
Exceptions, Exhibit X 6.
"SUBSTITUTE FIFTY CANADIAN" Ibid., May 2, trial
transcript vol. 1, 150.
"How shall I address you" Ibid., 164–66.

146 *"I move that all of the testimony"* Ibid., 168–69.
the jury convicted five "Five Men Found Guilty of Liquor
Plotting Charge," *Times-Picayune* (New Orleans), May 7, 1933.
two years in prison "Woodcock Move Clears Five Men in Rum
Plot Case," *Times-Picayune* (New Orleans), May 9, 1933.
"made an unusual impression" ESF autobiography, 88.
"a pretty government" Lyle, "Divine Fire," 174.
"a pretty middle aged woman" ESF autobiography, 70.
"a pretty young woman with a filly pink dress" Ibid.
"a pretty little woman who protects" Lyle, "Divine Fire," 174.
young enough to resent ESF autobiography, 70–71.
badly written Ibid.
Facing off against Edwin Grace ESF autobiography, 84–85.
"CLASS IN CRYPTOLOGY" Ibid., 82.

147 *She hoped the attention* Ibid., 83.
"He never put into words" ESF interview with Valaki,
November 11, 1976, transcribed January 10, 2012, 16.
"a certain grim look" Ibid.
A former artillery officer "About Henry L. Stimson," Stimson
Center, https://www.stimson.org/content/about-henry-l
-stimson.
"Gentlemen do not read each other's mail" Kahn, *The Reader*,
98.

148 *a new army codebreaking unit* Frank Rowlett, *The Story of
Magic: Memoirs of an American Cryptologic Pioneer* (Laguna
Hills, CA: Aegean Park Press, 1998), 6–33.
adjacent desks Ibid.
he popped his head in the vault "We Discover the Black
Chamber," ibid., 34–39.

148 *"Welcome, gentlemen"* Ibid.

149 *challenging them to solve different machines* Rowlett, *The Story of Magic,* 59–76.

Angooki Taipu A Craig P. Bauer, *Secret History: The Story of Cryptology* (Boca Raton, FL: CRC Press, 2013), 296–300.

Angooki Taipu B Ibid., 301–4.

Converter M-134 WFF, "Important Contributions to Communications Security, 1939–1945," 1, NSA.

150 *held secret for many years* Fischer, Willis and Panzer, "Memorandum Concerning a Bill for the Relief of William F. Friedman," August 21, 1950, NSA.

an enduring frustration Ibid.

evolved into the SIGABA WFF, "Important Contributions," 3–5.

Up to four rotors Timothy J. Mucklow, "The SIGABA/ECM II Cipher Machine: 'A Beautiful Idea,'" Center for Cryptologic History, National Security Agency, 2015.

10,060 SIGABA machines Ibid.

"the absolute security" WFF, "Important Contributions," 5.

"Never said a word to me" ESF interview with Pogue, 24.

"never opened his mouth" Ibid., 53.

151 *"was hugging me warmly"* Barbara Friedman to Ronald Clark, September 26, 1976, and "P.S.," October 6, 1976, box 14, folder 14, ESF Collection.

make himself a sandwich ESF interview with Pogue, 27.

"nothing more or less than exhaustion" Ibid., 46.

the ice pick lobotomy Michael M. Phillips, "The Lobotomy Files: One Doctor's Legacy," *Wall Street Journal,* http://projects.wsj.com/lobotomyfiles/?ch=two.

eventually, William did His psychiatrist starting in the 1940s was Dr. Zigmond Lebensohn of George Washington University, a junior colleague of Freeman. See Zigmond Lebensohn, "The History of Electroconvulsive Therapy in the United States and Its Place in American Psychiatry: A Personal Memoir," *Comprehensive Psychiatry* 40, no. 3 (1999): 173–81.

"Any story of my experiences" ESF, "A Cryptanalyst," *Arrow* (February 1928), box 12, folder 9, 531–34, ESF Collection.

152 *a letter years later to Barbara* ESF to Barbara Friedman, February 12, 1945, box 3, folder 26, ESF Collection.

Immortal Wife Irving Stone, *Immortal Wife* (Garden City, NY: Doubleday, 1944).

"to be a good wife" Ibid., 33.

"gray-haired" Ibid., 134.

"Everything that happened to John" Ibid., 140–41.

153 *"In many ways it parallels"* WFF to John Ramsay Friedman, August 13, 1945, box 4, folder 8, ESF Collection.
a square of red paper 1928 Friedman holiday card, Item 568, WFF Collection.

154 *"Friedman's Wishing Tree"* "Friedman's Wishing Tree," box 13, folder 9, ESF Collection.
scavenger hunts "The Single Intelligence Skool, Pap Problem No. 1," and other scavenger-hunt materials (envelopes, cryptograms, etc.), box 13, folder 9, November 6, 1938.
"The first item was a series of dots" Virginia Corderman to Vanessa Friedman, October 2, 1981, box 12, folder 15, ESF Collection.
Elizebeth designed a menu Ibid.

155 *The Crypto-Set Headquarters Army Game* Item 2097, WFF Collection.
A second prototype, Kriptor Item 2098, WFF Collection.
sent it to Milton Bradley WFF to ESF, November 29, 1938.
beguiled and frustrated Modernist fiction baffled WFF, but he collected it, read it, tried to make sense of it, thought the sentences were often beautiful, and corresponded with Joyceans, including J. F. Byrne, Joyce's real-life schoolmate and the inspiration for "Cranly" in *A Portrait of the Artist as a Young Man.* Byrne thought he had invented an unbreakable cipher that he called the "Chaocipher." WFF told him his cipher was worthless. See Sheldon, "The Friedman Collection: An Analytical Guide," 466, and Item 1405.1 in the WFF Collection.
with a burgundy binding Giambattista della Porta, *De furtivis literarum notis, vulgo de Ziferis* (Naples, 1563; forgery, London, 1591), Item 119, WFF Collection.

156 *"He came up to me one morning"* WFF, "Communications Intelligence and Security," 9.
American University in 1929 ESF's graduate-school course sheets, box 4, folder 23, ESF Collection.
his copy of the Voynich Manuscript Bauer, *Unsolved!* 83–86, includes a great account of the Voynich study group that WFF started at NSA in 1944 and his attempts to make sense of the weird manuscript using NSA technology.
"worked on the cryptogram" WFF speech to NSA, 1958, box 8, folder 4, ESF Collection.

157 *"for my work in the FBI"* FBI, "The Hoover Legacy, 40 Years After: Part 2: His First Job and the FBI Files," June 28, 2012, https://www.fbi.gov/news/stories/copy_of_the-hoover-legacy-40-years-after.
a professor friend Sheldon, "The Friedman Collection: An Analytical Guide," 434.

157 *a crimson warrior swinging an axe*   This image is pasted inside almost every book in the WFF Collection.

158 *her death in 1934*   WFF and ESF, *The Shakespearean Ciphers Examined* (London: Cambridge University Press, 1958), 208.
"I have no facilities"   Leroy Hennessey, "Twas Bill! Nay, Bacon! But Now E'en Fabyan Knows Not Who Did Shakespeare," *Chicago Evening American* (January 1922): box 14, "The Ideal Scrap Book," NYPL.
called Yardley an "ass"   George Fabyan to WFF, September 16, 1931, Item 734, WFF Collection.
Fabyan seemed glum and tired   Ibid., May 10, 1934.

159 *complained of a hernia*   Ibid., May 31, 1935.
"Always the same old, GF"   Ibid.
age of sixty-nine   Richard Munson, *George Fabyan: The Tycoon Who Broke Ciphers, Ended Wars, Manipulated Sound, Built a Levitation Machine, and Organized the Modern Research Center* (North Charleston, SC: Porter Books, 2013), 141–42.
a letter from an old Riverbank colleague   Cora Jensen to WFF and ESF, May 29, 1936, box 7, folder 18, ESF Collection.
It went dark   Munson, *George Fabyan*, 141–42.
far less money   Jensen to WFF and ESF.
$175,000 to his widow   "Will of Col. Fabyan Filed Tuesday Leaves $175,000 to Widow," 1936 newspaper clipping from Jensen, box 7, folder 18, ESF Collection.
Nelle died   John W. Kopec, *The Sabines at Riverbank: Their Role in the Science of Architectural Acoustics* (Woodbury, NY: Acoustical Society of America, 1997), 56.
Cumming was killed   Ibid.
$70,500   "Buy $800,000 Fabyan Estate as Playground," 1936 newspaper clipping from Jensen, box 7, folder 18, ESF Collection.
"hands it might fall into"   George Fabyan to WFF, July 29, 1935, Item 734, WFF Collection.

160 *wrote to his widow, Nelle*   WFF note in red pencil beginning "Colonel Fabyan's last letter to me," Item 734, WFF Collection.
ordered many records to be burned   ESF interview with Valaki, transcribed January 12, 2012, 2–3.
prevent embarrassing revelations   Ibid.
"A lie!"   WFF, annotated copy of Herbert O. Yardley, *The American Black Chamber* (Indianapolis: Bobbs-Merrill, 1931), Item 604, 43, WFF Collection.
"This is a patchwork"   Ibid., 44.
"Lies, lies, lies"   Ibid.

161 *unable to find a new job*   Kahn, *The Reader*, 104.

161 *"a glimpse behind the heavy curtains"* Yardley, *The American Black Chamber*, Foreword.

  there was no harm Ibid.

  31,119 copies Kahn, *The Reader*, 131.

  "A lovely girl dances" Yardley, *The American Black Chamber*, Foreword.

  "the beautiful blonde woman" Ibid., 90–119.

  "It did seem to me" Yardley, *The American Black Chamber*, 329.

  searched her desk Ibid., 331.

162 *"damned lie"* WFF copy of ibid.

  Yardley embellished Kahn, *The Reader*, 113.

  compressed time, invented dialogue Ibid., 117.

  startle adversaries WFF, "World War I Codes and Ciphers" (lecture, SCAMP, 1958), 22–23, NSA.

  "A lie! Which can be so proved" WFF copy of Yardley, *The American Black Chamber*, 45.

163 *four of his colleagues in the army* Ibid., front matter.

  Hollywood movies Kahn, *The Reader*, 173–86.

  Congress passed a law T. M. Hannah, "The Many Lives of Herbert O. Yardley," *Cryptologic Spectrum* 11, no. 4 (1981): 5–29.

  "Widespread interest in the romantic stories" "Extract from R.I.P. No. 98," April 5, 1943, 118–23, NSA.

  "I'll confess, Mrs. Friedman" ESF interview with Santry.

164 *"I never thought of my job"* Ibid.

  from the glassed-in control room ESF autobiography, 76.

  "Does the habit?" ESF interview with Santry.

165 *preferred speaking with female reporters* ESF autobiography, 71.

166 *code books designed for Chinese merchants* Chinese Telegraph Code book used in ESF's Gordon Lim case, 342.3, WFF Collection.

  "The whole deciphering science" ESF interview with Santry.

  intercepted letters and telegrams U.S. Customs Service, memorandum about the Ezra case, Frederick S. Freed, supervising customs agent, June 8, 1933, box 11, folder 10, ESF Collection; see also Leah Stock Helmick, "Key Woman of the T-men," *Reader's Digest* (September 1937): 51–55.

  Green Gang of criminal warlords ESF's Ezra case files list "Paul A. Yip" as the Shanghai contact of the Ezra twins. For background on Yip, see Kathryn Meyer and Terry Parssinen, *Webs of Smoke: Smugglers, Warlords, Spies, and the History of the International Drug Trade* (Plymouth, UK: Rowman & Littlefield, 1998), 159. The book describes Yip as "Green Gang member, opium trafficker, and double agent." The Ezra case files are in box 6, folder 25, ESF Collection.

166 *520 tins of smoking opium*   Freed memorandum, June 8, 1933.
   EZRA GANG FALLS IN TRAP   "Ezra Gang Falls in Trap of Woman
   Expert in Puzzles," *San Francisco Chronicle*, September 28 or 29,
   1933, box 18, folder 5, ESF Collection.
   "supplied the Federals"   Ibid.
167 *"We have to keep our ideas secret"*   Ibid.
   begged reporters to credit   "Woman Jails Dope Runners,"
   Universal Service, box 18, folder 5, ESF Collection.
   "The mystery-lure"   ESF to F. E. Pollio, February 14, 1938, box
   1, folder 9, ESF Collection.
   "She is entrusted"   Helmick, "Key Woman of the T-men."
   more than a million subscribers   Trusted Media Brands,
   "Expansion (1930s–70s)," http://www.tmbi.com/history/.
   "could lead to only one conclusion"   "Extract from R.I.P. No. 98."
   stormed into William's office   Ronald Clark, *The Man Who
   Broke Purple: Life of Colonel William F. Friedman, Who
   Deciphered the Japanese Code in World War II* (Boston: Little,
   Brown, 1977), 179.
168 *"I lost my tongue completely"*   Ibid., 180.
   Elizebeth flew to Vancouver   ESF Vancouver trip log.
   thirteen Luger pistols   A. H. Williamson, "Woman Translates
   Code Jargon in Assizes at Trial of Five Chinese," newspaper
   clipping, box 18, folder 6, ESF Collection.
   that were translated into English   Ibid.
   "select fully Wat list"   Ibid.
   "hams," "presses," and "tails"   "Woman Helps Canada Break Big
   Opium Ring," *New York Times,* February 8, 1938.
169 CANADA SMASHES   Staff Correspondent, "Canada Smashes Opium
   Ring with U.S. Woman's Aid," *Christian Science Monitor,*
   February 9, 1938.
   WOMAN TRANSLATES CODE JARGON   A. H. Williamson, "Woman
   Translates Code Jargon in Assizes at Trial of Five Chinese," box
   18, folder 6, ESF Collection.
   "careers unusual for their sex"   "These Women Make Their
   Hobbies Pay," *Look*, February 15, 1938, 46.
   printed fourteen pages   James W. Booth, "Lady Manhunter,"
   Detective Fiction Weekly, September 28, 1940, 60–73.
   a newspaper editor   A list of Booth's stories is available at the
   Crime, Mystery & Gangster Fiction Magazine Index maintained
   by Phil Stephenson-Payne, http://www.philsp.com/homeville
   /cfi/s116.htm#A2001.
   "PLEASE COOPERATE"   Theodore Adams to ESF (telegram), box 6,
   folder 4, ESF Collection.
   "Ad Absurdum!"   Ibid.

169 *she had taught her husband everything*   ESF autobiography, 73.
  regale him with stories   WFF to ESF, December 29, 1938.
  "When people introduce me"   Ibid.
170 *never wanted to see her name*   ESF to Mrs. T. N. Alford,
  October 19 (no year given, 1938 or 1939), box 1, folder 9, ESF
  Collection.
  Hawaii in October 1938   WFF, "Important Contributions to
  Communications Security, 1939–1945," 1, NSA.
  "highly secret communications"   Ibid.
  install and test them   Rowlett, *The Story of Magic,* 142.
  six bulky cipher machines   WFF, "Important Contributions."
  books of essays and poetry   WFF to ESF, January 6, 1939, ESF
  Collection.
  Nazi mobs in a thousand German towns   Martin Gilbert, "The
  Night of Broken Glass," in *Kristallnacht: Prelude to Destruction*
  (New York: Harper Perennial, 2006), 23–41.
  hacked apart a grand piano   Ibid., 47.
  her sister was simply overworked   Edna Dinieus to ESF,
  January 1, 1939, small blue binder.
171 *sometimes able to sneak on*   WFF to ESF, January 6, 1939, ESF
  Collection.
  got the sense his wife was struggling   Ibid., December 21, 1938,
  large blue binder.
  "I can almost forsee"   Ibid.
  a Christmas dance for passengers   Ibid.
  "simple personal habits"   "Munich Cardinal Praises Hitler's
  'Personal Habits,'" *Washington Post,* January 1, 1939.
  "far-reaching consequences"   Voices of the Manhattan Project,
  "Columbia University," Atomic Heritage Foundation and Los
  Alamos Historical Society, http://manhattanprojectvoices.org
  /location/columbia-university.
172 *the moon rose fat and white*   WFF to ESF, January 6, 1939, 26.
  He played shuffleboard   Ibid., November 29, 1938.
  he sent to Milton Bradley   Ibid.
  ran to thirty pages   Ibid.
  "I hope these sheets"   Ibid.
  love poems by Tennyson   Ibid., 21. The Tennyson poem he quoted
  was an erotic one, "The Miller's Daughter": "And I would be the
  necklace / and all day long to fall and rise / Upon her balmy bosom
  / With her laughter or her sighs." https://www.poetryfoundation
  .org/poems-and-poets/poems/detail/50267.
  "shelter'd here upon a breast"   William Makepeace Thackeray,
  "Song of the Violet," in *The Complete Works of William
  Makepeace Thackeray* (Boston: Houghton Mifflin, 1889), 291.

172 *"For I've known"*   WFF to ESF, January 6, 1939, 23.
173 *stood at the top deck's rail*   Ibid.
     One night when it was hot   Ibid.
     "The ocean," he said   Ibid., 19.

CHAPTER 1: GRANDMOTHER DIED

177 *"Lights out 'cause I can see in the dark"*   Fugazi, "Caustic
     Acrostic," recorded March–September 1997 on End Hits,
     Dischord Records No. 110, compact disc.
     At 4 P.M. on August 31, 1939   Heinz Höhne, *The Order of the
     Death's Head: The Story of Hitler's SS,* trans. Richard Barry
     (New York: Penguin, 2000), 260–66.
178 *"Grossmutter gestorben"*   Ibid.
     an excuse to start a war   Ibid.
     250,000 by 1939   "The SS (Schutzstaffel): Background and
     Overview," Jewish Virtual Library, http://www.jewishvirtual
     library.org/background-and-overview-of-the-ss.
     "the guillotine used"   Höhne, *The Order of the Death's Head,* 3.
     "enveloped in the mysterious aura"   Ibid., 1.
179 *a forty-three-year-old Catholic farmer*   Bob Graham, "World
     War II's First Victim," *Telegraph* (London), August 29, 2009,
     http://www.telegraph.co.uk/history/world-war-two/6106566
     /World-War-IIs-first-victim.html.
     "Attention, this is Gliwice"   Höhne, *The Order of the Death's
     Head,* 260–66.
     At dawn   Steven M. Gillon, *FDR Leads the Nation into War*
     (New York: Basic Books, 2011), 8–9.
     woke to a ringing phone   Ibid.
     "Well, Bill"   Ibid.
     Later that morning   Franklin Delano Roosevelt press
     conference, September 1, 1939, http://www.presidency.ucsb.edu
     /ws/?pid=15798.
180 *the possibility of direct Nazi attacks*   Richard L. McGaha, "The
     Politics of Espionage: Nazi Diplomats and Spies in Argentina,
     1933–1945" (Ph.D. diss., Ohio University, 2009), 392–93.
     destabilize foreign governments   Ibid.
     Roosevelt was convinced   Ibid.
     a 1940 speech to Congress   Franklin D. Roosevelt, "Message
     to Congress on Appropriations for National Defense," speech,
     May 16, 1940, http://www.presidency.ucsb.edu/ws/?pid=15954.
181 *within reach of Nazi bombing raids*   Frank Knox, "Our
     Heavy Responsibilities to the Nation," speech to the St. Louis
     Conference of the United States Conference of Mayors,

February 20, 1941, http://www.ibiblio.org/pha/policy/1941
/1941-02-20a.html.

181 *140,000 arriving* Stefan Rinke, "German Migration to Latin
America (1918–1933)," in Thomas Adam, ed., *Germany and the
Americas: O–Z* (Santa Barbara, CA: ABC-CLIO, 2005), 27–31.
A "bewildering abundance" Stefan Zweig, *Brazil: Land of the
Future*, trans. Andrew St. James (New York: Viking Press, 1941), 82.
"The glare of the sun" Ibid.
"some gigantic building site" Ibid., 214.

182 *open windows of brothels* Waldo Frank, *South of Us: The
Characters of the Countries and the People of Central and South
America* (New York: Garden City Publishing, 1940), 113–14.
two hundred in Argentina alone "Interrogation of Edmund Von
Thermann, German Ambassador to the Argentine from 1934 to
1942," RG 59, General Records of the Department of State, Entry
188, box 26, NARA.
"a fair sale for German Bibles" U.S. Bureau of Foreign and
Domestic Commerce, *Commerce Reports, Part 1* (Washington,
D.C.: Government Printing Office, 1915), 1011, https://books
.google.com/books?id=1eA9AQAAMAAJ.
spoke only German "German Immigration to Brazil."
"The German spirit" Stephen Bonsal, "Greater Germany in
South America," *The North American Review* 176 (January
1903): 58–67.
"this part of the world" "German Political Designs with
Reference to Brazil," *The Hispanic American Historical Review*
2, no. 4 (November 1919): 586–610, http://www.jstor.org/stable
/2505875.
"complete subservience" Thermann interrogation.
uniforms of green Victoria González-Rivera and Karen
Kampwirth, eds., *Radical Women in Latin America: Left and
Right* (University Park: Pennsylvania State University Press,
2001), 241–42.

183 *"mustachioed like Hitler"* David Sheinin and Lois Baer Barr,
eds., *The Jewish Diaspora in Latin America: New Studies on
History and Literaure* (New York: Garland Publishers, 1996), 210.
"Discipline, Hierarchy, Order" McGaha, "The Politics of
Espionage," 271.
Adolfo Hirohito Ibid., 272.
found much to admire Uki Goñi, *The Real Odessa: Smuggling
the Nazis to Perón's Argentina* (London: Granta, 2003), 37–38.
found it "embarrassing" Thermann interrogation.
"Knowing that war was going" McGaha, "The Politics of
Espionage," 98–99.

183 *"without much effort"*   Thermann interrogation.

184 *course in espionage tradecraft*   George E. Sterling, "The History of the Radio Intelligence Division Before and During World War II," unpublished manuscript, PDF file, http://www.w3df .com, 78–79.

actually looked like ink   Hedwig Elisabeth Weigelmayer Sommer interrogation by Boyd V. Sheets, RG 65, Classification 64 (IWG), box 211, 57, NARA.

shrunk documents   Ibid.

All This and Heaven Too   Rachel Field, *All This and Heaven Too* (New York: Macmillan, 1939). For the ESF and Allied intelligence aspects of the book, see Sterling, "History of the Radio Intelligence Division," 60, and Rose Mary Sheldon, "The Friedman Collection: An Analytical Guide," rev. October 2013, Marshall Foundation, PDF file, 345–46.

185 *fit in a suitcase*   Sterling, "The History of the Radio Intelligence Division," 78–79.

two kinds of suicide drugs   Sommer interrogation, 57.

its most capable Funkmeister   "Gustav Utzinger" is an alias—his birth name was Wolf Emil Franczok—but he was known in South America primarily as Gustav Utzinger, so I am using that name in the text. The bulk of the information about Utzinger's career comes from a postwar interrogation at the Wannsee Internment Camp in Germany. See Robert Murphy, "Reporting the Interrogation of Wolf Emil Franczok, Alias Gustav Utzinger," October 24, 1947, RG 65, Records of the Federal Bureau of Investigation, box 18, 64-27116, NARA. There are two sections of numbered pages in this document: an opening section of enclosures containing sworn statements by Utzinger, followed by a report of an interrogation.

186 *she freely shared anecdotes*   ESF speech to Mary Barteleme Club, Crystal Ballroom, Blackstone Hotel, Chicago, November 30, 1951, box 17, folder 10, ESF Collection.

"a vast dome of silence"   Ibid.

her husband's biographer   ESF interview with Clark.

"the world began to pop"   ESF interview with Pogue, 4.

were still with her   ESF interview with R. Louis Benson, January 9, 1976, obtained under the Freedom of Information Act from NSA; received October 2015; originally requested by G. Stuart Smith.

187 *chomping on the pipe*   "Robert Gordon and Elizebeth S. Friedman at a Desk," photograph, 1940, ESF Collection.

the cipher machines she kept   "History of USCG Unit #387."

reported to the chief   ESF interview with Benson.

187 *dread phone calls*   ESF autobiography, 78.
the unit began to analyze   Ibid., 77.

188 *first gunfight*   Bennett Lessmann, "The Story of the SS Arauca:
A Wartime Saga in Broward County," *Broward Legacy* 31, no. 1
(2011): 1–12.
"AM TRYING TO RUN"   "Arauca messages," box 6, folder 5,
ESF Collection.
"THE CRUISER HAS TRAINED"   Ibid.
"CRUISER NAMED ORION"   Ibid.
"THREE AMERICAN ARMY PLANES"   Ibid.
"Exciting, round-the-clock adventures"   ESF, "foreword to
uncompleted work," box 9, folder 11, ESF Collection. A six-page
manuscript with handwritten corrections by ESF.
same kinds of call signs   L. T. Jones, "History of OP-20-GU
(Coast Guard Unit of NCA)," October 16, 1943, RG 38, CNSG
Library, box 115, 5750/193, NARA.
bearing fixes   Ibid.

189 *"the goose that lays the golden eggs"*   WFF, "Communications
Intelligence and Security Presentation Given to Staff and
Students" (lecture, Breckinridge Hall, Marine Corps School,
April 26, 1960), 22.
"even more secret"   ESF interview with Benson.
adulterations of commercial codes   "History of USCG Unit
#387," 5–7.
"the pie circuit"   Ibid., 11.
giving away   Ibid., 8.

190 *"an entering wedge"*   Ibid., 62.
common German words   Ibid., 62–67.
anagram the letters   Ibid.

191 *eleven letters of the alphabet*   Ibid., 5–7.
Durchwalken   Ibid.

192 *"the following was produced"*   Ibid.
messages that used   For *The Story of San Michele* as a book
cipher, see Sterling, "History of the Radio Intelligence Division,"
80. For *Soñar la vida* and *O Servo De Deus,* see "History of
USCG Unit #387," 20.
One Nazi spy proposed   Hamburg to Rio, October 31, 1941, RG
457, SRIC, No. 3810.
a sophisticated process   Sterling, "History of the Radio
Intelligence Division," 60–61. See also "History of USCG Unit
#387," 67, and ESF's marginalia on newspaper clippings in Item
1006.1, WFF Collection.

193 *become the governess*   ESF's annotated, working copy of *All
This and Heaven Too,* Item 1006, WFF Collection, 15.

195 *753,506 . . .* Craig P. Bauer, *Secret History: The Story of Cryptology* (Boca Raton, FL: CRC Press, 2013), 255.

196 *first adopted them in 1926* "The Man, the Machine, the Choice," in David Kahn, *Seizing the Enigma: The Race to Break the German U-boat Codes, 1939–1943* (Boston: Houghton Mifflin, 1991).
Polish codebreakers were the first Bauer, *Secret History*, 256–83.

197 *dramatically more powerful* Ibid.
"search engines for the keys" Andrew Hodges, *Alan Turing: The Enigma* (London: Vintage, 2014), xviii.
one to five messages per day "History of USCG Unit #387," 216–30. All details from the coast guard's solution to this first Enigma machine are documented here.

200 *a linguist and scholar* Mavis Batey, "Knox, (Alfred) Dillwyn (1884–1943)," 2004, rev. ed. 2006, Oxford Dictionary of National Biography, http://dx.doi.org/10.1093/ref:odnb/37641.

201 *"This recovery of wiring"* "History of USCG Unit #387," 230.

202 *separate dining rooms* Kent Boese, "Lost Washington: Harvey's Restaurant," *Greater Greater Washington*, June 23, 2009, https://ggwash.org/view/2073/lost-washington-harveys-restaurant.
"Penalty of Leadership" "Biography of John Edgar Hoover," John Edgar Hoover Foundation, http://www.jedgarhooverfoun dation.org/hoover-bio.asp.
a Caesar salad Pamela Kessler, *Undercover Washington: Where Famous Spies Lived, Worked, and Loved* (Sterling, VA: Capital Books, 2005), 35–36.

203 *poured wine into the cryptologist's glass* ESF interview with Benson.
He got rid of them Michael Newton, *The FBI Encyclopedia* (Jefferson, NC: McFarland, 2003), s.v. "women agents," 374.
a secret dossier Curt Gentry, *J. Edgar Hoover: The Man and His Secrets* (New York: W. W. Norton, 2001), e-book, location 5837.
"dangerous than a man" Henry M. Holden, *FBI 100 Years: An Unofficial History* (Minneapolis, MN: Zenith Press, 2008), 37.
"don't tell 'em any secrets" Ibid.

204 *a public-relations disaster* Raymond J. Batvinis, *The Origins of FBI Counter-Intelligence* (Lawrence: University Press of Kansas, 2007), 10–26.
a lot of bad blood British Security Coordination, *The Secret History of British Intelligence in the Americas, 1940–1945* (New York: Fromm International, 1999), 468.
"that Jew in the Treasury" Gentry, *J. Edgar Hoover*, location 7845.
"J. Edgar Hoover is a man" British Security Coordination, *The Secret History*, 3.

204 *"It was once remarked"* Ibid., 468.

205 *"no attack is so unlikely"* Roosevelt, "Message to Congress on Appropriations."
"our teeming seaboard cities" Knox, "Our Heavy Responsibilities."
"the common defense" J. Edgar Hoover, "How the Nazi Spy Invasion Was Smashed," *The American Magazine* (September 1944): 20–21, 94–100.
a historic expansion FBI, "History of the SIS Division," vol. 1, NARA, three PDF files provided by Richard McGaha.
The first five SIS agents Ibid.
was actually Portuguese Transcript of John J. Walsh (former SIS agent) interview with Stanley A. Pimentel, May 19, 2003, National Law Enforcement Museum, 25–26, http://www.nleomf .org/museum/the-collection/oral-histories/john-j-walsh.html.

206 *The FBI had neither* British Security Coordination, *The Secret History,* 468.
She was to teach codebreaking ESF interview with Benson.
She read pacifist poetry All her life, Elizebeth wrote and typed copies of her favorite poems and kept them, including "Patterns" by Amy Lowell, about a young woman whose lover is killed in the First World War, and "Ultimatum for Man," a 1940 poem by Peggy Pond Church, a pacifist poet and schoolteacher in New Mexico whose land was taken by the government to build the Los Alamos facility for nuclear-weapons research. See box 11, folder 20, ESF Collection.

207 *"sleep is intermittent"* ESF to WFF (addressed to Munitions Building), June 1940, box 2, folder 5, ESF Collection.
what kind of airline Ibid., June 7, 1940.
fifty dollars Ibid.

208 *"this morass of debt"* WFF to ESF, June 4, 1940, box 3, folder 7, ESF Collection.
"not for some centuries" Ibid., June 10, 1940.

CHAPTER 2: MAGIC

209 *One day in September 1940* Frank Rowlett, *The Story of Magic: Memoirs of an American Cryptologic Pioneer* (Laguna Hills, CA: Aegean Park Press, 1998), 151–53.
called her Gene Ibid.
a professor at George Mason "Genevieve Grotjan Feinstein," NSA Cryptologic Hall of Honor, https://www.nsa.gov/about /cryptologic-heritage/historical-figures-publications/hall-of -honor/2010/gfeinstein.shtml.

209 *two patterns* Rowlett, *The Story of Magic*, 151–53. For a deft and clear technical explanation of Grotjan's insight and the larger process of breaking Purple, see the chapters on Purple in Craig P. Bauer, *Secret History: The Story of Cryptology* (Boca Raton, FL: CRC Press, 2013), 301–10.

her eyes were beaming Ibid.

210 *"Maybe I was just lucky"* Genevieve Grotjan Feinstein, NSA Oral History, May 12, 1991.

William agreed within seconds Ibid.

"Suddenly he looked tired" Ibid.

"The recovery of this machine" Ibid.

Someone went and got Cokes Ibid.

"sitting at his desk" Ibid.

211 *September 25, 1940* Jeffrey Kozak, "Marshall & Purple," Marshall Foundation, http://marshallfoundation.org/blog /marshall-purple/.

nothing at Purple's level WFF, "Contributions in the Fields of Communications Security and Communications Intelligence," undated, NSA.

a feat on par David Kahn, *The Codebreakers: The Comprehensive History of Secret Communication from Ancient Times to the Internet*, rev. ed. (New York: Scribner, 1997).

they demonstrated it Rowlett, *The Story of Magic*, 160–64.

"Last night your magicians" Ibid.

"By God" Ibid.

212 *"MAGIC Summaries"* Kahn, "One Day of MAGIC," in *The Codebreakers*, 1–67.

insisted on getting MAGIC raw David Stafford, "Churchill and Intelligence—Adventures in Shadowland, 1909–1953," *Finest Hour* 149 (Winter 2010–11), https://www.winstonchurchill.org /publications/finest-hour/finest-hour-149/churchill-and-intelli gence-adventures-in-shadowland-1909-1953.

"would be wiped out" George C. Marshall to Thomas E. Dewey, September 27, 1944, Papers of George Catlett Marshall, vol. 4: Aggressive and Determined Leadership, Marshall Foundation, http://marshallfoundation.org/library/digital -archive/to -thomas-e-dewey1/.

213 *He said hello* ESF interview with Pogue, 24; Rose Mary Sheldon, in discussion with the author, January 2015.

oily smoke rippling out Ulrich Steinhilper, *Spitfire on My Tail: A View from the Other Side* (Keston, UK: Independent Books, 2009), 306.

"The sky seemed full of them" Henry Steele Commager, *The*

Story of World War II, rev. Donald L. Miller (New York: Simon & Schuster, 2001), 38–41.

213 *telephones of Jewish families* "Historical Background: The Jews of Hungary During the Holocaust," Yad Vashem, http://www .yadvashem.org/yv/en/education/newsletter/31/jews_hungary .asp.

America had no standing Charles Lindbergh, "We Will Never Accept a Philosophy of Calamity," speech, Keep-America-Out -of-War rally, Chicago, August 4, 1940, http://www.ibiblio.org /pha/policy/1940/1940-08-04a.html.

"danger to this country" Charles Lindbergh, "Who Are the War Agitators?" speech, Des Moines, Iowa, September 11, 1941, http://www.charleslindbergh.com/americanfirst/speech.asp.

214 *British officers began to arrive* British Security Coordination, *The Secret History of British Intelligence in the Americas, 1940–1945* (New York: Fromm International, 1999), Introduction by Nigel West.

thirty-fifth and thirty-sixth floors Ibid.

Ian Fleming Jennet Conant, *The Irregulars: Roald Dahl and the British Spy Ring in Wartime Washington* (New York: Simon & Schuster, 2008), 84–86.

twenty-three-year-old Roald Dahl Ibid., xiv, 10–11.

He seduced actresses and heiresses Ibid., 99–126.

"I would do my best to appear calm" Roald Dahl, "Lucky Break," in *The Wonderful Story of Henry Sugar* (New York: Puffin, 2000), 201.

215 *"All I can say"* Conant, 30.

planted anti-Nazi information British Security Coordination, *The Secret History,* 66–87.

staged protests at rallies Ibid.

sending gorgeous female spies Ibid., 193–96.

in the Western Hemisphere Ibid., "Part VII: Counter-Espionage," 345–403.

one day in June 1941 Mark Riebling, *Wedge: From Pearl Harbor to 9/11: How the Secret War Between the FBI and CIA Has Endangered National Security* (New York: Touchstone, 2002), 3–15; John Pearson, *The Life of Ian Fleming* (London: Bloomsbury, 2013).

"a chunky enigmatic man" Ibid.; see also John Bryden, *Best-Kept Secret: Canadian Secret Intelligence in the Second World War* (Toronto: Lester, 1993), 66.

"no conception of offensive" Ibid., 67.

216 *In July 1941, Roosevelt* Thomas F. Troy, "Donovan's Original Marching Orders," *Studies in Intelligence* 17, no. 2 (1973): 39–67,

https://www.cia.gov/library/center-for-the-study-of-intelli
gence/kent-csi/vol17no2/html/v17i2a05p_0001.htm.

216 *"incomparably better"* British Security Coordination, *The
Secret History,* 471–72.
"The whole system" Ibid.
a husky, apple-cheeked colonel James Chadwick, "Frederick
John Marrian Stratton, 1881–1960," *Biographical Memoirs of
Fellows of the Royal Society* 7 (November 1961): 280–93.

217 *looked like Santa Claus* ESF interview with Benson.
The British operated Bob King, "The RSS from 1939 to 1946,"
November 22, 1944.
He wanted assistance "History of USCG Unit #387," Foreword.
others relied on "turning grilles" Ibid., 68–84.
made special devices and tools Jones, "History of OP-20-GU."

218 *"Fireside Chat" radio speech* Franklin Delano Roosevelt,
"Fireside Chat 16: On the Arsenal of Democracy," December 29,
1940, University of Virginia Miller Center, https://millercenter
.org/the-presidency/presidential-speeches/december-29-1940-fire
side-chat-16-arsenal-democracy.
36 minutes and 56 seconds Ibid.
Hitler responded Associated Press, " 'God With Us Up to Now,'
Hitler Says: Victory Sure in 1941 Army Men Are Told," *The
Brownsville Herald,* December 31, 1940.
climbed into the blacked-out streets Associated Press, "Charred
London Greets '41 With Cry 'To Hell With Hitler,' " *Washington
Post,* January 1, 1941.
January 4, 1941 Ronald Clark, *The Man Who Broke Purple:
Life of Colonel William F. Friedman, Who Deciphered the Japanese
Code in World War II* (Boston: Little, Brown, 1977), 158.
the Neuropsychiatric Section Mary W. Standlee, *Borden's
Dream: The Walter Reed Army Medical Center in Washington,
D.C.* (Washington, D.C.: Borden Institute, 2007), 214, 299–304,
334–36.

219 *keeping the mentally ill out of the army* William C. Porter,
"Psychiatry and the Selective Service," *War Medicine* 1 (May
1941): 364–71.
processing rather than healing Major M. R. Kaufman (physician
assigned to Neuropsychiatric Section at Walter Reed), "The
Problem of the Psychopath in the Army," in *Proceedings of
the Annual Congress of Correction of the American Prison
Association* 89 (1942): 128–38.
presumed to be less stressful William C. Porter, "The Military
Psychiatrist at Work," *The American Journal of Psychiatry* 98,
no. 3 (November 1941): 317–23.

220 *interfered with his ability to function*   Clark, *The Man Who Broke Purple*, 159.
"*the patient was isolated*"   Ibid.
"*mood swings*"   ESF to Ronald Clark, March 9, 1976, box 13, folder 30, ESF Collection.

221 *sailed across the Atlantic*   Robert L. Benson, "The Origin of U.S.-British Intelligence Cooperation (1940–1941)," *Cryptologic Spectrum* 7, no. 4 (1977): 5–8.
"*anxiety reaction*"   Ibid., 159.
discharged him on March 22   Ibid.
honorably discharged   Ibid.
temporary employee   " 'Court-martial' proceedings against William F. Friedman," box 13, file 14, ESF Collection.

CHAPTER 3: THE *HAUPTSTURMFÜHRER* AND THE *FUNKMEISTER*

223 *No code is ever completely solved*   ESF interview with Pogue, 79.

224 "*one of the most active*"   U.S. Department of Justice, Federal Bureau of Investigation, memorandum, *re: Siegfried Becker,* Francis E. Crosby to FBI Director, February 15, 1944, RG 65, box 18, 64-27116, NARA.
a "*deft Teutonic hand*"   Francis E. Crosby, "Memorandum for the Ambassador," February 4, 1944, RG 65, box 18, 64-27116, NARA.
He held the rank   Richard L. McGaha, "The Politics of Espionage: Nazi Diplomats and Spies in Argentina, 1933–1945" (Ph.D. diss., Ohio University, 2009), 22.
"*a sign of our loyalty*"   U.S. Department of Justice, Federal Bureau of Investigation, memorandum, *Subject: Johannes Siegfried Becker,* Francis E. Crosby to J. Edgar Hoover, November 22, 1944, including British translations of documents belonging to Becker, RG 65, box 18, 64-27116, NARA.
by direct or indirect steps   Crosby, "Memorandum for the Ambassador."
"*disliked his manner*"   "CSDIC Preliminary Interrogation Report on Heinrich VOLBERG," February 25, 1946, Records of the Security Service, KV2/89, TNA.
359,966   McGaha, "The Politics of Espionage," 211.
five foot ten   U.S. Department of Justice, Federal Bureau of Investigation, "Radio CEL, Albrecht Gustav Engels, Was., Et Al., Brazil—Espionage," RG 38, CNSG Library, box 77, NARA, 36.

225 *impregnating the wife*   McGaha, "The Politics of Espionage," 236.
grotesquely long fingernails   Charles F. Hemphill Jr., "Re:

Johannes Siegfried Becker," April 5, 1944, RG 65, box 18, 64-27116, NARA. "Subject has very unusual fingernails, in that they curve straight down over the tips of his fingers. HANS MUTH stated that this characteristic is so pronounced that they appear to be deformed."

225 *a Jewish retirement home* David Kahn, *Hitler's Spies: German Military Intelligence in World War II* (New York: Macmillan, 1978), 266.

five hundred people Ibid.

German for "informer" McGaha, "The Politics of Espionage," 185.

insufficiently ruthless Katrin Paehler, "Espionage, Ideology, and Personal Politics: The Making and Unmaking of a Nazi Foreign Intelligence Service" (Ph.D. diss., American University, 2002), 46.

according to an SS handbook Ibid., 215–16.

"no qualifications whatever" Theodor Paeffgen interrogation by Henry D. Hecksher, September 10, 1945, RG 65, box 183, NARA.

226 *Gestapo thug named Kurt Gross* Hedwig Elisabeth Weigelmayer Sommer interrogation by Boyd V. Sheets, RG 65, Classification 64 (IWG), box 211, 11, NARA; see also W. Wendell Blanke, "Interrogation Report of Karl Gustav Arnold," November 20, 1946, 13, NARA.

gladly told U.S. interrogators Ibid.

a Jewish man from Holland Ibid., 59–60.

"He was an intelligent person" Ibid., 22.

227 *trunkful of explosives* McGaha, "The Politics of Espionage," 231–32.

he abandoned the sabotage Sommer interrogation, 15.

carrying a revolver McGaha, "The Politics of Espionage," 185.

a broad-shouldered German U.S. Department of Justice, "Radio CEL," 59–61.

jittery mechanical engineer "The Starziczny Case," in Stanley E. Hilton, *Hitler's Secret War in South America 1939–1945* (Baton Rouge: Louisiana State University Press, 1999), e-book, location 1948; see also U.S. Department of Justice, "Radio CEL," 124.

lived with his Brazilian mistress Ibid.

"the only real professional" U.S. Department of Justice, "Radio CEL," 58.

228 *the SS Windhuk* Sommer interrogation, 17.

speaking to an American Utzinger interrrogation.

"acted according to his own lights" Ibid., 26.

229 *"in pursuit of his ends"* Utzinger interrogation, enclosure no. 3.

Vladimir Bezdek Vladimir Bezdek, Official Personnel Folder,

National Personnel Records Center, National Archives at
St. Louis, requested September 2016.

229 *read dictionaries in his free time*   Lekan Oguntoyinbo,
"Vladimir Bezdek: Retired WSU Professor, Linguist," *Detroit
Free Press,* May 19, 2000.
up and running in South America   U.S. Department of Justice,
"Radio CEL."

230 *an alphanumeric label*   "History of USCG Unit #387."
JOSE, or JUAN   Most of the Brazil decrypts from 1941 and
1942 are in RG 457, subseries SRIC, box 3, SRIC 1793–2591,
and box 5, SRIC 3723–3983, NARA; see also Hemphill Jr., "Re:
JOHANNES SIEGFRIED BECKER;" U.S. Department of
Justice, "Radio CEL."
He went by UTZ   Ibid.

231 *HUMBERTO was a piece of luck*   "History of USCG Unit
#387," 71.
"the greatest help to us"   "Final Report, British-Canadian-
American Radio Intelligence Discussions, Washington, D.C.,
April 6–17, 1942," RG 38, CNSG Library, Box 82, 5050/67,
NARA.
using book ciphers   "History of USCG Unit #387," 68.
switched to a grille-like cipher   Ibid.

232 *never sent useful information*   Ibid.
their own SIS filing system   FBI, "Subject: Frederick Duquesne,
Interesting Case Write-Up," March 12, 1985, eight PDF files.
The SIS used different serial numbers than the coast guard serial
numbers, but the texts of the messages are the same; for instance,
the handful of Mexico-to-Germany decrypts that Elizebeth
kept in box 6, folder 6 of her collection are identical to messages
included in the FBI/SIS Duquesne write-up.
"A considerable amount"   FBI, "History of the SIS Division,"
vol. 1, 288.
difficult for the coast guard   Jones, "History of OP-20-GU."
a massive spy investigation   Raymond J. Batvinis, "Ducase," in
The Origins of FBI Counter-Intelligence (Lawrence: University
Press of Kansas, 2007), 226–56.

233 *Sebold was secretly working*   Ibid.
The bureau reached out   "History of USCG Unit #387," 22–32;
Jones, "History of OP-20-GU."
to relay messages   Bativinis, "Ducase," in *The Origins of FBI
Counter-Intelligence.*
they were in an unknown code   Ibid.
Elizebeth broke it   "History of USCG Unit #387," 22–32.
Long Island to Hamburg   The coast guard called this radio link

Circuit 2-C and monitored it for the rest of the war. "History of USCG Unit #387," 22.

233 *"the greatest spy roundup"* Marc Wortman, "Fritz Duquesne: The Nazi Spy with 1,000 Faces," Daily Beast, February 26, 2017, http://www.thedailybeast.com/fritz-duquesne-the-nazi-spy -with-1000-faces.

234 *"gave birth to the popular cultural belief"* Batvinis, "Ducase," in *The Origins of FBI Counter-Intelligence,* 256.
"exposed the secret messages" Rose Mary Sheldon, "The Friedman Collection: An Analytical Guide," rev. October 2013, Marshall Foundation, PDF file, 345, text for Item 1006, WFF Collection.
too cavalier about publicity Jones, "History of OP-20-GU."

235 *disrupt their work* Diaries of Henry Morgenthau Jr., vol. 473, December 14–16, 1941, 37, Franklin D. Roosevelt Library and Museum website.
a Treasury staff meeting Diaries of Henry Morgenthau Jr., vol. 457, November 1–5, 1941, 237–64, Franklin D. Roosevelt Library and Museum website.
visible through a nearby window Peter Moreira, *The Jew Who Defeated Hitler: Henry Morgenthau Jr., FDR, and How We Won the War* (New York: Prometheus Books, 2014), 40.
at 10:45 A.M. Diaries of Morgenthau, November 1–5, 1941.
a birdlike man Moreira, *The Jew Who Defeated Hitler,* 85.
"She is very discontented" Diaries of Morgenthau, November 1–5, 1941.

236 *Harry Dexter White* James Nye, "Revealed: The Banker Who Shaped the Modern Financial World after WWII Was a Soviet Spy Who Wanted America to Become Communist," *Daily Mail* (London), March 5, 2013.
"I don't like to butt into this" Diaries of Morgenthau, November 1–5, 1941.
"But they knew" Ronald Clark, *The Man Who Broke Purple: Life of Colonel William F. Friedman, Who Deciphered the Japanese Code in World War II* (Boston: Little, Brown, 1977), 170.
Personnel started to stream in The most vivid recollection of the Munitions Building immediately after Pearl Harbor comes from John B. Hurt, the Japanese linguist on Friedman's team. Three pages, undated, 1944, NSA.

237 *1,177 crewmen* Wikipedia, s.v. "Attack on Pearl Harbor," last modified May 17, 2017, https://en.wikipedia.org/wiki/Attack _on_Pearl_Harbor.
wrote his will Hurt.
it would happen in Manila Ibid.

237 *a three-volume report*   WFF, "Certain Aspects of 'MAGIC' in the Cryptological Background of the Various Official Investigations into the Attack on Pearl Harbor," March 1957, NSA.
"cryptologic schizophrenia"   WFF, "Second Period, Communications Security" (lecture), NSA.

238 *a declassified NSA report*   ESF interview with R. Louis Benson, January 9, 1976, obtained under the Freedom of Information Act from NSA; received October 2015.
Elizebeth would never understand   Ibid.
"a date which will live in infamy"   "FDR's Day of Infamy Speech: Crafting a Call to Arms," *Prologue* 33, no. 4 (Winter 2001), https://www.archives.gov/publications/prologue/2001/winter/crafting-day-of-infamy-speech.html.

239 *Elizebeth's ration book*   ESF and WFF's Second World War ration books are in a black folder of letters given to the Marshall Foundation by John Ramsay Friedman.
appoint a new chief   ESF interview with Benson.
This upset her   Ibid.

240 *She got to know James Roosevelt*   Colin Burke, "What OSS Black Chamber? What Yardley? What 'Dr.' Friedman? Ah, Grombach? Or Donovan's Folly," http://userpages.umbc.edu/~burke/whatoss black.pdf.
a tall, irascible   Evan Thomas, "Spymaster General," *Vanity Fair* (March 2011), http://www.vanityfair.com/culture/2011/03/wild-bill-donovan201103.
James Roosevelt approached Elizebeth   Diaries of Morgenthau, December 14–16, 1941, 37.
Donovan reinforced the demand   William J. Donovan to Morgenthau, December 14, 1941, in ibid., 53.
Morgenthau grumbled   Ibid., 37.
three and a half weeks   ESF to Colonel Donovan, via Chief Liason Officer, Coordinator of Information, December 29, 1941, box 15, folder 14, ESF Collection.
She built it from scratch   Ibid.
a seethingly polite letter   Ibid.
defined by recklessness   Thomas, "Spymaster General."

241 *"My experience and observations"*   ESF to Colonel Donovan.
an honorary L.L.D.   ESF to Mrs. T. N. Alford, October 19, 1939, box 1, folder 9, ESF Collection.
Brazil had declared solidarity   Boris Fausto, *A Concise History of Brazil*, trans. Arthur Brakel (Cambridge, UK: Cambridge University Press, 1999), 228.
firing torpedoes at Brazilian ships   John Bryden, *Best-Kept*

Secret: Canadian Secret Intelligence in the Second World War (Toronto: Lester, 1993), 108–9.

241 *The positions of the ships* Ibid.

"Measures against members" Brazil to Germany, December 10, 1941, RG 457, SRIC, No. 2210.

242 *Operation Drumbeat* Bryden, *Best-Kept Secret*, 108–9.

the ruthless U-boats Ibid.

MARCH 14, 1942 South America to Germany, March 14, 1942, RG 457, SRIC, No. 2418.

an FCC listening station Rhode Island Radio, "Radio Intelligence Division," http://www.61thriftpower.com/riradio /rid.shtml.

8,398 American servicemen Eric Niderost, "Voyages to Victory: RMS Queen Mary's War Service," Warfare History Network, January 16, 2017, http://warfarehistorynetwork.com /daily/wwii/voyages-to-victory-rms-queen-marys-war-service/.

MARCH 7, 1942 South America to Germany, March 7, 1942, RG 457, SRIC, No. 2414.

MARCH 8 South America to Germany, March 8, 1942, RG 457, SRIC, No. 2413.

243 *MARCH 12* South America to Germany, March 12, 1942, RG 457, SRIC, No. 2418.

MARCH 13 Ibid.

MARCH 14 South America to Germany, March 14, 1942, RG 457, SRIC, No. 2419.

one million Reichsmarks Niderost, "Voyages to Victory."

able to take evasive maneuvers ESF wasn't the only Allied codebreaker who noticed that the *Queen Mary* was in peril; British and Canadian agencies solved similar messages. It was like multiple witnesses reporting the same crime to 911. See Bryden, *Best-Kept Secret*, 121.

"hiding place" Santiago to Hamburg, March 5, 1942, RG 457, SRIC, No. 3739.

31 degrees Celsius Brazil to Hamburg, March 7, 1942, RG 457, SRIC, No. 3799.

"Throughout country" Brazil to Germany, March 16, 1942, RG 457, SRIC, No. 3831.

the docked Swiss ship Brazil to Hamburg, March 17, 1942, RG 457, SRIC, No. 3821.

guessed that authorities Jones, "History of OP-20-GU."

244 *"You'll blow the house up!"* John Humphries, "The Man From Brazil," in *Spying for Hitler: The Welsh Double-Cross* (Cardiff: University of Wales Press, 2012), 199–211.

March 15, 1942 Leslie B. Rout Jr. and John F. Bratzel, "Climax of the Espionage War in Brazil: 1942–55," in *The Shadow War: German Espionage and United States Counterespionage in Latin America during World War II* (Frederick, MD: University Publications of America, 1986), 172–222.

245 *Engels assumed*   Ibid.
"MEYER CLASEN"   Brazil to Germany, March 18, 1942, RG 457, SRIC, No. 3964.
That day Engels was arrested   Rout and Bratzel, "Climax of the Espionage War in Brazil."
West believed   Ibid.
Robert Linx drove around Rio   George E. Sterling, "The History of the Radio Intelligence Division Before and During World War II," unpublished manuscript, PDF file, http://www.w3df.com, 85.

246 *verbatim copies*   Ibid.; see also Jones, "History of OP-20-GU."
"delivered the complete information"   J. Edgar Hoover, "How the Nazi Spy Invasion Was Smashed," *The American Magazine* (September 1944): 20–21, 94–100.
tougher on the prisoners   Rout and Bratzel, "Climax of the Espionage War in Brazil."
fill in a handful of missing words   Jones, "History of OP-20-GU."
went on a hunger strike   Rout and Bratzel, "Climax of the Espionage War in Brazil."
looked the other way   Ibid.

247 *one of the radio transmitters*   Sommer interrogation, 23.
three long letters out of prison   Rout and Bratzel, "Climax of the Espionage War in Brazil."
"Warning"   Germany to Chile, March 23, 1942, RG 457, SRIC, No. 3809.
a pack of cigarettes arrived   C. F. Hemphill, "Osmar Alberto Hellmuth," January 1, 1944, RG 65, box 18, 64-27116, NARA.

CHAPTER 4: CIRCUIT 3-N

249 *"Flight, fight, or neurosis"*   Ronald Clark, *The Man Who Broke Purple: Life of Colonel William F. Friedman, Who Deciphered the Japanese Code in World War II* (Boston: Little, Brown, 1977), 258–59.

250 *"heebeegeebees . . . hbgbs*   WFF, "Bletchley Park Diary," ed. Colin MacKinnon, http://www.colinmackinnon.com/files/The_Bletchley_Park_Diary_of_William_F._Friedman_E.pdf.
three consecutive days   ESF to Barbara Friedman, May 22, 1942, box 3, folder 22, ESF Collection.
friends knocked on their door   Ibid.
"Artiste de Boudoir"   WFF to ESF, telegram beginning "YOUR

RENOWN," May 1942, box 1, General Correspondence, ESF Collection.

250 *"Doctor of Successful Marriage"* WFF to ESF, telegram beginning "BOARD OF OVERSEERS," May 1942, box 1, General Correspondence, ESF Collection.

251 *leftist political causes* Barbara Friedman to WFF, undated, box 4, folder 8, ESF Collection.
"I hope you will let nothing interfere" WFF to Barbara Friedman, October 11, 1944, box 3, folder 21, ESF Collection.

252 *Arlington Hall* Jennifer Wilcox, "Sharing the Burden: Women in Cryptology During World War II," Center for Cryptologic History, NSA, 2008.
Many were women Ibid.
secret cryptology courses Patricia Ryan Leopold, in discussion with the author, January 2015. See also Craig Bauer and John Ulrich, "The Cryptologic Contributions of Dr. Donald Menzel," *Cryptologia* 30, no. 4 (2006): 306–39. DOI: 10.1080/01611190600920951.
guarded by U.S. Marines ESF, "foreword to uncompleted work."
operated the bombes Wilcox, "Sharing the Burden."

253 *"rolling down one's legs"* Martha Waller, in discussion with the author, via e-mail, March 2015.
the Star of David "Star of David; Badges and Armbands," National Holocaust Centre and Museum, UK, https://www.nationalholocaustcentre.net/star-of-david.
Exactly as she had feared Jones, "History of OP-20-GU."
"thereafter completely changed" Ibid.
"the matter got out of hand" "R.I.P. No. 98, Appendix II, American Measures Against Communications Intelligence Publicity," April 5, 1943, RG 457, Friedman Collection, Entry UD-15D19, box 22, NARA, 400–401.

254 *also taken aback* F. H. Hinsley and C. A. G. Simkins, *British Intelligence in the Second World War, Vol. 4, Security and Counter-Intelligence* (London: Her Majesty's Stationery Office, 1990), 149.
"Rather shattering" [Redacted] to J. M. A. Gwyer, MI5, April 26, 1943, KV2/2845, TNA.
jurisdictional squabbles "R.I.P. No. 98, Appendix II," RG 457.
without their approval Ibid., 384.
in the Western Hemisphere Ibid., 394–97.
"more informal" BSC, 472.
"sotto voce" Ibid., 473.

255 *for tighter control* Jones, "History of OP-20-GU."
a weeklong conference "Final Report, British-Canadian-American Radio Intelligence Discussions."

255 *On the day she explained* "Brief of Minutes, Committee B, Method of Obtaining W/T Intelligence From Intercepted W/T Traffic, Including D/F Bearings," British-Canadian-American Radio Intelligence Discussions, Washington, D.C., April 8, 1942, RG 38, CNSG Library, Box 82, 5050/67, NARA. See also "Recordings of Final Report, British-Canadian-American Radio Intelligence Discussions, Washington, D.C.," April 6–17, 1942, envelope no. 2, list of April 8, 1942, speakers, RG 38, CNSG Library, Box 82, 5050/68, NARA.

fewer than twenty cryptanalysts ESF interview with Benson.

weirder, harder stuff "History of USCG Unit #387," 15.

ZUM NEUE JAHR Ibid., 95–96.

256 *a partial list* All of these circuits are described in "History of USCG Unit #387."

diplomats and even military officers L. T. Jones, "Memorandum to Op-20-G, Subj: Clandestine Radio Intelligence," September 7, 1944, obtained under the Freedom of Information Act from NSA; received October 2015; originally requested by G. Stuart Smith.

257 *just another communications channel* Ibid.

"when wireless is perfectly applied" John B. Kennedy, "When Woman Is Boss," interview with Nikola Tesla, *Collier's,* January 30, 1926, http://www.tfcbooks.com/tesla/1926-01-30.htm.

"a miscellany" Jones, "Memorandum to Op-20-G."

about the Nazi grasp Ibid.

the shapes were parallelograms "History of USCG Unit #387," 37–38.

others like labyrinths Ibid., 199.

"explosives from cacao" Argentina to Berlin, October 22, 1943, Serial CG3-2213, RG 38, CNSG Library, box 79, 3824/3, NARA.

258 *baby girl, Jutta* Hamburg to Iceland, June 1, 1944, RG 457, SRIC, No. 3687.

"My dear JOHNY" Berlin to Argentina, November 11, 1943, Serial CG3-2348, RG 38, CNSG Library, box 79, 3824/3, NARA.

Lieutenant Jones did that ESF interview with Benson; Jones, "History of OP-20-GU."

Jones did that, too ESF interview with Benson.

sometimes quarreling Ibid.

"one of the workers" Ibid.

$4,200 a year ESF payroll slip, July 1945, listing her previous salaries and government classification, Personnel Folder.

Her initials, ESF Germany to ?, 1942, RG 457, SRIC, No. 3648.

Her handwriting appeared See, for instance, Berlin to Argentina, April 6, 1944, Serial CG4-4142, RG 38, CNSG Library, box 79, 3824/3, NARA, with a handwritten stapled

note by ESF that begins "Comment," or see one of the messages
from the *Jolle* supply ship to Argentina, June 27, 1944, Serial
CG4-5077-A, RG 38, CNSG Library, box 79, 3824/2, NARA.

259 *a characteristic burst*   Berlin to Argentina, May 30, 1944, Serial
CG4-4847, RG 38, CNSG Library, box 79, 3824/3, NARA.
Berlin writes that a man named Curt "had his leg in a cast as a
result of a bombardment of BERLIN when he was going down
the stairs carrying a young lady on his back." ESF wrote in red
pencil, "neat trick."
described the meeting   Government Code and Cypher School,
memorandum, *CLANDESTINE*, Major G. G. Stevens to
D.D.(S), December 24, 1942, HW14/62, TNA.
at unexpected times   Ibid.
a device he called a "snifter"   George E. Sterling, "The History
of the Radio Intelligence Division Before and During World
War II," unpublished manuscript, PDF file, http://www.w3df
.com, 19–20.

260 *October 10, 1942*   David P. Mowry, "Cryptologic Aspects of
German Intelligence Activities in South America during World
War II," Series IV, vol. 11 (2011), Center for Cryptologic History,
National Security Agency, 85–86.
called it Circuit 3-N   "History of USCG Unit #387," 231.
chipped in with clues   Ibid.; Mowry, "Cryptologic Aspects."
twenty-eight encrypted messages   Ibid.

261 *1511 Calle Donado*   Utzinger interrogation, 4.
legitimate paying clients   Ibid.
girlfriend worked at AMT VI   In FBI memos exchanged after
Utzinger's arrest in August 1944, Bureau officials write that
before Utzinger left Germany for South America, he asked
friends to look after a woman named Hilde Burckhardt, who
told a roommate that she and Utzinger both worked for AMT
VI. See Federal Bureau of Investigation, memorandum, Subject:
"Gustav Utzinger, with aliases," James P. Joice Jr. to John Edgar
Hoover, October 5, 1945, RG 65, Classification 64 (IWG), box 14.
Also, in one of the Circuit 3-N decrypts, "Blue Eye" talks about
"participating in the construction of several directional shortwave transmitters," strongly suggesting that she was with AMT
VI. See Berlin to Argentina, October 26, 1943, Serial CG3-2236,
RG 38, CNSG Library, box 79, 3824/3, NARA.
signed the messages "blue eye"   Berlin to Argentina, October 26,
1943, Serial CG3-2236.
called herself "the Ahnfrau"   Berlin to Argentina, February 8,
1944, Serial CG4-3535, RG 38, CNSG Library, box 79, 3824/3,
NARA. The coast guard codebreakers weren't entirely sure how

"the Ahnfrau" was related to "Luna"—they sometimes wrote "grandmother?" or "mother?" or "wife?" next to her name on the decrypts—but I am fairly certain, from the context of the decrypts, that she was Utzinger's grandmother.

261 *"celebrating your birthday"* Berlin to Argentina, October 28, 1943, Serial CG4-2837, RG 38, CNSG Library, box 79, 3824/3, NARA.

to brush his teeth Berlin to Argentina, April 29, 1944, Serial CG4-4447, RG 38, CNSG Library, box 79, 3824/3, NARA.

262 *built transmitters* Ibid., 3.

in the public square Ibid., 4.

ancient Krupp cannon Ibid., 3.

under extreme pressure Ibid.

Waldo Frank toured Argentina Waldo Frank, *South American Journey* (New York: Duell, Sloan and Pearce, 1943).

noticed "a spawn" Ibid., 76.

a pro-Nazi newspaper Ibid., 83–85, 128.

"a very uncertain grip" Ibid., 213.

before he could escape, five cops Ibid., 217.

263 *only listening posts* Utzinger interrogation, 4–5.

a small farm Arthur F. Carey, "Gustav Edward Utzinger, with Aliases, Espionage," August 15, 1945, RG 65, Classification 64 (IWG), box 14.

code name "Boss" McGaha, "The Politics of Espionage," 189.

on a Spanish ship Carey, "Gustav Edward Utzinger."

he had brought an Enigma Utzinger interrogation, 5.

very sensitive information Ibid., 4.

264 *a mythical radio organization* Utzinger interrogation, enclosure no. 3.

Red, Green, and Blue "Camp 020 Interim Report on the Case of General Friedrich Wolf," October 1945, RG 59, Entry 1088, box 26.

"seasoned collaborators" Argentina to Berlin, October 14, 1943, Serial CG3-2179, RG 38, CNSG Library, box 79, 3824, NARA.

"I am teaching my boys" Argentina to Berlin, January 20, 1943, Serial CG3-896, RG 38, CNSG Library, box 80, 3824/4, NARA.

another useful channel Sommer interrogation, 38.

two Enigmas Utzinger interrogation, Sommer interrogation. Dietrich Niebuhr gave Utzinger one Enigma and Becker gave him at least one other before the Red Enigma arrived via the wolf courier system.

a Liliput "History of USCG Unit #387," 212.

265 *a cheerful progress update* Argentina to Berlin, February 28, 1943, RG 38, Serial CG3-933, CNSG Library, box 80, 3824/4, NARA.

"Old boy, now we are off" Berlin to Argentina, February 28,

1943, RG 38, Serial CG3-860, CNSG Library, box 80, 3824/4, NARA.

265 *Nazi censors*   Earl R. Beck, *Under the Bombs: The German Home Front, 1942–1945* (Lexington: University Press of Kentucky, 1986), 35.

difficult to find toothbrushes   Ibid., 24.

patrons stealing glasses   Ibid.

the Battle of Stalingrad   Joseph Goebbels, "Nation, Rise Up, and Let the Storm Break Loose," February 18, 1943, German Propaganda Archive, Calvin College, http://research.calvin.edu /german-propaganda-archive/goeb36.htm.

a fourteen-year-old girlfriend   Goñi, *The Real Odessa*, xxiii.

a secret lodge   Ibid., 20–22.

266 *a small notebook*   "Summary of Traces, BECKER Siegfried," address book, July 11, 1945, KV2/89, TNA.

an alluring figure   Utzinger interrogation, 8.

an informal deal   Ibid., 2–5.

"Hitler's struggle"   Goñi, *The Real Odessa*, 22.

a stack of text   The coast guard documented its solution of the Green Enigma in "History of USCG Unit #387," 230–61. For an account of both the coast guard and British solutions for this machine, see Philip Marks, "Enigma Wiring Data: Interpreting Allied Conventions from World War II," *Cryptologia* 39, no. 1 (2015): 25–65. DOI: 10.1080/01611194.2014.915263.

267 *a unit called Intelligence Service Knox*   Hinsley and Simkins, *British Intelligence, Vol. 4*, 182.

at about the same time   Marks, "Enigma Wiring Data;" see also "History of USCG Unit #387," 262.

a G-model Enigma   Ibid.

a "less superior Enigma"   ESF interview with Clark.

"There was much celebration"   ESF interview with Benson.

268 *"enciphered with LILY"*   "History of USCG Unit #387," 212.

somewhat more complex   Marks, "Enigma Wiring Data."

frequencies of certain juxtapositions   "History of USCG Unit #387," 212–15.

never solve the Blue   Ibid., 262.

269 *"We have antenna"*   Argentina to Berlin, January 18, 1943, Serial CG3-921, RG 38, CNSG Library, box 80, 3824/4, NARA.

on the second floor   ESF, "foreword to uncompleted work."

"Our work was compartmented"   George Bishop to Vanessa Friedman, September 22, 1981, box 12, folder 14, ESF Collection.

a few of the SPARS   ESF Personal History Statement.

her home for tea   WFF to Barbara Friedman, January 16, 1944, box 3, folder 24, ESF Collection.

269 *"I think I was mesmerized"*   Waller, in discussion with the author, via e-mail, March 2015.

270 *a personal diary of the trip*   WFF, "Bletchley Park Diary."
at least fourteen letters   ESF to WFF, May 31, 1943, box 2, folder 7, ESF Collection.
"a momentary return for you"   Ibid., May 16, 1943.

271 *a corsage of violets*   Ibid., April 27, 1943.
cigarettes in the evenings   Ibid., May 9, 1943.
110 degrees   ESF, "foreword to uncompleted work."
rolled out extensive reforms   "History of USCG Unit #387," 156.
Nazi cryptographer, Fritz Menzer   David P. Mowry, "Regierungs-Oberinspektor Fritz Menzer: Cryptographic Inventor Extraordinaire," *Cryptologic Quarterly* 2, nos. 3 and 4 (1983–84): 21–36.
"Procedure 62"   "History of USCG Unit #387," 195–202.
"Procedure 40"   Ibid., 203–6.
5-by-5 square of letters   Ibid.

272 *volume of traffic abruptly rose*   Utzinger interrogation.
fifteen different messages   Sommer interrogation, 27.
"Roseman, Morgenthau and Frankfurter?" Berlin to Argentina, September 18, 20, 21, Serial CG3-1949, RG 38, CNSG Library, box 79, 3824/3, NARA.
"The Fisher Co."   Berlin to Argentina, November 21, 1943, Serial CG3-2477, RG 38, CNSG Library, box 79, 3824/3, NARA.

273 *deposing the old regime*   McGaha, "The Politics of Espionage," 269.
the Nazis called him "Godes"   Argentina to Berlin, July 14, 1943, Serial CG3-1586, RG 38, CNSG Library, box 80, 3824/4, NARA.
code name "Moreno"   Argentina to Berlin, May 12, 1943, Serial CG3-1788, RG 38, CNSG Library, box 80, 3824/4, NARA.
"the interests of the Axis powers"   Argentina to Berlin, July 24, 1943, Serial CG3-1582, RG 38, CNSG Library, box 80, 3824/4, NARA.
"ready in every respect"   Argentina to Berlin, May 12, 1943, Serial CG3-1788, RG 38, CNSG Library, box 80, 3824/4, NARA.
"USA is considered greatest enemy"   Argentina to Berlin, August 15, 1943, Serial CG3-1658, RG 38, CNSG Library, box 80, 3824/4, NARA.

274 *"V-men residing here"*   Argentina to Berlin, July 24, 1943.
"completely in our camp"   Argentina to Berlin, February 28, 1943, Serial CG3-858, RG 38, CNSG Library, box 80, 3824/4, NARA.
Bolivian minister of mines   Utzinger interrogation, enclosure no. 5.
Elias Belmonte   Utzinger interrogation, 8.

274 "Final objective" Argentina to Berlin, August 28, 1943, Serial CG3-1893, RG 38, CNSG Library, box 80, 3824/4, NARA.
a secret weapons deal McGaha, "The Politics of Espionage," 296–338.

275 *"An agent will depart"* Argentina to Berlin, July 14, 1943, Serial CG3-1608, RG 38, CNSG Library, box 80, 3824/4, NARA.
anything quite this exciting "Interim Report on the Case of Osmar Alberto Hellmuth," RG 65, 64-27116, NARA.
red mustache Ibid.
was "easy prey" Ibid.
an upscale neighborhood Goñi, *The Real Odessa*, xxiii.
spoke often, in secret Utzinger interrogation, 8. In the decrypts, Becker and others refer to Perón by name as well as his group, which they called "The Colonels Lodge."

276 *he would board a ship* "Interim Report on the Case."
"with good prospects" Ibid.

277 *"lack of imagination"* Argentina to Berlin, July 15, 1943, Serial CG3-1445, RG 38, CNSG Library, box 80, 3824/4, NARA.
"such an easy time!" Argentina to Berlin, May 12, 1943, Serial CG3-1702, RG 38, CNSG Library, box 80, 3824/4, NARA.
it was almost time Utzinger interrogation, enclosure no. 4.
The Chilean's descriptions Argentina to Berlin, December 11, 1943, Serial CG3-2746, RG 38, CNSG Library, box 81, NARA.

278 *sixty kilograms of gifts* Argentina to Berlin, October 8, 1943, Serial CG3-2103, RG 38, CNSG Library, box 80, 3824/4, NARA.
"HELLMUTH enjoys" Argentina to Berlin, October 7, 1943, Serial CG3-2125, RG 38, CNSG Library, box 80, 3824/4, NARA.
Osmar Hellmuth sailed "Interim Report on the Case."
"Luna is probably Gustav Utzinger" Argentina to Berlin, June 13, 1944, Serial CG4-4991, RG 38, CNSG Library, box 79, 3826/2, NARA.
"bloc in South America" Argentina to Berlin, January 6, 1944, Serial CG4-2907, RG 38, CNSG Library, box 81, 3824/4, NARA.

279 *in the middle of the night* "Interim Report on the Case."

280 *called Camp 020* Oliver Hoare, ed., *Camp 020: MI5 and the Nazi Spies* (Richmond, UK: Public Record Office, 2000).
Colonel Robin Stephens Gilbert King, "The Monocled World War II Interrogator," Smithsonian.com, November 23, 2011, http://www.smithsonianmag.com/history/the-monocled-world-war-ii-interrogator-652794/.
prisoners told credible stories Ian Cobain, "How Britain tortured Nazi PoWs," October 26, 2012, *Daily Mail* (UK), http://

www.dailymail.co.uk/news/article-2223831/How-Britain-tor
tured-Nazi-PoWs-The-horrifying-interrogation-methods-belie
-proud-boast-fought-clean-war.html.

280 *"I am speaking with authority"* "Interrogation of Hellmuth at
Camp 020 by Lieut. Colonel Stephens," November 17, 1943, RG
65, box 19, 64-27116, NARA.

281 *"possessed, almost arrogant"* Hoare, ed., *Camp 020*, 267.
"must necessarily deteriorate" "Interim Report on the Case."
"Sargo," he told his captors Ibid.

CHAPTER 5: THE DOLL LADY

283 *wearing out from overuse* "History of USCG Unit #387," 215,
262.
"Enigma arrived via RED" Ibid., 262.
"birthday surprise for LUNA" Ibid.
a right-wing Bolivian general Richard L. McGaha, "The
Politics of Espionage: Nazi Diplomats and Spies in Argentina,
1933–1945" (Ph.D. diss., Ohio University, 2009), 284–92.

284 *had set the coup into motion* Ibid.; Becker's Bolivian friend,
Elias Belmonte, was the link between the Nazi/Argentine group
and the Bolivian coup plotters. See also the Berlin to Argentina
and Argentina to Berlin decrypts sent between January and April
1944 in RG 38, CNSG Library, Box 79.
the "first fruit" Argentina to Berlin, January 17, 1944, Serial
CG4-3174, RG 38, CNSG Library, box 81, 3824/4, NARA.
"We extend hearty wishes" Argentina to Berlin, December 28,
1943, Serial CG4-2758, RG 38, CNSG Library, box 81, 3824/4,
NARA.
had discussed its delivery "History of USCG Unit #387," 262.
decided to double the crypto Ibid., 263–66.

285 *to write a punch-card program* Ibid., 270.
able to hammer Argentina J. Lloyd Mecham, *The United States
and Inter-American Security, 1889–1960* (Austin: University of
Texas Press, 1965), 214–15.
sent a secret cable Washington to ISK, February 19, 1944, cable
no. CXG204, HW 19, Records of the Government Code and
Cypher School: ISOS Section and ISK Section, subseries 361,
TNA.

286 *transmitted the wiring details* Washington to ISK, February 24,
1944, cable no. CXG228, HW 19/361, TNA.
they had just solved Red themselves ISK to Washington,
February 20, 1944, telegram no. CXG636, HW 19/361, TNA.
"We urgently need reports" Argentina to Berlin, January 26,

1944, Serial CG4-3780, RG 38, CNSG Library, box 81, 3824/4, NARA.

286 *a new commander*   Sommer interrogation, 5.

"a leak in your courier organization"   Berlin to Argentina, February 21, 1944, Serial CG4-3831, RG 38, CNSG Library, box 81, 3824/3, NARA.

"go on in full revolutions"   Berlin to Argentina, January 26, 1944, Serial CG4-3780, RG 38, CNSG Library, box 81, 3824/3, NARA.

287 *"Dear DARK EYE"*   Berlin to Argentina, March 2, 1944, Serial CG4-3736, RG 38, CNSG Library, box 81, 3824/4, NARA.

"Here, life goes on"   Berlin to Argentina, April 5, 1944, Serial CG4-4132, RG 38, CNSG Library, box 81, 3824/3, NARA.

vulnerability to chemical warfare   Berlin to Argentina, February 4, 1944, Serial CG4-3785, RG 38, CNSG Library, box 81, 3824/3, NARA.

He wrote a story   Sunday Express Correspondent, "Britain Smashes South America Spy Ring," *Sunday Express* (London), January 30, 1944.

288 *"Security Calypso"*   C. H. Carson to Mr. Ladd, memorandum, *Subject: Osmar Alberto Hellmuth,* February 27, 1945, with attached "Security Calypso" lyrics by Young Ziegfield, RG 65, box 20, 64-27116, NARA.

289 *a peeved memo to Hoover*   U.S. Department of Justice, Federal Bureau of Investigation, memorandum, *re: Osmar Alberto Helmuth* [sic], *Memorandum for the Ambassador,* D. M. Ladd to FBI Director, December 16, 1943, RG 65, box 18, 64-27116, NARA.

"The memorandum was written"   Crosby to FBI Director, February 15, 1944; Crosby, "Memorandum to the Ambassador."

290 *to search their files*   FBI, memorandum, *Memorandum No. 205, Series 1944, Memorandum for All Legal Attaches,* John Edgar Hoover, September 30, 1944, RG 65, box 19, 64-27116, NARA.

291 *direction-finding automobiles*   George E. Sterling, "The History of the Radio Intelligence Division Before and During World War II," unpublished manuscript, PDF file, http://www.w3df .com, 91–92.

Becker assured him   Utzinger interrogation, enclosure no. 4.

beneath a chicken coop   Carey, "Gustav Edward Utzinger;" Rout and Bratzel, *The Shadow War.*

eardrum destroyed   Argentina to Berlin, March 25, 1944, Serial CG4-3971, RG 38, CNSG Library, box 81, 3824/4, NARA.

"He fell on February 19"   Argentina to Berlin, March 22, 1944, Serial CG4-3945, RG 38, CNSG Library, box 81, 3824/4, NARA.

292 *die for National Socialism*   Argentina to Berlin, April 9, 1944, Serial CG4-4174, RG 38, CNSG Library, box 81, 3824/4, NARA.

292 *captured and hanged* Sommer interrogation, 38.
  sent him an ominous message Berlin to Argentina, February 8,
  1944, Serial CG4-3535.
  an intriguing piece of news Argentina to Berlin, March 22, 1944,
  Serial CG4-3890, RG 38, CNSG Library, box 81, 3824/4, NARA.
  Velvalee Dickinson whirled around John Jenkisson, "The FBI
  vs. New York Spies," *New York World Telegram*, June 22, 1945.
293 *five suspicious letters* ESF to Edward C. Wallace (U.S. Attorney,
  Southern District of New York), April 1, 1944, box 7, folder 1,
  ESF Collection.
  she had fallen into debt Jenkisson, "The FBI vs. New York Spies."
  he called the supervisor R. A. Newby to D. M. Ladd, March 14,
  1944.
294 *"According to Mr. Wallace"* Ibid.
  "performed in this connection" Ibid.
  "ADVISE AS TO SUBMISSION" U.S. Department of Justice, Federal
  Bureau of Investigation, teletype, *Kin. Velvalee Dickinson*, New
  York to Director, March 18, 1944. Obtained under the Freedom
  of Information Act from FBI; received December 2015.
  "Concerning the project" U.S. Department of Justice, Federal
  Bureau of Investigation, memorandum, *Subject: Velvalee
  Dickinson*, J. Edgar Hoover to SAC, New York, March 23, 1944.
  Obtained under the Freedom of Information Act from FBI;
  received December 2015.
  "My dear Mr. Wallace" ESF to Wallace.
295 *an advantage, not a disadvantage* Hoover to SAC, New York,
  March 23, 1944.
  "the woman spy of this war" George Kennedy, "The War's
  No. 1 Woman Spy," *West Sunday Star*, August 20, 1944, box 7,
  folder 2, ESF Collection.
296 *"Who are all these people?"* "Doll Woman Enters Guilty Plea in
  Censor Case; Faces Ten Years," *New York Times*, July 29, 1944.
  "except that it's larger" Kennedy, "The War's No. 1 Woman Spy."
  "the frustration of childlessness" Ibid.
  "one of the cleverest woman operators" J. Edgar Hoover,
  "Hitler's Spying Sirens," *The American Magazine* (December
  1944): 40–41, 92–94.
297 *"He is hidden"* Argentina to Berlin, April 6, 1944, Serial
  CG4-4163, RG 38, CNSG Library, box 81, 3824/4, NARA.
  "The enemy succeeded" Argentina to Berlin, August 11, 1944,
  Serial CG4-5629, RG 38, CNSG Library, box 81, 3284/4, NARA.
  arrested by the Argentine federal police Utzinger interrogation,
  enclosure no. 4.
  Juan Perón himself appeared Ibid., 4.

297 *"the final chapter"*   Sommer interrogation, 32.

298 *"Technical advantages"*   Rout and Bratzel, *The Shadow War,* 454.

299 *a seven-page story*   J. Edgar Hoover, "How the Nazi Spy Invasion Was Smashed," *The American Magazine* (September 1944): 20–21, 94–100.
   a fifteen-minute film   "Battle of the United States," *Army-Navy Screen Magazine* 42, Steven Spielberg Film and Video Archive, United States Holocaust Memorial Museum, https://collections .ushmm.org/search/catalog/irn1003973; see also https://www .youtube.com/watch?v=pdMTRjRvqGk for an uncut version.
   director Frank Capra   Mark Harris, *Five Came Back: A Story of Hollywood and the Second World War* (New York: Penguin Books, 2014), 233.

300 *the Friedman family Christmas card*   ESF and WFF, "B U L L E T I N ** 1944 ** F R I E D M A N," box 4, folder 6, ESF Collection.
   "Bill, Will, Billy"   Ibid.

301 *thirty thousand tons of bombs*   "1945, Summary of Air Operations, January," in Royal Institute of International Affairs, *Chronology and Index of the Second World War, 1938–1945* (1947; repr., London: Meckler, 1975), 317.

302 *223 planes*   Randall Hansen, *Fire and Fury: The Allied Bombing of Germany, 1942–1945* (New York: NAL Caliber, 2008), 260.
   found melted together   Ibid., 263.
   arrested in Buenos Aires   U.S. Department of Justice, Federal Bureau of Investigation, personal and confidential memorandum by special messenger, *Subject: Johannes Siegfried Becker, Buenos Aires,* John Edgar Hoover to Frederick B. Lyon (chief, Division of Foreign Activity Correlation, Department of State), April 21, 1945, RG 65, box 20, 64-27116, NARA.
   confiscated an address book   "Summary of Traces, BECKER Siegfried."
   He gave statements   Utzinger interrogation, 14.
   "delicacies and champagne"   Ibid., 21.

303 *hardly enjoyed Becker's privileges*   Ibid., 12–14.
   He no longer feared   Ibid., enclosure no. 4.
   war criminals into Argentina   Uki Goñi, *The Real Odessa: Smuggling the Nazis to Perón's Argentina* (London: Granta, 2003), xxiii, 107.
   an angry mob invaded   United Press, "Rebels Slay President, Seize Power in Bolivia," *Washington Post,* July 22, 1946.
   William got bronchitis   ESF to Barbara Friedman, February 9, 1945, box 3, folder 26, ESF Collection.
   extra-long time getting dressed   ESF to Barbara Friedman, February 22, 1945, box 3, folder 25, ESF Collection.

304 *"evil influences"*   ESF to Barbara Friedman, April 12, 1945, box 3, folder 26, ESF Collection.
a family friend died   WFF to Mrs. A. J. McGrail, November 29, 1945, attached to a "Harvard Honor Roll" sheet filled out by WFF and listing the accomplishments of Colonel A. John McGrail, NSA.
held her arm   ESF to Barbara Friedman, May 3, 1945, box 3, folder 27, ESF Collection.
"where I am, when I am"   Ibid.
"Remember, darling"   Ibid., May 11, 1945.
305 *"We have difficulty believing"*   Ibid., May 9, 1945.
"It's absolutely terrifying"   Ibid.
slept with the windows open   Ibid., June 18, 1945.
sent her a poem   Ibid., June 4, 1945.
a ninety-day assignment   Ibid., July 4, 1945.
"the last great secret"   Randy Rezabek, "TICOM: The Last Great Secret of World War II," *Intelligence and National Security* 27, no. 4 (2012): 513–30, http://dx.doi.org/10.1080/02684527.2012.688305.
306 *the military air terminal*   ESF to Barbara and John Ramsay Friedman, July 14, 1945, typed letter to her children, box 3, folder 27, ESF Collection.
She thought he looked handsome   Ibid.
waited until the C-54 took off   Ibid.

CHAPTER 6: HITLER'S LAIR

307 *rumbled up the twisting incline*   WFF diary of touring postwar Europe, dictated July 26, 1945, signed September 2, 1945, thirteen-page typescript in large blue binder, Marshall Foundation, 3–4.
a castle on a mountaintop   Ibid.
acoustics research   Army Security Agency, "European Axis Signal Intelligence in World War II as Revealed by TICOM Investigations and by Other Prisoner of War Interrogations and Captured Material, Principally German," May 1, 1946, NSA, 37–44.
The Allies wanted to know   Ibid.
308 *character in a murder mystery*   WFF diary of postwar Europe, 4.
Vierling's prototypes   "European Axis Signal Intelligence."
a brief supper of hot dogs   Ibid.
growing spookier   Ibid.
six TICOM teams   Randy Rezabek, "The Teams," TICOM Archive, http://www.ticomarchive.com/the-teams.
"a heavy feeling of sadness"   WFF diary of postwar Europe, 1.
309 *ate a C-ration*   Ibid., 4.
like LEPIDOPTERA   Ibid., 2.

309 *"The destruction to be seen"* WFF diary of postwar Europe, 8.
his own cryptologic publications TICOM discovered a copy
of his classic paper "The Index of Coincidence" that had been
translated into German from French. See Rose Mary Sheldon,
"The Friedman Collection: An Analytical Guide," rev. October
2013, Marshall Foundation, PDF file, 90.
for his own library Item 167.3, WFF Collection.
a grim duty WFF to ESF, October 6, 1917.

310 *"Zionism is only one"* WFF to Barbara Friedman, March 15,
1945, box 4, folder 8, ESF Collection.
did not sleep well WFF diary of postwar Europe, 5.
rode in the Army staff jeep Ibid., 5–7.
"He gave us a little speech" Ibid., 5.

311 *loved to entertain friends* Heike Görtemaker, *Eva Braun: Life
With Hitler* (New York: Alfred A. Knopf, 2011), 216.
"I think it is too bad" WFF diary of postwar Europe, 7.
"I shall have it made" Ibid., 6. The paperweight isn't part of
William's collection at the Marshall Foundation; no one seems to
know what happened to it.
this bowl of smoking ice ESF to WFF, July 26, 1945, box 2,
folder 8, ESF Collection.

312 *"At 1535 a visit with Dr. Turing"* WFF spiral-bound diary of
his 1945 England trip, box 13, folder 13, ESF Collection, 21.
stripped Turing's security clearance Andrew Hodges, *Alan
Turing: The Enigma* (London: Vintage, 2014), 574–664.
an apparent suicide Ibid., 614–15.
country village of Beaconsfield WFF diary of 1945 England trip, 16.
three high-value German POWs WFF to commanding general,
Army Security Agency, "Report on Temporary Duty, ETO,"
October 1, 1945, NSA.
unbreakable all the way Ibid.

313 *a burlesque show* WFF diary of 1945 England trip, 8.
asked him to tell stories Ibid., 36.
asleep and dreaming Ibid., 27–28.
"nearly as big as Dallas" Peter J. Kuznick, "Defending the
Indefensible: A Meditation on the Life of Enola Gay Pilot Paul
Tibbets Jr., *The Asia-Pacific Journal* 6, no. 1 (2008), http://apjjf
.org/-Peter-J.-Kuznick/2642/article.html.

314 *a sex dream about Enid* WFF diary of 1945 England trip, 28.
martinis in befuddled silence Ibid., 32–33.
"renewed call to surrender" Ibid.
listen to radio bulletins ESF to WFF, August 12, 1945, box 2,
folder 8, ESF Collection.

315 *"end the war P.D.Q."* Ibid., August 7, 1945.

315 *watch some tennis*   WFF diary of 1945 England trip, 37–38.
"A day we will remember!"   WFF to ESF, August 10, 1945, box 3, folder 9, ESF Collection.
"war will be a fact"   Ibid., August 14, 1945.
praising his son's vocabulary   WFF to John Ramsay Friedman, August 13, 1945, box 4, folder 8, ESF Collection.

316 *Elizebeth stayed in*   ESF to WFF, August 15, 1945, box 2, folder 8, ESF Collection.
"Bobbie, darling"   Ibid.
"The O.S.S is starting"   Ibid., September 4, 1945.
Elizebeth heard a radio interview   Ibid., August 16, 1945.

317 *a cold Sunday*   Ibid., undated (late August 1945).
came over with his wife   Ibid.
"I find it hard to tell you"   WFF to ESF, August 19, 1945, box 3, folder 9, ESF Collection.
She realized how tricky   ESF to WFF, August 26, 1945, box 2, folder 8, ESF Collection.
"Dearest"   Ibid.
"very soon"   WFF to ESF, August 29, 1945, box 3, folder 9, ESF Collection.
"I LOVE YOU!"   Ibid.
General Douglas MacArthur   Douglas MacArthur, "General MacArthur's Radio Address to the American People," September 2, 1945, https://ussmissouri.org/learn-the-history /surrender/general-macarthurs-radio-address.

318 *Elizebeth heard Truman say*   ESF to WFF, September 4, 1945, box 2, folder 8, ESF Collection.
"It is our responsibility"   Harry S. Truman, "Radio Address to the American People After the Signing of the Terms of Unconditional Surrender by Japan," September 1, 1945, http:// www.presidency.ucsb.edu/ws/?pid=12366.
no point in having fought a war   ESF to Barbara Friedman, April 12, 1945, box 3, folder 26. She wrote that now Americans needed to fight "for a truly international post war world."
She wondered if William heard   ESF to WFF, September 4, 1945, box 2, folder 8, ESF Collection.
"You are the dearest and best"   Ibid.
Around September 12   WFF, "Report on Temporary Duty."
Elizebeth opened the door   As best I can tell, neither Friedman ever wrote about this exact moment of reunion. I admit I'm inferring it from what WFF writes in his letters about the schedule of his trip home and the timing of his arrival.

319 *sorted through the voluminous files*   ESF, "foreword to uncompleted work."

319 *a detailed technical account* "History of USCG Unit #387."
historians of codebreaking Ibid., "Foreword."
Five copies were printed Ibid. Two copies went to the navy's
OP-20-G, one to the Coast Guard brass, one to the Army, and
one to British intelligence.
destroy the rest ESF, "foreword to uncompleted work."
"government tombs" Ibid.
prepared to leave Ibid.
The navy forced her Ibid.
At the end of her final workday ESF, "foreword to uncompleted
work."

320 *"to that particular form"* Ibid.
"thrilling records" Ibid.
On August 14, 1946 H. L. Morgan (Acting Chief, Civilian
Personnel Division, USCG) to ESF, August 14, 1946, box 6,
folder 8, ESF Collection. This folder also contains the envelope in
which the Reduction in Force letter arrived, and on the front and
back of the envelope, ESF wrote a note explaining that it was her
idea to eliminate her own job.

321 *show by the Amazing Dunninger* ESF to Barbara Friedman,
December 3, 1944, box 3, folder 25, ESF Collection.
both debunker and illusionist Wikipedia, s.v. "Joseph Dunninger,"
last modified March 9, 2014, http://www.geniimagazine.com
/magicpedia/Joseph_Dunninger.
"came away with theories" ESF to Barbara, December 3, 1944.

322 *alive and kicking* WFF and ESF, *The Shakespearean Ciphers
Examined* (London: Cambridge University Press, 1958), 9.
asked the Friedmans Ibid., 161–63.
the following message Ibid.

323 *"IN HER DAMP PUBES"* Ibid., 258. They were renowned for
their scientific insights and their serious feats of codebreaking,
but this is one of those passages that shows how the Friedmans
were also very funny, in a delicate and savage and wonderfully
idiosyncratic way.
"great natural gifts" Ibid., 205.
"a sincere and honourable woman" Ibid., 264.
"found in her texts" Ibid.
"was therefore at the mercy" Ibid., 265.
"whose work on the question" Ibid., ix.
they thanked Fabyan, too Ibid.
"give the devil his due" ESF to Mrs. Percival White, March 28,
1958, box 1, folder 23, ESF Collection.
"Vile creature" ESF interview with Valaki, transcribed
February 16, 2012, 5.

323 *an accident*   ESF, "Pure Accident."

324 *She tries to imagine herself*   ESF must have done this when she wrote the book, because there's a passage by her describing "the gradual crystallization" of her opinions about the Bacon Cipher project in 1916 and 1917. The rest of the book is written in the first-person plural "we" but this passage uses the singular "I." WFF and ESF, *Shakespearean Ciphers*, 211.

EPILOGUE

327 *otherwise ordinary Tuesday*   Transcript of Donald F. Coffey oral history interview with NSA, November 4, 1982.
Scattered clouds   Weather Underground, "Weather History for KDCA, December 30, 1958," https://www.wunderground.com /history/airport/KDCA/1958/12/30/DailyHistory.html.
at least three men   Possibly four men. S. Wesley Reynolds, NSA director of security, writes that he visited the Friedman home with an NSA man named Cook and a third man from the attorney general's office. Coffey would make four. See S. Wesley Reynolds memo, January 2, 1959, RG457, Entry UD-15D19, "Reclassification of Friedman Articles," box 57.
a rented truck   Coffey oral history.

328 *a Defense Department order*   Garrison B. Coverdale to William G. Bryan (undated), RG457, Entry UD-15D19, box 57; see also Rose Mary Sheldon, "The Friedman Collection: An Analytical Guide," rev. October 2013, Marshall Foundation, PDF file, 5.
forty-eight items   Ronald Clark, *The Man Who Broke Purple: Life of Colonel William F. Friedman, Who Deciphered the Japanese Code in World War II* (Boston: Little, Brown, 1977), 252; see also "Inventory of the material taken from Friedman's house," RG457, Entry UD-15D19, box 57.
"went berserk"   Coffey oral history.
denied this   Ibid.
wrote in a memo   S. Wesley Reynolds memo.
The ciphers were obsolete   Sheldon, "Analytical Guide," 7.
"The NSA took away from me"   Clark, *The Man Who Broke Purple*, 252.

329 *silent rage*   Coffey oral history.
"The mad march of red fascism"   J. Edgar Hoover, "Speech Before the House Committee on Un-American Activities," speech, March 26, 1947, http://voicesofdemocracy.umd.edu /hoover-speech-before-the-house-committee-speech-text/.
"psychic giddiness"   Zigmond Lebensohn to Ronald Clark, May 10, 1976, box 1, folder 38, ESF Collection.

329 *unable to work or solve puzzles*   Ibid.

a rope and a noose   John Ramsay Friedman to Ronald Clark (undated), box 14, folder 14, ESF Collection.

"a tree to hang myself"   Sheldon, "A Very Private Cryptographer," 15.

proponent of electroshock therapy   Lebensohn, "Electroconvulsive Therapy . . . A Personal Memoir."

The first course of shocks   Lebensohn to Clark, May 10, 1976.

330 *"was almost elated"*   Ibid.

"Anxiety kept her figure slim"   Murial Pollitt to ESF, October 6, 1981, box 12, folder 15, ESF Collection.

to get the pen moving   Ibid.

"I found it an outlet"   ESF to Anne [?], October 24, 1951, box 1, folder 17, ESF Collection.

stayed active   Ibid.

"That part of my life is over"   Ibid., October 13, 1951.

suitcase full of lantern slides   Ibid.

at least fifteen mutilated sheets   ESF speech to Mary Bartelme Club. Her draft of the speech begins with an unnumbered page called "Introduction" followed by fourteen more pages. The last page ends in midsentence; the concluding pages appear to have been lost or destroyed.

agonized about what to say   ESF to Anne [?], November 8, 1951, box 1, folder 17, ESF Collection.

331 *pink ballroom at the Blackstone Hotel*   Irene Powers, "Benefit Fetes Aglitter with Holiday Spirit," *Chicago Tribune,* November 18, 1951.

wasn't free to talk   ESF speech to Mary Bartelme Club, "Introduction."

"Perhaps you may think"   Ibid., 1.

showed slides of code messages   Ibid., 1–7.

two and a half hours   ESF to Anne [?], December 26, 1951, box 1, folder 17, ESF Collection.

a luncheon at Cambridge   ESF, "foreword to uncompleted work."

"As befits a woman"   Ibid.

sheet of lined yellow paper   ESF, "Notes for 'Foreword, 1959," box 17, folder 20, ESF Collection. This is the seven-page handwritten draft of the "foreword to uncompleted work" typescript.

"FOREWORD"   ESF, "foreword to uncompleted work."

332 *President Truman established*   James Bamford, *The Puzzle Palace: A Report on NSA, America's Most Secret Agency* (New York: Houghton Mifflin, 1982).

332 *the most secret of agencies*   Ibid.
"mostly nonsensical"   WFF to Roberta Wohlstetter,
September 17, 1969, box 14, folder 12, ESF Collection.
"secrecy virus"   Clark, *The Man Who Broke Purple*, 252.
333 *a nice ceremony*   "Ceremony Honoring William F. Friedman,"
Arlington Hall Post Theatre, October 12, 1955, box 14, folder 12,
ESF Collection.
feared the NSA   WFF undated letter, box 14, file 12, ESF
Collection.
"Frightening to be alone"   Clark, *The Man Who Broke Purple*,
258–59.
"a great desire to live"   WFF to Wohlstetter, September 17, 1969.
His feet swelled so much   ESF to John Ramsay Friedman,
March 20, 1967, box 4, folder 2, ESF Collection.
334 *taking notes on his condition*   ESF daybook, 1969, box 20, ESF
Collection.
The doctor stayed at the house   Ibid.
"My beloved died at 12:15"   Ibid.
"Dear heart be courageous"   ESF Collection.
"the greatest brain of the century"   Joe Mauborgne to ESF,
telegram, box 14, folder 1, ESF Collection.
"His effect on world history"   Herman Wouk to ESF,
November 3, 1969, box 14, folder 1, ESF Collection.
"Our business now involves"   Juanita Morris Moody to ESF,
November 7, 1969, box 14, folder 2, ESF Collection.
335 *"Woman's Privilege Card"*   Cosmos Club "Woman's Privilege
Card," November 14, 1969, box 17, folder 24, ESF Collection.
She designed his tombstone   ESF sketch of WFF's tombstone,
box 13, folder 31, ESF Collection.
specified that certain letters   Ibid. The *a*- and *b*-forms are clear
on ESF's sketch, and she also included a more detailed tracing of
that specific line of text. Later, ESF explicitly told Ronald Clark
that "WFF" is the cipher message; see ESF to Clark, October 7,
1976, box 15, folder 4, ESF Collection. See also Elonka Dunin,
"Cipher on the Elizebeth and William Friedman tombstone at
Arlington National Cemetery Is Solved," http://elonka.com
/friedman/index.html.
an emotional thank-you letter   John Ramsay Friedman to
Eugene McCarthy, November 12, 1969, box 293, Eugene J.
McCarthy Papers, Minnesota Historical Society.
336 *Immediately after his funeral*   ESF letter to family and friends,
January 28, 1970, box 13, folder 31, ESF Collection.
a sense of duty   Ibid.
entice a first-rate historian   Ibid.

336 *paid for a typist*   Ibid.

six-hour days   ESF to Ronald Clark, June 12, 1974, box 15, folder 1, ESF Collection.

"entertained like a queen"   Ibid.

got her on tape   ESF interview with Marshall staff.

337 *had not indexed*   The only guide to the thousands of documents in the ESF Collection is an eighteen-page "Container List" that lists the names of the folders in the twenty-two boxes but does not describe their contents.

asked her to inspect   ESF to David Kahn, undated two-page letter on carbons, box 15, folder 2, ESF Collection.

The NSA men's chorus   "Dedication Ceremony for the William F. Friedman Memorial Auditorium," program, May 21, 1970, box 14, folder 12, ESF Collection.

338 *a competent account*   ESF to Marshall Foundation, July 14, 1977, box 15, folder 4, ESF Collection.

her savings dried up   ESF to Stuart and Mabel, April 30, 1974, box 15, folder 1, ESF Collection.

"There is just one thing"   ESF typewritten diary, February 7, 1967, box 3, folder 20, ESF Collection.

She gave an interview   Connie Lunnen, "She Has a Secret Side," *Houston Chronicle,* May 24, 1972.

"In a few years"   WFF and ESF burial wishes, box 16, folder 23, ESF Collection.

arteries failed   Maureen Joyce, "Elizabeth Friedman, U.S. Cryptanalyst, Pioneer in Science of Code-Breaking Dies," *Washington Post,* November 2, 1980.

Washington Post   Ibid.

New York Times   Alfred E. Clark, "E.S. Friedman, 88, Cryptanalyst Who Broke Enemy Codes, Dies," *New York Times,* November 3, 1980.

her ashes were scattered   John Ramsay Friedman eulogy at Arlington National Cemetery, November 1980, box 6, folder 26, ESF Collection.

339 *"beacon of hope"*   Barbara Osteika (ATF historian, Department of Justice), in discussion with the author, April 2015.

An FBI cryptanalyst   Jeanne Anderson (FBI cryptanalyst, Cryptanalysis and Racketeering Records Unit), in discussion with the author, via e-mail, September 2015.

briefed U.S. leaders   "Juanita Moody," NSA Center for Cryptologic Heritage, Hall of Honor, https://www.nsa.gov /about/cryptologic-heritage/historical-figures-publications /hall-of-honor/2003/jmoody.shtml.

Ann Caracristi   "Ann Caracristi," NSA Center for Cryptologic

Heritage, Hall of Honor, https://www.nsa.gov/about/crypto
logic-heritage/historical-figures-publications/women/honorees
/caracristi.shtml.

339 *brief, verifiably true comments*   Sheldon, "Analytical Guide."
See the entries for Items 658, 1006, and 1006.1.

340 *"There are plenty of mysteries"*   ESF interview with Valaki,
transcribed February 21, 2012, 8.
first joined the agency   Valaki obituary.
"Well, thanks again, Mrs. Friedman"   ESF interview with
Valaki, transcribed January 12, 2012, 5.

341 *"Girl cryptanalyst and all that"*   Ibid.
Valaki shut off the recorder   Ibid., 6.
"You mean to say"   ESF interview with Valaki, transcribed
February 21, 2012, 8.
"I'll bet no two women"   ESF interview with Valaki, transcribed
January 10, 2012, 8.
the women laughed   Ibid. The transcript reads, "((Both laugh.))"

INDEX

ABOUT THE AUTHOR

JASON FAGONE is a journalist who covers science, technology, and culture. Named one of the "Ten Young Writers on the Rise" by the *Columbia Journalism Review*, he writes for the *San Francisco Chronicle* and has written for *GQ*, *Esquire*, *The Atlantic*, the *New York Times*, *Mother Jones*, and *Philadelphia* magazine. Fagone is also the author of *Ingenious: A True Story of Invention, the X Prize, and the Race to Revive America* and *Horsemen of the Esophagus: Competitive Eating and the Big Fat American Dream*. He lives in San Francisco, California.

PRAISE FOR *THE ONLY GOOD INDIANS*

"Scary good. Stephen Graham Jones is one of our
most talented and prolific living writers."
—Tommy Orange, author of Pulitzer Prize finalist *There There*

"Fans of Stephen King's *It* and Peter Straub's
Ghost Story should find plenty to love."
—Silvia Moreno-Garcia, bestselling author of *Mexican Gothic*

"Jones boldly and bravely incorporates both the difficult
and the beautiful parts of contemporary Indian life into his story,
never once falling into stereotypes or easy answers but also not
shying away from the horrors caused by cycles of violence."
—Rebecca Roanhorse, bestselling author of
Trail of Lightning and *Black Sun*

"I like stories where nobody escapes their pasts
because it's what I fear most."
—Terese Marie Mailhot, *New York Times*
bestselling author of *Heart Berries*

"*The Only Good Indians* is the most American horror novel I've ever read."
—Grady Hendrix, *New York Times* bestselling author of
The Southern Book Club's Guide to Slaying Vampires

"Stephen Graham Jones is a literary master who happens to write horror,
and you've never read a book quite like *The Only Good Indians*."
—Tananarive Due, American Book Award winner
and author of *The Good House*

"*The Only Good Indians* is equal parts revenge thriller,
monster movie, and meditation on the inescapable undertow
of the past. A gripping, deeply unsettling novel."
—Carmen Maria Machado, National Book Award finalist
and author of *Her Body and Other Parties*

PRAISE FOR

THE
ONLY
GOOD
INDIANS

BY STEPHEN GRAHAM JONES

"One of 2020's buzziest horror novels."

—*Entertainment Weekly*

"Gritty and gorgeous."

—*New York Times*

"Subtly funny and wry at turns, this novel will give you nightmares. The good kind, of course."

—Buzzfeed

"Fans of Stephen King's *It* and Peter Straub's *Ghost Story* should find plenty to love in this tale of friends who are haunted by a supernatural entity they first encountered in their youth."

—Silvia Moreno-Garcia,
bestselling author of *Mexican Gothic*

"*The Only Good Indians* is equal parts revenge thriller, monster movie, and meditation on the inescapable undertow of the past. A gripping, deeply unsettling novel."

—Carmen Maria Machado,
National Book Award finalist and
Guggenheim Fellow and author of
Her Body and Other Parties

"The best yet from one of the best in the business. An emotional depth that staggers, built on guilt, identity, one's place in the world, what's right and what's wrong. *The Only Good Indians* has it all: style, elevation, reality, the unreal, revenge, warmth, freezing cold, and even some slashing. In other words, the book is made up of everything Stephen Graham Jones seemingly explores and, in turn, everything the rest of us want to explore with him."

—Josh Malerman,
New York Times bestselling author of
Bird Box and *A House at the Bottom of a Lake*

"*The Only Good Indians* is scary good. Stephen Graham Jones is one of our most talented and prolific living writers. The book is full of humor and bone-chilling images. It's got love and revenge, blood and basketball. More than I could have asked for in a novel. It also both reveals and subverts ideas about contemporary Native life and identity. Novels can do so much to render actual and possible lives lived. Stephen Graham Jones truly knows how to do this, and how to move us through a story at breakneck (literally) speed. I'll never see an elk or hunting, or what a horror novel can do, the same way again."

—Tommy Orange, author of
Pulitzer Prize finalist *There There*

"A heartbreakingly beautiful story about hope and survival, grappling with themes of cultural identity, family, and traditions."

—*Library Journal* (starred review)

"Stephen Graham Jones is one of our greatest treasures. His prose here pops and sings, hard-boiled poetry conspiring with heartbreakingly alive characters."

—Sam J. Miller, Nebula Award–winning author of
Blackfish City

"How long must we pay for our mistakes, for our sins? Does a thoughtless act doom us for eternity? This is a novel of profound insight and horror, rich with humor and intelligence. *The Only Good Indians* is a triumph; somehow it's a great story and also a meditation on stories. I've wondered who would write a worthy heir to Peter Straub's *Ghost Story*. Now I know the answer: Stephen Graham Jones."

—Victor LaValle, author of
The Ballad of Black Tom and
The Changeling

"*The Only Good Indians* is a masterpiece. Intimate, devastating, brutal, terrifying, yet warm and heartbreaking in the best way. Stephen Graham Jones has written a horror novel about injustice and, ultimately, about hope. Not a false, sentimental hope, but the real one, the one that some of us survive and keeps the rest of us going. And it gives me hope that this book exists and is now in your hands."

—Paul Tremblay, author of
A Head Full of Ghosts and
The Cabin at the End of the World

"Jones hits his stride with a smart story of social commentary—it's scary good."

—*Kirkus Reviews* (starred review)

"Jones . . . has written a masterpiece. The book is . . . as instinctive and essential as it is harsh. Despite the blood and bleakness, *The Only Good Indians* is ultimately also about hope and the promise of the future. . . . Read it."

—*Locus* magazine

"Jones has written a chilling and original story of revenge set in contemporary Indian Country that had me staying up late turning pages as fast as I could. The book is bloody and brutal at times, but also intimate, heartbreaking, and ultimately hopeful. Jones boldly and bravely incorporates both the difficult and the beautiful parts of contemporary Indian life into his story, never once falling into stereotypes or easy answers but also not shying away from the horrors caused by cycles of violence. I highly recommend."

—Rebecca Roanhorse,
New York Times bestselling author of
Trail of Lightning and *Black Sun*

"Stephen Graham Jones is a literary master who happens to write horror, and you've never read a book quite like *The Only Good Indians*. It's so sharp that it cuts: from laser-precise details, characters, and prose to its unique sensibility and imagination steeped in Native American life. This book is so scary that you'll hear noises while you're reading. Keep the lights on and hang on for the ride."

—Tananarive Due,
American Book Award winner and
author of *The Good House*

"Packed tight with language you can chew, deeply original, and stinking with the fear-inducing funk of folk horror, *The Only Good Indians* is the most American horror novel I've ever read. Without Stephen Graham Jones, horror would be a hell of a lot more boring."

—Grady Hendrix,
New York Times bestselling author of
The Southern Book Club's Guide to Slaying Vampires

"Jones spins a sharp, remarkable horror story out of a crisis of cultural identity. . . . Jones's writing is raw, balancing on the knife-edge between dark humor and all-out gore as he forces his characters to reckon with their pasts, as well as their cultures. This novel works both as a terrifying chiller and as biting commentary on the existential crisis of indigenous peoples adapting to a culture that is bent on eradicating theirs. Challenging and rewarding, this tale will thrill Jones's fans and garner him plenty of new readers."

—*Publishers Weekly* (starred review)

"*The Only Good Indians* mauled me. I like stories where nobody escapes their pasts because it's what I fear most. Everyone's worst deed, if they're somewhat decent, is usually always there looming in their peripheries. What Stephen Graham Jones does for me, is create new possibilities for Indigenous story-makers."

—Terese Marie Mailhot,
New York Times bestselling author of
Heart Berries

THE
ONLY
GOOD
INDIANS

THE
ONLY
GOOD
INDIANS

a novel

STEPHEN GRAHAM JONES

SAGA PRESS

LONDON SYDNEY **NEW YORK** TORONTO NEW DELHI

SAGA)) PRESS

AN IMPRINT OF SIMON & SCHUSTER, INC.

1230 AVENUE OF THE AMERICAS, NEW YORK, NEW YORK 10020

First Saga Press paperback edition January 2021

SAGA PRESS and colophon are trademarks of Simon & Schuster, Inc.

For information about special discounts for bulk purchases,
please contact Simon & Schuster Special Sales at 1-866-506-1949
or business@simonandschuster.com.

The Simon & Schuster Speakers Bureau can bring authors to
your live event. For more information or to book an event, contact
the Simon & Schuster Speakers Bureau at 1-866-248-3049
or visit our website at www.simonspeakers.com.

Interior design by Michelle Marchese

Manufactured in the United States of America

7 9 10 8

Library of Congress Cataloging-in-Publication Data

Names: Jones, Stephen Graham, 1972– author.
Title: The only good Indians : a novel / Stephen Graham Jones.
Description: First edition. | New York : Saga Press, [2020]
Identifiers: LCCN 2019032510 (print) | LCCN 2019032511 (ebook) |
ISBN 9781982136451 (hardcover) | ISBN 9781982136475 (ebook)
Subjects: GSAFD: Horror fiction.
Classification: LCC PS3560.O5395 O55 2020 (print) |
LCC PS3560.O5395 (ebook) | DDC 813/.54--dc23
LC record available at https://lccn.loc.gov/2019032510
LC ebook record available at https://lccn.loc.gov/2019032511

ISBN 978-1-9821-3645-1
ISBN 978-1-9821-3646-8 (pbk)
ISBN 978-1-9821-3647-5 (ebook)

For Jim Kuhn.
He was a real horror fan.

This scene of terror is repeated all too often in elk country every season. Over the years, the hunters' screams of anguish have rocked the timber.

—Don Laubach and Mark Henkel, *Elk Talk*

WILLISTON, NORTH DAKOTA

The headline for Richard Boss Ribs would be INDIAN MAN KILLED IN DISPUTE OUTSIDE BAR.

That's one way to say it.

Ricky had hired on with a drilling crew over in North Dakota. Because he was the only Indian, he was Chief. Because he was new and probably temporary, he was always the one getting sent down to guide the chain. Each time he came back with all his fingers he would flash thumbs-up all around the platform to show how he was lucky, how none of this was ever going to touch him.

Ricky Boss Ribs.

He'd split from the reservation all at once, when his little brother Cheeto had overdosed in someone's living room, the television, Ricky was told, tuned to that camera that just looks down on the IGA parking lot all the time. That was the part Ricky couldn't stop cycling through his head: that's the channel only the serious-*old* of the elders watched. It was just a running reminder how shit the reservation was, how boring, how nothing. And his little brother didn't even watch normal television much, couldn't sit still for it, would have been reading comic books if anything.

Instead of shuffling around the wake and standing out at the family plot up behind East Glacier, everybody parked on the logging road behind it so they'd have to come right up to the graves to turn their cars around, Ricky ran away to North Dakota. His plan was Minneapolis—he knew some cats there—but then halfway there the oil crew had been hiring, and said they liked Indians because of their built-in cold resistance. It meant they might not slip off in winter.

Ricky, sitting in the orange doghouse trailer for that interview, had nodded yeah, Blackfeet didn't care about the cold, and no, he wouldn't leave them shorthanded in the middle of a week. What he didn't say was that you don't get cold-resistant because your jackets suck, you just stop complaining about it after a while, because complaining doesn't make you any warmer. He also didn't say that, first paycheck, he was gone to Minneapolis, bye.

The foreman interviewing him had been thick and windburned and sort of blond, with a beard like a Brillo pad. When he'd reached across the table to shake Ricky's hand and look him in the eye while he did it, the modern world had fallen away for a long blink and the two of them were standing in a canvas tent, the foreman in a cavalry jacket, and Ricky already had designs on that jacket's brass buttons, wasn't thinking at all of the paper on the table between them that he'd just made his mark on.

This had been happening more and more to him the last few months. Ever since hunting went bad last winter and right up through the interview to now, not even stopping for Cheeto dying on that couch.

Cheeto hadn't been his born name, but he had freckles and orange hair, so it wasn't a name he could shake, either.

Ricky wondered how the funeral had gone. He wondered if right now there was a big mulie nosing up to the chicken-wire fence around all these dead Indians. He wondered what that big mulie saw, really. If it was just waiting all of these two-leggers out.

Cheeto would have thought it was a pretty deer, Ricky figured. He had never been a kid to get up early with Ricky to be out in the trees when light broke. He hadn't liked killing anything except beers, probably would have been vegetarian if that was an option on the rez. His orange hair put enough of a bull's-eye on his back, though. Eating rabbit food would have just got more dumb Indians lining up to put him down.

But then he'd died on that couch anyway, not even from anybody else, just from himself, at which point Ricky figured he'd get out as well, screw it. Sure, he could be this crew's chain monkey for a week or two. Yeah, he could sleep four to a doghouse with all these white boys, the wind rocking the trailer. No, he didn't mind being Chief, though he knew that, had he been around back in the days of raiding and running down buffalo, he'd have been a grunt then as well. Whatever the bow-and-arrow version of a chain monkey was, that'd be Ricky Boss Ribs's station.

When he was a kid there'd been a picture book in the library, about Heads-Smashed-In or whatever it was called—the buffalo jump, where the old-time Blackfeet ran herd after herd off the cliff. Ricky remembered that the boy selected to drape a calf robe over his shoulders and run out in front of all those buffalo, he'd been the one to win all the races the elders had put him and all the other kids in, and he'd been the one to climb all the trees the best, because you needed to be fast to run ahead of all those tons of meat, and you needed good hands to, at the last moment after sailing off the cliff, grab on to the rope the men had already left there, that would tuck you up under, safe.

What had it been like, sitting there while the buffalo flowed down through the air within arm's reach, bellowing, their legs probably stiff because they didn't know for sure when the ground was coming?

What had it felt like, bringing meat to the whole tribe?

They'd almost done it last Thanksgiving, him and Gabe and Lewis and Cass, they'd *meant* to, they were going to be those kinds of Indians for once, they had been going to show everybody in Browning that this is the way it's done, but then the big wet snow had come in and everything had gone pretty much straight to hell, leaving Ricky out here in North Dakota like he didn't know any better than to come in out of the cold.

Fuck it.

All he was going to hunt in Minneapolis was tacos, and a bed.

But, until then, this beer would work.

The bar was all roughnecks, wall-to-wall. No fights yet, but give it time. There was another Indian, Dakota probably, nursing a bottle in a corner by the pool tables. He'd acknowledged Ricky and Ricky had nodded back, but there was as much distance between the two of them as there was between Ricky and his crew.

More important, there was a blond waitress balancing a tray of empties between and among. Fifty sets of eyes were tracking her, easy. To Ricky she looked like the tall girl Lewis had run off to Great Falls with in July, but she'd probably already left his ass, meaning now Lewis was sitting in a bar down there just like this one, peeling the label off his beer just the same.

Ricky lifted his bottle in greeting, across all the miles.

Four beers and nine country songs later, he was standing in line for the urinal. Except the line was snaking all back down the hall already, and the last time he'd been in there there'd already been guys pissing in the trash can and the sink both. The air in there was gritty and yellow, almost crunched between Ricky's teeth when he'd accidentally opened his mouth. It wasn't any worse than the honeypots out at the rig, but out at the rig you could just unzip wherever, let fly.

Ricky backed out, drained his beer because cops love an Indian with a beer bottle in the great outdoors, and made to push his way out for a breath of fresh air, maybe a fence post in desperate need of watering.

At the exit the bouncer opened his meaty hand against Ricky's chest, warned him about leaving. Something about the head count and the fire marshal.

Ricky looked past the open door to the clump of roughnecks and cowboys waiting to come in, their eyes flashing up to him but not asking for anything. It was the queue Ricky would have to mill around in to wait his turn to get back in. But it was starting to not really be his decision anymore, right? Inside of maybe ninety seconds, here, he was going to be peeing, so any way he could up the chances of being someplace where he could do that without making a mess of himself, well.

He could stand in a thirty-minute line to eyeball that blond waitress some more, sure. Ricky turned sideways to slip past the bouncer, nodding that he knew what he was doing, and already a roughneck was stepping forward to take his place.

There wasn't even any time to stiff-leg it over beside the bar, by the steaming pile of bags the dumpsters were. Ricky just walked straight ahead, out into the sea of crew cab trucks parked more or less in rows, and on the way he unleashed almost before he could come to a stop, had to lean back from it because this was a serious fire-hose situation.

He closed his eyes from the purest pleasure he'd felt in weeks, and when he opened them, he had the feeling he wasn't alone anymore.

He steeled himself.

Only stupid Indians brush past a bunch of hard-handed white dudes, each of them sure that seat you had in the bar, it should have, by right, been theirs. They're cool with the Chief among them being the chain monkey, but when it comes down to who has an eyeline on the white woman, well, that's another thing altogether, isn't it?

Stupid, Ricky told himself. *Stupid stupid stupid.*

He looked ahead, to the hood he was going to hip-slide over, the bed of the truck he hoped wasn't piled with ankle-breaking equipment, because that was his next step. A clump of white men can beat an Indian into the ground, yeah, no doubt about it, happens every weekend up here on the Hi-Line. But they have to catch his ass first.

And now that he was, by his figuring, about three fluid pounds lighter, and sobering up fast, no way was even the ex–running back of them going to hook a finger into Ricky's shirt.

Ricky grinned a tight-lipped grin to himself and nodded for courage, dislodging all the rifles he couldn't keep stacked up in his head, rifles that were *actually* behind the seat of his truck back at the site. When he'd left Browning he'd taken them all, even his uncles' and granddad's—they were all in the same closet by the front door—and then grabbed the gallon baggie of random shells, figuring some of them had to go to these guns.

The idea had been that he was going to need stake-money when he hit Minneapolis, and rifles turn into cash faster than just about anything. Except then he'd found work along the way. And he'd got to thinking about his uncles needing to fill their freezer for the winter.

Standing in the sprawling parking lot of the roughneck bar in North Dakota, Ricky promised to mail every one of those back. Would he have to pull the bolts, though, mail them in separate packages from the rifles, so the rifles wouldn't really be rifles anymore?

Ricky didn't know, but he did know that right now he wanted that pump .30-06 in his hands. To shoot if it came to that, but mostly just to swing around, the open end of the barrel leaving half-moons in cheeks and eyebrows and rib cages, the butt perfect for jaws.

He might be going down in this parking lot in a puddle of his own piss, but these grimy white boys were going to remember this Blackfeet, and think twice the next time they saw one of him walking into their bar.

If only Gabe were here. Gabe liked this kind of shit—playing cowboys and Indians in all the parking lots of the world. He'd do his stupid war whoop and just rush the hell in. It might as well have been a hundred and fifty years ago for him, every single day of his ridiculous life.

When you're with him, though, with Gabe . . . Ricky narrowed his eyes, nodded to himself again for strength. To fake it anyway— to try to be like Gabe, here. When Ricky was with Gabe, he'd always want to give a whoop like that too, the kind that made it where, when he turned around to face these white boys, it'd feel like he was holding a tomahawk in his hand. It'd feel like his face was painted in harsh crumbly blacks and whites, maybe a single finger-wide line of red on the right side.

The years can just fall away, man.

"So," Ricky said, his hands balled into fists, chest already heaving, and turned around to get this over with, his teeth clenched tight so that if he was turning around into a fist it wouldn't rattle him too much.

But . . . no one?

"What the—?" Ricky said, cutting himself off because there *was* something, yeah.

A huge dark form, clambering over a pearly white, out-of-place 280Z.

Not a horse, either, like he'd knee-jerked into his head. Ricky had to smile. This was an elk, wasn't it? A big meaty spike, too dumb to know this was where the people went, not the animals.

It blew once through its nostrils and launched into the truck to its right, leaving the pretty sloped-down hood of that little Nissan taco'd up at the edges, stomped all down in the middle. But at least the car had been quiet about it. The truck the elk had slammed into was much more insulted, screaming its shrill alarm loud enough that the spike grabbed onto the ground with all four hooves. Instead of the twenty logical paths it could have taken away from this sound, it scrabbled up across the loud truck's hood, fell off into the between space on the other side.

And now that drunk little elk was banging into another truck, and another.

All the alarms were going off, *all* the lights going back and forth.

"What is into you, man?" Ricky said to the spike, impressed.

The feeling didn't last long. Now the spike was turned around, was barreling down an aisle between the cars, Ricky right in its path, its head down like a mature bull—

Ricky threw himself to the side, into *another* truck, setting off *another* alarm.

"You want some of me?" Ricky yelled to the elk, reaching over into the bed of a random truck. He came up with a jawless oversized crescent wrench that would be a good enough deterrent, he figured. He hoped.

Never mind he was outweighed by a cool five hundred pounds.

Never mind that elk don't *do* this.

When he heard the spike blow behind him he turned already swinging, crashing the crescent wrench's round head into the side mirror of a tall Ford. The big Ford's alarm screamed, flashed every light it had, and when Ricky turned around to shuffling hooves behind him, it wasn't hooves this time, but boots.

All the roughnecks and cowboys waiting to get into the bar.

"He . . . he—" Ricky said, holding the wrench like a tire beater, every second truck in his immediate area flashing in pain, and showing the pounding they'd just taken. He saw it too, saw *them* seeing it: this Indian had got hisself mistreated in the bar, didn't know who drove what, so he was taking it out on every truck in the parking lot.

Typical. Momentarily one of these white boys was going to say something about Ricky being off the reservation, and then what was supposed to happen could get proper-started.

Unless Ricky, say, wanted to maybe *live*.

He dropped the wrench into the slush, held his hand out, said, "No, no, you don't understand—"

But they did.

When they stepped forward to put him down in time-honored fashion, Ricky turned, flopped half over the 280Z he *hadn't* trashed, endured a bad moment when somebody's reaching fingers were hooked into a belt loop, but he spun his hips hard, tore through, fell down and ahead, his hands to the ground for a few overbalanced steps. A beer bottle whipped by his head, shattered on a grille guard right in front of him, and he threw his hands up to keep his eyes safe, veered what he thought was around that truck but not enough—his hip caught the last upright of the guard, spun him around, into *another* truck, with *another* stupid alarm.

"*Fuck you!*" he yelled to the truck, to all the trucks, all the cowboys, just North Dakota and oil fields and America in general, and then, running hard down a lane between trucks, hitching himself ahead with more mirrors, two of them coming off in his hands, he felt a smile well up on his face, Gabe's smile.

This is what it feels like, then.

"Yes!" Ricky screamed, the rush of adrenaline and fear sloshing up behind his eyes, crashing over his every thought. He turned around and ran backward so he could point with both hands at the roughnecks. Four steps into this big important gesture he fell out into open space, kind of like a turnrow in a plowed field, caught his left boot heel on a rock or frozen clump of bullshit grass, went sprawling.

Behind him he could see dark shapes vaulting over whole truck beds, their cowboy hats lifting with them, not coming down, just becoming part of the night.

"White boys can move . . ." he said to himself, less certain of all this, and pivoted, rose, was moving again, too.

When the footfalls and boot slaps were too close, close enough he couldn't handle it, knew this was it, Ricky grabbed a fiberglass dually fender, used it to swing himself a sharp and sudden ninety degrees, into what would have been the truck's long side, what should have been its side, but he was sliding now, he was going under, leading with the slick heels of his work boots.

This was the kind of getting away he'd learned at twelve years old, when he could slither and snake.

The truck was just tall enough for him to slide under, through the muck, his momentum carrying him halfway across. To get across the rest of the truck's width, he reached up for a handhold, the skin of his palm and the underside of his fingers immediately smoking from the three-inch exhaust pipe.

Ricky yelped but kept moving, came up on the other side of the truck fast enough that he slammed into a beater that didn't have an alarm. Two truck lengths ahead, the dark shapes were pulling their best one-eighty, casting left and right for the Indian.

Duck, Ricky told himself, and disappeared, ran at a crouch that felt military, like he was in a trench, like shells were flying. And they might as well be.

"There he is!" a roughneck bellowed, and his voice was far enough off that Ricky knew he was wrong, that they were about to pile onto somebody else for ten or twenty seconds, until they realized this was no Indian.

Ten trucks between him and them finally, Ricky stood to his full height to make sure it wasn't that Dakota dude catching the heat.

"I'm right here," Ricky said to the roughnecks, not really loud enough, then turned, stepped through the last line of trucks, out into the ditch of the narrow ribbon of blacktop that had brought him here, that ran between the bar's parking lot and miles and miles of frozen grasslands.

So it was going to be a walking night, then. A hiding from every pair of headlights night. A cold night. Good thing I'm Indian, he told himself, sucking in to get the zipper on his jacket started. Cold doesn't matter to Indians, does it?

He snorted a laugh, flipped the whole bar off without turning around, just an over-the-shoulder thing with his smoldering hand, then stepped up onto the faded asphalt right as a bottle burst beside his boot.

He flinched, drew in, looked behind him to the mass of shadows that were just arms and legs and crew cuts now, moving over the trucks.

They'd seen him, made his Indian silhouette out against all this pale frozen grass.

He hissed a pissed-off blast of air through his teeth, shook his head once side to side, and straight-legged it across the asphalt to see how committed they might be. They want an Indian bad enough tonight to run out into the open prairie in November, or would it be enough just to run him off?

Instead of trusting the gravel and ice of the opposite shoulder, Ricky took it at a slide, let his momentum stand him up once his boot heels caught grass, then transferred all that into a leaning-forward run that was going to have been a fall even if he hadn't caught the top strand of fence in the gut. He flipped over easy as anything, the strand giving up its staples halfway through, just to be sure his face planted *all* the way into the crunchy grass on the other side.

Ricky rolled over, his face to the wash of stars spread against all the blackness, and considered that he maybe should have just stayed home, gone to Cheeto's funeral, he maybe shouldn't have stolen his family's guns. He maybe should have never even left the rez at all.

He was right.

When he stood, there was a sea of green eyes staring back at him from *right there*, where there was just supposed to be frozen grass and distance.

It was a great herd of elk, waiting, blocking him in, and there was a great herd pressing in behind him, too, a herd of men already on the blacktop themselves, their voices rising, hands balled into fists, eyes flashing white.

INDIAN MAN KILLED IN DISPUTE OUTSIDE BAR.

That's one way to say it.

THE HOUSE
THAT
RAN RED

FRIDAY

Lewis is standing in the vaulted living room of his and Peta's new rent house, staring straight up at the spotlight over the mantel, daring it to flicker on now that he's looking at it.

So far it only comes on with its thready glow at completely random times. Maybe in relation to some arcane and unlikely combination of light switches in the house, or maybe from the iron being plugged into a kitchen socket while the clock upstairs *isn't*—or is?—plugged in. And don't even get him started on all the possibilities between the garage door and the freezer and the floodlights aimed down at the driveway.

It's a mystery, is what it is. But—more important—it's a mystery he's going to solve as a surprise for Peta, and in the time it takes her to drive down to the grocery store and back for dinner. Outside, Harley, Lewis's malamutant, is barking steady and pitiful from being tied to the laundry line, but the barks are already getting hoarse. He'll give it up soon enough, Lewis knows. Unhooking his collar now would be the dog training him, instead of the other way around. Not that Harley's young enough *to* be trained anymore, but not like Lewis is, either. Really, Lewis imagines, he deserves some big Indian award for having made it to thirty-six without pulling into the drive-through for a burger and fries, easing away with diabetes and high blood pressure and leukemia. And he gets the rest of the trophies for having avoided all the car crashes and jail time and alcoholism on his cultural dance card. Or maybe the reward for lucking through all that—meth too, he guesses—is having been married ten years now to Peta, who doesn't *have* to put up with motorcycle parts soaking in the sink, with the drips of Wolf-brand chili he always leaves between the coffee table and the couch, with the tribal junk he always tries to sneak up onto the walls of their next house.

Like he's been doing for years, he imagines the headline on the *Glacier Reporter* back home: FORMER BASKETBALL STAR CAN'T EVEN HANG GRADUATION BLANKET IN OWN HOME. Never mind that it's not because Peta draws the line at full-sized blankets, but more because he used it for padding around a free dishwasher he was bringing home a couple of years ago, and the dishwasher tumped over in the bed of the truck on the very last turn, spilled clotty rancid gunk directly into Hudson's Bay.

Also never mind that he wasn't exactly a basketball star, half a lifetime ago.

It's not like anybody but him reads this mental newspaper.

And tomorrow's headline?

THE INDIAN WHO CLIMBED TOO HIGH. Full story on 12b.

Which is to say: that spotlight in the ceiling's not coming down to him, so he's going to have to go up to it.

Lewis finds the fourteen-foot aluminum ladder under boxes in the garage, Three Stooges it into the backyard, scrapes it through the sliding glass door he's promised to figure out a way to lock, and sets it up under this stupid little spotlight, the one that all it'll do if it ever works is shine straight down on the apron of bricks in front of the fireplace that Peta says is a "hearth."

White girls know the names of everything.

It's kind of a joke between them, since it's how they started out. Twenty-four-year-old Peta had been sitting at a picnic table over beside the big lodge in East Glacier, and twenty-six-year-old Lewis had finally got caught mowing the same strip of grass over and over, trying to see what she was sketching.

"So you're, what, scalping it?" she'd called out to him, full-on loud enough.

"Um," Lewis had said back, letting the push mower die down.

She explained it wasn't some big insult, it was just the term for cutting a lawn down low like he was doing. Lewis sat down opposite her, asked was she a backpacker or a summer girl or what, and she'd liked his hair (it was long then), he'd wanted to see all her tattoos (she was already maxed out), and within a couple weeks they were an every night kind of thing in her tent, and on the bench seat of Lewis's truck, and pretty much all over his cousin's living room, at least until Lewis told her he was busting out, leaving the reservation, screw this place.

How he knew Peta was a real girl was that she didn't look around and say, *But it's so pretty* or *How can you* or—worst—*But this is your* land. She took it more like a dare, Lewis thought at the time, and inside of three weeks they were a nighttime *and* a daytime kind of thing, living in her aunt's basement down here in Great Falls, making a go of it. One that's still not over somehow, maybe because of good surprises like fixing the unfixable light.

Lewis spiders up the shaky ladder and immediately has to jump it over about ten inches, to keep from getting whapped in the face by the fan hanging down on its four-foot brass pole. If he'd checked *The Book of Common Sense* for stunts like this—if he even knew what shelf that particular volume might be on—he imagines page one would say that before going up the ladder, consider turning off all spinny things that can break your fool nose.

Still, once he's up higher than the fan, when he can feel the tips of the blades trying to kiss his hipbone through his jeans, his fingertips to the slanted ceiling to keep steady, he does what anybody would: looks down through this midair whirlpool, each blade slicing through the same part of the room for so long now that . . . that . . .

That they've carved *into* something?

Not just the past, but a past Lewis recognizes.

Lying on her side through the blurry clock hands of the fan is a young cow elk. Lewis can tell she's young just from her body size—lack of filled-outness, really, and kind of just a general lankiness, a gangliness. Were he to climb down and still be able to see her with his feet on the floor, he knows that if he dug around in her mouth with a knife, there wouldn't be any ivory. That's how young she still is.

Because she's dead, too, she wouldn't care about the knife in her gums.

And Lewis knows for sure she's dead. He knows because, ten years ago, he was the one who made her that way. Her hide is even still in the freezer in the garage, to make gloves from if Peta ever gets her tanning operation going again. The only real difference between the living room and the last time he saw this elk is that, ten years ago, she was on blood-misted snow. Now she's on a beige, kind of dingy carpet.

Lewis leans over to get a different angle down through the fan, see her hindquarters, if that first gunshot is still there, but then he stops, makes himself come back to where he was.

Her yellow right eye . . . was it open before?

When it blinks Lewis lets out a little yip, completely involuntary, and flinches back, lets go of the ladder to wheel his arms for balance, and knows in that instant of weightlessness that this is it, that he's already used all his get-out-of-the-graveyard-free coupons, that this time he's going down, that the cornermost brick of the "hearth" is already pointing up more than usual, to crack into the back of his head.

The ladder tilts the opposite way, like it doesn't want to be involved in anything this ugly, and all of this is in the slowest possible motion for Lewis, his head snapping as many pictures as it can on the way down, like they can stack up under him, break his fall.

One of those snapshots is Peta, standing at the light switch, a bag of groceries in her left arm.

Because she's Peta, too, onetime college pole vaulter, high school triple-jump state champion, compulsive sprinter even now when she can make time, because she's *Peta*, who's never known a single moment of indecision in her whole life, in the next snapshot she's already dropping that bag of groceries that was going to be dinner, and she's somehow shriking across the living room not really to catch Lewis, that wouldn't do any good, but to slam him hard with her shoulder on his way down, direct him away from this certain death he's falling onto.

Her running tackle crashes him into the wall with enough force to shake the window in its frame, enough force to send the ceiling fan wobbling on its long pole, and an instant later she's on her knees, her fingertips tracing Lewis's face, his collarbones, and then she's screaming that he's stupid, he's so, so *stupid*, she can't lose him, he's got to be more careful, he's got to start caring about himself, he's got to start making better decisions, please please *please*.

At the end she's hitting him in the chest with the sides of her fists, real hits that really hurt. Lewis pulls her to him and she's crying now, her heart beating hard enough for her and Lewis both.

Raining down over the two of them now—Lewis almost smiles, seeing it—is the finest washed-out brown-grey dust from the fan, which Lewis must have hit with his hand on the way down.

The dust is like ash, is like confectioner's sugar if confectioner's sugar were made from rubbed-off human skin. It dissolves against Lewis's lips, disappears against the wet of his open eyes.

And there are no elk in the living room with them, though he cranes his head up over Peta to be sure.

There are no elk because that elk *couldn't* have been here, he tells himself. Not this far from the reservation.

It was just his guilty mind, slipping back when he wasn't paying enough attention.

"Hey, look," he says to the top of Peta's blond head.

She rouses slowly, turns to the side to follow where he's meaning.

The ceiling of the living room. That spotlight.

It's flickering yellow.

SATURDAY

On break at work—he's supposed to be training the new girl, Shaney—Lewis calls Cass.

"Long time, no hear," Cass says, his reservation accent a sing-song kind of pure Lewis hasn't heard for he doesn't know how long. In response, Lewis's voice, smoothed down flat from only ever talking to white people, rises like it never even left. It feels unfamiliar in his mouth, in his ears, and he wonders if he's faking it somehow.

"Had to call your dad to get your number," he says to Cass.

"What happens when you move away for ten years, yeah?"

Lewis shifts the phone to his other ear.

"So what's going?" Cass asks. "Not calling from jail, are you? Post office finally figure out you're Indian, what?"

"Pretty sure they know," Lewis says. "It's the first checkbox."

"Then it's her," Cass says with what sounds like a grin. "*She* finally figured out you're Indian, enit?"

What Cass and Gabe and Ricky had told him when he was running off with Peta was that he should get his return address tattooed on his forearm, so he could get his ass shipped back home when she got tired of playing Dr. Quinn and the Red Man.

"You wish she'd figure it out," Lewis tells Cass on the phone, turning to be sure Shaney, his shadow for the day, isn't standing in the break room doorway soaking all this in. "She even lets me hang my Indian junk on all the walls."

"Like *Indian*-Indian," Cass says, "or Indian just because an Indian owns it?"

"I called to ask a question," Lewis says, quieter, closer.

Of Gabe and Ricky and Cass, Cass was always the one he could dial down to "serious" easiest. Like the real him, the real and actual person, wasn't buried as deep under attitude and jokes and bluster as it was with Ricky and Gabe.

Not that Ricky, being dead, really has a telephone number anymore.

Shit, Lewis says inside.

He hasn't thought about Ricky for nearly ten years now. Not since he heard.

The headline flashes in his head: INDIAN MAN HAS NO ROOTS, THINKS HE'S STILL INDIAN IF HE TALKS LIKE AN INDIAN.

Lewis breathes in, covers the handset to breathe out, so Cass won't hear it across all these miles.

"Those elk," he says.

After a long enough time that he can be sure Cass knows exactly the elk he's talking about, Cass says, "Yeah?"

"Do you ever . . ." Lewis says, still unsure how to say it, even though he ran it through his head all last night and all the way in to work. "Do you ever, you know, *think* about them?"

"Am I still pissed off about them?" Cass fires right back. "I see Denny on fire on the side of the road, you think I stop to piss on him?"

Denny Pease, the game warden.

"He's still on the job?" Lewis asks.

"Running the office now," Cass says.

"Still a hard-ass?"

"He fights for Bambi," Cass says, like that's still in circulation all this time later. It was what they all used to say about the game crew: anytime Man was in the forest, all the wardens' ears perked up, and their citation books flapped open.

"Why you asking about him?" Cass says.

"Not him," Lewis says. "Just thinking, I guess. Ten-year anniversary, I don't know."

"Ten years in, what, a week?" Cass says.

"Two," Lewis says, shrugging like he doesn't mean to have it all figured down this precise. "It was the last Saturday before Thanksgiving, wasn't it?"

"Yeah, yeah," Cass says. "Last day of the season . . ."

Lewis winces without any sound, closes his eyes tight. The way Cass dragged that last part out all suggestive, it's the same as reminding Lewis it *wasn't* the last day of the season. Just, the last day they'd been able to all get together to hunt.

But it was also, as it turned out, the last day of their season in a different way, he guesses.

He shakes his head three times like trying to clear it, and tells himself again that no way did he see that young elk on the floor of his living room.

She's dead, she's gone.

To pay for her, even, the day before he left with Peta, he'd taken all the packaged-up meat of her and gone door-to-door down Death Row, giving it to the elders. Because she'd come from the elders' section—the good country saved back for them up by Duck Lake, so they could fill their freezers from the field instead of the IGA—because she'd *come* from there, it was all full circle and Indian, hand-delivering the meat to their doors. Never mind that Lewis couldn't find any of his meat stamps, had to use Ricky's little sister's stamps. So, instead of STEAK or GROUND or ROAST, all the butcher paper the young elk was wrapped in had a black raccoon handprint on it, because that was the only one she had that wasn't a flower or a rainbow or a heart.

But no way could that elk be coming up from thirty stew pots all this time later, walking a hundred and twenty miles south to haunt Lewis. First because elk don't do that, but second because, in the end, her meat had got where it was meant to get, he hadn't even done anything wrong. Not really.

"Gotta go, man," he tells Cass. "My boss."

"It's Saturday," Cass says back.

"Rain nor sleet nor weekends," Lewis says back, and hangs up more abruptly than he means, holds the phone on its cradle for a full half minute before lifting it back.

He dials in the number Cass's dad gave him for Gabe. It's Gabe's dad's number, actually, but Cass's dad had looked out the window, said he could see Gabe's truck over there right now.

"Tippy's Tacos," Gabe says after the second ring. It's how he always answers, wherever he is, whoever's phone. There was never a place called that on the reservation, as far as Lewis knows.

"Two with venison," Lewis answers back.

"Ah, *Indian* tacos . . ." Gabe says, playing along.

"And two beers," Lewis adds.

"You must be Navajo," Gabe says right back, "maybe a fish tribe. If you were Blackfeet, you'd want a six with that."

"I've known some Navajo can flat put it away," Lewis says, a deviation from the usual routine, like bringing it down, breaking it. For maybe five seconds Gabe doesn't say anything, then, "*Lewdog?*"

"First try," Lewis says, his face warming just to be known.

"You in jail?" Gabe asks.

"Still a comedian," Lewis tells him.

"Among other things," Gabe says back, then, probably to his dad, "It's Lewis, remember him, old man?"

Lewis doesn't hear the reply, but does hear a basketball game cranked up high enough to be blasting through the whole house.

"So what's up?" Gabe asks when he's back. "Need bus fare home, what? If so, I can hook you up with somebody. Little light at the moment myself."

"Still hunting?"

"That'd be under the 'Among Other Things' category, wouldn't it?" Gabe says.

Of course he's still hunting. Denny'd have to work 24/7 to write up even half of what Gabriel Cross Guns poaches on a weekly basis, and the rangers over in Glacier would have to work even harder to find his tracks, going back and forth across the Park line, the return prints a couple hundred pounds deeper than the ones sneaking in.

"How's Denorah?" Lewis says, because that's where you start after this long.

Denorah's Gabe's daughter by Trina, Trina Trigo, has to be twelve or thirteen by now—she was walking around already when Lewis left, anyway, he's pretty sure of that.

"My finals girl, you mean?" Gabe says, finally all the way into this call, it feels like.

"Your what?" Lewis asks all the same.

"You remember Whiteboy Curtis from Havre?" Gabe asks.

Lewis can't dredge up Whiteboy's actual last name—something German?—but yeah, he remembers: Curtis, the baller, this naturally gifted farm kid who was born for the court. He didn't see it all with his eyes, he *felt* the game through his feet like radar, and didn't even have to think to know which way to cut. And he had that basketball on a string, one hundred percent. Only thing kept him from going college was his height, and that he insisted he was a power forward, not a stop-and-pop sharpshooter. At high school height, sure, someone just six-two could crash in, dominate as a power forward. And he had some jumps, too, could rise up and flush it—only in pregame, with a lot of setup, but still. In the end, though, he wasn't built like Karl Malone, but like John Stockton. Just, he couldn't accept that, had the idea he could go inside at the next level, bang his way through the bigs, not be a pinball bouncing off them. Insisting he was that power forward, he'd lost so many teeth he looked like a hockey player, last Lewis had heard. And the concussions weren't exactly doing anything good for his short-term memory. It would have been better for the rest of his life if he'd never figured he could play.

Still?

"He had that jump shot," Lewis says, seeing it again, the way Whiteboy Curtis would just hang and hang, wait for everyone else to sink back down before releasing the ball so perfect, his eyes laser-guiding it up and up, and, finally, in.

"Denorah's like that," Gabe whispers, like the best secret ever. "Just, *better*, man. Serious. Browning's never seen nothing like her."

"I should come watch her play," Lewis says.

"You should," Gabe says. "Just, don't tell Trina I told you to come. Maybe don't even talk to her. If she looks at you? She does, maybe cut your hair, change your name, jump on a ship."

"She still out for blood?"

"Woman can hold a grudge," Gabe says. "Got to give her that."

"For no reason, of course," Lewis says, leaning back on the usual lines again.

"So to what do I owe this call, Mr. Postman?" Gabe says then, being all fake formal. "I forget to put a stamp on something, what?"

"Just been a while," Lewis says.

"It was a while eight, nine years ago," Gabe says. "You're talking to me, man."

A lump forms in Lewis's throat. He tilts his face back, closes his eyes.

"I was just remembering when Denny—"

"Fucked us permanent?" Gabe cuts in. "Yeah, something about that maybe rings a bell or two . . ."

"You ever been back there by Duck Lake again?" Lewis asks.

"You have to have an old-timer with you," Gabe says. "You know that, man. How long you been gone again?"

"I mean where it happened," Lewis says. "That drop-off place."

"That place, that place, yeah," Gabe says, driving a nail into Lewis's heart. "It's haunted, man, didn't you know? Elk don't go there anymore, even. I bet they even tell stories around the elk campfires, right? About what went down that day? Shit, we're legends to them, man. The four boogeymen—the four *butchers* of Duck Lake."

"Three," Lewis says. "The three boogeymen."

"They don't know that," Gabe says.

"But you really think they might remember?" Lewis asks, just hanging it all out there at last.

"*Remember?*" Gabe says, the smile one hundred percent there in his voice. "They're fucking elk, man. They don't really have campfires."

"And we killed them all anyway, yeah?" Lewis says, blinking the heat from his eyes. looking around again for Shaney.

"What's this about?" Gabe says then. "You still missing that crappy knife, what?"

Lewis has to strain to dial back to what Gabe's saying: that trading-post knife he'd bought, with the three or four interchangeable blades, one of them a weak little saw, for the breastbone and pelvis.

"That knife was a piece of shit," Lewis says. "If you find it, lose it again fast, yeah?"

"Will do," Gabe says, his voice far from the phone for a moment, basketball pouring into his end of the line. "Hey, we're watching a—"

"I got to get gone, too," Lewis says. "Nice hearing your dumb-ass voice again, though."

"Shit, I should charge by the minute," Gabe says, and ten, twenty seconds later the line's dead again, and Lewis is standing there with his shoulder against the wall, tapping the handset into his forehead like a drumstick.

"Should I be taking notes, Blackfeet?" Shaney asks from the doorway.

Lewis hangs the phone up.

Shaney's Crow, so calling him "Blackfeet" is this running joke, their tribes being longtime enemies.

"Something Peta said last night," Lewis lies, always trying to be sure to remind Shaney about his wife, and then say something about her again, just to be sure. Not because he's the ladies' man of the USPS—there isn't one—but because him and Shaney are the only two Indians at this station, and for the last week, ever since Shaney passed the background check and hired on, everybody's been doing that thing they do with armchairs or end tables when they match: trying to push him and her together over in the corner, leave them there to be this perfect set.

"Something your *wife* said?" Shaney asks, Lewis sliding past her, leading them back to the big sorting machine. He flicks it on to continue this lesson.

"We've got this crap light in the living room," he tells her. "Won't come on when it's supposed to. She thinks it might be a short in the wall. Was calling a guy I know who does electrician stuff on the side."

"On the side . . ." Shaney repeats, and nudges an envelope this way into the sorting machine instead of that way.

Lewis tracks that fast piece of mail up into the belly of the beast and shakes his head with wonder when nothing catches, nothing crumples.

Shaney grins a mischievous grin, bites her lower lip in at the end of it.

"Next time," she says, and hip-checks Lewis.

He rolls with it, doesn't push back, is miles and years away.

MONDAY

Duckwalking backward on his stripped-down, double-throaty Road King that's about to find its lope, Lewis clocks Jerry already at the edge of the post office's parking lot, hanging his loose right hand down by the rear wheel of his custom Springer, his index and middle fingers waggling in an upside-down peace sign before they curl up into his fast fist. Lewis has no idea what it means, never rode with a real and actual gang like Jerry did in his Easy Rider youth, but it must mean something like *This way* or *All clear* or *Smoke 'em if you got 'em*, because Eldon and Silas throttle in right behind him, leaving Lewis to watch the back door like always, even though where they're headed is *to* Lewis's new place way the hell over on 13th.

Pecking order's pecking order, though, and Lewis, even though this is his fifth year slinging mail, is still the new guy. Being last, though, that means that when Shaney comes running out the side door, Lewis's bitch seat is what she jumps onto, barely making it.

Her hands fall perfect to his hips, her front to his back, and very much right there.

"Hello?" he says, throttled down and wobbling.

"I want to see too," she says, shaking her head and loosening her hair.

Yeah, this is exactly what Peta needs to clock pulling into the driveway.

Still, Lewis grabs the next gear, falls in line, having to goose it to stay with.

Why they're all going to his place is because Harley, at nearly ten years old, has taken to jumping the six-foot fence like a young dog, a fact Eldon says he'll only believe when he sees it. So, he's going to see it. They all are, including, now, Shaney.

Third in line is Silas, on his rattletrap scrambler that's not good past fifty, but gets kind of fun at seventy-five, if near-death experiences are your thing. Eldon, snapping at Jerry's heels, is on his slammed bobber, which he can only swing because he lives close to the post office, can walk in if the weather's bad, so doesn't need to keep and insure a truck or a car. Of the four of them he's the only one not married, too, which frees up some funds, for sure. Jerry tells him to just wait, though, it'll happen—"They'll drop sooner or later," haw haw haw. At fifty-three, Jerry's the oldest of them, and comes complete with the silver handlebar stache, freckled-bald head, ratty ponytail, and icy blue eyes.

Silas is pretty much mute, and might even have some Indian in him somewhere, Lewis thinks. Not enough to have been Chief before Lewis earned that title, but . . . maybe as much Indian blood as Elvis had, however much that is? Like, enough to fill up a pair of blue suede shoes? Eldon claims to be Greek and Italian both, which is maybe a joke Lewis doesn't quite get. Jerry doesn't claim to be anything other than in constant need of another beer.

It's good to have found them, after losing Gabe and Cass and Ricky.

Well, after having left them.

No headlines about this. It's just the same old news as ever.

The five o'clock traffic they slip past on River all cranes a bit to keep Shaney in view ten or twenty feet longer. Meaning her button-up flannel's probably untucked and flapping, threatening to come off altogether.

Great.

Wonderful.

Lewis shouldn't have said anything about Harley, he knows. It would be better just to be headed home alone, to maybe sink a few free throws in the driveway before Peta's back. But—Harley, right? He's not just not-young, he's actually pretty damn old for a dog his size, has been hit twice on the road, one of those by a dump truck, and he's been shot once, in the hip. And that's just what Lewis knows about. There've been snakebites and porcupines and kids with pellet guns and all the usual dog fighting that any dog's going to get up to.

No way should Harley be able to clear that fence. No way should he even have a reason to try. Still, four times now Lewis has found him out in the road, and Peta's found him twice.

He *must* be jumping, maybe scrabbling a bit to get all the way over.

And Lewis should have kept it to himself.

Except?

Thinking and thinking about the young elk who couldn't have been on his living room floor, Harley barking it up outside, Lewis had finally made what felt like a connection between the two. Could Harley have been barking at her, at the elk? Can he see her *without* a spinning fan? Has she been there all along, these past ten years?

Worse, if Harley can sense her, then is that what's been driving him over the fence? Maybe it's not about getting *to* all the dogs in heat out there or whatever. Maybe it's about getting *away* from the house.

Never mind that the lease is for twelve months and they lose the deposit if they pack up, disappear.

"Hold on," Lewis says back to Shaney, and rolls the throttle back to shoot across the river, go weightless a bit over the train tracks on the other side, avoid the way they always rattle his teeth not once but twice—one for each rail.

Shaney does a whoop from the thrill of it and Lewis gears down for the slow turn onto 6th, gets all the way into fourth for the straight shot down American Ave, taking the lead because none of these jokers have been to his new place. Three fast turns later, maybe taking them a bit fast like to test Shaney, it's his driveway.

"This is it," Lewis says into the sudden silence of no panheads, no V-twins.

Jerry and Eldon and Lewis all cock their bikes over, but Lewis waits for Shaney.

"Oh yeah," she says, placing her hands on his back and pushing off the seat all at once, a dismount Lewis is glad he won't have cycling through his head for the rest of the week.

"So where's this great flying dog already?" Jerry croaks.

"Close to your bedtime, Granddad?" Eldon says, just out of arm's reach but going boxer-light on his feet anyway.

Silas grins up at the front of the house, settles on a high window, Lewis thinks. He studies it, too. It's just his and Peta's bedroom window, no curtains yet.

"Well, mailboy?" Jerry says again.

It's what he calls everybody, Lewis is pretty sure. Probably because names have started slipping out of his head.

Lewis does the code for the garage, makes a show of tapping his shoes at the door, then ushers them all into the World Famous Jumping Dog Show.

"He just started doing it," he says on the way through the kitchen, walking backward like a proper tour guide. "I always thought he had some wolf in him maybe, along with sled dog or pit fighter. Now I'm thinking kangaroo."

"Snow kangaroo," Jerry says, the leathery skin around his eyes crinkling.

Silas snickers, running the tips of his fingers along the top of the table and then looking at them like for grime.

"Dog needs what's on the other side of the fence, that's a dog'll learn to jump," Shaney informs them all.

Jerry says something about this through his stache but it's lost, and anytime Lewis asks him for a repeat Jerry just waves it off.

"Where's the little lady?" Eldon asks, clapping his big hand onto the back of the couch.

"Making the big bucks," Lewis says, miming the bright orange wands Peta parks planes with, using them both to direct this little tour group to his right, his right.

"Aren't any big bucks in Great Fa—" Eldon starts to say, doesn't get to finish, some of Peta's more lacy underthings suddenly drying on the back of a chair.

"Divert, divert," Lewis tells him, waving him back with his make-believe wands but smiling.

Still, "*Nice*," Shaney says to Lewis and only to Lewis about the showy bra when she passes.

Past her, thankfully, Silas has liberated the housing for the Road King's headlight assembly from the kitchen table, is holding it up to peer through it.

"Still looking for hard bags?" he asks.

"Got some?" Lewis says back, and flips the latch up on the sliding door. "What color?"

"Because everything else matches so well?" Eldon cuts in.

"Foul, foul," Lewis calls, shining his wands on him because evidently he's a ref now.

No, his bike doesn't all match yet. But it will. He's going to Pinocchio it up from the rolling skeleton it is now into the real bike it wants to be. The hard bags are what Peta's insisting on, since, in a skid, they take the heat from the asphalt, keep the flesh and muscle on your leg and hip. Lewis tried telling her that only matters for riders who lay it over, but that pretty much just warranted a glare, not even a halfway grin.

"Anything about Silas's scoot there suggest he's got extra parts of any color laying around?" Jerry says over his shoulder. "Any extra pieces, he just tacks them on, don't he?"

Silas's bike right now is mid-transformation, somewhere between a cafe racer and a twelve-year-old's drawing of his dream bike, but he has to grin and shrug about this, because it's true.

Lewis twirls the sawed-off broomstick up from the sliding door's track, swishes the glass back dramatically, and presents the backyard to these unbelievers, letting them go first so they can see there's no trickery involved.

How he knows Harley will be there instead of running wild from yard to yard, it's that he hooked Harley's chain to the rusted baling wire of the laundry line before work, like every morning. Last he saw, Harley was running back and forth, he had a water pan, some shade, some grass, a clueless look on his face—everything a dog could need. The laundry line isn't a permanent solution, but it's solution enough until Lewis finds some hog-wire panels for the top of the fence.

"Maybe he's a pole vaulter like his momma," Eldon calls back from the uneven deck.

Lewis has bragged to them all about Peta, and Jerry and Silas have even met her a couple of times when it was raining and she had to pick him up in the truck.

"Or an escape artist," Silas adds.

Lewis steps out after them, parts them to see from pole to rusted pole of the laundry line, and he's right: no Harley. Also, no baling wire running between the poles anymore, to clip wet clothes to.

"I'm going to kill that dog," he says, stepping out farther to make sure Harley's not just standing there watching the house, which is when Shaney, over at the next corner of the back of the house, finishes that thought: "Think you're a little late for that, Blackfeet."

With her lips she shows Lewis where to look, and from the way she's not joking, he gets a flash of warning, can feel regret washing up into his throat.

It's Harley. He's hanging by his chain from the top of the fence, eyes open but not seeing anything, gouges and furrows clawed into the fence because it took a while for him to strangle out, evidently.

"Well, shit," Jerry says.

Harley was the first gift Peta ever got Lewis, nine years ago. One of her other aunts' dogs had thrown a litter, and the dad was supposed to have been a real scrapper, and Lewis had already been talking about how the last good rez dog he'd had, he'd been a kid, and a horse in the parade had kicked it in the head while Lewis was grubbing for candy with the rest of the kids. So Harley, he'd been perfect, almost made Great Falls feel like home that first year—they grew into it together. And now he's dead on the chain Lewis tied him up with.

"Sorry, man," Eldon says, studying the high-dollar boots he always changes into for riding.

"Looks like he almost made it," Shaney says for all of them, meaning they believe Lewis about Harley having found springs in his legs late in life.

"Stupid dog," Lewis says, keeping it short because he doesn't trust his voice not to break into pieces, choke him up.

And then one of Harley's hind legs twitches once, exactly in rhythm somehow with the way that elk on the living room floor blinked her eyes. The elk that wasn't dead on the floor of Lewis's living room, that wasn't *alive* on the floor—that wasn't there at all.

Lewis's response to Harley being sort of alive isn't the right response, isn't the response he's proud of: he sucks air in and steps back, almost falls on his ass.

Of the five of them it's Silas who dives forward to hug on to Harley, lift him up, get the pressure off his throat. Jerry reaches up with a meaty paw, unhooks the chain from the top of the fence, and Shaney's already guiding Harley's bloody collar up over his head, being careful of his ears.

Silas turns around, Harley cradled in his arms, and Lewis pulls his eyes away for just a moment, finds himself watching Shaney, who's maybe going to step forward, hug Harley to her, but then she's jerking back all at once, startled from the wall of sound suddenly rushing at them all.

Eldon grabs Lewis's shoulder like to pull him out of the way, or use him to push off of, and even Jerry looks up faster than his walrus-looking self usually does.

The whole backyard is shaking and loud and fast and dangerous, the kind of sensory trauma where Lewis is pretty sure that, if there were a sprinkler rainbowing a wall of water back and forth, that iridescent sheet of color would collapse, turn to mist.

It's the train that runs behind this neighborhood twice a day, what Peta calls the Thunderball Express. It's why her and Lewis can swing rent on a place with a ceiling this high. It's also why Harley can't be getting out of the backyard anymore.

Lewis looks up at the coal and graffiti smearing past, sees tomorrow's headline in his head: ONCE-LOCAL MAN CAN'T EVEN TOUCH HIS OWN DYING DOG.

Sometimes the headlines get it right. And the story on 12b this time, it's accompanied by a small out-of-focus black-and-white photograph Lewis's mind takes on reflex, because he can't really deal with it in the moment, what with the train screaming past, tearing its necessary hole in the world: Harley's mouth yawning open, flashing teeth, snapping back at the source of what he thinks is the cause of all this pain.

Silas jerks his face away right as the bite's happening, right when Harley's teeth have hooked into the skin of his cheek, but that just makes it worse, really.

TUESDAY

Lewis is using little short tear-offs of masking tape to outline a certain dead animal on the carpet of the living room floor. It's to prove that it couldn't have happened, that she wouldn't have even fit right there. That's what he's telling himself anyway.

He's got the couch shoved back, Peta's grandmother's antique coffee table pushed the other way. Peta's family isn't old-money Great Falls—is there any such thing?—but they've been here in one way or another since about when the original reservation was staked out.

She's in the garage with Harley, on the nest of sleeping bags and blankets she pulled together for him when she walked home from the make-do park-n-ride two streets over, found Lewis and Shaney and Eldon on the back porch dribbling water into Harley's mouth. Jerry was gone in the truck, delivering Silas to the hospital, his face packed in towels.

After they'd pulled away, Jerry driving easy, one hand on the wheel, the other keeping Silas upright, Eldon said it figured that a mailman would get it from a dog, right?

Right.

According to Peta, who spent most of her childhood nursing dogs and cats and baby birds, Harley could still go either way. Silas was never in that kind of danger—though, before he left, Lewis could see yellowy teeth through the flapped-open cheek skin.

Jerry says Lewis shouldn't hold it against Harley. He didn't know what he was doing. When the whole world hurts, you bite it, don't you?

Harley's nest of sleeping bags and blankets were meant to be the insulation around the sweat lodge Lewis had planned for the backyard, but screw it. Maybe they still will be. Maybe, next year, wrapped in heat and darkness and steam, Lewis will dip some water out of the bucket and tip a little out for Harley. In memory of, all that.

You can do it for dogs the same as people, he's pretty sure. And, if not, some old chief gonna step down out of the sky, slap his wrist?

Lewis tears off another longish rectangle of masking tape, sticks it to the carpet in front of the couch, then peels it up and sticks it again, trying to get the slow turn down from the belly to the front of the back leg just right. Thing is, these re-stuck sections of tape all curl up after a few minutes, like retracting from the shape Lewis is forcing them to be part of.

The rear hoof is just starting to come together when Peta steps back in with the dishrag over her shoulder, the bottle of goat milk in her hand, and for a slice of an instant she's a mom, tired from one in diapers, one just balancing around on wobbly legs. But that's another life than this one, Lewis reminds himself. She doesn't want kids, was up-front about that even those first couple of weeks in East Glacier. Not because Lewis is Indian, but because she thinks her pre-Lewis self made enough bad decisions of the chemical variety that any kids she had would have to pay that tab, so they'd be starting out with the world stacked against them already.

The headline kicks up in Lewis's head on automatic, straight out of the reservation: not the FULLBLOOD TO DILUTE BLOODLINE he'd always expected if he married white, that he'd been prepping himself to deal with, because who knows, but FULLBOOD BETRAYS EVERY DEAD INDIAN BEFORE HIM. It's the guilt of having some pristine Native swimmers—they probably look like microscopic salmon, even though the Blackfeet are a horse tribe—it's the guilt of having those swimmers cocked and loaded but never pushing them downstream, meaning the few of his ancestors who made it through raids and plagues, massacres and genocide, diabetes and all the wobbly-tired cars the rest of America was done with, those Indians may as well have just stood up into that big Gatling gun of history, yeah?

"How's he?" Lewis says, tipping his head to the garage.

"I think it's helping," Peta says, holding the goat milk up.

According to one of the luggage guys at the airport, you can bring a parvo puppy back with goat milk. Harley's not that kind of sick, but if goat milk can keep a puppy alive when its insides are turning to slurry, then surely it can do something for a dog that spent most of yesterday dying and then coming back, right?

It makes as much sense as anything.

At some point, though, and Lewis hates hates hates this, at some point, and soon, it's going to come down to a rifle, and Harley's last walk, or carry, whatever.

It won't be because Harley was a bad dog. It'll be because he was the best dog.

It'll have to be the same rifle from ten years ago, too. He'll drive up to the reservation to bum it off Cass, even—it's the one he used for that young cow elk. The cow elk he's tracing out on the carpet with a hundred torn-off pieces of masking tape.

"Need some help?" Peta says about this little project.

Any other person, any other woman, any other wife of a stupid husband who's trying to hide from his dying dog by outlining an elk on the living room floor in masking tape, she'd tell him to quit messing her house up, to quit wasting tape, to be sure and clean up every bit of that when he's done.

Peta works her way down beside Lewis, takes the roll of tape, tears off squares, and holds them up on her fingertips for when he's ready.

Her theory about what he saw is that, the same way you can put lights on the spokes of a bike and they'll gel into a picture at speed and hold that blurry glowing image, there must have been some random pattern of light and dark dust on the back of the fan blades. They produced a kind of blob in all that spinning, and Lewis just took it to his guilty place: that young elk.

About which, he hasn't told her the whole story.

She's vegetarian, and not for health reasons, but ethical ones. More nights than not, he's eating potatoes or tofu or beans. And that's fine. Every middle-aged Indian needs a diet exactly like this. So, Peta *would* listen to the whole story, sure, and make the right noises, hold her eyes in a way that meant she was getting it, but it would hurt her to hear it, and she'd have to go down to the high school, run around and around the track to try to stay ahead of that story. It's better not to tell her all of it, then, not to burden her with it, scar her memory up. Who knows, even? She might just stand up after hearing it, walk away, not come back.

Twenty minutes later, maybe an hour, Lewis has the shape of the cow elk more or less roughed out, emphasis on the "rough."

He stands to see it from higher up, and has no clue how the bow-and-arrow Blackfeet did it back in the day. The horses they drew into ledgers or onto the sides of lodges weren't anatomically accurate—neither's this—but they did suggest a sort of intimacy with the shape, with the form, that this masking-tape elk doesn't even come close to. It's more like somebody told Lewis about an elk than that he ever saw one in real actual life.

Peta covers her mouth with her hand to keep from laughing, and Lewis has to smile as well.

"Looks like a five-year-old tried to trace a giant sheep, doesn't it?" he says. "While he was working on his third beer of the morning."

Peta collapses onto the pushed-back couch, pulls her legs up under her, adds, "But the sheep kept kicking, trying to get away."

"Sheep don't know anything about art," Lewis says, and falls into the couch beside her.

From this position, he of course ends up looking through the fan from the underside, and then at the little spotlight that's dead again up there. It's a mystery he's resigned to never solving. Some lights you never figure out, and shouldn't even try to.

"What next, then?" Peta asks.

For maybe thirty seconds Lewis doesn't answer, then, "It's stupid," he finally says.

"What?" Peta says. "You mean like climbing up a shaky ladder alone in the middle of the day and almost cracking your head open?"

Point.

After stopping to say hey to Harley, tell him that Eldon's covering the morning shift, Lewis walks the tall aluminum ladder around the side of the house again.

"It was here," Peta says, positioning the ladder just shy of directly under the fan.

"How can you tell?" Lewis asks.

To show, she pivots around to the other side, braces her feet wide to lower the ladder down until the red plastic cap at the top fits perfectly into the wedge slammed into the wall on the other side of the living room.

"Oh," Lewis says. "Think we're getting the deposit back?"

"Security deposits are overrated," Peta says back, and maybe she really is Indian, right?

"Wait," Lewis tells her, and retreats to the garage, comes back from the chest freezer with the trash bag that's gone-with through six rental houses and one never-finished basement.

It's maybe going to smell. But maybe not.

"We still have that?" Peta says.

Lewis tries to open the bag but it's more like peeling a plastic tamale, the bag's so old. Inside, kind of making his heart swell, is the hide he promised that young elk to use someday, to make everything she went through worth it.

The story he told Peta was that it was snowing thick, and she'd looked like a full-grown cow, not a teenager. That he never would have pulled the trigger if he'd seen her right.

It's not a complete lie. Just, it's not the complete truth, either.

Lewis swallows the memory down, gets back to whatever this is he's doing: re-creating the scene of the crime? No. More like staging the accident all over again. With, this time, props.

"Is it still . . . ?" Peta asks about the tight bundle of rolled elk hide, hair still on.

Lewis shrugs, doesn't know about the hide, if it's still all one piece, or crumbly. It has a lot of nicks and holes, he knows, because, first, he's a crap skinner, but second, that trading-post knife he was using only held its edge for like three minutes.

Should he thaw it before unrolling it? Would the microwave work? Would he ever be able to eat anything warmed in there again?

"I'll just—" he says, and ceremonially sets the hide down in the middle of the masking tape. It looks like a fat, hairy burrito, and Lewis has to focus to keep from coughing, because that'll turn into a gag, and he doesn't want to be rude to her memory.

"That's probably good enough," Peta says, sitting back and eye-balling the hide, the tape, the whole setup.

"Well, then," Lewis says, one foot already on the lowest rung, one hand up higher.

"The fan's at the same speed?" she asks.

"I haven't messed with it," he says. "You?"

She shakes her head no, nods for him to go, that she's watching.

"I was on this rung," he narrates, using his hand to touch where his foot was, and then he's going up, up.

He waits until the spinning blades of the fan are at his chest again to look down through them. At Peta, on the couch. At a dead elk made of masking tape, with a hairy burrito for a gut sack.

"Maybe it's the light," Peta says, and unfolds from the couch, backs into the edge of the living room, where she was standing when Lewis started his big slow-motion fall. "Am I making a shadow?" she calls up to him. She turns the hall light on and off behind her, keeping her feet in the same place.

"You had a bag," Lewis tells her, still holding on to the chance that this might work, that there might be an explanation.

"O-kay . . ." she says, not as confident in this bag-possibility as he is, but all the same, she bounds into the kitchen to dig one up.

While she's gone Lewis looks over the top of the fan, at the gouge the ladder left in the wall of the living room. The new wound in the house.

Moving there like an afterimage, like it was left behind, is just trying to creep past without being seen, he's ninety percent sure there's the shadow of a person up against that wall. A thin shadow, just for a flicker of a moment.

A woman with a head that's not human.

It's too heavy, too long.

When it turns as if to fix him in its wide-set eyes, he raises his hand to block her vision, to hide, but it's too late. It's been too late for ten years already. Ever since he pulled that trigger.

WEDNESDAY

What wakes him the next morning is . . . a basketball? Dribbling?

Lewis rolls out of bed and into the closest sweats, has to hold them up with his left hand all the way down the stairs—the dryer ate their drawstring back when they were brand-new.

There's definitely someone dribbling a basketball in the driveway.

Lewis steps down from the kitchen into the garage, says to Harley, "Who is it, bub?"

Harley thumps his heavy tail once against a Star Wars sleeping bag but that's all he can manage.

A neighbor kid, maybe? Did the former tenants tell all the kids on this street that they can come over anytime, shoot hoops?

If so, cool. Lewis needs someone to play with who's at his skill level. Playing against Peta—doing anything athletic with her—is a study in shame, pretty much. Even grabbing her waistband when she slides past, pushing her in the back when she's laying it up, he never can hit twenty-one before she does. He can never even get to *ten* before she wins.

Lewis bumps the garage door button with the side of his fist, his face pre-hard because that's what you do in what could be a trespassing situation, what could be the former tenant, drunk, weaving back to the home he sort of remembers.

Slowly—it's an old, heavy door—the sneakers out there become legs, then shape into a woman, then become . . . *Shaney*?

She spins around, making room against an imaginary defender, and comes back, rises into a fallaway that scissors her legs in the air, the back one touching down just as the ball banks in, smooth as butter. She shags the rebound, claps the ball between her hands like heads-up, and passes it across, a clean bounce right on target.

Lewis catches it because the other option is getting popped in the gut with it.

"I wake you, early bird?" she says like a challenge.

"Day off," Lewis says.

"To spend with him," Shaney says back, going to Harley now that the door's open.

She cups his wide head in her hand, draws her nose to his, and squeezes her eyes shut, keeps them like that.

"You smell it, don't you?" Lewis says.

"He's dying," she says, massaging his notched ears.

She rolls into a sitting position on the unassembled sweat, says about Harley and all his scars, "He's an old warrior, isn't he?"

"You come just to see him?" Lewis asks, trying not to make it sound confrontational. She hears it anyway.

"Your wife wouldn't want me here, right? White girls of red men are always the most jealous of my kind."

"Your kind?" Lewis says, though he kind of already knows.

"Indian, unattached, an ass like this," Shaney goes on. "I know Jerry says I'm bad news."

The headline back on the reservation: BASEBALL BASEBALL BASEBALL.

"What's her name about anyway?" Shaney asks. "She a white-girl tortilla, or all against wearing animal skins, what?"

"Peta with an *e*, not an *i*," Lewis recites, falling through Peta's own explanation. "She was supposed to have been a boy, her dad's name is Pete, so he put an *a* on his own name, handed it down."

Shaney nods like she can track that, sure, and when she threads her bangs away Lewis clocks that her left eye's all bloodshot, and that—has he ever even seen her forehead?—the skin above her eyebrow on that side's drawn tight and bumply, like from sudden contact with a dashboard, or an aerosol can exploding in a burning pile of trash.

The eye, though. Bad date last night, Lewis has to think. Either that or the wrong boyfriend. He doesn't ask, tries not to be too obvious about looking. Which pretty much means he telegraphs his thoughts word for word across to her, he knows.

"Anyway, I came over for a book, Mr. Library," she says, shaking her hair back over her forehead and eye. "Not to jump your bones. Call her up, tell her that, I'll wait. My day off, too, yeah?"

Lewis looks at her about this, about the book thing, because usually this kind of lead-in is the setup for some joke. Reading about wizards and druids at the mall, or werewolves and vampires being detectives, it doesn't exactly bump a thirty-six-year-old's cool meter up. And if anybody knew centaurs and mermaids are sometimes part of it? Or demons and angels? *Dragons?*

Keep those book covers folded all the way back, Lewis knows.

Except here's a girl actually asking to see them.

Even Peta doesn't really understand the fascination, the compulsion, the draw. How, camping, he always tucks a paperback or two in his pack, each inside its own separate ziplock bag. She's a super-athlete, though. She was always running too fast or jumping too high to pick up reading. It's nothing bad about her.

Keep saying that, Lewis tells himself.

Keep saying that and dribble out from under the garage, into the bright open sky. It's that kind of November day.

"Anything particular?" Lewis says back to Shaney without looking, all his fascination on the rim so he can come up onto his toes to shoot. What he has planned is a trip to the bank, to show off, to match the shot she just made look so easy, but then he has to collapse that idea at the last instant to keep his sweats on his body. Nothing underneath.

"Nothing I haven't seen before," Shaney says. "Tall Indian choking on the court, I mean."

The ball's bouncing through the junk lumber behind the goal. Lewis picks through barefoot to retrieve it, finds an even worse way back to the concrete pad.

"Court intrigue or heroic quests to save the realm?" Lewis asks. "Ships or horses, elves or—"

"I don't know, something *exciting*," Shaney says. "First in a series, maybe? Nice long series. Something to keep me busy all night."

Can she ever just talk about one thing?

"You being serious?" Lewis asks, chest-passing the ball across to her slow enough that she rips it from the air, like disgusted with the weakness of that lackluster pass. Her billowy T-shirt snags the grabby ball when she whips it across her midsection, though, so, pissed off about that, she bounces the ball high and wrenches her arms around behind her to tie her shirt, take up the slack, then catches the ball back. Peta, in this situation, usually tucks the front of her shirt up under her sports bra, but that isn't really an option for Shaney today, Lewis can tell.

"Oh," she says, following the look Lewis doesn't mean at her stomach.

It's a long ragged scar up and down, not side to side and low like a C-section. It's an open-heart-surgery scar, just, too low for the heart, and with an ugly, uneven ridge of scar tissue. Is this and whatever happened to her forehead and eye a matching pair? One really bad night instead of a lot of pretty sucky ones?

Lewis wants to ask about that car crash, or ask if the baby made it, or if they got the guy did this, except what if she was the only one to make it through that wreck? What if the baby didn't make it? What if that guy's still out there, carrying no scars himself?

"Say it," Shaney says about the scar, "go ahead, I've heard them all. Did I go to the emergency room or the butcher?"

She axles the ball between her two bird-fingers, rotating it with her thumbs, keeping it between Lewis and her midriff . . . as much bluff as she has, she still doesn't much want him looking, he can tell.

"Can hardly even see it," he one hundred percent lies. "Did— did everything . . ."

Her eyes flit up to the goal, and her non-answer is all the answer he needs, and her story comes together in his head, glued together from all the other stories he knows: she was young, the emergency room doc was a reject from the American medical system, so she ran from that tiny grave as far as she could, which ended up being about one tank of gas away from her reservation.

"Sorry," Lewis says. Not for seeing, but for whatever happened.

"We're from where we're from," she says back. "Scars are part of the deal, aren't they?"

Lewis steps out onto the court proper, wading into this game.

"So you really want a book?" he asks, still sure this is some complicated joke.

"I *read*, yeah," she says like insulted, shrugging one shoulder, dribbling up to him and then turning around in invitation. One thing all guy ballers can learn from how the girls play, it's that trick right there: giving the defender your ass, so you can protect the ball, slash out either way around them. Problem is, guys always think it's an ego thing, that it's a bigger coup to face up, lock eyes, then juke them any the hell way. And maybe it is. But the guys get their pockets picked more, too.

Shaney presses right up to Lewis, dribbling far from her body so he can't reach around.

This would be a bad time for Peta to come strolling up, he knows. He might as well be leaning around a barely halter-topped girl in the bar who's pretending not to know how to play pool. Peta won't be walking up, though. She's not off work for hours yet, and even then, it's a ten-minute hike in from the park-n-ride, her work duffel slung over her shoulder, her ear protection slung around her neck, the world so quiet to her, probably, after planes throttling up around her all day.

Peta.

Lewis vows to keep her name in his head for the next few minutes.

Shaney leans right like she's going to use her left to throw the ball ahead, a long dribble that gets her into layup territory if she glides a bit, underhands it, has that kind of touch, but then she's twisting left already, and Lewis, like always, like with Peta, falls for it. Shaney eels past, has had a coach really make her get her footwork right, and the net's already spitting the ball down.

"I had to hold my pants up," Lewis calls out.

"No you didn't," Shaney says, and bounces the ball into the garage, well away from both Harleys: the dying dog and the parked Road King. "Now, book me, officer."

It takes a second for Lewis to get that. And he's fully aware she just smuggled "handcuffs" into his head. He leads her in all the same, one hand clutching his sweats, and when he makes the turn up the stairs, Shaney's still back at the kitchen table.

"Blackfeet?" she's saying.

She's almost touching the rolled elk hide on the table, and she either just said her name for Lewis, or she's asking if this hide is from the reservation.

"What?" Lewis says, stopped with his non-sweats-hand gripping the newel post at the turn upstairs.

"I didn't know," she says, looking across to him with new eyes. "You're a—you're a bundle holder? They let it come all the way down here?"

From the look on his face, she explains: "It's like a pipe holder. Just, with a bundle."

"Oh, that's just—" Lewis begins, doesn't finish. "I didn't really grow up traditional."

"Think tradition found you just the same," she says, impressed, and almost touches the outermost brown hairs, then draws back like afraid of what might happen, what might pass from this Blackfeet bundle to her Crow self.

It's just an elk hide, Lewis doesn't say. Mostly because now she's drifted over to the couch, can see the masking-tape insult to all elk on the carpet of the living room. She looks from it to him then back, and, without saying anything, she's there, has the masking tape, is tearing off a few long strips, affixing them to the side of the couch. They look like long, careful shavings of wood, curling up.

Lewis doesn't say anything, just steps over like caught, a hundred possible explanations swirling through his head, all of them built to fail.

Moving deliberately, Shaney applies the long strips to the carpet, not adding to the elk, but giving it some insides—that inward-going tube of an arrow Lewis has always seen on lodges and in ledgers, that ducks back from the mouth to the stomach, for reasons he's never had clue one about. Why would the esophagus and stomach be more important than the heart, the liver?

"Now it's right," Shaney says.

It is. It was a smushed sheep before. Now it's . . . not so much a young cow elk, but a shape that somehow represents a young elk better than even an actual young elk, lying right there.

"How'd you know?" Lewis asks.

"You asking that because I'm a girl?"

"It was just a blob with legs," Lewis tells her.

Now she's looking to the ladder, to the nothing happening on the ceiling.

"My books are—" Lewis tries, but this isn't a library visit anymore.

"Why do this here, by the couch?" she asks just generally, coming back around to fix him in her thirsty eyes. She opens her hand to the masking tape elk, leaves her fingers spread like that.

"Wherever I did it, that could be the question," Lewis says, stalling.

"But you did it *here*, not anywhere else," Shaney says back, not pushing this time, but eliciting.

"It's stupid," he says, sitting down on the third step of the stairs. "Just something I thought I saw the other day."

She leans back onto the arm of the couch, her eyes still locked on him, says, "Which was?"

"It's not like in the books," Lewis says. "When you—when you see something that doesn't fit, like."

"Like a werewolf digging through your trash," she completes for him, hauling his current book up from the coffee table and showing him the cover, which is . . . a werewolf digging through a dumpster, trash strewn all over the alley.

Lewis nods, even more caught, his hands cupped over his mouth, his breath hot on his palms.

Is he really about to tell her? Does the hot girl from work get to know what his wife doesn't?

But she knew how to finish that elk on the floor, didn't she? That has to mean something. And—Lewis hates himself for saying it, for thinking it, but there it is: she's Indian.

More important, she's asking.

"It was the winter before I got married," he says. "Six—no, *five* days before Thanksgiving, yeah? It was the Saturday before Thanksgiving. We were hunting."

"We?" Shaney prompts.

"Guys I grew up with," Lewis says with a shrug, like they're not the real focus. "Gabe, Ricky, Cassidy—*Cass*."

Shaney nods like he's doing good so far, and looks over to the masking tape elk again, kind of for both of them it feels like, and then Lewis is talking, is confessing, is saying it all out loud for the first time, which must mean it really happened.

THAT SATURDAY

The sky was spitting these hard little snowballs that kept catching in Lewis's girly eyelashes that he always thought were maybe just normal eyelashes.

"Wearing mascara now, princess?" Gabe asked all the same, bumping over into him. "Gonna bat your eyes, bring all the big bulls to your door?"

"You should talk," Lewis said, lifting his chin to Gabe's own frosted eyelashes.

Off-rez, people always used to default-think that Lewis and Gabe were brothers. Gabe, at six-two, had always been a touch taller, but otherwise, yeah, sure. In John Wayne's day Lewis and Gabe would have been scooped up to die in a hail of gunfire, would have been Indians "16" and "17," of forty. Cass, though? Cass would have been more the sitting-in-front-of-the-lodge type, the made-for-the-twentieth-century type, maybe even already wearing some early version of John Lennon shades. Ricky, he'd be Bluto from *Popeye*, just, darker; put him in front of a camera, and all he could hope to play would be the Indian thug off to the side, that nobody trusts to remember even half a line. Of Lewis and Gabe and Cass, though, he was the only one who could struggle out a sort-of beard, if he made it through the itchy part, and didn't have a girlfriend at the time. "Custer in the woodpile" was the excuse he would always give, smoothing his rangy fourteen hairs down along his cheeks like Grizzly Adams.

Gabe leaned across to Lewis, making smoochy lips, saying, "A little flirting would probably work better than what we're—" but then Cass, ahead of them at the truck, raised his left hand, silencing them.

"What you got?" Ricky asked, coming back.

He was always ranging out to the side, sure they were just missing a whole herd, that all the elk were single-filing it past just out of sight, ducking their heads down so their racks wouldn't crest over the snow.

"Shh," Cass said, coming down to one knee to read sign like a real Indian.

Tracks.

Elk had been nosing into the bed, probably remembering that some trucks carry hay, and hay never gets *all* the way gone. Not without elk that are tall enough to lean over the side of the truck, that have long enough necks to even get under the toolbox for every last straw.

"Heavy guys," Gabe said, lowering down to insert a trigger finger into the deep hoofprint. He had some complicated method where a bull weighed this much if it was up to his second knuckle, that much if it was halfway past that, but Lewis never bought it.

"Told you they were up here," Ricky said, looking all around like these elk might be turned around at the tree line like a stupid whitetail, to twitch their tails and watch.

"Up here" wasn't *high*-high, snowmobile or horse country, but halfway there, anyway, just down from Babb, over toward Duck Lake. With the weather moving in, the elk should have been filing down from the timber, to wait the big snow out. The idea was to meet them halfway.

"This is some *bull*shit," Cass said, his usual call, and Ricky responded with his obligatory line, "Literally," toeing over a fresh black mound, the pellets more tapered at one end, not both. Nine times out of nine, that'll mean "bull," not "cow."

"They're playing with us," Gabe said, reseating his rifle strap on his shoulder.

"Catch me if you can," Lewis said for the bulls and then lined up on the walking-away tracks, his eyes going downhill with them, downhill, to—

"Shit," Cass said, turning around to kick snow.

"They know," Ricky said with a chuckle, impressed.

"Tricksy, tricksy . . ." Lewis said, smacking his gum too loud, and Cass cut his eyes over at him, not sure he'd heard that word right but not wanting to ask, either.

Gabe didn't say anything, just kept watching where the big bulls had gone—where they *were*.

"Anybody pack some grey braids in with their bear kit?" he finally said with his trademark grin, the one that usually ended up either getting beat in by the end of the night or looking out through bars. Sometimes both. A hundred years ago he would have been the guy always trying to get a raiding party together to sneak over the line, have some fun, come hell-for-leather home in the morning with half of America massed up right behind.

"*No*, man," Ricky said, his eyes hot so he could really mean this, really drill it in. "If we get caught over there, it's—"

"Then let's not get caught, what say?" Gabe said, looking from face to face like polling a jury.

"We can't," Lewis said to Gabe about the off-limits section. "Ricky's right, if Denny catches us again, then he'll—"

"It's not fair though," Cass whined, flicking something off the end of his finger and watching it fly. "That section's reserved for elders, but what if none of the elders are even hunting it, right?"

"Old guys get up early," Gabe threw in, like just seeing this brilliant point. "If they were going to hunt their section today, they'd already have been and gone. We'll just be cleaning up the ones they weren't going to shoot. No big. Cassidy's right."

"Cass," Cass said.

"Whoever he is today, he's right," Gabe corrected, setting his feet to take Cass's elbow.

It wasn't that the elders' section was all the way off-limits, it was that only elders—plus one and only one—could use trucks to get in and out. Anybody younger was supposed to hoof it, which would be a two-hour walk in at least, and it was already an hour and a half after lunch, with the sun going down just after four, and taking the thermometer with it.

"Elders aren't the only ones with empty freezers," Cass said with an obvious shrug. "Anyway, it's my truck. You three bail, I take the heat."

When Ricky didn't say anything, Lewis just looked away, down to the elders' section again.

It *was* some good-ass country around Duck Lake, no two ways about that. And Gabe knew every logging road, every two-track, every old game trail that'd been widened out by four-wheelers and chain saws. And it does suck to be the only Indian without an elk.

"Last day of the season . . ." Gabe pled to all of them.

Technically it wasn't, but it was the last time they could come out for a whole Saturday like this together. There would still be lunch breaks on their own, though, eating and driving down some road somebody maybe saw an elk walking alongside. There would still be being late *to* work because of a set of deep tracks crossing from ditch to ditch. But Lewis heard what Gabe was saying, what he was arguing: the last day of the season, the rules are different.

Anything goes. Whatever fills your freezer. You've put in enough days out in the cold and the snow that you feel like the elk owe you, almost.

Included in that are any moose or mulies you might jump along the way.

"Shit," Lewis said, because he could feel himself starting to cave.

"That's back where you found Junior, enit?" Cass said to Ricky, but Ricky was watching the trees again, always seeing an ear twitch where there were no ears.

Cass was talking about when Ricky found Junior Big Plume floating facedown in Duck Lake, and was reservation-famous for the weekend.

"Shut up," Ricky said, his hunting face all the way on, which was pretty much just a cigar-store Indian mask. Still, Cass let it drop.

Gabe took advantage of the silence to take a long read of all the faces, all the eyes, all the weak, weak spines. "Well, the elk aren't going to shoot themselves, gentlemen," he finally said, satisfied with what he'd seen, evidently. He hauled his rifle around to clear the chamber, Cass's rule since the new hole in the front floorboard of his truck, the hole Gabe insisted Cass would thank him for come summer, which is right where Lewis would like to freeze-frame that day, just stop it completely, hang it on the wall, call it "Hunting" or "Snow" or "Five Days Before Turkey and Football."

But he can't. The rest of the day was already happening, had already pretty much happened right when Gabe kept looking down-hill, to where he said the elk were.

"Was he right?" Shaney asks, her legs tucked under her to the side like a traditional.

Lewis chuckles a sick chuckle, says, "About the elk not shooting themselves?"

Shaney nods, and Lewis looks away, says Ricky was right too.

"About what?" Shaney asks.

"Getting caught."

Since Cass's squarebody crew cab didn't have a winch, each time he couldn't tell where the road was and slogged out into the soft stuff, everybody had to pile out again, take turns on the stretched-out come-along, the other two digging with planks and trying to do some magic or other with the jack, one person behind the wheel to feather the accelerator and work the shifter, keep the truck rocking back and forth.

Four separate times at least, certain death loomed, but either that wobbly high-lift sliced down into fluffy snow instead of crunchy skull, or the come-along hook snapped back over the cab of the truck, instead of through any faces.

It was so funny even Lewis was laughing.

It didn't feel like anything could go wrong.

Sure, yeah, he wanted an elk and wanted it bad, but all the same, this was what hunting is about: you and some buds out kicking it through the deep snow, your breath frosted, your right-hand glove forever lost, your Sorrels wet on the inside, Chief Mountain always a smudge on the northwest horizon, like watching over all these idiot Blackfeet.

At least until they got to where it happened.

It was a steep hill, maybe a half mile in from the lake. The big snow was already crowding in, pushing the wind ahead. That's the only thing Lewis has to explain how the elk didn't hear Cass's Chevy struggling through the snow. The squirrels had been chattering about it, the few birds that were still out were annoyed enough to glide to farther-and-farther-away trees, but the elk, maybe because of that wind in their faces, they were oblivious, just trying to chomp whatever they could, since it was all about to be buried.

Looking back, Lewis tells Shaney that the one thing that could have maybe saved them was some horses, the wild ones that were always showing up in the least expected places all over the reservation, their eyes wide and crazy, their manes and tails shaggy and tangled. If four or five of them had pounded through on some important horse mission or another, that might have spooked the elk, or at least got them listening closer, smelling harder, paying better attention.

But there were no horses that day. Only elk. What happened was the same thing as had been happening the last half mile: Cass lost the road again, in spite of Gabe guaranteeing that it turned here, here, *here*. Instead of trying to back up and find the road again, though, Cass drove *into* this wrong direction, his foot deep in the pedal, the wheels already churning for purchase, the only thing keeping the truck going its own sagging momentum.

"Going for the record, going for the record here . . ." Gabe said, lifting his butt off the seat like he was the thing weighing them down, and, in the back, Ricky rocked forward, trying to help the truck along. Sitting beside him, Lewis wondered what the penalty was just for *being* in the elder section. But he knew: nothing, so long as you're not rifled up. If you are carrying, though? Denny throws the key away.

"We're gonna make it, we're gonna make it!" Cass said, one hand on the wheel, the other to the four-wheel-drive shifter to tap the transfer case into high should they be so lucky as to need it. What he was doing, not exactly on purpose, was driving from one part of an S-curve in the road to the other, snow flying every which direction, the tires spinning great white rooster tails of it up and over, some of it probably not even landing, just hitching a ride on the wind, to sift down over Cutbank or Shelby—

somewhere so far away from this as to, right then, just be a legend, pretty much.

"Shit, shit," Lewis said, hooking a second hand through the grab strap, straightening his legs against the floorboard even though he knows that's the wrong way to take a jolt. It was just instinct to brace himself, though. Three times already they'd barely missed a lichen-shadowed boulder left behind by some glacier twenty thousand years ago. They had to be owed one right in the grill-teeth sooner or later, right?

Instead of that granite stop sign, what they almost drove onto and down *into* was open space.

Cass didn't have to hit the brake, he just had to stop gunning the truck forward.

"What the hell?" Ricky said, not able to see from the backseat. Lewis, either.

The engine sputtered out, dropping them into a vast silence.

"Good one," Cass said, disgusted, trying to clear his side of the windshield, finally cranking his window down instead, and Lewis was just thanking any gods tuned in right now: they should be a smoking wreck at the bottom of this drop-off.

"Shh, shh," Gabe said to them all then, and leaned forward over the dash, looking down and down.

And then.

"What?" Shaney asks.

Then Gabe reached over for his rifle, his fingers coming into place on the pistol grip one by delicate one, like all four at once might be too loud.

What Lewis remembers clearest about the next sixty seconds, maybe closer to two impossible minutes, is the way his heart clenched in his chest, the way his throat filled with . . . with terror?

Is that what too much joy and surprise can ball up into, when it comes at you all at once?

There was the instant sweat, his head full of sound, his eyes letting in too much light for his head to process. It was like . . . he doesn't have words for it, really. "That fight or flight rush," he tells Shaney, only, running wasn't even a distant option. It was what he'd always imagined war to be like: too much input all at once, his hands acting almost without his say-so, because they'd been waiting for this moment for so long, weren't going to let him miss it.

Gabe either.

He popped the handle of his door, rolled down into the snow smooth as anything, his rifle whipping out after him.

Following his lead, nobody said anything, just fell in, Ricky coming out his door, Cass trying to jam the truck into Park so it wouldn't roll over the rock lip it was teetering on.

The door on Lewis's side opened like a whisper, like fate, and when he committed his right foot down to the powdery surface that ended up being two feet deep, he just kept falling, his chin stopping a hand's width into the powder the front tires had churned up. His forward motion never faltered, though. He crawled ahead like a soldier, pulling with his elbows, his rifle held ahead to keep the barrel clear.

And—that was when the frenzy washed over him.

He'd seen big herds in the Park, over at Two Dog Flat, had seen them in spring over by Babb, bounding across the road at night, but this many huge perfect bodies against all that stark white was something he'd never seen this close before. At least, not with a rifle in his hands, and no tourists around to snap pictures.

Gabe's rifle going off was distant, was down at the other end of some long, long tunnel.

Lewis, knowing that this was how you got to be a good Indian, finally remembered how to jack a round in. Once it was seated, he pulled that Tasco up until it cupped his right eye, and he was firing now as well, and firing again, just waiting to pull the trigger until he could see brown in the crosshairs. Just anywhere *near* the crosshairs—how could he miss?

He couldn't.

Three rounds, then he was rolling over, digging in his pants pocket for shells, and the elk, trained in the high country, the sharp drop in front of the truck throwing the sound every which way, their first instinct was to crash uphill, to what was supposed to be safety.

On the other side of the truck Ricky was screaming some old-time war whoop, and Gabe maybe was too, and so was Lewis, he thinks.

"You couldn't hear if you were or not?" Shaney asks.

Lewis shakes his head no, he couldn't.

But he does remember Cass standing behind his opened door, his rifle stabbed through the rolled-down window, and he's just shooting, and shooting, and shooting, only stopping to thumb another round in, and another, one of them launching onto the dash and clattering the whole way across, hissing down into the snow by Lewis.

"We could have fed the whole tribe for a week on this much meat," Lewis says, his eyes hot now. "For a month. For the whole winter, maybe."

"If you were that kind of Indian," Shaney says, getting what he's saying.

"There's more," Lewis says, finally looking to the masking-tape elk on his living room floor.

In the hollow deafness after all this, the four of them stood there on that rocky ledge, the snow skirling past, the weather almost on them, and Gabe—he always had the best eyes—counted nine huge bodies down there in the snow, each probably pushing five hundred pounds.

Cass's Chevy was a half-ton.

"Shi-*it*," Ricky said, breathing hard, smiling wide.

This was the kind of luck that never happened, that they had only ever heard about. But never like this. Never a whole herd. Never as many as they could bring down.

"Okay there?" Gabe said across to Lewis, and Cass reached up with the side of his finger, dabbed at Lewis's right eye.

Blood.

Before, when he'd had scope-eye as a kid—when the scope had recoiled back into his eye orbit—he'd felt that shock wave move in slow motion from the front of his head to the back. It makes your brain *fluid* for a slowed-down moment, leaves you scrambled, and because of that you can never remember what exactly it is you're doing to make this scope-eye happen. Except the obvious: pressing it right up to your eye, pulling the trigger.

This time Lewis remembered every shot, the lead-on-meat slap of every slug, but never even felt the force of that sudden recoil going from front to back through his head.

Five years after this a dentist will look at his X-rays and trace out the bone evidence of this trauma around his right eye and ask was it a car wreck, maybe?

"Almost," Lewis will tell him. "But it was a truck."

The last time he saw Cass's Chevy, it was up on blocks by a barbed-wire fence over at a high place north of Browning, the windshield caved in, the hood yawning open like a long scream. The engine must have been good enough, otherwise it would have still been there. The wheels and tires had been yanked as well. After Lewis left that part of the country for what he secretly knew was going to be forever, the first cinder block holding that truck up went soon enough, he imagines, a rust-coated brake drum crumbling down through that stony grey, making the truck look like a horse kneeling, and after that it would have been fast. The land claims what you leave behind.

That day with all the elk, though, back then the Chevy was still on its first or maybe second life, was young and hungry, was telling the four of them it could carry as many elk as they could pile in. Realistically, even just three elk in the back of a half-ton is pushing it, is going to have that truck sitting down on its springs, the nose pointing at the sky, the front brakes useless.

And that was if the stupid come-along was going to cooperate, help get those heavy bodies up the slope, and if four Indians without gambrels or cherry pickers could somehow get the second and third elk in on top of the first.

"And that's when it started snowing heavy," Lewis tells Shaney, touching his face with his fingertips like feeling those cold dabs again.

She doesn't say anything, is just watching, soaking all this in. Not because she wants to know, Lewis doesn't think, but . . . is it more like she knows he needs to say it? To have told *some*one, at least?

"Old-time buffalo jump!" Gabe called out then, and he vaulted right the hell off the ledge, slid on his ass down into that jumble of dead and dying elk.

Lewis and Ricky and Cass picked their way down after him, unsheathed saws and knives, and got to work. Inside of five minutes it was clear this was going to be a haunches-only affair—already big wet flakes were finding their way into the red body cavities, dissolving instantly against that steamy heat. But pretty soon the flakes were going to start winning that little war, no longer melting but piling up, making these carcasses look like giant stuffed animals slit open, all their batting leaking out.

Gabe and Cass doubled up on the one bull that fell, trying to keep his cape intact, since Gabe knew a back-alley taxidermy guy who'd do a mount for meat, so long as he picked the cuts. Ricky was on a monologue about how this Thursday, Thanksgiving, was going to be an *Indian* holiday this year, with the four of them bringing in a haul like this.

"Thanksgiving Classic," Gabe said, giving what just happened a proper name.

Cass whooped once, setting that name in place.

Lewis whipped his hand over his head about the cow he'd just dressed out in record time—it's a rodeo thing, is deep in his DNA—and moved on to the next, the young cow, but when he dropped to his knees to make that first cut from her pelvis to her sternum, she found her front legs, tried to climb up out of the snow.

Lewis fell back, called over to Cass for a rifle. He never once looked away from this young elk, though. Her eyes, they were—don't elk usually have brown eyes? Hers were more yellow, almost, branching into hazel at the edges.

Maybe it was because she was terrified, because she didn't understand what was happening. Just that it hurt.

The shot that brought her down had caught her midway through the back, from the top, and taken her spine out. So her rear legs were dead, and her insides were going to be a mess as well.

"Whoah, whoah," Lewis said to her, feeling more than seeing Cass's rifle plunk into the snow just short of his right leg. He felt down for it, that young elk still struggling, blowing red mist from her nostrils, her eyes so big, so deep, so shiny.

"And I couldn't find a *shell*," Lewis says to Shaney. "I thought I was out, that I'd used my last up by the truck, when everything was crazy."

"But you had one," Shaney tells him.

"Two," Lewis says back, looking down at his hands.

This close he didn't need the scope or a sight.

"Sorry, girl," he said, and, careful of his swelling-up eye, lined the barrel up, pulled the trigger.

The sound was massive, rolling up the slope and then crashing back down.

The young elk's head flopped back like it was on a hinge, and she sank into the snow.

"Sorry," Lewis said again, quieter, so Cass couldn't hear.

But it was just hunting, he told himself. It was just bad luck for the elk. They should have bedded down with the wind in their favor. They should have pushed through to some section the hunters didn't have access to—that *trucks* can't get to, anyway.

After the shot, Lewis looked behind him for some buckbrush or something to hang the rifle from, but then a sound brought him back to the young elk.

The sound was the crust of snow, crunching.

She was staring at him again. *Not* dead. Her breath was raspy and uneven, but it was definitely there, somehow, when no way should it have been. Not after having her back broken, half her head blown to mist.

Lewis took a long and involuntary step back and fell, jammed the butt of the rifle down just ahead of his ass so he could be sure where the barrel was going to be, because he didn't want it driving up under his head, trying to separate his jaw from his face.

She was trying to stand again was the thing, never mind that the top of her head was missing, that her back was broken, that she should be dead, that she had to be dead.

"What the hell?" Cass called over. "My rifle's not *that* off, man."

He laughed, leaned back down into the big cow he was insisting was his. Lewis had his right leg straight out in the snow, was feeling in all his pockets for one more shell, please.

He found it, ran it into the chamber, working the bolt back and forward to be sure the cartridge seated right. This time, talking to the young elk the whole while, promising her that he was going to use every bit of her if she would just please *die*, he nestled the barrel right against her face, so the bullet would come out the lower back of her skull, plow into her back where she'd already been shot once.

Her one yellow eye was still watching him, the right one haywire, the pupil blown wide, looking somewhere else, someplace he couldn't see without turning around.

"So that's where I put the barrel this time," Lewis says to Shaney. "I figured—I don't know. That first shot must have glanced off her skull, right? Looked worse than it was. So this time I didn't want to give it any chance to bounce off. The eye could be like a tunnel in—*in*to her."

Shaney doesn't blink.

"You were gonna be a tough one, weren't you?" the Lewis of back then said, his lower lip starting to tremble, and then he pulled the trigger.

Cass's rifle bucked free of his one-handed grip and the young elk fell *again*, and what he was saying in his head, what he was telling himself even though he was Indian, even though he was this great born hunter, what he was telling himself to make this okay, to be able to make it through the next minute, and the next hour, it was that shooting her, it was just like putting one in a hay bale, it was just like snapping a blade of grass in a field, it was like stepping on a grasshopper. The young elk didn't even know what was happening, animals aren't aware like that, not in the same way people are.

"You believed that, too, didn't you?" Shaney says.

"For ten years," Lewis says back. "Until I saw her again, right over there."

"Still dead?" Shaney asks, sitting on the second step of the stairs now, her hand on the knee of his sweats, and that's how she is and that's how Lewis is when Peta walks in the front door.

FRIDAY

When the Thunderball Express slams past at 2:12 in the morning, Lewis's half-asleep mind turns those slamming wheels into thundering hooves, going up and down faster and faster into some muck, until he sits up hard from . . . *what*?

His knee-jerk thought is that the train threw one of those grey rocks into the fence, knocked another slat out, left it spinning on the cross-board like a cartoon, but it wasn't the train at all, he realizes. It was a . . . *chain*? He sits up when that word connects to what's right under the bedroom: the garage. The sound that shook him awake was the garage door chain in its long greasy track, the little motor grinding and pulling.

And the garage door would only be going up because someone pushed the button. And Peta's not on her side of the bed, is she?

Lewis sits up, his feet to the carpet now, his head trying to swim back to functional. When he can balance enough, he steps into those same useless sweatpants, feels his way across the bedroom to stumble down the stairs, the same ones Peta caught him and Shaney on. Doing nothing, but still, right? Lewis made sure Shaney left with an armload of books, a whole series, to prove to

Peta why she'd been there, but the whole time, stacking them up in Shaney's arms, it felt like an overcorrection, like trying to hide a body on the lawn by covering it with eight other bodies.

And Peta bought it, too, that was the thing. There was that moment when it all could have gone the other way, sure, but the reason it didn't, she told him later, it was his eyes. Lewis hadn't even been there on the stairs, not really.

All the same, Lewis knows it might have hurt Peta less for Shaney to have been just stepping back into her jeans. That would have been better than Lewis telling another woman something so intimate, so personal, so private. And that he'd been telling the elk story to another *Indian*, which Peta could never be no matter how fast she ran, no matter how high she jumped, that was maybe the final cut. The deepest, anyway. The one she was probably still nursing.

Thursday, the little him and Peta saw each other, she was cordial but not really herself. Not like she had something to say so much as she didn't have anything to say. And now she's gone from the bed at two in the morning, when she's a sleeper who treasures every minute she can get until that five o'clock alarm.

Coming down the stairs, holding on to his sweats, Lewis over-shoots the step-off at the bottom, falls ahead into the living room, tight-wiring it around the edge of the masking tape, and, because it's not on any human schedule, that stupid spotlight in the ceiling flickers. From Lewis's stumbling footfalls? Is *that* what makes it come on? Or are the aftershocks from the garage door what's flickering it on?

Before the mystery of that spotlight, though, there's the mystery of the wife.

Lewis pulls the door to the garage in kind of reverently, and the light built into the garage door motor is still on because it's only been about forty seconds since the door came up. Sitting on the pad of concrete out at the soft edge of that light, her knees to her chest, arms hugging them, white-blond hair cascaded down her back, is Peta in her sleep shirt, her ankle socks, her breakfast scrunchy around her left wrist, ready for cereal.

She's been crying. Lewis doesn't have to see her face to know. He can tell just from the curve of her back.

He steps down onto the cold smooth concrete of the garage to pick his way through to her, and that's when he sees it.

Harley, but not. Not anymore.

That good dog from his childhood, that got its head kicked in by a horse? That's the closest thing to this. To what's happened to Harley. Except that was just one fast kick, there and back in a snap, so nobody watching the parade even knew what had happened for a breath or two.

This—this is a dog that got ran down *by* a horse, and the horse had some serious score to settle, went down and came up over and over, slamming that dog into nothing, into a red chunky smear, teeth here, flash of bone there, matted fur all in and among.

The vomit's coming up and out of Lewis's mouth before he even realizes it. It's hot, thin, and already on his hands, like keeping it from splashing down onto the floor of the garage is suddenly the most important thing. Once he feels it stringing out between his fingers, he really gags, is shambling for the outside, losing everything under the basketball goal, his sweats around his ankles.

It's probably not a good look, but Peta's not looking, either. When it's over he yanks his stupid sweats up and pushes his forehead into the flaky paint of the basketball pole just to have something solid to hold on to.

"I don't understand," he says.

"He's dead," Peta says, kind of obviously.

Yeah, *but*, Lewis doesn't say.

If the door was only open *four inches*, like he's been leaving it for air circulation, then . . . then: "What could have done that?"

Peta looks over from her grief hole, says, "Should we call your coworker about it?"

Lewis deserves that, he knows. "Coworker" is what he was calling Shaney for the few minutes after hustling her out the door with her armload of books.

"No, let's not," he says, cleaning his hands in the dirt. "I don't even have her number."

Except, he realizes, he does, doesn't he? In the work directory thing, the brand-new one.

"What could have even done that to him?" he says, settling down beside Peta.

She scooches over like giving him room. Like there's limited space on this two-car-wide pad of dribbled-in, oil-stained concrete.

No: like she doesn't want to be touching him.

"It wasn't his fault," Peta says, staring off into the nothing all around, "he was just a dog," which—is that an answer, really?

Without meaning to, Lewis clocks her socked feet.

No blood, no gore.

No hooves, either.

But the door, it was only up *four inches*. And Peta had to *raise* it to come sit out here and think. The only explanation is that it was either one of the two of them stomping Harley, or it was some-one . . . some*thing* else.

Lewis cranks around, heart thudding in his chest, and stud-ies the dark cave of the garage for a tall, top-heavy form standing flat up against a wall, hidden just nearly enough, her yellow eyes drinking the light in.

It wasn't horse hooves that did that to Harley, he's sure. It was an elk. How he knows is that it's after midnight, meaning it's tech-nically *Saturday* now—it's exactly one week before the ten-year anniversary of the Thanksgiving Classic.

"I don't know if we should be here anymore," he says.

Peta doesn't look over.

"All new houses take some getting used to," she says, always the rational one. "Remember that one with the attic?"

The place Lewis was so sure was haunted. The one where he nailed a board across that attic door up in the ceiling, in case anything wanted to crawl out, stand by the bed on his side. Or on any side. *Indians are spooky* had been his explanation to Peta. It's pretty much all he's got now, too.

"I can't sleep," he says.

"You were sleeping a few minutes ago."

"Why are you up?" Lewis asks, watching the side of Peta's face.

"Thought I heard something," she says, shrugging one shoulder.

"Harley?" Lewis says, because it's the obvious thing.

"The stairs," Peta says, which instantly sucks all the heat from Lewis's body.

He breathes in, breathes out long and shaky.

"I didn't tell you the whole story about that . . . *hunting* thing because I didn't want you to have to have it in your head," he says.

This gets Peta looking over. A mealy-mouthed excuse like this, it deserves her full attention.

"You don't like to hear stuff about . . . animals," Lewis adds.

"It's about you," she says back without hesitating. "It's who *you* are."

"I didn't tell her the end of it," Lewis says, his voice barely more than a creak.

Peta's still watching him. Waiting.

"You sure you want to know?" he asks.

"Who are you married to?" she says back. "Her, or me?"

Lewis nods, taking that hit, and wades into it one more time, starting with how, when he split that young elk open, when he carved into that elk who didn't know when she was dead, what spilled out into the snow were her milk bags. They were light blue, muscular and veiny, the ductwork still attached and ready.

She was too young to be pregnant, probably couldn't have carried full term all the way to spring, and it was too early for a calf to be this far along anyway, but still—*that's* why she was fighting so hard, he knew, and still knows. It didn't matter that she was dead. She had to protect her baby.

And that baby, that embryo or fetus, that *calf*, it was still rounded like a bean in there, its head shape ducked down into its chest like it was going to look up at him from its mother's gore, like it was going to wobble up onto four spindly legs, walk away, grow bigger but never actually develop, so it'd end up being a seven-hundred-pound big-eyed, smooth-skinned fetus, always looking for its dead mother.

When Cass wasn't watching, which was the whole time, Lewis used the butt of the rifle to scrape a hole in the frozen dirt, and nestled that unfinished, only-wriggling-a-little-bit elk calf into the ground, covered it as best he could, and then—never mind that storm swirling in, dumping load after load of snow—insisted on dressing this young elk mother out right, and all the way.

Nothing was going to spoil. No part of her could go to waste.

To do it right, he hacked a thick branch from the brush, split her sternum with just the knife—she wasn't even old enough to need the saw—then cracked the pelvis like prying a butterfly's wings apart, jammed the branch into her chest to prop her open. To be sure to get every bit of her ruptured guts, the last little bit of her lungs, he even crawled in like a kid with his first elk, scooping and pushing, and when he finally rolled out, dislodging the branch, Gabe was standing there watching.

"Just the hindquarters today, Super Indian," he said with a smile, a big brown leg Fred-Flintstoned over his shoulder, the black hoof cupped in his hand, blood dripping down the back of his jacket.

Lewis didn't take Gabe's bait. Just kept working.

The next part of his promise to the young elk was the skinning, which was a job he really needed to hang her up for, from a stout rafter in a shop, the radio playing over on the workbench. What he *had* was a trading-post knife that was too sharp at first, then too dull, and by the end of the job Gabe and Ricky and Cass were all standing there watching, the snow coating their shoulders, not even melting in their hair anymore.

And Lewis was maybe crying by then, he admits to Peta. He doesn't say it for pity, just because not saying that part would feel like lying.

"What did Gabe and them say about that?" Peta asks, her hand to his forearm now, because he's trying not to cry *here*, he's trying not to be that stupid, that needy.

"They were my friends," Lewis says, sputtering now and trying to just keep it in. "They didn't—they didn't say anything."

Peta reaches up to his forehead, delicately removes a flake of paint from the basketball pole, and then pulls him to her chest, her palm to his cheek, and this, her, it's home, and it's not haunted, not even a little. This is where he wants to live forever.

But he *still* hasn't said it all, about that day.

What he didn't get to before turning into a blubbering excuse for a grown-up is the four of them struggling that young elk up the hill, finally just using the junk come-along as a cable after digging the truck out, never mind that this ledge is the exact point on the reservation where all the wind gathers to sweep down like the end of the world.

Against all rationality, and though each step uphill takes about twenty steps in total, the young elk makes it all the way up, and in one piece, the four of them sweating in the freezing air. And neither Gabe nor Ricky nor Cass even asks Lewis why this is so important. They don't blame him, either, when Denny Pease is waiting by the truck on his game warden four-wheeler, looking from face to face like impressed that they thought they could get away with something of this magnitude on his watch. It's just as well. The snow's too deep, is coming down too fast. Without Denny's radio to call in more help, the truck doesn't make it out, Gabe and Cass and Ricky and Lewis don't get found until spring, and Lewis never meets Peta, never gets Harley, never goes to work for the post office, never builds his Road King.

The condition Denny lays down that day, it's that the four of them can either throw their honorable kills back down that slope and pay the fine for what they've done here, multiplied by *nine*, not counting any elk that ran off shot, are out there dying now, or they throw all this meat back down that slope and then cash out once and for all, never hunt on the reservation again. Happy early turkey day, turkeys.

It's a small price to pay, really. It's not like Lewis has the nerve for shooting big animals anymore. Not after having gone to war against the elk like that. That craziness, that heat of the moment, the blood in his temples, smoke in the air, it was like—he hates himself the most for this—it was probably what it was like a century and more ago, when soldiers gathered up on ridges above Blackfeet encampments to turn the cranks on their big guns, terraform this new land for their occupation. Fertilize it with blood. Harvest the potatoes that would grow there, turn them into baskets of fries, and sell those crunchy cubes of grease back at powwows.

Even after taking Denny's option two—I will hunt no more forever—all Lewis could think about, standing there, was that young elk he'd spent so much time on. She was freezing solid on the ground between them all, skinless, surrounded by sawed-off legs.

"Can we at least keep her?" he asked Denny, Gabe already rushing to the ledge to sling his elk haunch out into open space, the storm swallowing it whole.

Like a ritual, Cass stepped forward, hurled his leg down the slope, and then Ricky did it, his going the highest before disappearing, all five of them tracking its descent until they couldn't.

Denny looked over to Lewis about his question, then down to the young elk, all her muscles showing, a hole blown in her back, her head mostly gone, and Lewis, in the driveway well after midnight, shudders against Peta. Not because Denny shrugged a what-the-hell about the young elk, but because Harley's dead, isn't he? And not just dead, but killed, in a way that had to be terrifying. It should have been *Lewis* under those flashing elk hooves, Lewis knows, it should have been *him* paying for that young elk. Not Harley.

"I don't understand what's happening," he says into Peta's chest, his hand gripping hard on her leg. All her running muscles are still there, are there forever, probably.

"Something must have got in," she says back, about Harley.

She's right, of course, but the real question isn't what got in, it's *when* did it get in.

Lewis's breath hitches and he stands all at once, faking resolve, is already at the side of the house, digging for a shovel.

Peta stands at the edge of the concrete watching, her elbows in her hands, parentheses of concern around her eyes.

"I'm sorry," she says. About all of it—the elk, Harley. Maybe even Shaney.

When Lewis comes in thirty minutes later, Harley poured into the hole he dug that was bigger than it needed to be, a blanket and a sleeping bag or two packed in all around him to keep him warm, he steps out of those useless sweatpants, balls them into the kitchen trash, and what he sees nestled in there is a ball of crumbling masking tape.

Peta peeled the elk up from the living room floor. *Good*, he tells himself, standing there naked, chest heaving, *good*.

Just, it doesn't feel good.

SATURDAY

To keep his hands busy, maybe occupy his mind if he's lucky, Lewis puts the Road King up on the stand, is going to take it down to the frame again, clean and detail every bolt thread, check and double-check every connection, blow every line out twice, make it brand-new, cherrier than cherry.

He's just gone what he thinks might be a full five minutes without images of Harley cycling through his head when two of Great Falls' best show up, doing that thing where they park their patrol car across the bottom of the driveway. Lewis keeps on tracing the throttle cable he's tracing, like that's the only thing he's interested in today. The cops walk up spaced out wider than a single shotgun blast. The reason for their spacing, Lewis knows, is that he's sitting in the dark of the garage, shirt off, hair over his face, and he didn't walk out to meet them, is making them come to him.

"That really your name?" the first officer asks.

"Like, on purpose?" the second adds.

"What's this about?" Lewis says, hands in clear view up in the frame of the Road King. Though of course, should they pop him in the back with their .40-calibers just because, then their report could be all about how it looked like he had a gun tucked up under the tank.

"This is about your killer dog," the second officer says.

"What'd he do?" Lewis asks.

"According to emergency room personnel," the first officer states, looking at his notebook like reading it but of course he's not, "a dog at this address bit a man on the face."

"Silas," Lewis says with a shrug. "That's between me and him, isn't it?"

"Not when the hospital gets involved," the second officer says. "We have to see if the animal qualifies as a menace, a threat to public safety."

"Aren't you people cops, not dog cops?" Lewis says, standing, each of the cops taking a tactical step back, their right hands suddenly loose by their sides.

"We're the *police* asking to see your *dog*," the first officer says, that thing rising in his voice that isn't so much saying this call can go bad, but that he's kind of hoping it will.

"You really want to see him?" Lewis asks.

Where he leads them is the tamped-down grave on the back side of the fence, close to the tracks. He explains to them that he buried Harley there because he liked to bark at the train. They ask what happened to him. Instead of telling them that an elk from back home followed him all the way down here, is apparently on this big revenge arc, and instead of telling them option *two*, which is that there was something in this house before he even got here, and it's using his own memories and guilt against him, Lewis just shrugs.

What he also doesn't tell them is that there's always the chance that he's just flat-out losing it, here. That all the bad medicine from that hunt's built up all these years, has turned into something that's messing with his head from the inside. Or maybe that repetitive scope-eye he got that day in the snow was worse than he thought, it kicked something loose in his brain, something that's just now blooming.

"Do you not want to tell us because you put the dog down yourself, and don't want us checking serial numbers on any unregistered weapons?" the first officer asks.

"You don't have to register hunting rifles," Lewis says. "Do you?"

"You shot a deer rifle this close to other houses?" the second officer says, real concern in his eyes.

"Elk rifle," Lewis corrects. "And no, I didn't shoot it this close to other houses. He hung himself from the fence, trying to get over."

"'He,'" the second officer prompts.

"Harley," Lewis fills in.

"Like your bike," the second officer says.

Lewis doesn't dignify this.

"The one you're taking apart as well," the second officer says.

"What are you saying?" Lewis asks.

"What are you doing?" the first officer says right back.

Lewis thrusts both his hands up through his hair and like that both cops have drawn, are in that shooting crouch they like.

Going slow, finger by finger, Lewis lowers his hands back to his sides.

Dealing with cops is like being around a skittish horse: No sudden movements, nothing shiny or loud. Zero jokes.

Still, Lewis leans forward, shakes what hair he has to show there're no weapons there.

"You'll dig him up?" the second officer asks, holstering his pistol.

"I can," Lewis says, toeing at the loose dirt of the obvious mound.

"You're supposed to have a permit to bury on private property," the first officer says. "Otherwise everybody buries their pets down at the park, or in their neighbor's lawn, because they don't want to mess their own grass up."

Lewis looks up to the tracks on their long spine of gravel-skinned earth, says, "Think BNSF cares?"

"We'll make the necessary inquiries," the first officer says, his pistol holstered now as well.

"Fine," Lewis says.

"We might need to see the animal, too," the second officer says. "To confirm."

"That he's dead?" Lewis asks.

"That you're not hiding him," the second officer says.

"Unless you want to grant us permission to inspect your house," the first officer says.

Lewis hiccups half a laugh out, shakes his head no-thank-you to that home inspection. Just on principle.

"He was a barker," he says, about Harley. "You'd have heard him when your car pulled up."

"We'll be back soon," the first officer assures Lewis. "Either with the proper documents to perform a comprehensive search, or with the railroad's reply."

"Or to dig my dead dog up," Lewis adds, stupid Indian that he is.

"That, too," the second officer says, and then the three of them are walking back to where this started.

Lewis sits back down on the purple milk crate beside the Road King.

"Shouldn't you be at work?" the second officer says in farewell, his chrome shades already back on.

Lewis shrugs, is leaned down into the frame of the Road King again, rolling a vacuum tube for pliancy.

"Anything you want to tell us, sir?" the first asks.

"I miss my dog, yeah," Lewis says, and like those were the magic words, the patrol car is easing away. They're coming back, though. Because cops are exactly what Lewis needs in his life. His thinking is already twitchy enough without having to act all law-abiding for them.

He goes inside for a sandwich, eats it standing over the sink so as not to crumb the place up, and when he comes back to the Road King, two of the books he loaned Shaney are on the purple crate, like she showed up in the garage while he was leaning against the kitchen counter shaking Fritos into his mouth from the bag. He picks the two paperbacks up, studies the spines. The first two in the series. He grins a little, maybe the first time he's smiled in two or three days. He wishes he could go back, read them all over again for the first time. He wishes he had the concentration for reading at all, right now.

Instead, what he can't stop thinking, it's why now? Why did this elk, if it is an elk, why did she wait so long to come for him? Was it so he could have time to cobble a life together, get people and things he cares about, so she could take them away the same way he took her calf from her? But why start with him, not with Gabe and Cass? Not that he wishes any kind of ill on them, but if she's from the reservation, well: That's where they are, right? Why trek all the way down here first? And Ricky doesn't factor in, since he died just a few months after Lewis left, and it wasn't anything out of the ordinary, just another Indian beaten to death outside a bar.

The only part that makes sense, Lewis supposes, it's starting with the dog. It's what serial killers and monsters always do, since dogs bark the alarm, dogs *know* there's some shape standing over there in the shadow.

But how, right? *How* did she do it?

Does she, like, inhabit random people to do her bidding or something? Could she have snagged some kid bopping down the road after lights-out, gotten him or her to weasel through that four-inch space under the garage door and go to town on Harley with a mallet?

But none of Lewis's mallets have heads that big. And what happened to Harley *looked* sort of like elk hooves.

Lewis stands, studies the garage with this possibility in mind. What else could have done this to Harley, right?

"Post driver," he says, drifting over to it in the corner. It's the kind you need two hands and your whole body for. The kind that should be over in the corner with the T-posts, since that's what it goes with.

Lewis doesn't want to lift it, see the business end, but he has to.

It's clean, pristine, even has delicate flakes of rust-backed paint just hanging on, flakes that no way could have stayed there while crushing a dog to death with ten or twenty heavy blows.

A quart paint can, maybe? Those are hand-sized, could do the job. Lewis inspects all of them as well, even the ones that are light, obviously dried up. Nothing.

"You're being stupid," he tells himself, and sits down hard on the concrete step leading up to the kitchen door, pulls the mallet from its hook on the side of the rolling tool chest, bounces its butt between his feet.

The mallet is clean, too. For a mallet.

Of *course* an elk can't "inhabit" a person. That person would fall over onto all fours and probably instantly panic. Unless she's like that shadow he saw in the living room. Woman body, elk head, no horns.

That's all he's really got, though: a shadow he probably saw wrong, and something he thought he saw through the spinning blades of a fan.

Those two cops would love him to come in with that for evidence, or as explanation.

And thinking about it isn't making it go away. Lewis chuckles at himself, shakes his head, and drops the mallet handle-first into the closest receptacle, which is one of the cheapo rubber boots Peta always keeps by the door of all the houses they've rented since moving out of her aunt's basement. The boots are big enough that either Lewis or her can pull them on, wade out into the snow for the mail, then slip them off, not track slush in.

Lewis is so used to these boots by now that he doesn't even see them, not until they're in motion. From the mallet he dropped into the right one.

He pushes away just on instinct—boots don't move on their *own*—then comes back closer to be sure he's seeing what he's seeing.

The motion is ants. The boots are coated in small black ants, summer ants, even though Thanksgiving's just next week. Halloween ants, maybe? Is that a thing? If not, it should be, once he realizes what the ants are after: Harley. What's left of him, smushed into the tread of the rubber soles, smeared on the toes because it wasn't all stomping, evidently. There was some kicking, too.

Lewis shakes his head no, please no, and backs away, out into the light, then feels his way around to Harley's grave. Lewis is breathing deep but he's not going to cry. He's a stoic Indian, after all. When he was a kid, he thought that was the fancy word for "stone-faced," which he figured was some connection to Rushmore, since he knew it wasn't supposed to look like that.

That was back when he was stupid, though. Before now, when he's even stupider.

Except—what he's thinking now. No.

Last night out here in the dark, when he kept asking Peta about Harley, about how that could have happened. Could she have just been seeing a dog that finally died from its injury? Had Lewis seen something completely different? Did his questions even make sense to her?

He thinks back through what he can dredge up of her answers, to see if they track with a Harley that *wasn't* stomped to mush.

On his knees back by the fence, he claws the earth open with his desperate fingers, breathing fast. He pulls up the blanket with the ducks on it and the sleeping bag that tapers at the foot end, and one Star Wars sleeping bag deeper and it's going to be Harley.

But then Lewis doesn't pull that field of stars away.

Does he even really want to know? Will seeing Harley smashed to pulp prove anything about who was wearing those boots? Is that even for sure dead dog *in* the tread of those boots? What if Peta was hauling the trash out from the kitchen and it burst, and she had to wade through it? Halloween ants would go for that just the same, wouldn't they? If there is such a thing as Halloween ants?

What he's really afraid of, though, he knows, it's that Harley will just be dead from being strangled by his collar, from hanging on the fence.

And now the ground is trembling with Lewis's chest. The train's coming. The train's always coming.

Lewis closes his eyes against the screaming wheels and the sparks, but then a rock chip catches him on the arm and he falls away slapping at it, and that's when he finally looks at the train rushing past. Not at the different-colored cars, not at the graffiti smearing past at sixty miles per hour, but at the space *between* the cars, that space that's full full full full, then, for a flash, for a slice of an instant, empty.

Only, it's not.

Standing out there in the yellow grass is a woman with an elk head, and—no, no.

Lewis stumbles forward, the train cars whipping just past his face now.

Is she wearing a thick brown jacket with reflective stripes? Like the kind the ground crew of an airport wears?

"It can't be," Lewis says, and the moment the train's gone he's scrambling up onto the hot tracks, but of course the grass over there is just grass again, like nobody was even there.

SUNDAY

For once, Lewis wishes he were at work. Because the only other option is faking sleep until Peta's gone to her shift. For maybe thirty seconds he felt her standing in the doorway with her morning coffee, watching the mound of covers he was trying to make rise and fall like completely normal, non-mechanical breathing, but at least that kept her from asking him why he was acting like such a weirdo last night, cooking alone out on the grill like some backyard warrior, then staying in the garage with the Road King until late.

Are you pissed about something? she might have asked, if she'd seen an eyelid flicker.

The answer he had ready was *No*, and *Harley*, but then she bought his sleeping ruse, he guesses.

Well. If she's Peta she bought it.

If she's something else, well.

The dots he's trying and trying not to let connect in his head are that Peta showed up *on* the reservation, didn't she? And it was the exact summer *after* the Thanksgiving Classic, when he was all busy flipping the whole place off with both hands, denying it his sacred presence from here on out. Maybe that's the answer for why this is starting with him, not Gabe or Cass: because he was the first to leave.

As for the case against Peta, or for her *not* being Peta, it doesn't help that she's a vegetarian, either, Lewis has to admit. Which is what you call a person who doesn't eat meat. What you call an animal that doesn't eat meat, that's "herbivore."

Elk are herbivores. Grass-eaters. Vegetarians.

And: maybe she wasn't lying when she said it was her past making her not want to have kids? Only, the past she meant was the one where she already lost a calf?

Moments after the front door shuts and locks, Lewis covers his face with his pillow, screams into it.

If it's her or if it's not her, either way, one fact is that *somebody* in those boots stomped Harley dead. And he definitely for sure saw a woman with an elk head through the boxcars flashing past— maybe even flashing past at the same flicker rate as the ceiling fan?

It's too much to hold in his head all at once.

He loves Peta, and also he's terrified of her.

Worse, there's no proof either way. No way to tell.

Lewis smushes the pillow up his face, pushes it behind him, and, with his ears cleared now, hears a telltale creak on the stairs. As if, say, someone didn't actually just leave. As if someone just shut the door and locked it from the *inside*.

Slowly, deliberately, the most distinct footfalls he's ever heard are coming up the stairs, but each light *thunk* is preceded by a draggy *whisk*. Because an elk will feel forward with its hooves, right? Find the leading edge of the next step before actually taking it?

Lewis rolls over fast, his bare back to the door, and stares at the curtainless window, trying to memorize each waver and imperfection in the glass so he can clock the reflection when it comes. And his right ear, the one that's up, dials down as sensitive as he can get it. The kind of sensitive that can hear a large set of nostrils breathing his scent in, should that happen.

A tear spills from his left eye, soaks into the pillow.

Is she there now? And if she is, then which her is it? Peta-Peta, or Peta with an elk head?

When one of the ripples in the window glass finally smears with color, with motion, Lewis breathes in deep, says, "Hey, forget something?"

No answer.

When he breathes his next breath in, it's thready, unsteady, isn't a breath he can trust not to explode into a scream.

"Or was it—?" he says, rolling over fake-groggily like he has an end to that sentence.

The doorway's empty.

Lewis closes his eyes, opens them, doesn't let himself rush to the window to see who, or what, might be walking away.

It's too early for this shit.

He brushes his teeth and pees at the same time, spits into the toilet and sort of on his hand, and makes his way downstairs, taking the steps slow, trying to memorize each creak. It's hopeless, though. Each stairstep makes one sound in the middle, a completely different sound eight inches over. Of course.

At the kitchen table he stands before the elk bundle—the hairy burrito—for maybe thirty seconds, finally pushes a finger into it. It's mushy and rough at the same time, smells like some soft cheese that was on the table at a party once, that he knew better than to eat.

"Cheese," though. Now he's thinking cheese.

It'll wreck his digestion, but, figuring that's the least of his concerns right now, he makes a grilled cheese for breakfast, just staring down into the yeast-craters in the toasting bread while it cooks in the pan.

Because he's such a good and considerate husband, he eats it over the sink. Either that or he's kind of scared of the living room. His big irrational fear now is that that spotlight in the ceiling's not broken, it's just waiting for him to be the right kind of alone, so it can shine down like a UFO beam, a woman with an elk head materializing in it. Or it can shine down on Peta when she's standing right there, that light showing her true form.

Which is just *Peta*, Lewis insists to himself, raising the last triangle of the grilled cheese high to slam it down through the rubber flaps of the disposal, as if a grilled cheese can be the deciding gavel. It sort of works, but the crust breaks away on the backswing, lets that last good bite go flying. Trying to imagine either him or Peta finding a moldering hunk of bread and, worse, cheese behind some jar or can next week, he flips the light on, hunts the lost bite down.

Instead of finding it, he sees a paperback on top of the refrigerator. Not either of the two Shaney left in the garage, which he already put up, but the third in the series. Already. Beside it is Peta's thermos, the one she takes to work, that she can never find, that she evidently didn't find this morning, either. The thing with her is she's tall, so when she comes home she sets stuff in the first place she sees, which is generally somewhere high. The top of the refrigerator, this time. The book must have been out front somewhere, the porch, maybe.

"You can bring them back all at *once* . . ." Lewis says to the idea of Shaney, taking the book down from its perch, and, as if by design, the guilty last bite of grilled cheese is right there behind it. Lewis pinches it up like it's gross, like he wasn't just eating it ten seconds ago, and delivers it to the sink, the headline scrolling across the back of his forehead: INDIAN MAN FIRST IN HISTORY TO PICK UP AFTER HIMSELF.

He smirks about that, kind of stupidly proud, and takes the stairs two at a time up to the linen closet that's his new bookshelves.

Before filing the book back in the stack, he fans the top corner to make sure Shaney isn't a page-folder. She isn't—this alone pretty much means she's a good person—but he does catch something. He flips through again, slower, and doesn't see whatever it is again until . . . the inside of the back cover.

"Seriously?" he says.

She *writes* in books, apparently. In *borrowed* books. In pencil, and light like she maybe meant to erase it, but still, right?

Screw it, Lewis tells himself. Who cares? They're mass market, not collectors', and the story's still the same, and it's not like she's putting happy faces or question marks in the actual margins. But now Lewis has to flip through to be sure there's none of *that* going on, and, even while doing it, he's wondering if he's doing it to have something to rib her about at work. Which sounds a lot like flirting.

But it's not that, he insists. This is book-policing. And it's his book anyway. He can look through it all he wants.

Flipping through, he finds himself sitting down against the wall, back in the world of this story again. It's the series about that stone the elves didn't want, but didn't want *found*, either, because it can destroy the whole world. So they hide it in a magic fountain. Which is also, Lewis forgets just how, twinned with the wishing well in the mall that one doofus works in . . . Andy? Yeah, Andy. Of course "Andy." Andy the This, Andy the That. And then all the magical creatures are trolling the mall for the stone their magical radar is telling them is here somewhere. It's kind of hilarious, and has more sex scenes than really make sense either for such an actiony story or for a story in a place as public as a mall, but that's magical creatures for you.

"Did you like it, at least?" Lewis mumbles, flipping to Shaney's notes.

She's not jotting stuff down about the novel, though. She's still thinking through the masking tape on Lewis's living room floor.

What makes this elk so special? is the first note.

Under it she's drawn three lines, like giving herself room to figure this out. But they're blank.

Peta could have answered it, though. Because Lewis told her: this young elk had been pregnant, and farther along than she should have been for November. He thought that was what gave her fight, but what if—what if every great once in a while an elk *is* special, right? What if there are wheels within wheels up there on the mountain, where ceremony used to take place? Was that unborn elk supposed to, Lewis doesn't know, grow some monstrous rack, be a trophy for some twelve-year-old's first kill? Was it supposed to be the big elk an old man chooses not to shoot on his last hunt? Was it supposed to clamber up onto a certain stretch of blacktop, wait for headlights to crunch into it? Was it supposed to find new and safer grass for the herd? Was it not even about the calf, but the mother?

What process had Lewis broken by popping this elk back in illegal country?

"You're thinking crazy," he tells himself, just to hear it out loud. He's right, though. These are the kind of wrong thoughts people have who are spending too much time alone. They start unpacking vast cosmic bullshit from gum wrappers, and then they chew it up, blow a bubble, ride that bubble up into some even stupider place.

Elk are just elk, simple as that. If animals came back to haunt the people who shot them, then the old-time Blackfeet would have had ghost buffalo so thick in camp they couldn't even walk around, probably.

But they killed them fair and square, Lewis hears, and in Shaney's voice, he thinks. Probably because he's reading her writing.

Shaney's next question, still in her voice, is, *Why now?*

Lewis is the only one who maybe knows this answer.

It has to do with her meat, doesn't it? All that meat he gave away door-to-door on Death Row, which is where the closest-to-death of the elders get to live.

It's not out of reach to think that one of those elders he gave the meat to is still alive. Some of them old cats can sit in the same chair for ten, twenty years. Or—

That's it, Lewis knows all at once, sitting up from the wall, his face muscles tense from this certainty.

One of those elders *was* still alive . . . until last week, or last month.

This has to be it.

One of those elders finally kicked last week, and way in the back of her freezer, frozen to the side of it after all these years, there was one last packet of that meat left. Because it was locked in ice, her old fingers could never pry it free, and the reason none of her kids or grandkids ever mixed it into Hamburger Helper or cooked it up with taco seasoning, it was that raccoon stamp.

If you don't know the story of the meat, if that elder couldn't remember the kind young man assuring her it was elk, then what you think, with a black footprint like that on the white paper, it's that somebody's ground up some raccoon from somewhere—the road south, probably—and left it in this freezer like a joke, like a dare.

No, nobody ate it. Nobody would.

But now, with that elder dead, another family's getting that house, right? Which means new furniture, new appliances. Out with the old freezer, in with the new one.

The meat finally thaws, gets tossed. For the birds, for the dogs. And that last packet of meat, it was Lewis's one chance, wasn't it? He'd promised the young elk that none of her would go to waste. But now some had.

That's why now, Shaney. Shit.

The moment that packet of raccoon-printed meat hit the ground, started to thaw, the ground hatched open back in the elders' hunting section. What clambered out, just like a monster movie, was the ghost elk, the one he'd had to shoot three times.

At first she's wobbly on her legs, but with each step south, her hooves are more sure.

Peta didn't stomp Harley, *she* did, this ghost elk. After . . . what let her in, though?

"That I was still thinking of her?" Lewis says in the hallway.

His memory of that young elk, his guilt over her, that was the tether she pulled herself back with, wasn't it? That's why she's starting with him, not Gabe, not Cass: because they don't remember her. She's just one of a thousand dead elk to them.

It's all making sense now. Without Peta here to talk him down, this is all making *perfect* sense, actually.

Shaney's last note is only half formed, still in its egg of parentheses: *(ivory?)*

Shit. Of course. Lewis stands, paces back and forth, slapping the paperback against his thigh, pushing his other hand through his hair.

Shaney knows elk as well as he does, doesn't she?

The thing with elk is that their canines, they used to be tusks, thousands of years ago. They're shorter these days, but are still ivory. That's why they look good polished up and sewn to a traditional dress. If Lewis and Gabe and Ricky and Cass had been thinking that day in the snow, their pockets would have been clinking with elk ivory, to trade back in town.

What Shaney is saying with that *(ivory?)* is that there's a way to tell who the ghost elk is.

Check the teeth.

Lewis holds his left hand out. It's shaking. He drops the paperback, grabs his left wrist with his right hand to steady it. When that doesn't work he retreats to the garage again. By the time Peta gets home hours later, he's got the Road King down to smaller and smaller pieces, arrayed all around him like an exploded diagram from a repair manual.

She stands out there under the goal and studies him—he can feel her. She's studying him and she's trying to make sense of all these parts, all this grease and oil. All this effort. A husband who doesn't go to work anymore, and refuses to talk about it.

Finally she sets her duffel down, tosses her ear protection on top—she's superstitious about them, won't leave them at work—and uses her foot to flip the basketball up to her hands.

She spins it backward, dribbles it twice.

"Real leather," she says, impressed. "Where's it from?"

Lewis looks from it to the street, realizes he has no idea. Shaney was just shooting with it. He never stopped to ask if she'd brought it.

Like he can say that now.

He shrugs.

"Make it take it?" Peta says, chest-passing the ball at him, which is practically a skills challenge, what with how he has the garage booby-trapped. But that's Peta.

Lewis catches the ball the same as he had to when Shaney zinged it at him—what is it with the women in his life? After rocking back and almost falling off the purple crate, he chocks the ball under his arm to clean his hands on his pants legs and eases out under the floodlight with her, dribbling once like testing is this ball up to his exacting standards.

"Eleven?" she says, coming around to place herself between him and the rim, palms up, eyes ready even though she has to have been awake going on twenty hours.

No way is she anything but herself.

Lewis smiles his best Gabe-smile and looks up to that orange rim like asking Peta if she's sure about this, asking if she really thinks she's ready for what he's about to bring.

She is, and then some.

They trade buckets until they're both slick with sweat, and Lewis never stops to think that she's going easy on him, that she's letting him feel like he has a chance.

What he's doing specifically, for the first time in days it feels like, is *not* thinking, just stopping on dime after dime, faking to make airspace, running through the tall grass and junky lumber for the ball again and again. It's one hundred percent exactly what he needs, and would have never known to ask for.

By the end he's laughing, she's laughing, and then they're in each other's slick arms, and he's guiding her back onto the mound of blankets and sleeping bags whose sweat lodge dreams are over, and the door is lowering over them as they add their clothes to the pile, and the world is kind of perfect.

MONDAY

It's lunch by the time Lewis gets the drive belt back onto the Road King. The bike is still a skeleton, but now it's a skeleton with an engine that cranks and can blur the spokes of its rear wheel. No forks or bars, no seat or pegs, and the throttle is just a cable, but this is something, he tells himself. It's a sign that he's climbing back to his life. Once the Road King's back together enough, then he can finally ride in to work, provided he still has a job. There's a lot of steps to getting fired from a federal job, though.

Lewis might can stop them by sitting in the big office and taking his medicine, promising to be a model employee from here on out, offering to take all the shit details, covering for whoever needs it, coming in on holidays, on snow days, whatever it takes.

The best excuse he'll have for his prolonged absence will be Harley, but that's also the most embarrassing excuse: a dog. Is he really that fragile? Is he going to lodge a complaint about "Chief" next? Will his new name around the post office be "Kid Gloves"?

It's not supposed to be easy, though, he tells himself. It's not supposed to be easy or comfortable or fun or any of that. And if it ends up with him keeping his job, maybe even getting his own route someday, well, then it'll have been worth it. Sorry, Harley. You deserved better. But right now it's about proving to Peta that he's not stalling out in the middle of his life. Right now it's about showing her that she's not going to have to carry him from here on out. She would, Lewis knows, she'd do it for as long as she could, but the strain would show. She's pretty much superhuman, and would never complain, but they're supposed to be a team, too. He delivers the mail, she brings the planes in, and they meet at the end of the day over tofu and beans, compare notes, and, like last night, work out their kinks on the court. And also on the floor of the garage.

The more Lewis thinks about it, tightening this bolt, wiping that smudge off, no way can it be Peta, right? If she were really this "Elk Head Woman" he's making up, which isn't even a Blackfeet thing so far as he knows, then why would she have saved him from bashing his head in on the hearth when he fell off the ladder? Answer: she wouldn't have. That would have been just what she wanted. And it would have been the perfect accident, giving her an excuse to go up to the reservation for the funeral, look across the grave to Gabe and Cass right there, like they know it's their turn.

No, Peta's Peta, Lewis decides.

But the only reason he thought she wasn't?

Shaney.

He stands to find a cable clip he knows is in the garage *some-where*, probably within arm's reach, and manages to kick a housing into the torque wrench he has leaned against the front tire, and it all clatters down like misshapen dominoes and Lewis just stands there, unable to take his frustration out on any of this because that'll make even more of a mess.

Shaney, though.

What if that elder whose freezer got cleaned out didn't die a week or two ago, but a month or two ago? Shaney could have heard it hit the ground, clawed up from the pile of weathered bones at the bottom of that steep slope, and walked on wobbly legs down across half of Montana, finally got steady enough that she was able to stride right through the front door of the post office, fill out a federal application.

And it makes sense that when that elk took two-legged form, human form, that person might have skin to match Lewis's own. He doesn't know why he didn't see this before.

The clincher comes when he's back in the house, cleaning a grimy washer in the sink, trying to use the dish brush as little as possible, since Peta doesn't like grease showing up in the kitchen. Which Lewis completely understands. So he's just brushing the washer lightly, with the very tip of the blue bristles, not the white ones.

Walking back to the garage turning the washer back and forth to see if it's possible anymore to tell which side had been facing out, his heel on the floor shakes the house in the way it needs for that spotlight in the living room to jiggle on in his peripheral vision.

Lewis freezes, half afraid to look directly at it, since he's trying hard for this episode of his life to be over.

For it to really be over, though, he has to prove he's not scared anymore, doesn't he? He makes himself look.

On cue, like it's shy, the light sucks back up into the bulb.

Lewis stamps his heel on the kitchen floor again. Nothing.

"Such bullshit," he says, shaking his head at the stupidity of all this, haunted houses and ghost elk and Crow women, and then, not wanting to in the least but making himself, because he *doesn't* need to be handled with kid gloves, he looks up, through the spinning blades of the fan instead of at them. The angle from the kitchen is bad, so he feels safe, but still, there's the chance he's going to see a woman-shape up in the far corner, trying to skitter out of view.

Nothing.

He lets himself breathe out, rubs the coolness of the washer against his chin, and tracks down from the fan to . . . where he saw her first. The couch, the carpet there.

"Oh shit," he says, letting the washer fall and not bothering to go after it.

Why didn't he see this the other day? It's so obvious.

When—when Peta was helping him re-create the conditions of what he thought he'd seen, he'd climbed the ladder, he'd looked down through the fan blades at the carpet, and . . . he'd also looked down at Peta, on the couch, looking back up at him, not judging, not having to suppress her smile, just playing along with her losing-it, spooky-Indian husband.

That's not the important part, though. The important part is that, through the revealing blades of the fan, through that flickering that strips away fake faces or whatever, she'd been *herself*.

"I'm sorry," Lewis says to her. For avoiding her the other night. For even allowing the possibility that it could have been her.

It was never her. That's just what he was supposed to think— that's what Elk Head Woman wants, for him to tear down his own life. That way she doesn't even have to do anything, can just sit back and watch.

She's devious like that.

And . . . using the same logic he used to indict Peta—that she'd shown up the summer right after the Thanksgiving Classic— Lewis realizes it can't be any accident that the day Shaney showed up at his house, delivered by *him*, Harley was already most of the way dead.

If he hadn't been, then he would have gone for her throat, wouldn't he have? He would have ripped her mask off, shown her for what she was.

Shit.

It's like rebuilding a carburetor. You seat that last jet and then realize this thing can breathe on its own now.

Like to confirm that, when Lewis steps outside to just not be in the house for a minute or two, the fourth book in the series is there, balanced on top of a stray beer can so anybody leaving or coming home has to trip on it, find it.

Lewis stares down at the book for maybe twenty seconds before nudging it over with his toe, like there might be something ready to spring up from the can. The book falls open-faced on the rough concrete, the can doing its tinny roll for a foot or so, until it finds a pebble to stop against.

Lewis kneels, saves the book from the scratchy concrete, and flips to the inside back cover first thing, for the next note.

It's blank, doesn't even have erased pencil marks.

Lewis studies the street and across the street, and both ways as far as he can see.

Shaney must be close, right?

Nothing moves, though. No big ears flick back in the trees, no large black eyes blink, no hooves shift weight.

Lewis takes the book back inside, sits at the kitchen table with it, and interrogates the cover, the spine, flips through the pages.

"Why the books?" is where he finally lands. That's the part that doesn't track, that's the thing niggling at his mind, trying to poke holes in his theory, his suspicions. If—if Shaney *is* this young elk reborn or not yet dead or come back for unfinished business, then why is she interested in a fantasy series about a jewelry store worker trying to save the world from itself?

Lewis scans the back cover to be sure he remembers this one right. Yep. It's the installment where the usher at the movie theater, the one who's more than she seems, figures out that the food court is actually the front gate of the fairy prison. It's maybe the best book in the series, really. It ends with Andy on a woolly

mammoth, rampaging through the makeup and perfume department of one of the pricey department stores, and then it has that
epilogue—rare for an installment in a series—where the dwarves
discover carbonated drinks, and you can tell from the glimmer in
their eyes that this is going to be trouble.

None of which applies in any remote way to the Thanksgiving
Classic, to hunting in general, to the post office, to Gabe and Cass,
to Peta. At least not in any way Lewis can track.

He shuts the book, ditches it on the stairs so he can grab it next
trip up.

So it's Shaney, then. If it's not Peta, and it's not, then Shaney's
the main and only one left. And maybe she didn't crawl up from
that killing field up on the reservation, maybe she had a whole
real life before . . . before she stopped being herself, opened her
eyes, and looked around with a different set of instincts. Maybe
she was up in Browning for Indian Days, or maybe she clipped an
elk on the interstate down here, or maybe she just signed on to the
wrong job, took a cigarette break on the wrong stoop by the load-
ing dock, breathed in more than smoke.

The how doesn't really matter. What does is that she's coming
for him. And that she's been trying to set Peta up, meaning Peta's
a target as well.

Lewis shakes his head no about that.

This ends here. Or, it ends where he wants it to end, not where
Shaney wants it to.

To be sure, though, beyond a shadow, he has to somehow get
Shaney into his living room one more time. He has to get her in
his living room *while* he's up on the ladder, so he can look down at
her through the spinning blades of the fan.

And it's probably best to do this when Peta's not around. Meaning tomorrow, when she's at work.

It's leaving himself vulnerable, Lewis knows—it won't look

good if Peta walks in on them alone in the house *again*—but if Peta's piddling in the kitchen while Lewis relays some made-up work news or whatever, then Shaney's defenses will be up, and her real face might not show.

No, her real *head*.

But what will get her over here tomorrow, right?

The one thing Lewis maybe has going is that he's half sure that on Tuesdays, Shaney doesn't go in until noon. And that he, of course, won't be going in either. For the moment, there's more important things.

He paces and paces the house, looking in every corner for a reason to get Shaney over here. Something work? Indian? Basketball? Should he act like he's ready to go the next step with her? Does he need some help keeping his sweats up? Would she even be interested, or has he been reading her completely wrong?

No, he finally figures out—not that he's been reading her wrong or right—but, none of those options. There's a better way to get her into his living room.

Silas.

He was the gift Shaney didn't know she was giving him.

That's the way it always is in fantasy novels, isn't it? The evil wizard or dastardly druid builds his own doom into his plan, like he knows he really shouldn't be doing this, or like there's some magical realm rule where he has to leave a single scale off the dragon's belly, to give the puny crew a one-in-a-thousand chance.

Harley biting Silas is that missing scale, that chink in the armor, that one chance he has in a thousand. Lewis thinks it through once, twice, and nods the third time through, when there aren't any red flags.

This can work.

He digs around for the work directory, can't put hands on it, so just calls the post office, talks Margie out of Shaney's number because he's coming back to work but his bike's all taken apart, and she lives out by him, he can catch a ride with her.

Ten digits later, Shaney's phone is ringing.

"Blackfeet?" she creaks after his *Hello?* Lewis listens for rustling in the background to see if she's alone.

"Hey," Lewis says. "Silas is back in the mail room, isn't he?"

"Frankenface?" she says.

Lewis winces, is responsible for that.

"We were kind of short handed," Shaney goes on, coughing like a smoker.

"Sorry about that," he says.

"What already?" Shaney says. "It's my day off, man."

"Thought you were off Wednesday," Lewis says like the question it is.

"Missing two people, schedule's fried," she says back.

"But you go in tomorrow, right?" Lewis asks. "Silas, his build, his bike—he was supposed to get a bracket from me the other day, right? When Harley . . . you know?"

"He's not supposed to ride for two weeks," Shaney recites. "Wind can blow his stitches out."

"But he can still tinker around in the garage," Lewis says. "This'll help, it'll be good."

"A bracket?" Shaney repeats.

"For his headlight," Lewis tells her. "I was thinking you could bring it up there to him, maybe."

Long pause, in which Lewis imagines Shaney shying away from the bright window by her bed.

"Think it's going to take more than a new headlight to get that bike in shape," she finally says.

"It's a start," Lewis says.

"I'll have to leave an hour *early* . . ." Shaney says, playing up the groan.

"Thank you, thank you," Lewis says, and gets off the phone before she can tell him to leave it on the porch like she's been doing with the books.

For two hours, then, Lewis wears out the carpet, going back and forth, gesticulating, orchestrating, framing ideas between his held-out fingers, trying to figure everything through from each possible angle. He tries to work on the Road King, get it closer to ready to maybe *really* go back to work, but his mind is too jumpy, won't settle down. By midafternoon he's got the tape measure out, is duct-taping a free-throw line on the driveway. His rule is he can't quit until he makes three in a row, no iron, no backboard, but about fifty tries in he's counting trash, because, he tells himself, this is basketball, and basketball is made of trash. Still, he can swish two easy enough, just, the third always back-irons out at some crazy angle, like the world is laughing at him. Maybe it's for the best, though. This way he'll still be shooting when Peta walks up, and maybe they can pull a replay of last night, happy ending and all, he'll even use the same stupid joke about it not mattering they don't have any protection out in the garage—Indians like to go bareback anyway, yeah?

Just like always, Peta will grin and pull his mouth to hers.

It's looking good for a repeat—well, him being out there when she walks up, anyway—except then she's calling, is having to cover some no-account's shift again. Some no-account like Lewis, she doesn't say. Some no-account who just doesn't show up, doesn't even call in.

"Fine, cool, great," Lewis tells her, the phone going from side to side of his head because he can't find the best way to hold it, can't figure out what to do with his hands in the moment, or this whole day. But it *is* good, her pulling an extra shift, maybe edging into some overtime. Money's about to be tight once his next check's docked, if there even is a next check, so any way to earn a bit more's pretty much just what the doctor ordered.

"Love you," Lewis says into the phone. "Anything I can do, bring, be?"

It's his usual sign-off.

"Just yourself," Peta says like always, and they hang up together.

An hour later he's warming a can of chili for dinner, eating it with a whole tube of crackers, all the crumbs raining down into the sink. By nine he's nodding off at the kitchen table, and by ten he's in bed trying to read the fourth book in the series, a longneck leaning cold against his right side.

Peta says it's a bad habit to get into, drinking at lights-out, that your body can forget how to go to sleep on its own, and Lewis is sure she's right, but she's not here, either, and this whole "sleeping" thing seems to always be about to happen instead of actually happening, anyway. The pages gather in his left hand, thin out in his right. He'd forgotten how fun this mall is, and what a good contrast this magic is with the commerce going on around it. It's cool how the seasons and the decorations are always changing, kind of giving a theme or motif to each installment, and it's hilarious how these pagan characters both recognize and are seriously insulted by all these holidays—especially the elves. But they're insulted by everything.

Inside an hour, finally, gratefully, Lewis starts to lose the line between reading and sleeping. Because Peta isn't here to reach across, guide the book off his chest, careful to save the page, Lewis tells his distant index finger to be a bookmark, and then, right before the nothing of sleep, he asks himself what's he going to do when Shaney's elk head shows under the fan, and he's up on the ladder.

He'll know at last, but what will he *do*?

He mutters an answer but his lips and mouth and voice belong to someone else at this point of almost-sleep, and his ears can't make out the words, quite.

He feels himself chuckle with satisfaction all the same.

TUESDAY

Lewis is standing on a five-gallon bucket, looking over the fence at what used to be Harley's grave. Something's dug him up, scattered the blankets and sleeping bags across the train tracks.

Lewis insists on that: something dug him up. Otherwise Harley struggled out of the dirt himself and walked up the railbed, that Star Wars sleeping bag holding on until it snagged on the grabby edge of one of the wooden ties.

Peta was already gone by the time he got up, so she didn't have to see this. What's she made of, Lewis wonders for the fiftieth time, that she can crawl in at one in the morning, be gone again before sunup? Can't they just shut the airport down, let her catch a couple more hours' sleep? But it's good she's not here, too.

Shaney's coming over.

Lewis eats toast and a candy bar for breakfast, candy bar first because it makes the toast taste better.

Something definitely dug Harley up. That's the only way it could have gone. Coyotes, probably, but a badger will scavenge as well. He doesn't want to picture the long-headed shape of a woman out there on her knees at three in the morning pulling the ground open, but not wanting to have to see it just cleans the focus on that image up, pretty much.

Lewis is puking into the sink before he even realizes he's gagging. It's not from nerves about Shaney, he tells himself. It's from thinking about what Harley must look like now.

When he's done throwing up he turns the disposal on and a chunk spits back up at him and he falls down in the kitchen trying to get away.

"You're doing great," he tells himself from the floor. "You're completely ready, Blackfeet."

It's the first time he's ever called himself that.

He hand-over-hands his way back to the kitchen table, where at least he probably can't fall down. Not as far down, anyway. Well, not from as high.

His fingers busy themselves smoothing down an unruly tuft of the elk hide.

It still doesn't smell right, but it doesn't smell like cheese anymore, so that's some kind of progress, right?

10:40 already. If Shaney's shift starts at noon, and for some reason stopping by here means leaving an hour early, which has to be a lie, then she should be here in the next ten, fifteen minutes, Lewis calculates.

Enough time to unroll that hide. Not for the nicks he knows he left all over and through it, so it would probably only be good for a few pairs of gloves, nothing of size, but because . . .

Maybe some elk *are* special, right?

What if it wasn't that she was carrying a calf early? Or, what if she was carrying that calf early because she needed to get it birthed before . . . before some Gabe or Cass or Ricky or Lewis poached her in late spring, or some shed-hunter popped her with the handgun he only carries for bear?

What if she needed to get that calf out because she was already scheduled to die, so she could get skinned?

In the museum, behind glass, there's an old winter count, drawn on . . . it's probably buffalo, Lewis imagines. But why not elk?

And who's to say it's all drawn, either?

It could be that, back when, the people would bring any hides or skins that looked different in to the old-time version of a postal inspector. Because maybe some hides, some skins, right when they peel back from the meat, there's already some markings there, right? A starting point, maybe. A story of things to come. Pictures of the winter yet to come.

That day in the snow, the Thanksgiving Classic, there'd been too much blood and hurry to wipe the skin clean.

But there's time now.

Lewis clears the table and unrolls it delicately, like parchment.

The back side of the skin is black with freezer burn or something, Lewis isn't sure. He tries to wipe it away with paper towels but it's in the pores like ink, which he guesses either blows his big theory or proves it, only what this skin is tattooed with is a storm so bad it eats the world.

"Little late," Lewis says down to the young elk. Could have used this kind of warning about 1491 or so.

There is something, though. In the last roll, which would have been the first when it was being rolled up, is that trading-post knife that he thought he'd lost.

He'd put it in here, really?

For what reason?

Lewis extracts the knife. The blade that's on is the short skinning one with the curved nose. The handle still fits perfect in his hand, too, which is why he bought it in the first place. Oh, the adventures he thought they were going to have.

Instead, that was the last day he ever hunted.

He sits back in his chair, studies the ladder he's already got set up half under the fan, already test-leaned over to see if it lines up with the dent in the wall. *Thank you, Peta.* Even when she's not here, she's saving him.

10:55. Shaney should be here by now.

Lewis stands, studies the living room all over again to see what he's forgetting.

Nothing he can think of.

When things are simple, there's not a lot to keep in mind.

He crosses to the front door, cocks it open, then comes back to the living room, looking from the Road King's headlight bracket on the floor and up to the fan, confirming the angle one last time. It's dead-on. The elk was right here.

And she's about to be again.

Lewis puts one foot on the lowest step of the ladder and reaches up for the red-handled screwdriver on the fourth step, at eye level.

He can't just be standing on a ladder for no reason, right?

It's 11:05 before tires crunch up in front of the house.

"Okay, then," Lewis says, and nods to himself, steps up the ladder until the spinning blades of the fan are at his hips again.

His angle down onto the headlight bracket on the floor is perfect.

Shaney doesn't step on the beer can on the porch, just knocks on the door. It creaks in with her knocking, because Lewis left it slightly ajar.

"Blackfeet?" she calls.

"In here," Lewis calls back, the screwdriver handle clamped between his lips muffling his words, the strain of keeping both hands busy at the little spotlight compressing his breath.

"Say what?" Shaney calls back, leaning in, it sounds like.

"In here!" Lewis says, louder, hopefully clearer.

She steps in timidly, like this might be a trap.

"What the hell are you doing up there?" she says from the edge of the living room.

"Stupid light," Lewis says, and trying to speak around the screwdriver means losing grip *on* that screwdriver. It tumbles down behind him, bounces into the corner.

"Smooth move, Ex-Lax," Shaney says with a smile.

"There's the bracket," Lewis says, nodding down to it, and then realizes that, without a screwdriver, why's he up on the ladder anymore? Why isn't he climbing down *for* the screwdriver?

But he can't leave this step of the ladder—he has to look down through the fan, see Shaney's real self.

His hand, moving on its own almost, rounds to his back pocket, comes back with the knife from the elk hide. He looks at it like just seeing it, doesn't remember having grabbed it.

With this round-nosed skinning blade, though, it's a putty knife, an art knife, the widest flathead. He swallows it into his hand, works it up into the space between the spotlight's buried can and the crumbly ceiling.

"Need help?" Shaney asks, and Lewis looks to her, shakes his head, and finally sees her for the first time. She's dressed for work, just normal clothes, same flannel as ever, but her hair's still done from, he guesses, the night before. It's spiral-curled but still forever long, over half her face.

No, Peta does not need to come home, find Shaney here looking like this.

But it's fake, too, Lewis reminds himself. She's showing him what he wants to see, she's making herself up like this specifically to get to him.

"So when are you coming back?" Shaney asks. "There's a pool at work, you know?"

"Tomorrow," Lewis says, straining with the fake adjustment he's doing on the light. "Day after."

"Make it Friday and I'll split the take," Shaney says.

"What's the pool up to?" Lewis asks, because he's no dumb Indian.

Shaney just smiles, nods across to the headlight bracket, says, "Frankenface will know what this is?"

"His name is Silas," Lewis says.

"Past tense," Shaney says, and, stepping down into the living room at last, she reaches over to the bank of switches, turns the fan off, finding the control first try.

Lewis's heart drops. His face goes numb.

She knows exactly what he's doing.

"I'm saving your life," she says, and is to the couch now, is squatting down, knees together, to collect the bracket.

Lewis zeroes in on her through the fan but the blades are already sagging, losing their speed, the rate of flicker slowing, slowing.

Through them like that, Shaney's just Shaney.

"No, no, turn it on," Lewis says, pleading, holding hard onto the ladder. "The—the switches. When the fan's off, the power comes on for this light."

"What kind of bullshit wiring job is that?" Shaney asks, looking from the fan to the light in disbelief. But, using up the rest of the wishes Lewis has left for his whole life, the bulb in the spotlight flickers the slightest bit. Just the filament glowing on for a moment, but it's enough.

Lewis looks from it down to Shaney and she shrugs, cradles the bracket, careful of the bolts he intentionally left hanging, and crosses to the switches, turns the fan back on. It whirs back up like sad to have been turned off after such a long and constant run.

"Oh, hey," Lewis says, pointing with the knife to the carpet in front of the couch. "That fall out?"

The fan pushes Shaney's hair across her face but she clears it, looks down to the bracket, touches the three bolts in the outer ring, shrugs up to Lewis.

"There, there," he says, still pointing, and she steps across, her legs coming into the tunnel of vision the spinning blades are carving, but then, her face just out of it, she looks up, says, "You trying to see down my shirt, Blackfeet?"

She looks down to her own chest and, instead of pinching the flannel together at her throat, she pops it out, lets it slap back, then looks up to Lewis with a little bit of devil to her eyes.

"No, no," Lewis says, coming a step down the ladder to see her face through the blades, but, at this angle, at this getting-there speed, it's just Shaney.

Shit.

Shit shit shit.

But still, that she knew not to step into that space? That she knew to turn the fan off?

"If it's a loose *connection*," she says, "you have to, like, jam something in right there." She points with her free hand to the spotlight, and because he's the one faking this repair job, Lewis has to play along.

She steps ahead, to see the light around the fan, and Lewis takes a step higher, high enough to work the blade of the knife up beside the can.

Just like she said, the little spotlight glows on, holds steady.

"Pay me later," she says, already turning around, drawn by the elk hide on the table.

Lewis climbs down, follows her.

"What happened to it?" Shaney asks, still almost touching it but not quite.

"Neanderthals," Lewis says, the funniest joke.

Shaney just looks up at him, squinting.

In the fourth book, the one she just returned, there's Neanderthals slouching around the mall with their heavy spears and heavy brows, and it's kind of the running joke. Every time there's the next version of "Cleanup on aisle nine," Andy shakes his head, spits air, and says, *Neanderthals*, like they were put in the mall specifically to ruin his life.

Lewis swallows hard, a thrill rushing through him, all the way to his fingertips.

"You're kind of weird, you know?" Shaney says.

"Me and Andy, yeah," Lewis says.

While it might be possible to somehow forget "Neanderthals," each installment in the series is Andy the Something: *Andy the Water Bringer*, *Andy the Giant Slayer*, *Andy the Unemployed*. The fourth one is, of course, *Andy the Mammoth Rider*. No way could she not know who he's talking about, here.

Shaney holds Lewis's eyes for a moment like checking if he's for real, then makes the turn to the front door.

"Wait," Lewis says, his heart pounding in his chest, his face heating up with possibility. "That's the wrong one," he blurts, making this up as he goes.

Shaney looks down to the headlight bracket she's holding.

"I think about anything will match his bike," she says.

"I've got the right one," Lewis says. "It's just out here, I just saw it . . ."

Shaney just holds him in her stare, like waiting for him to draw the curtain on whatever charade or joke this is going to be.

"You can go out this way," Lewis says, stepping past her for the garage door, not giving her a chance to say no.

He grinds the big door up, which brings the light on, then he just stands there inspecting all the parts strewn across the concrete and boxes and old towels.

"What happened?" Shaney says, impressed.

"It always gets like this," Lewis tells her, trying not to let on how fast things are cycling through his head. It's like that flicker rate he needed to see through to the real, it's behind his *eyes* now.

He's afraid to look directly back at Shaney. Instead he jogs three steps ahead, knocks on the basketball to get it dribbled up.

"Forgot your best friend the other day," he says, and underhands it across to her, hardly even a pass at all. Instead of one-handing it to her hip like he knows she can do, she steps to the side, lets the ball lob on past, still watching him like trying to figure him out.

"Oh, oh, you'll like this," he says, stepping across to the Road King.

"I've got to get to work, Blackfeet," Shaney says back, trying to pick a path to freedom.

"Just wait, wait," Lewis says, and comes around to the other side of the bike, the side without the purple crate. He crosses the poles, throws some sparks, but pulls the current before the engine can start up, so that it sounds like a failure, like it choked down. "Shit, shit," he says, shaking his hand like he got burned, and leans over to look inside. "Oh, of course," he aha's, and, without looking up, does his right hand to pull Shaney over.

She approaches slowly, uncertainly.

"I've heard a motorcycle before," she tells him.

"New pipe," Lewis tells her, and is still guiding her closer, closer. "You've got to—" he says, finally looking up. "Here, here," he says, taking the headlight bracket from her, setting it down wherever on his side. "This vacuum hose right here, just plug it so I can turn the engine over. Should start right up."

Shaney inspects the Road King like for safety, says, about the rear wheel and all the dangerous possibilities there, "How about just show me when you've got it together, yeah? I promise to be real impressed."

"I want you to tell Silas," Lewis says then, like embarrassed to be admitting it. "Don't tell him this part, it's still secret, but I ordered him the same exhaust. As, you know, apology. Just tell him how throaty this sounds."

"I don't have to actually hear it to—" she begins, but Lewis is already leaning over, guiding her finger to the open end of the vacuum tube, its junction right there, which would be so much easier to stopper it with.

Is this the first time he's touched her skin to skin? It might be, he thinks.

There's no sparks, no flood of memories or accusations rushing in, no replay of four Indians pouring lead down a slope.

"You're making me late, Blackfeet," Shaney says, and Lewis, coming back to his side, realizes he *can* see down her shirt for a flash.

She sees him seeing, says, "All you had to do was ask, yeah?"

"No, it's not like—" Lewis starts, and then they both turn to the basketball, slowly rolling down the slope of the garage like the biggest, softest, most drunk pinball.

"This better not get me dirty," Shaney says over the wide tank, her eyes locked on his, like she means the exact opposite.

"I know who you are," Lewis says back right as the engine cranks over, and that narrows her eyes because she couldn't have heard that right, and, just like that last day hunting, this is the moment where he could back out, this is where he could stop this from happening. He could break the connection, kill the bike, make like he said something else—*Watch out for your hair*, yeah. That would be the perfect thing to have said.

Except that's exactly what he doesn't want her doing.

She killed Harley, he makes himself remember. She killed Harley and she's trying to turn Peta against him. And the final way he can tell she is what she is, the thing that gave her away for sure, even more than the basketball, it's that she doesn't know about Andy the Mammoth Rider. The books were just props to her, just excuses to be over here, turning the wheels of her plan. If she really *had* read book four, then she would have ridden that emotional roller coaster Lewis and the rest of the world had to ride after Andy "died" at the end of book three, and then was a no-show for the first half of book four. He wasn't pulling a Gandalf, though, was just trapped in the fizzy world inside the fountain drink dispenser, waiting for the right set of circumstances to get born into the world again—which turned out to be a mammoth one of his ancestors had driven over a cliff, except that mammoth had fallen into a pool that was right where the fountain was. So, when that mammoth fell *again*, like happens when time is a cycle, the elves carved Andy up from its belly. At first he was just a skinny mammoth fetus, but then he grew into himself over the course of a single day, rode that dead mammoth's mate right through the makeup and perfume department, became the true champion and savior he was always meant to be. It wasn't the kind of return a person forgets, especially if she just read it.

"You know who I *am*?" Shaney says over the scream of the *un*muffled four-stroke, still holding that dummy vacuum tube shut with the pad of her index finger, and that's all she gets out before this thing is happening.

The drive belt is on Lewis's side of the bike—Shaney isn't stupid, *Elk Head Woman* isn't stupid, she would have clocked that danger, that little conveyor belt of instant death—but he started it in first gear, meaning that naked rear wheel he's already got in place, *it's* spinning as well, is an instant silver blur.

It takes maybe half a second for those chrome spokes to grab her long spiral curls, crank her head both up and to the side, her neck obviously cracking. But her hair's still pulling, still winding into the spinning spokes, the *flickering* spokes. An instant after her neck breaks, the top of her head scalps off and her forehead tilts loosely down into the rear wheel, the spokes shearing skull as easy as anything, carving down into the pulpy-warm outside of her brain. It's greyish pink where it's been opened, and kind of covered with a pale sheath all around that, the blood just now seeping into the folds and crevices.

Lewis backs off the throttle, lets the starter wires go.

Silence. Just that bare wheel winding down. Shaney's throat is still sucking air in, her eyes locked on Lewis, calling him *traitor*, calling him *killer*, calling him "Blackfeet" one last time. Then she falls back, slumping into the sleeping bags and random parts, her left foot twitching, a line of saliva, not blood, threading down from the corner of her mouth. But there is bright red aerated blood—a spattery stripe bisecting the garage, going from floor to wall to ceiling then down the other wall again. It's a line between who Lewis used to be and who he is now.

He stands, pushes the button on the wall.

It's time to lower the door on all this.

STILL TUESDAY

Lewis never built the sweat he wanted, but if he stands in the up-stairs shower long enough that it's all steam, he can pretend, can't he? The blood and brains Shaney splashed on his face swirls down the drain, is gone forever.

Her little yellow Toyota truck is still pulled up outside, but once he's clean he can drive it wherever, walk back, no witnesses. *Yes, Officer, she stepped by for a part, but she left with that part. The proof of that is—hey, that part's not here, right?*

Easy as that.

Now the next ten years of his life can start, finally. Payment came due for that young elk, for all *nine* of those elk—ten if the unborn calf counts—but, this far from the reservation, he just managed to duck paying it.

As for what to do with Shaney herself, his first instinct is to bury her with Harley, but the cops are going to be digging there before too long, and the noon train's coming through anyway, and he doesn't need an audience for that kind of work.

No, for once in his life he's going to be smart about a thing. And he's not even really a killer, since she wasn't even really a person, right? She was just an elk he shot ten years ago Saturday. One who didn't know she was already dead.

Still, his soapy hand, when he raises it into the stream of hot water, it's trembling, won't stop trembling. Twice so far he's ripped the shower curtain to the side, sure he saw a shadowed figure standing out there, sure he'd heard a door creak, or footsteps. *Hooves.*

It's just nerves, he tells himself. Any first-timer would be having the same exact panic attack now.

He puts his face into that scalding water some more, promises himself not to crank his brain up, but it cranks up anyway, can't stop dwelling on how could it be that Shaney didn't seem comfortable around a basketball, didn't automatically catch it like any real player would have to, just let it slip past like an object, not a thing she'd sweated countless hours over.

But, *could* she still be that same player she was before? Did she sidestep that pass because she was holding a delicate bracket? Did she dodge it because, unlike him, she's not obligated to catch a ball she didn't ask to be thrown?

It doesn't matter. What does is that she didn't know the books.

Lewis steps out, towels off, his shape blurry in the mirror.

She didn't know about the books, he repeats in his head.

Meaning?

Meaning she was Elk Head Woman.

Because?

Because she was lying.

That means she's a monster?

Lewis squats down in the hall, his face in his hands, his head shaking back and forth to resist this line of reasoning.

No, he finally has to admit to himself.

It doesn't mean *for sure* she's that monster, but added together with the basketball being so alien to her, and her knowing where to stand in the living room, and to turn the fan off, and, and: What about how she wouldn't touch her own hide on the kitchen table?

Lewis stands nodding.

That, yeah.

She could have been lying about the books just for an excuse to break up his marriage, because that's what she does, that's the *human* thing she just does, but touching the skin she'd worn the last time she was alive, that probably would have made her relive her first death all over again, wouldn't it have?

Lewis nods. It would have, yes. Definitely.

Oh, and also: She was lying about having to leave an hour early, wasn't she? What she really wanted was more time alone over here *before* work. And Lewis can prove this.

He calls work again, gets Margie on the line again.

"You must really want to hear my voice," she tells him.

"Shaney," he says, switching ears like that can get across how much he doesn't have time for small talk, "she—she never showed up, but if I run I think I can catch her, but I don't know her address— the flower farm, right?"

It's stupid, nobody lives over there, it's probably not even zoned for residential, but it was the first sort of close place he could dredge up.

Margie's silence means she's weighing this for the bullshit it is.

"Please, please, Jerry'll kick my ass if I'm not there," Lewis adds, bouncing up and down like that can help make his case.

It's that easy to get an address.

Moments later, still in his towel, he has the fold-out map of Great Falls spread over the back of the elk hide, his hair dripping all over the red and blue lines.

"No way," he says when he finally finds Shaney's place.

She *did* have to leave an hour early, because she *does* live all the way on the other side of town—Gibson Flats. That's not even really Great Falls at all, is it? But, at the same time, she *did* really leave with the books. And they *have* really been showing back up, haven't they?

Lewis plunks down into a chair, his eyes lost.

Finally, an explanation bubbles up.

It's thin, it's anorexic, but: What if she read the first chapter or two of book one, and it wasn't for her, was just stupid elves in trench coats, halflings at the hot dog stand, so she drove the whole long-ass way over, dropped all ten books off on Lewis's porch?

That would explain her not knowing about the story, the people in it.

But, then, who found that stack? Who's been parsing them out one here, two there? And why?

To make you do what you did, Lewis hears in his head, in a colder voice than his own.

He stands breathing hard, shaking his head no.

She *was* Elk Head Woman. She had to be. She was—she's the only Indian in his life down here, right? Lewis *sees* other Indians out and about, but that's always just a nod without stopping. No, if it's going to be anybody, it's her.

There is one last way to tell, though, isn't there? One way she wrote in the back of the third book?

Lewis goes to the corner of the living room behind the ladder, comes back with the red-handled screwdriver.

Next, the garage, the mound of sleeping bags and blankets.

He pulls them back from Shaney, tries to close her eyes that won't stay closed.

Not gagging even a little, he threads her scalped hair up from her mouth where he'd stuffed it, thinking that had to be some old-time Indian shit. That she didn't fight back when he was burying her in the sleeping bags and blankets, that meant it worked, the hair in the mouth.

This next part, though. Getting her hair into the spokes of the Road King, that was easy in comparison. This is a lot more involved.

But Lewis has skinned out he doesn't know how many elk and deer. Once even a moose, right? He's even delivered an embryo or a fetus up from a pregnant young elk, one he didn't tell Peta was still kind of struggling in its thin, veiny bag.

He can do this.

First is opening her mouth with his fingers, then it's forcing his hand in as deep as he can and pulling down hard, breaking the jaw at the crunchy-wet hinge so he can have the kind of access he needs. So he can see her top row of teeth.

Elk Head Woman told him what to look for, didn't she? She told him how he'd know her.

(ivory)

He positions the screwdriver between a canine tooth and the next one over and jams the red handle with the heel of his hand, driving it in deep enough to lever out the canine he needs, bloody root and all. Because she's fresh, the tooth doesn't want to let go.

It does, though. Along with the one he was using for leverage.

Lewis rattles them in his hand, considers it lucky to have accidentally pulled two. That way he can compare them: normal and ivory.

Except both of these are the same.

He gets some carb cleaner, sprays them down because there's got to be ivory in there somewhere. When there's not, he closes his eyes, falls to his knees on the sleeping bags.

Next he's laughing to himself, and sort of crying.

Work didn't *have* to be so short-staffed, did it?

NATIVE AMERICAN MAN SINGLEHANDEDLY TAKES DOWN USPS.

He's trying to work a grin up about that headline when he finds his eyes fixed on the stomach of Shaney's flannel shirt. Because that's not where the blood and damage is, it's as safe a place as any to concentrate on, maybe even better than most. But . . . no. No no no.

He can *untuck* that shirt if he wants, can't he? He can untuck it, pull it up to check if she's got that long vertical knot of scar tissue. If she does, if she *was* field-dressed, then—then she was definitely Elk Head Woman.

Unless she got butchered on an operating table. Unless some drunk IHS doctor scarred her for life, made her into a woman always trying to get eyes to focus on her chest, not below that.

Lewis shakes his head no, doesn't want to have to do this, doesn't want to have to know either way. What if she doesn't have that scar at all, right?

Still, he owes it to her, even has his hand to her stomach, his fingers bunching the flannel up into his palm, is *going* to look around, face this truth. On the count of three. Now on this next count of three.

What saves him—who, *who* always saves him—it's Peta.

The front door opens, then closes.

Shit.

Fast, fast, he gets Shaney buried again. He can still make this work. Peta doesn't have to know. That's hydraulic fluid splashed all over the ceiling and walls. It smells like opened body because of Harley.

It takes thirty seconds to stop hyperventilating, and then another full minute to clear his eyes.

Nodding to himself for strength or something like it, Lewis walks into the kitchen, ready to jerk his head up like surprised by Peta being here, unloading her lunch box. She's not at the counter like always, though. To find her he has to look up, and up.

She's . . . now *she's* on the ladder?

"You figured it out!" she says, her whole face a smile, like the last two or three back-to-back shifts don't even matter, suddenly.

"What?"

"The whole *thing's* loose," she says, and wiggles the knife jammed in alongside the light.

The bulb flickers on, goes back off.

A warm smile crosses Lewis's face.

He did fix it. This is the perfect gift, the best surprise. He is a good husband.

Smiling like it was nothing, he walks past the table, into the living room, and only slows when he catches the way Peta's looking him up and down, slow and unsure.

"What?" he says, and only then does he look down to his hands, wrist-deep in Shaney's blood. It's probably splashed onto his chest and face, too, from the tooth extraction, and it's not hydraulic fluid—the Road King doesn't *have* this much hydraulic fluid.

The red against the white of the towel is unmistakable.

"Are you ok—?" Peta says, eyes fixed on him, stepping down the ladder without looking, probably concerned that he's hurt himself bad enough to go into shock, and that's how it happens, that's *why* it happens: because she's worried he might be cut somewhere. The left toe of her work boot that she's no longer paying attention to, thick for safety, misses the next rung, and her other foot was already shifting, and she knows better than to clutch onto the sides of the ladder because that just means bringing it down with her, and there's already one gouge in the wall.

Just from instinct, probably, her hands slash up, to try to find something to hold her.

What she finds is the handle of the knife jammed in alongside the spotlight's can. She brings it down with her, and, instead of bursting across the room to tackle her into the wall like a good husband would, Lewis stands there clutching his towel, watching this happen in what feels like the slowest motion ever.

Anybody but Peta would fall down into the rungs, get tangled up, break their fall with their basic awkwardness.

She used to be a pole vaulter, though.

She knows to push off, to arc away.

It's beautifully executed, and, being her, she even manages to fling the knife to the side so it won't impale her on landing like Lewis guesses he was expecting, since everything else is already going wrong.

Peta's used to falling onto huge mats, though. Not the sharp brick corner of the fireplace hearth, back-of-the-head first.

The cracking sound of her skull opening up is distinct, and permanent, and looking away doesn't help Lewis process it, or accept it.

Just like with Harley, he doesn't rush up to hold her in these last moments.

He just stares, shocked.

Her body spasms and her breath hitches for maybe ten seconds, her eyes locked on him like trying to communicate something, like . . . like trying to relive the last ten years with him? Like, now she can go back to sitting at that picnic bench in East Glacier and start the two of them all over again, live it right up until now. What *was* she drawing that day? Was she drawing her dream house, complete with fireplace and a little apron of brick before it, "hearth" labeled above it? Did she know all along this was what was going to happen, but then did it anyway, because these ten years were worth it?

"Peta," Lewis finally says, seconds and a lifetime too late.

The corners of her mouth grin just a little and then, like has to happen, her hips die down with the rest of her, like electricity's leaving her body, is running into the ground or wherever it goes.

Lewis is still just standing there.

Peta's blond hair is staining red, spreading out into the pale carpet. It's not a stain he's getting out. No, they're not getting the security deposit back.

"Hey," he says when he can, once he's sure it's too late, once he's sure she's not going to be answering.

Her pupils are fixed and dilated, her mouth open in a way she'd never let it hang in life.

Ten years, Lewis says to himself.

They made it ten years.

That's a pretty good run, isn't it? For an Indian and a white woman, especially when she outclassed him so much, and when he had all the usual baggage?

And—and maybe he *was* thinking right all along, he tries hard to believe. Maybe she appeared the summer after the Thanksgiving Classic for a reason—because she wanted to move from place to place with him, get him to invest his whole life in her, then stage a grand death scene like this, one he'd never be able to shake, would always be running from.

Wouldn't that be the best revenge? Death is too easy. Better to make every moment of the rest of a person's life agony.

Like with Shaney, though, there's a way to check. A way to know.

Lewis steps up onto the ladder to pull the knife from the wall where it stuck. Because the way he broke Shaney's jaw open let her teeth cut into his wrist like a bite from the wrong side of the grave, for Peta he grabs her chin from the outside, sets his knee against her forehead, and cracks the hinge that way.

Her teeth come out so much easier. All of them, like they were just waiting, were hardly in there at all. Maybe that's a difference between white people and Indians?

Lewis lines all her teeth in a rough crack of grout between the bricks of the hearth.

None of them are ivory either.

He sits back, hugs his legs to him, resting his chin in the cradle between his knees.

This is a thing he did, a thing he's definitely done.

Planes are probably going to be crashing into the terminal for months now, and mail's going to be piling up on the dock at the post office.

In addition, two women are dead who probably didn't have to be.

Lewis stares into what would be the fire if the chimney weren't boarded up—the lease says no open flames, only gas grills—and then he has no real choice but to smile when the light in the ceiling flickers on, even without its can being pressed to the side.

It's shining its bright little spotlight down on Peta. On her . . . stomach? Her belly?

Because everything means something now, what Lewis flashes on is that up-and-down scar that either is or isn't on Shaney's stomach in the garage.

It's a scar he knows for sure Peta doesn't have. But still, he's thinking of it for a reason, isn't he? Or, the world showed it to him in the driveway that day for a reason.

Soon enough, that reason pushes against the tight fabric of Peta's uniform shirt.

Something's struggling under there. Something's moving.

It's like—it's like when Andy was trapped in the belly of that dead mammoth, but for Lewis, for this, it's not a mammoth in there, is it?

"Indians like it bareback," he hears himself recite, and chuckles about it.

Some of his salmon *were* pretty good swimmers.

Sure, it's only been, what? Two nights? That's plenty, though. Nine months would be a luxury, an indulgence, would take so long he'd probably forget. And anyway, Peta would be all falling apart by then.

Forty-eight hours to gestate feels just about perfect, to Lewis.

He can even see some tiny limb pushing against Peta's skin now. Something in there suffocating, drowning, fighting to live.

His plan, only half formed but that's how he does it, had been to stand after a few more minutes, stand and feel his way outside, flop over the fence, sit between the two rails of the train tracks, wait for the Thunderball Express to come, deliver his judgment at sixty miles per hour, its air horn filling the whole world with sound.

But this is something new, something unexpected, something wondrous.

Lewis never thought he might be a father.

There's still hope, isn't there?

This can all still work out.

Using the same dull knife he used to pry Peta's teeth out, the same one he used to carve open that young elk ten years ago, he slits the tight skin of Peta's swelling-up belly.

A thin brown leg stabs up and he grabs on to it, traces it to its terminus.

A hoof, a tiny black hoof.

Lewis nods about the rightness of this, pulls that leg gently, his other hand ready.

Two days later, his elk calf wrapped in its mother's ten-year-old hide, he wakes under a rock ledge that's partway between the rent house in Great Falls and the reservation he still calls *his* reservation.

Shaney's yellow Toyota truck is two or three miles back on the plains, tucked just in from the gas station where he's pretty sure someone called him in, from all the headlines that made the rounds Wednesday: NATIVE MAN ON KILLING SPREE, TWO DEAD SO FAR, BABY MISSING.

The story on 12b is him, here, sleeping in the cold like the bow-and-arrow days. The story on 12b is how he looks up into the huge white flakes drifting down out of the sky, just like the Thanksgiving Classic.

He raises his face to those cold wet flakes, closes his eyes, holds this elk calf he's been calling his daughter close to his chest. She hasn't been growing as fast as Andy, and she hasn't moved since stabbing that leg up into the air, but she will, he knows. He just has to get her home, to land she knows, to grass she remembers. He'll watch her grow for the rest of the year, keep the coyotes and wolves and bears away, and, when she can, he'll let her go on her own, stand there crying from sadness, from happiness. And then it'll all be over. Indian stories always hoop back on themselves like that, don't they? At least the good ones do.

Lewis smiles, pulls her tighter, breathing heat down onto her thin ears, and on the ridge above him there's four men sighting down along the tops of their rifles. He looks up to them, his lips moving, trying to explain to them what he's doing, how this can work, how it's not too late, how they don't need to do this, it's not like the papers have been saying, he's not that Indian, he's just him, locked into the steps of this story but finding his way now, finally, making it all work.

When they shoot is when he finally feels what he's been waiting for, what he's been gambling on, praying for: long delicate legs against his chest, kicking once, twice, again. That small head nuzzling up into his neck, and the long lashes of big round eyes brushing his cheek as they open, and then closing against the mist of blood that, for the moment, is its whole world.

Her name in Blackfeet is Po'noka.

It means elk.

THREE DEAD, ONE INJURED IN MANHUNT

Four Shelby men were attacked last night, following the apprehension of the fugitive Lewis A. Clarke (see Wednesday's edition), who had apparently been fleeing back to his tribe's ancestral reservation. Clarke was the main suspect in the brutal murders of both his wife and a federal coworker.

Reports indicate this group of four hunters had been out all day, aiding in the search for Clarke. Representatives for the highway patrol say that while armed citizen patrols of this sort might seem helpful, staying off the roads is actually more beneficial to recovery and apprehension efforts.

Reports that these four Shelby men were actually the ones to find Clarke, now deceased, are unconfirmed.

According to sources at the hospital, who were able to speak to the lone survivor before surgery, the four Shelby men had in the back of their truck both Clarke and the deer or elk calf he had apparently been carrying for reasons unknown.

At some point in the drive back to town, according to this survivor, someone stood from the bed of the truck while it was moving. It was a girl of twelve or fourteen, Indian. Presumably she had climbed into the truck earlier, when it was going west.

When the driver of the vehicle slowed to keep her from falling or blowing out, and alerted his three cab mates to her, the survivor says the girl "rushed forward over the toolbox" and "through the rear window" into the cab, which is where the eyewitness testimony ends.

Anyone finding an Indian teenager perhaps hitchhiking or loitering is advised to alert the authorities.

Names of the dead and injured are being withheld until families of the slain can be notified.

More on this story as it develops.

SWEAT LODGE MASSACRE

FRIDAY

The way you protect your calf is you slash out with your hooves. Your own mother did that for you, high in the mountains of your first winter. Her black hoof snapping forward against those snarling mouths was so fast, so pure, just there and back, leaving a perfect arc of red droplets behind it. But hooves aren't always enough. You can bite and tear with your teeth if it comes to that. And you can run slower than you really can. If none of that works, if the bullets are too thick, your ears too filled with sound, your nose too thick with blood, and if they've already gotten to your calf, then there's something else you can do.

You hide in the herd. You wait. And you never forget.

What you do after you've made your hard way back into the world is stand on the side of the last road home, wrapped in a blanket torn from a wrecked truck, your cold feet not hard hooves anymore, your hands branching out into fingers you can feel creaking, they're growing so fast now. The family of four that picks you up is tense and silent, neither the father nor the mother nor the son saying anything with their mouths, only their eyes, the infant just sleeping. They make room in the backseat because if they don't stop, someone else will, and the father driving the car says that that never ends well for starved-down fourteen-year-old Indian girls wearing only one thin blanket.

You're fourteen, then. Already.

Just a few hours ago you're pretty sure you were what he would have called "twelve." An hour before that you were an elk calf being cradled by a killer, running for the reservation, and before that you were just an awareness spread out through the herd, a memory cycling from brown body to brown body, there in every flick of the tail, every snort, every long probing glare down a grassy slope.

But you coalesced, you congealed, you found one of the killers about to spark life into the body of another, a life you could wriggle into, look out of. He had to be groomed first, though, groomed and cornered and isolated.

It was so easy. He was so fragile, so delicately balanced, so unprepared to face what he'd done.

You settle into the soft fragrant backseat of the car taking you the rest of the way home. The father behind the wheel turns the radio knob constantly backward and forward, looking for a song that may not exist, and the mother beside him, holding the new calf—baby, baby baby *baby*—to her chest, she's staring out the side window, maybe at all the dry grass slipping by.

The boy in the backseat with you smells like chemicals. They steam up from his skin and his eyes are wet and mad behind his waist-long hair, and in him you can sense all his ancestors before him, and you're surprised he doesn't recognize you for what you are.

You say something to him in your own language, this mouth and teeth and throat and tongue not made for the shape of these words, and the boy just stares at you for a long moment, says, "What are you?" and then rotates in the seat, away from you.

So he does sense you. Just, not in a way he can acknowledge.

Good, good.

If you can't interact with him, then interact with the grassland sweeping past with so little effort. Lean over to the center of the backseat to see the white mountains rolling up. It feels like you're rushing, like you're stretched out and running. You can't help but smile a bit from the feeling, from the velocity. It's your first smile in this face. On the last hill down to the town, though, your smile droops when the necessary memory rises, from seeing the train tracks from far away and above like this.

The memory is an old one, not your generation but a few before, a thing that happened right there to the south, just past where the last fence is. The memory is of how the herd came down here in the night. How they found good grass closer and closer to the buildings, where nothing much ever grazed, and then they kept eating and eating, bloating their sides out because they needed this to get through the winter that was coming.

But then the hunters stepped out onto their porches, saw tall brown bodies in the waving yellow, and reached back inside for their rifles.

They approached on their bellies all morning, and the herd knew they were there, their smell so tangy, their crawling so loud, but the grass was good and the horizon was open on the other side from the hunters. The herd could run as one when they needed to run, could dig in with their hooves, bunch their haunches and burst away, move like blown smoke across the rolling prairie, collect in a coulee they knew. The water that ran through that rocky bottom was already trickling through their heads. From its taste they knew exactly where it came from in the mountains, and its whole story getting here.

They didn't know about trains, though. Not like the hunters did.

When the locomotive and all its boxcars thundered through, the scent hot and metal, it was as if those tracks in the grass had stood up. They became a flashing moving wall of sparks and wind no elk could run through (one tried) and the screeching and tearing of those great metal wheels covered the boom of the hunters' rifles firing again and again, until the sound of the rifles and the sound of the train were the same sound, and in the backseat of this impossibly fast car you reel from the acrid taste of this memory, causing the chemical boy in the backseat to pull away from you even more, but it was fair, what happened that day, and it had been the herd's own fault.

You run when you first taste hunters on the air, don't you? When you first even *think* that might be their ugly scent. One more pull of grass isn't worth it. Even if it's good and rich. Even if you need it worse than anything.

Knowledge of this day lodged in the herd, got passed down like what headlights meant, like how those blocks of salt aren't for elk tongues in the daytime, like how the taste of smoke means to walk somewhere else slowly, head down, feet light. The price of the knowledge about trains had been high and the coming winter harder, as less hooves means more wolves, but the herd didn't feed down near town anymore, and they didn't trust the metal tracks anywhere they encountered them, knew they could stand up into a sudden wall.

Instead they kept to the high country, the lonely places where the air tasted of trees and cold and the herd, the places the trucks never lurched into.

Until one did.

In the backseat of this speeding car you thin your lips, remembering that day as well.

An elk mother, cornered, will slash with her hooves and tear with her mouth and even offer the hope of her own hamstrings, and if none of that works, she'll rise again years and years later, because it's never over, it's always just beginning again.

The father lets you out in the parking lot of the grocery store, where you've told him with this new voice that you can call your aunt, but you're really only there to dip into another car that's not even locked. You stand from that car with a duffel bag of clothes, never mind the starving dogs circling around now, snarling and tearing at the air, the wiry hair at their spines maned up, tails tucked over their soft parts.

You snap your teeth back at them, watch them roil with spit-flecked rage, writhe with desire they want you so bad, want you gone even worse.

Town is such a funny place, isn't it?

You can't rid yourself of the annoyance of these dogs without drawing more annoyances, though. But you won't be here long.

It's another thing the herd's always known: never stay in one place. Keep moving, always moving.

First, though, one of their calves is sitting in eighth-grade geography—*girl*, girl girl girl, not "calf." And this girl has this certain father you remember, and that father, he has a friend you remember as well, from looking up a long snowy slope, their monstrous forms black against the sky.

For them, ten years ago, that's another lifetime.

For you it's yesterday.

THE GIRL

Her name is Denorah. Her dad used to tell her she was supposed to have been *Deborah*, since that was the name of one of her dead aunts, but his handwriting had never been so good, and then he'd smile that sharp-at-the-right-side smile that had probably been killer in high school, a hundred thousand beers ago.

Her dad is Gabriel Cross Guns. He's the one who shot the hole in your back, took your legs away.

In Denorah's eighth-grade geography right now, six days before Thanksgiving turkey, Mr. Massey is saying that all of the details aren't in, that it might have been highway patrol that shot that Native American man just off the reservation, that it doesn't have to have been vigilantes or militia, even though this state is stacked deep with the second, all of them hoping to be the first.

"*Native American?*" Tone Def says back to Mr. Massey. "I thought he was Blackfeet."

"Tone Def" is Amos After Buffalo's hip-hop name.

The class falls in behind him against Mr. Massey. Not because they care, but because it's fun to harass the white teacher.

Tone Def Amos—the name he's earned—stands up from the

desks, calls out about is there really some big *difference* between state troopers and idiots with guns anyway, at which point Christina or one of them over by the window says back that this dead Indian who nobody on the reservation really remembers anyway, he killed his wife and gutted the baby from her and pulled all her teeth out, didn't he? Does it matter who shot him down for that? And now voices are rising and more kids are standing and a couple are crying already just from the tragedy and drama of it all and there's probably not going to be any geography going on today.

Denorah pages her spiral notebook over to a blank page and tries to think if she really even remembers this Blackfeet who got shot. His name, sure. His name was a joke, except this joke had been by the dead Indian's parents, who probably came up with it down the hall in history class. But it's hard to tease what she actually remembers apart from what she's been told twenty or thirty times.

There *is* this dim sort-of image of her dad and Cassidy, as he likes to be called now even though it's a girl name. They're crossing the living room before dawn on a Saturday, and Denorah's sleeping on the couch, just a kid, not even in kindergarten yet. At the door are two not-yet-dead Indians: the joke-named one just shot down over by Shelby, and Ricky Boss, who she's pretty sure died from getting beat up outside a bar over in North Dakota. Unless that was somebody else. But she's pretty sure about the North Dakota part.

Anyway, Denorah remembers this early morning living room not because it was the Saturday before Thanksgiving, and not because her real dad and Cassidy were being too loud with their hot-hot coffee. It's not because the knocking at the door woke her from sleeping on the couch, either. Her real dad long-stepped past her to keep the door quiet, Cassidy right behind. It was Ricky Boss and that other dead one, Lewis, with rifles slung over their shoulders and sleep in their eyes. The only reason Denorah has any of this left ten years later, now, it's because when Ricky Boss was slurping from his coffee, the steam like a veil in front of his eyes, he was looking straight through it at her on the couch, like he knew what was going to happen that day, hunting, and wanted instead to just stay inside, drink the rest of his coffee.

It was something she would have tried to draw, once upon a time. When she used to draw.

It had started in sixth grade, the drawing, two years ago, before she got serious about basketball. It was right after the museum visit. A class project. It didn't matter that they were all drawing just in spiral notebooks. Miss Pease, who's her aunt now, explained how, back when, ledgers were the spiral notebooks of the day.

Denorah would never admit it no matter what, but she'd believed Miss Pease that day. Sitting in the second row, she hadn't even had to close her eyes to see the picture of an old-time lodge, inside it all manner of things for sale: beaver pelts, pipes, braids of sweetgrass, hunks of boiled buffalo meat with brown-colored ropy string through them (for hanging on pegs), pounded-flat strips of pemmican (yuck), beaded bags like at the trading post for tourists, where the flaps are big to show off the beadwork, and, way back in the corner, a stack of blank ledgers. She knew she just had to put her finger on the fast-forward button of that picture, keep it pressed until that lodge grew shoulders, squared up into a building, a store, one with a school supplies aisle. Now the ledgers are spiral notebooks, just like Miss Pease was saying.

It felt magical in class that day, opening her spiral up to that new blank page, that modern-day ledger. She imagined it being in a museum, even pictured a class of sixth-graders single-filing it up to the glass display case someday, to see how the old ones used to do it, back when spiral notebooks were everywhere, back in that handful of years when Indians only had reservations, before they got all of America back.

The assignment was for class to draw their favorite holiday. It was supposed to be Christmas or Thanksgiving or the powwow from the summer like everybody else was doing, but what Denorah drew is from when her sister's team went to regionals in basketball the year before, the holiest day of all in her family, even though they weren't completely a family yet, since her mom and new dad were only dating.

This is the day her big sister, Trace, who's her new dad's daughter he already had, scored ten just in the first quarter, then eight in the second, and came back from halftime to go six for twelve, then in the fourth, when it was down to the wire and the whole gym was stomping and screaming, when the other team had figured out how to swing a double-team over to her if she so much as touched the ball, she passed out of it every time, to whoever had to be free, and racked up nine crushing assists in a single quarter. The other side was chanting, *Indians go home, Indians go home*, but Trace was home, all of this was home, no place more than a basketball court in the last thirty seconds.

What Denorah drew at the bottom right of the blue-lined page she'd quartered up into panels that day in sixth grade, it was her sister at the end of that game, the one free throw she had in the whole second half, a technical for illegal defense, only the way Denorah drew her arms, it isn't proper form at all. Her big sister's got her arms up and out like she's holding a bow, like the ball balancing in the cup of her outstretched fist is an arrow, one she's aiming up at the whole world.

She rode that historical free throw, that game, that *win*, to a full-ride four-year college scholarship down in Wyoming, and Denorah talks to her every week on the phone, big sis to little one, no "step" between. When she was done with her ledger art that day, when it got Miss Pease's "B–" scrawled top-right—"Is this really *Indian*, D? Shouldn't you do something to honor your *heritage*?"—

she mailed it to Trace, careful to fold along the panel lines, and Trace said Denorah got it just right, that's just how it happened, thank you thank you, Denorah should keep practicing, she's better than any twelve-year-old has any right to be, she'll tell her coach, she'll make her listen, this is A-plus work. A-plus *plus*.

It's two years after that B minus, though.

Denorah hasn't drawn probably since last summer. Not since her hands got big enough to make a basketball look like it's going one way when it's really going a completely different way.

She really is good. That's not just something her sister tells her. Coach says it after practice, to Denorah's new dad, and when he's home from his busy time in a couple of months, he's promised they're going to go up to the gym every night and run drills, work on her left-side attack—so long as she keeps her grades up. Because they don't just hand scholarships out.

New dad: "And why do you need to be sure to go to college?"

Same daughter: "Because you can't eat basketballs when you grow up."

Even though secretly she kind of thinks she can.

But still, she won't get good enough to live on basketball if she doesn't run all those drills, and she doesn't get to run those drills if she doesn't maintain a solid B average.

In the left-hand margin, Denorah pencils her grades in with the lightest pencil. It's her way of reminding herself that they're not stable, that they can change in an instant, with the least quiz:

> Pre-Algebra: B−
> Biology: C+
> English: B+
> Geography: A
> Athletics: AAA+
> Health: ?

So, health, once that six-week unit kicks in, can make all the difference, Denorah maths out. She draws three hearts by "Health" like hit points and shades in the first one, half of the second one. She pretends the red margin line on the left side of the page is a pole and draws some dramatic shade coming down from it. She thinks about how a good point guard holds the defender's eyes with her own, to keep that defender from watching the ball. She remembers back when spiral notebooks were Big Chief tablets, and how she used to think those came from Chief Mountain, and that her reservation was the only reservation that got them. She tunes in to Mr. Massey, trying to defend both the highway patrol *and* the Shelby vigilantes, trying to flip this turtle of a discussion onto its back to see the real issues scratched on its belly, but there's nothing there Denorah hasn't heard in her first three classes already, so she closes her spiral notebook and holds it closed, looks out the window at the storage trailer with the big dent in its side, from when a senior tried to push it over with his dad's truck and got expelled, joined up with the fire crews, and burned up before he even would have graduated.

But . . . what?

There's a figure standing there in the ragged shade of that storage trailer. A pair of eyes that blink once and resolve into a blank face so much like Denorah's, the long hair not braided, a bright white scrimmage jersey, gym shorts, knee socks—*My workout gear from the car?* Denorah thinks, leaning closer to the glass to see.

You look right back at her, your hair lifting all around your shoulders.

She doesn't know you yet, no.

She will.

DEATH ROW

Gabriel Cross Guns, right before lunch.

While his daughter he hasn't seen in going on two weeks is flinching back in her seat in geography, drawing her teacher's attention, he's raising a dusty rifle from the front closet of his dad's living room and trying not to make a big production of it.

The rifle's an old Mauser that his dad used to load with bird shot, so it could be a real mouser. The baseboards of the living room are pitted and scored from it, and there's a crater near Gabe's right eye that's not acne, but a ricochet that he was never sure if it was a pellet or salt or a shattered fragment of mouse bone or splinter of wood or what, just that it stung and it was close enough to his eye that he'd slapped at this sudden dab of pain without thinking, and probably just succeeded in pushing it in deeper, meaning he's got some salt or lead or some baseboard or some rodent in his face. It's a dot in his life he's always touching. It makes him feel like Cyclops from the X-Men, like he can place his finger to that dot, that button, that release, and glare a ruby optic blast at whatever he wants, blow it so deep into next week that nobody'll ever catch up with it.

He hasn't read comic books for years now, though, is only thinking about them from a couple of weekends ago, from a smoky couch he was sitting on that he's pretty sure Ricky's little brother died on back when—drowned, technically. Gabe was sitting on the couch because he'd woke on it with his boots on, and when he sat up he'd had to carefully extract his arm from deep between the cushions, down where the lumpy-flat mattress was folded over three times.

His hand had come up from that with a comic book, and he'd wondered if it was worth anything at the pawnshops in Kalispell, and thinking about pawning shit got him thinking about that old Mauser his dad always said he was going to sell if he needed some quick cash. It's supposed to be a historical gun, from World War I or World War II, that he'd inherited from one of *his* uncles, who got it on the actual battlefield.

Gabe wonders if shooting years of bird shot through the barrel at mice has worn the rifling down. To find out, he jacks the breech open, holds the butt to the light coming in through the front window, and looks down the barrel from the front.

As if he can tell whether the rifling's worn down or if it's at factory specs, yeah. What was he thinking? It's old anyway, right? Being wore down is what an eighty- or whatever-year-old rifle that's maybe from an actual German in an actual war is supposed to be, right? Anyway, the crispness of the rifling won't be what sells this rifle. What sells this rifle will be this goofy forestock that tapers up nearly to the end of the barrel and has what looks to be hand-carved checkering scratched in.

Gabe shoulders the rifle all at once, tracks an imaginary antelope bounding from right to left.

"*Lead it, lead it . . .*" he says, left eye closed, right sighting, then comes to a sudden stop on his dad's bored-with-this face.

His dad palms the rifle away, runs the bolt back to be sure there's no live round.

"Think I'm stupid?" Gabe says, turning sideways to get past his dad to the refrigerator.

"You can't have my uncle's war trophy," his dad says.

"Don't want it, it's too old," Gabe says back, twisting the top off a stubby bottle of V8. He doesn't like the way it coats his mouth like cold spaghetti sauce or the way it clumps down his throat like throw-up he's having to swallow, isn't even that fond of the way it pools in his stomach and boils in his gut, but, technically, it's not food, and he's supposed to be fasting today, for tonight's sweat. The rocks are already all heating up in the fire. By dusk they'll be crawling with red heat, ready to shatter if the handler isn't careful, and—Gabe hasn't told Cass this yet, and he probably won't tell Victor Yellow Tail, who's laying down a cool hundred for the sweat—but these particular rocks, they're from a scattering of old tipi rings he found way back in Del Bonito in August. Meaning they won't be the first Blackfeet to use them, ha. Maybe it'll make them better or hotter or someshit, right?

Anything helps.

It's not the first sweat he's thrown, but it's the first one he's thrown in honor of a friend just gunned down the day before.

As to what Lewis had been doing to get himself shot, that's the big mystery. Going crazy from marrying a Custer-haired woman, Gabe figures but knows better than to say out loud—yet. Give it a few months. Give it a few months and that'll be the joke going through the reservation.

The best jokes are the ones that have a kind of message to them. A warning. This one's warning would be to stay home. To not go postal.

It's what Gabe thinks he's maybe going to do right now, with his dad following his every step like Gabe's sixteen again, is only swinging by to thieve anything not bolted down.

"For recycling," his dad says, about the plastic bottle Gabe just banked into the white trash can by the back door.

"Oh yeah," Gabe says, casing the interior of the fridge some more, "Indians use every part of the V8, don't we?"

His dad grunts, settles the Mauser into the corner by the door, and canes across the linoleum, reaches into the trash for the clear plastic bottle.

Gabe shuts the refrigerator in frustration.

"How long since you've even shot a mouse?" he says. "That rifle's just sitting there. You know it."

"What are you wearing that for?" his dad says back.

The black bandanna tied high on Gabe's left arm, with the knot on the outside because that makes it look more like a headband, just, on his arm.

Gabe stands to his full height, always feels more traditional when his back's straight like a ramrod—well, when it looks like he's got a stick up his ass, anyway.

"You heard about Lewis?" he says to his dad. "You remember him, Lewis?"

His dad lowers his face as if rattling the right tape into some slot in his head, then comes up with an old-man smile, says, "Little Meriwether?"

"Still not funny," Gabe drolls. "Highway patrol shot him yesterday, yeah? Here, here, here," making up the bullet holes as he goes.

He watches his dad for a flicker of reaction but instead his dad says, "Didn't he die already once?"

"What? No. That's . . . you're thinking of Ricky, Dad. Ricky Boss?"

"Boss Ribs Richard," his dad says, putting faces with names.

"Lewis was trying to come home at last," Gabe says.

"For Thanksgiving moose?" his dad says with a smile.

Oh yeah: Turkey Day's not even a week out, is it?

"Is everybody wearing those on their arms?" his dad asks, circling his own left bicep with his hand.

"He was my friend, Dad. Cass is wearing it, too."

"Just the two of you, then?"

"Lewis had been gone a long time already."

Gabe's dad looks out the kitchen window, at the wall of the house right beside his, maybe. Who knows what old men look at?

"Do mice even come out in winter?" Gabe asks.

"It was my uncle Gerry's rifle," his dad says back.

"He's not coming back for it, Dad."

"He used to shoot prairie dogs with it," his dad goes on, a smile ghosting his lips up. "But only the ones who were wearing those German helmets."

Gabe has to spin away from this.

"His wife is dead, too," Gabe says. "Lewis's, I mean."

"Field-dressed her," his dad adds in.

So. The headlines are even circulating over here on Death Row, then. Great. Wonderful. Perfect.

"They don't know what happened yet," Gabe says.

"Meriwether . . ." his dad says then, casing the refrigerator himself now, probably taking inventory for whatever Gabe might have palmed. "He was selling that raccoon meat that one time, wasn't he?"

"I don't even know why I come over here," Gabe says, brushing past his dad, pushing through the front door he'd hung himself one many-beers day. It's not his fault it's crooked, though. Whoever framed the doorway must not have had a square. Or maybe it's whoever poured the foundation's fault. Or whoever came up with the whole idea of "doors."

He roars his truck to life and backs out without looking, finds the three unbroken teeth in his transmission that still allow first gear, and touches two fingers to his eyebrow, saluting his dad bye, if he's even watching.

Two houses later he pats the Mauser nosed down into the floorboard on the passenger side. His dad didn't even see him snag it when Gabe had brushed past. In court-mandated substance-abuse counseling once—completely unnecessary, but slightly better than ninety days in lockup—Neesh had explained to the ten little Indians in group about counting coup. How that's what all of them were sort of already doing, did they know that?

Twenty bored eyes looked back at him.

Counting coup, he explained, using his ancient-old hands to form each word, act out what he was saying, counting coup was running up to the baddest enemy and just tapping them, then getting clear before that enemy could bash you with anything.

That, he claimed all reverentially, was what each person there in group had already done: rushed up to overdoses, to freezing while doped up, to crashing a car because of impaired reflexes, to vomiting in their sleep and drowning—addictive behavior was the *big*-time enemy, couldn't they see? And the fact that they were all here meant they'd already ran up to it, had already counted coup on it, and gotten away with their lives. The question now was whether they would come back to the tribe proud of how close they'd got, or if they'd go back again and again, until the enemy got its hook into them all the way, left them in a ditch somewhere.

Gabe has always remembered that. "Counting coup." It's kind of what he lives by, isn't it? With wives and girlfriends, with jobs, with the law, with how much gas is left in the tank, and now with this: he'd counted coup on his dad, had actually been brushing right past him while his other hand snaked the rifle, swinging it forward with his left leg until the butt could catch on the steel toe of Gabe's boot, just like Denorah's foot had back when he'd taught her cowboy dancing.

But he knows better than to think about her.

Not because he doesn't want to, but because he won't stop, will have to go out and find something *to* stop his thinking. Either that or show up at Trina's front door again, apologizing, begging, asking her to give Den something from him. Maybe just a bottle of Sprite with a chance to win under the cap.

At which point the lecture comes, about how he can't keep showing up like this, about how she's at practice but don't go back there, about how don't call her that, it's *Denorah*, not Den, okay?

Better yet, don't call her at all.

Gabe pats the rifle, rolls it over so the hand-carved checkering won't rub away against the seat.

Snow swirls across the blacktop and, screw it, Trina can't tell him what to do. D's his daughter, too, isn't she?

Gabe hauls the steering wheel over, takes the turn that leads down to the school. Just to drive by. She knows his truck. Everybody knows his truck. They should invite him to drive it slow in the parade, let him rain candy out the window.

On the way down to the school, though, his mind churning through who might even have cartridges for a rifle this old and weird, he sees a girl walking on the opposite side of the street, away from school.

"D?" he says, letting his foot off the gas.

She's wearing a scrimmage jersey and shorts, probably for the game tomorrow, but has her hair down, like Denorah never wears it since she got all serious about ball.

That can't be her, can it?

Gabe coasts past and just glances over. In case it's not her, the last thing he needs is word getting out that Gabriel Cross Guns is creeping on the junior high set.

Is it her, though? And, shouldn't she be cold?

Right when he's cranking his window down to see better, you raise your face, level your eyes at him through your black hair blowing everywhere, and this is the first time you've seen him since that day, the air full of sound, your nose breathing in just blood, your calf gasping inside you, your legs gone.

Don't look away.

Make him be the one to break eye contact.

Listen to his truck accelerate away.

It doesn't matter that he saw you now, either. The next time he lays eyes on you, you'll be taller, different, better. Already these stolen clothes are getting tight.

SEES ELK

Cassidy is changing his name again.

From here on out, as long as it lasts, he'll be *Cashy*, he thinks.

It's payday in the Thinks Twice household. Or, the Thinks Twice camper, anyway. Not that Thinks Twice is his born name either, it's just what his auntie Jaylene always called him like to remind him what to do, but, in his head anyway, it kind of stuck.

In addition to his check, cashed and still in big bills, Gabe's sliding him forty just for rehabbing his old sweat and keeping the fire stoked all day. In the old days, which means up until last month, forty dollars extra would have pretty much turned itself into a cooler of beer. Just, *poof*, Indian magic, don't even need any eagle feather fans or a hawk screeching, just look away long enough for it to happen.

Since Jo, though, Cassidy is a new man. Gainfully employed— even more legal once he takes the driving test—home by an hour after dark most nights, up with the sun like there's this big long string tied between him and it. Who'd have thought a Crow would be the one who finally stepped in, saved his lame ass? Never mind all the sage around the camper, all her smudging. There *was* something bad following him, Cassidy finally had to admit, but it wasn't anything Indian. Or, well, it was pretty Indian, he guessed: a bench warrant. But it wasn't even for anything bad, was just an unpaid ticket, which can happen to anybody.

Still, he can tell Jo's waiting for the other shoe to drop. At the high school basketball games he hauls her to so she can start knowing everybody, he can tell she's always looking around, watching for a ten- or fifteen-year-old with pale eyes like his, even though he guarantees her he never slipped any past the goalie, that he would have heard *all* about it if he was owing anybody child support.

What he figures is that he's shooting blanks, just like all the Indians when they're fighting John Wayne, and what he blames for *that*, or thanks, is uranium in the water. Gabe and Ricky and Lewis had grown up down in Browning, which, the water's not perfect there, but you can usually drink it anyway. Cassidy has been living on his dad's place in East Glacier most of the time, though, where the water's cloudy with who knows what-all. It is kind of weird, though, he's always thought. Of him and Gabe and Ricky and Lewis, they've altogether only had one kid? He figures Lewis and that blonde he'd run off with might be waiting till the time was right like white people do, or maybe she already had some before Lewis, didn't want any more, but that *Ricky* never threw any kids before he died—it's not like he was ever careful, right? The only one of them to leave a kid behind so far, though, it's Gabe, and that was, shit, what? Fourteen years ago already? It's really been that long since him and Trina? At least he did it right, though. Denorah Cross Guns can flat-out *ball*. She's the one Jo's always standing up for at the games and telling to shoot, shoot, that this game is hers if she wants it.

She's right, of course, the girl has Trina's drive on the court, not Gabe's interest in what's happening behind the gym, but still, the first time Jo came up out of her seat like that, not looking around for permission or to see if she was the only one who could see the magic happening at the top of the key, Cassidy knew she was going to be all right here. It's so stupid, too: Jo was just a random Crow girl he got to talking to at the powwow last summer—well, her and her cousin, whatever her name was. They'd been doing a thing of standing up in front of tourists' cameras right when they were about to take the perfect shot of the grand entrance, and they weren't doing it to protect the Blackfeet or anything, they were just doing it for the hell of it. Cassidy had liked that, had scooted over to stand up with them, and then, before he even really knew it was happening, he was driving down to her rez every other weekend, then every weekend, then every day if he could get away. And then, after her big fight with her mom, and after her cousin moved down south, taking away Jo's couch, Cassidy was coming back with Jo's stuff mounded in the bed of his truck, a borrowed horse trailer dragging behind.

So it was all on accident, him and Jo, but at the same time it feels like it was meant to be. It's like he walked into the best thing ever, and all he was doing was screwing around at the powwow. But maybe that's the way it works when it's real?

Cassidy rolls the doubled-over pad of cash in his front pocket and considers going into the camper to wake Jo just to be sure she's there, that he isn't just dreaming all this, but . . . she works nights, is a stocker, the one and only Crow to ever work at the new grocery so far—she needs her sleep, he knows. Instead he goes to the camper of the old truck, where the dogs sleep. They won't miss the pile of sleeping bags and blankets and old jackets for one night, will they? For one sweat?

Maybe he'll buy them forty dollars of dog food.

Well, twenty.

He hauls an armful of the blankets out and down and drops them to the dirt, finds a corner here, an edge there, a sleeve poking out like it's reaching up to get saved. One by one Cassidy separates them all and shakes them out, then carries them to the frame of the old sweat. The poles are still good, are from some kids' tent he guesses, propped up with forks of tied rebar at the four points. Not four for any bullshit Indian reason, but because, first, the day Cassidy had put this together, he'd only been able to scavenge eight pieces of rebar, two per prop, X'd like he was doing, and, second, the tent frame was made to hold a kid's *tent*, not forty-odd pounds of dog bedding.

Too, it's good that Gabe wants to do this in winter. In summer, because of Cassidy's great idea to dig the floor of the sweat down a foot and a half, it's sometimes soggy. Frozen like this, though, it'll be perfect. Anyway, it'll be good to sweat the past year out. Reset, like. The old-time Indians had it right, Cassidy figures.

At the frame he tightens all the shoestrings holding the poles and rebar together, then shakes each blanket and sleeping bag and jacket out before draping them across the white plastic skeleton of the sweat, saving his prison brother's old Army coat for the flap. The sleeping bag with the silvery insides flashing in the sun spooks the horses and Cassidy notes this, reminds himself to keep this one free, that maybe it'll scare the magpies, too. Last summer they stole threads from his favorite shirt when it was hanging on the line between the camper and the pens. Somewhere back in the trees there was a nest with colorful accents, he figured, which was great and all, but not at the cost of his favorite shirt. Now, to protect Jo's clothes that he should probably be taking down so they don't smell like smoke, he's got Christmas tinsel strung up all down the wire. So far the magpies just think it's pretty. They squawk their thanks to him for dressing the place up, keeping things interesting.

When the sweat's together—it looks like an igloo made from homeless dudes—Cassidy goes to the tack shed, roots around, comes back with a mallet and enough junky tent stakes to be sure no flaps blow open tonight except the door flap. But the door flap is a greasy old BDU jacket with a rock in its pocket to keep it down, so that should be good.

Next, what should have been first: sweeping out the floor. If he'd done it before, he could have used a normal broom. Now he has to use the broken-off head of a broom, and a tray like from a cafeteria. Zero clue where that's from, but it works.

Slapping the tray on the side of the pen spooks the horses again, too.

"What's with y'all fraidy-cats today?" Cassidy says to them.

The paint whinnies back, stomps her front hoof like trying to pull the ground between them closer, and Cassidy ducks back into the lodge for one last sweeping. A little extra dirt doesn't matter to him, but it's the Yellow Tail kid's first sweat, so he's probably going to have his face right down by the ground, just trying to breathe. *Heat rises, kid. Way it is. Sorry.* Maybe this'll clean him right up, though. It's a different kind of getting baked, right?

"Here all night," Cassidy calls out to the horses and the dogs, and parade waves to them all. The paint swishes her shampoo-commercial tail back at him. The dogs are pretending to be guarding some other camper, it looks like. One they'd be more proud to be associated with.

Cassidy turns around to take his world in. For miles around there's just yellow grass and crusty snow and, in the folds of the hill where seeds can blow and water can flow, clumps of trees. The only thing keeping this from being 1800 or all the centuries before are the utility poles hitching the power cable out to the camper. Well, he supposes the camper's not very pre–white people, either. Or the horses.

He's always kind of wondered about the dogs, though. Back when, dogs would sometimes pull little travois, wouldn't they? He's pretty sure he's seen drawings. But, wouldn't those dogs have pretty much just been domesticated wolves? At the same time, though, all the dogs living on the streets in town, they may have started out as Saint Bernards or Labradors or Rotts or whatever, but, to grit out the winter, to fight it out over every scrap, they bush their coats up, they bare their teeth first thing, and their ears aren't as floppy as whatever line of lapdog Frisbee-catchers they come from. It's like, living like they do, it's turning them back into wolves.

Case in point: Cassidy's three women, each faster to snap at your hand than the other. The black one with the blaze, Lady-bear, is the mom of the two others. There used to be a boy dog he called Stout, because he was, but the problem with the males is that they're never content to hang around the camper. Stout went out on self-imposed patrol one day, which Cassidy figured was really just looking for women in heat or something to fight, and he must have found one or the other, because the next time Cassidy saw him was when him and Jo were trotting the horses a couple miles off, just killing the afternoon. Stout was a mat of ratty hair and a few bones.

"Always wondered where he got off to," Jo'd said, her paint dancing and spinning under her.

"Not far," Cassidy said.

That was probably only a couple of months after she'd moved in, when he was still trying to prove to her that he was a Real Indian. Exhibit one: I ride my own horses on the same land my ancestors did.

Whatever. It had worked, Cassidy supposes. Never mind she was twice the rider he was, and probably three times the Indian.

Not that she could come into the sweat tonight. If it was just him and her, sure, always, forever, please. Gabe had read in one of his books that women and men didn't mix in the sweats, though, and anyway: the kid. Cassidy still remembers his own first sweat. It was bad enough sitting in all that dark heat with a bunch of naked uncles. Add a woman into that mix—especially one like Jo: two unfair inches taller than Cassidy, curvy, solid, long black hair—and it wouldn't have been ceremonial anymore, it would have been about how, *Look, I'm tough, this heat isn't anything, I can take it longer than any of these old-timers.*

He probably never would have sung, either, if there'd been a woman there. Yeah, sweats shouldn't be like the bar, he figures.

But now that it was built, they could heat up the rocks whenever they wanted, get clean, screw whatever book Gabe had. What, were the Indian police going to thunder down from the sky on lightning bolts, write Cassidy up for letting a woman into the sacred sweat lodge?

If they did, he'd ask them about their dogs, maybe. And also what they did for water for the sweat, back in the pre-bucket days.

In town, Cassidy could just run a hose over, snake it under a blanket, spray the rocks when they needed more steam. This far from town, though, all high and lonely, water was in the five-hundred-gallon tank behind the pens, and cost a tank of gas to drag all the way here.

What did the old Indians use?

Probably they built their sweat lodges by creeks, Cassidy figures, or where the snow was melting downhill. *His* solution is . . . that old green and white cooler that doesn't have a lid anymore, that he's been using for the dogs to drink from.

"Sorry," he says to them, tumping it over.

Miss Lefty bats her tail once against the dirt in response. Her name is Miss Lefty because that's a funny name for a dog.

Cassidy scoops the cooler into the tall snow living in the shade behind the horses' lean-to. Their ears are all directed right at him.

After working the cooler into the sweat, all that's left is a shovel for Victor, tonight's designated rock handler. Cassidy jogs around back of the camper, not sure where he last saw the shovel but sure he can't use the wide flat one they muck out the stalls with. He's not a hundred percent on board with keeping every part of a ritual intact, but he is against the shitty end of that shovel being anywhere near him.

Jo's around back of the camper, it turns out, wetting the air needle with her lips to push it into the basketball, shoot a few thousand on the little court maybe a camper's length past the outhouse. Really it's just the foundation left over from a house that used to be here, that blew away. All Cass had to do was grind down all the water pipes level with the concrete and screw an old backboard to the utility pole the tribe left behind, that he spent a whole day digging the hole for, and another day getting to stand up straight.

Jo's sitting on the little weight bench Cassidy liberated from one of his kid cousins, her foot stepping on the bicycle pump's base to keep it from tipping over into the dirt.

She jabs the needle in, tries to hold the ball between her knees while she works the air pump's squeaky plunger. Cassidy wants to step over, help with either the ball or the pump, but one thing about Jo is that she'll either do it herself or she'll go down trying.

"Couldn't sleep," she says, checking the pressure.

"Who needs sleep when there's basketball?" Cassidy says.

"Macaroni's cooking," Jo says, tilting her head inside.

"Hot dogs?" Cassidy asks.

"Once you cut them in," she says back, then, about the lodge: "Who for?"

"Lewis," Cassidy says. "You know—that guy. One I grew up with, got himself shot yesterday?"

"That how y'all do it up here, throw a sweat? This a Blackfeet wake or something?"

"Just a remembering. It was Gabe's idea."

"Gabe," Jo says, flat as it's possible to say a name.

"Also there's this kid," Cassidy adds.

Jo nods, meaning he doesn't need to take her through it again: Victor Yellow Tail's kid needing something traditional to maybe ground him, keep him from burning out on dope and 90 proof.

"Almost forgot," Cassidy says, "payday," pulling the cash up from his pocket enough to show some thick green.

"My man."

Soon enough Jo's involved trying to grease the bicycle pump's plunger, so Cassidy steps into the camper to cut hot dogs into the macaroni, cube some Velveeta in—smaller the better, so they'll melt. He walks a bowl out to her, spoon already in, the cheese thick enough that the spoon isn't even tapping into the side of the bowl.

"Needs ketchup," she says after her first bite.

They're sitting by the fire now, away from the smoke.

The dogs still haven't moved, know this lunch isn't for them. The horses are lined up against the fence, their jaws on the top rail, tails swishing like cats.

"We should run them guys tomorrow," Cassidy says about her paint and the sorrel. The mouse-colored gelding isn't broke enough yet, maybe never will be.

Jo looks over, waits until Cassidy's chewing a big bite before saying, "Aren't you not eating today or something?"

Cassidy chews, swallows, says like a question, "I did skip breakfast, yeah?"

"You never eat breakfast."

"But especially not today."

Jo shakes her head, spoons another bite in, says, "Where you gonna stash that drug-dealer roll?"

"The safe, I figure," Cassidy says, and as one they look over at the truck he dragged up a few months ago. It wasn't stealing, he assured Jo, it wasn't even salvaging—it was *his* truck. Just, he'd walked away from it for a few years. But it had been a good pony once upon a time, deserved to rust into the ground close to people instead of out by itself.

The safe he's talking about is a powder-black thermos he's got shoved up inside a rotted-out glasspack up under the truck, rotted out because when he'd pulled the engine forever ago, he'd left the headers mouth-up to all the rain and snow the sky could funnel down into them. The result is an exhaust system no one would ever try to spirit away, as it would crumble in their hands. Besides, anything worth stealing off that truck, making it work on some other truck, it was taken years ago.

And, an actual safe, tucked under the bed in the camper, or in a high cabinet, or hidden in some clever cutout? That would be exactly what anybody breaking in would key on, carry off to crack open in their buddy's shop. It's not like him and Jo can be at the trailer every hour of the day, and, way out and lonely like they are, the dogs and horses are no way keeping anybody from taking a pry bar to the flimsy door.

But no one would ever look in a trashy old glasspack barely hanging on under a truck with no engine, no wheels, only one vent window, its four drums up on cinder blocks. Already there's six hundred dollars in that thermos in large bills, and, smushed way down at the bottom under that green, a little medicine pouch with a secret ring in it, for Jo.

She struggles another apparently too-dry bite down and hands her bowl to Cassidy, says, "I'm gonna teach you good taste if it takes forever," and steps inside for the ketchup.

Forever, Cassidy repeats, liking that, spooning another bite in. It's not *that* dry.

The paint at the fence shakes her head from too many horse thoughts rattling around in there and Cassidy shakes his head just the same, trying to get a rise out of her. She's smart enough it works sometimes.

Not this time.

She's looking past Cassidy.

He turns, stands slow, dropping his bowl and Jo's both.

"Holy shit," he says, having to move side to side to stay standing, from the dogs rushing this spilled lunch.

He doesn't care about it anymore.

Spread out behind him, just down the slope from the camper, are probably eighty, ninety elk. Maybe a hundred.

They're all looking right back at him, not a single tail flicking, not one eye blinking.

Cassidy swallows hard, wishing more than anything for his rifle.

The name he was born with wasn't Cassidy Thinks Twice, even though that's what he's doing now—*Where's my gun, where's my gun?*—but Cassidy Sees Elk.

Names are stupid, though.

Pretty soon he won't even need his.

FOUR THE OLD WAY

Standing up at Ricky's grave behind the old lodge, sharing an after-lunch beer with him, tipping a bit out for Cheeto, too, what the hell, it's not giving alcohol to minors if the minor's in the ground, Gabe is still thinking about the basketball girl he saw not wearing a jacket, walking in the snow by the school.

What he's talked himself into is that it couldn't have been Denorah. Den's tight-laced like her mom, wouldn't be walking around with her hair just flying around her head like some Indian demon. And it had been school hours anyway, right? One rule about sports that Gabe's pretty sure still holds, it's that truancies mean you can't play. A one-for-one kind of system, each truancy keeping you on the bench for a game, even though there's so many more school days than there are games. It's what Gabe blames for never being the basketball star he's sure he could have been.

He shakes his head no again, that it couldn't have been her. That he's not a bad dad for not having stopped to get her warm, hike her up to wherever she was going. Just, what that means, he supposes, it's that she was some other baller, out in shorts in the cold. Meaning he's just not a good Indian.

But bullshit to that, too.

Gabe cashes his beer and hooks the neck of the bottle into one of the chicken-wire squares of the Boss Ribs family fence.

What if Den's having some big war with Trina, though, right? They are alike, Denorah's like a little clone of the girl Gabe knocked up fourteen years ago—fifteen, really—but that little clone, she's got some Cross Guns in her veins, too. What this means, Gabe knows, it's that she's going to reach an age where she'll want to take the world in her teeth and shake until she tears a hunk of something off for herself. And then, whether it's good or bad, whether it's a scholarship or a five-year bid in state or two kids in as many years, she'll sit in the corner by herself and chew it down, dare anybody to say this isn't exactly what she wanted.

She's going to be like him, he knows. She's got that in her. She didn't get that smile of hers from Trina, anyway. Gabe's seen it when she plays, in spite of the restraining order. The order's not about staying five hundred feet away from Denorah, or even from Trina, though he kind of self-imposes that one for purposes of self-preservation, it's about not coming to any more home basketball games. Because of boisterousness, which is just cheering. Because of fighting, which wasn't his fault. Because of public intoxication, which was only that one time.

With the right jacket and hat and sunglasses, though, he can still slip in with the visitors, so long as he doesn't draw attention to himself. He's pretty sure Victor, the tribal cop slipping him cash for the sweat tonight, has seen him over there all incognito, but Gabe keeps his hands in his pockets and he doesn't explode up out of his seat each time Denorah pulls a no-call, so Victor lets this sleeping dog lie.

It's not easy being quiet, though.

Denorah is a special player, that once-in-a-generation kind. Yeah, he's her dad, but everybody else says it, too, even that newspaper guy. She's got everything her big stepsister had, but Trace, she does it at college with all the basics she's had drilled in, which, as far as Gabe can tell, has pretty much stamped all the Indian out, left it on the practice gym floor.

Den's got those basics down pat, can walk that line in practice day after day just like her coach wants. When the game's down to it, though, when the buffalo chips are down, as Cass used to say when he was still just Cass, when two defenders are ganged up on this little Indian girl straight out of Browning, that's when she'll smile a smile that Gabe has to smile with her.

It's that hell-for-leather look, that come-at-me look, that let's-do-this look.

Instead of passing out of a double-team like she's been told to, what Den'll do is back off these two defenders, look from one to the other, and then get her dribble and her feet just out of sync enough to throw them off balance, giving her room to split them.

Second game of the season, she even threw the ball between a tall girl's legs and caught it before the second bounce, cut for the basket straight as any arrow ever let loose.

That was the game they had to escort Gabe out, disinvite him from the rest of the season. The reason he got kicked out was because her coach had benched her for showing off. For being Blackfeet. It was like—it was like what Gabe had read about in that one book. Those two Cheyenne from the old days who got caught by the cavalry, sentenced to death, but asked if they could die like they wanted.

Sure, the stupid Custers said.

The way those two Cheyenne wanted, it was to die on their horses, with all those soldiers shooting at them as they ran past.

Only, they did it once, and made it through all the bullets.

And then again.

Finally they had to walk slow, give those plowboy soldiers a chance.

That's what Coach had done to Den: made her slow down, when she was faster than any of them, fiercer than them all.

Gabe figures he should maybe slip through town on the way to Cass's, look D up, make sure everything's good, be sure that wasn't her out walking in the cold.

It's the day before their first scrimmage, isn't it? There's really one place she'll be.

"She's good, man," he says down to Ricky, cracking the top off a second beer, killing this one all at once, like old times.

He hooks it into the fence alongside the first one. They look like two bottles in the side of a hamster cage. One for him, one for Ricky.

Gabe opens a third, considers it, the white chill swirling up from that brown neck.

"So ask Lewis what the hell, if you see his ass," he says to Ricky, tipping the first swig out for him, for them, for all the dead Indians. But for Lewis first.

He wasn't the best of them, was maybe the stupidest of them, really, always with his nose in a book, but that didn't mean the staties needed to pop him like that.

But—Gabe squints his face up, tracks a cloud scudding along the treetops, the sky grey and forever behind it—but why would Lewis have been carrying a dead little elk around with him? At first Gabe knew he had to have heard it wrong, but then the paper confirmed it: when the truck Lewis's body had been in back of had wrecked, there'd for sure been an elk calf thrown in there beside him, because he'd been *carrying* it, and it might be some Indian evidence, or evidence of Indians, who knew.

Just, when the emergency responders piled onto the scene of the accident, they were interested in dead *people*, not dead animals that had probably been in the ditch already, as far as they knew. By the time they realized it was evidence, had come back for the calf, the coyotes had probably dragged it off, made themselves a fine meal.

Good for them.

It doesn't explain what Lewis had been doing with the calf in the first place, though.

The only thing Gabe has to sort of explain it is how sentimental or whatever Lewis had got over that skinny elk, that day they'd jumped that herd out in the elder section.

Lewis had known they were just sawing haunches off and running, but he'd insisted on taking all of her, even her head, which he'd just have to throw away in town anyway, right? He'd skinned her too, hadn't he, and carried that rolled-up wet hide under his arm like a football, like he was some Jim Thorpe wannabe? Like this really *was* some bullshit Thanksgiving Classic? Gabe nods, can see it again, Lewis struggling up that long hill, all the odds against him.

What he told them was that he needed her head because he wanted her brains, to tan that hide with. Like he knew anything about leather. Like that hide hadn't been thrown out years ago, like every other hide any of them had ever saved. Like she'd even *had* all her brains in her head at that point.

Good job, Lewis.

Gabe tips beer number three up to him and drinks it down, hooks it in the fence in line with the others. One bottle for Ricky, one for Lewis, and one for himself. They chime against each other once, then still.

He shakes another one up from the little cooler, for Cass, even though he's seeing Cass in a few, here. Four's too many for three-thirty in the afternoon, on zero food, but screw it. He'll sweat it all out come nightfall, and then some.

Was that what Lewis had actually been smuggling home, that old hide? Had Lewis made it down to the city with it, kept it frozen all this time, and had he been trying to bring it back to the reservation now? Did the cops just not know the difference between a thawed-out, ten-year-old skin and an elk calf? Had they shot him enough times that it was all guesswork?

But *why*?

Was Lewis going to hand-deliver it to Denny at the Game Office, say he'd done his penance, could he please please hunt on the reservation again?

You don't have to ask, though, Gabe tells Lewis. *You just have to not get caught.*

The last ten years of his ban, which he guesses is short for *banishment*, he's taken probably twice again as many elk as they popped that day. Well, enough that when he was stashing some of the meat in his dad's freezer in the garage a few months ago—it was that little one-horn that had still been in velvet—he'd had to make room by cleaning all the old stuff frozen to the walls of that freezer.

The reservation dogs had eaten well that night.

Gabe had watched them until they were done, paper and all, and then nodded once to them because they owed him now, didn't they?

They knew. They would remember.

"Tell me when trouble's coming for me," he'd said to them. "You'll know."

And then he'd laughed. Like now.

Gabe has to wait until he can wipe his smile away to drink the top off the fourth beer. He checks the time.

He's just waiting until he's sure where Denorah'll be. Who he'll call *Den* if he wants, and *D* if she's on defense, *Killer* if she's not.

He hooks the bottle in the fence alongside the others, just like it's the old days, the four of them always together, and leaves a drink sloshing in it for the dead Indian hamsters. Halfway to the truck he turns back to the grave, undoes the black bandanna from his arm, and ties it into the fence as well, like a prayer he doesn't know how to say with words. It's about Lewis, though. And Ricky. And how they all used to be.

Easing back down the logging road in first, just riding the transmission, he crunches the clutch and brake in, leans over the dash to be sure he's seeing what he's seeing.

He has to set the parking brake, get out to be sure.

Elk tracks in the snow. A big cow, just walking up the road like following him up to see Ricky, but a heavy cow, too. Gabe points his trigger finger down into the hoofprint and wonders if it's a small horse with elk feet, carrying a rider.

He stands, looks ahead for anything this ridiculous, but the road hooks back to the right almost immediately.

Still. This is a good sign, isn't it? Strong medicine, like Neesh used to say? The sweat's going to be good for the kid tonight. It's going to be good for all of them.

Gabe climbs back into the truck, eases down the two-track, has his eyes on the road enough that he doesn't catch the flash of black in the rearview mirror when a full-grown woman in a too-tight scrimmage jersey steps out of the trees, into the bed of the truck, her long black hair swirling in after her.

It's where hunters carry the animals they shoot, isn't it? It's where they put you, ten years ago. Don't smile too much about this, just work your way under the toolbox.

The night is almost here. It's the one you've been waiting for.

OLD INDIAN TRICKS

It's about form, yes, Coach is right about that, everybody knows that, but what Denorah's big sister taught her, forwarding and re-winding through hours of tape, is that it's also about using the exact same form every single time you step up to the stripe.

And that form, that ritual, it's not just the second and a half or two seconds of your free throw, either.

It starts with how you toe up to the line. For Denorah it's right foot first, right up to the paint, then backing off about a shoelace width because the point will get erased if your feet are illegal. If you're a good shooter that's not the end of anything, as you usually get a do-over in junior high, but if you missed and one of your tall girls hauled the rebound down, has position to slip it back in for an easy two, then, well, then you've screwed things up, haven't you?

So, out at the pad of concrete at the back of her family's two acres at the edge of town, where her new dad's rigged up some floodlights for summer, where she had to measure out and paint the charity stripe herself, Denorah shoots and shoots and shoots, never mind that she can see her breath in the cold.

Eighty-six out of one hundred, then seventy-nine, which gets her breathing hard and mad, then an even ninety.

Because the scrimmage this week is Saturday night—tomorrow night—and there's no practice the day before a game so everyone can have fresh legs and be mentally prepared, today is for free throws.

They've never been Denorah's weak point, but she's also never had a game where she swished every one. So there's room for improvement. Like with Trace, her whole future might come to rest on making one point the easy way, with the whole gym thundering and crumbling around her, the floor beneath her shoes trembling, sweat pouring into her eyes.

Coach never has them do free throws at the front of practice, but the back, so they'll know what it's like to make themselves have good form when they're exhausted, just want to fling the ball up there, say a prayer after it.

But Denorah needs fresh legs for tomorrow, so she's compensating by trying to get a cool five hundred shots in before dark. Or to get five hundred either way, whether she gets to slope down to the house for dinner or ends up shooting right through it again.

Toe up to the line with the right foot, back off a touch for safety, then work the left up until it's dead even with the right. Spin the ball back with the lines going from thumb to thumb and dribble twice, fast and hard with the right hand, using the whole shoulder, elbow straightening out each time. Catch it, look up to the rim, bend the knees, back straight, ass out, and push up with the front of the thighs, extend with the right arm, left hand just there to keep the rock steady, the calves pushing right at the end, when the middle finger of that right hand is gripping on to the rubber of the valve hole, imparting the perfect spin.

Swish, swish, swish.

The machine girl is on automatic, she's locked and loaded, doesn't even need to concentrate anymore. Fouling her in the act of shooting is the same as putting a pair of points up on the scoreboard for her team.

"Bring it," Denorah says, lining up again.

The only thing she regrets about basketball, that baseball and football and even golf have over it, it's that those players all get to wear war paint under their eyes.

What Coach tells them in the locker room before every game, it's that their war paint is on the *inside* of their faces, it's in how they hold their faces, it's in how they look the other girls in the eye and don't look away. Dribbling and passing and shooting are just the parts of the game that get recorded in the stats. There's also who wants it worse.

To steel herself against the kind of bullshit Indian teams always get hurled at them when the game's close, Denorah tries to inoculate herself with all the bullshit that the other side of the gym will be chanting.

> *It's a good day to die.*
> *I will fight no more forever.*
> *The only good Indian is a dead Indian.*
> *Kill the Indian, save the man.*
> *Bury the hatchet.*
> *Off the reservation.*
> *Indian go home.*
> *No Indians or dogs allowed.*

Her sister heard them all in her day, perverted on spirit ribbons, usually illustrated, too. Shoe-polished on the windows of buses, the big one was always, *Massacre the Indians!*

Bring it, Denorah says in her head, and drops another through the net. If the only good Indian is a dead one, then she's going to be the worst Indian ever.

Her promise to herself for tomorrow, win or lose, is to be back out here again right after the game, working on any shot she should have made but didn't.

They don't just give those scholarships away.

Denorah shags the rebound, jogs back to the line without stopping to spin, bank one in from the block. From what would be the block if she'd measured that out too, painted it.

Her dream is to somehow extend this pad of concrete, so there can be a three-point line.

Someday.

Not today.

Today it's just free throws.

Swish, swish, a sound in the grass behind her but she can't look, it's probably just Mom, home early from work, and . . . rattle, bounce, clunk.

Denorah starts to spin around to whoever made her miss but reminds herself at the last moment that *she* made herself miss, that *she* let her concentration flag, that *she* didn't follow through on the ritual.

"Hey, Finals Girl," an unhilarious male voice calls out from behind her, moments after a truck engine's been turned off.

Finals Girl.

It's what her real dad calls her when she's on the court, ever since she was his lucky charm when she was four and he was watching her in June, during the NBA finals.

She looks back with just her head.

He's sitting in his truck, window down, one arm patting the side of the door like it's a horse he's riding, not a pickup he's picked back through the grass with.

"The septic's there," Denorah says, nodding to the greasy grass by the scattered pipes he must have just driven on one side of or the other.

"That's why I got four-wheel drive," her dad says, pushing the shifter up into place. "The slop."

He's been drinking. She can tell from his eyes. They're loose in his head, too happy for this early in an afternoon.

"Just wanted to wish you luck for tomorrow," he says.

Denorah looks around for the ball, walks a beeline across to it.

"You're not supposed to be here," she says, and he's already cutting her off by rolling his hand, saying, "Your new and important dad doesn't think I'm a good influence, blah blah blah . . ."

Denorah toes up to the stripe, brings the left foot even, her back to him.

He won't get out of his truck. Not when he might be needing to make a hasty exit.

"You're going to kill them tomorrow night," he says. "We're doing a sweat tonight, to, you know. Like, help the team."

His talking is good, Denorah's telling herself. He's like a whole crowd chanting, *Indians go home.*

Just noise.

Swish.

"That's the ticket," he says, patting his door again in token applause.

"You and who's doing this sweat?" Denorah asks, chancing a look back to him while snabbing the ball from the tall grass.

"Cass," her dad says, then, "you want to keep a song in your head when you're shooting free throws, and always let go on the same beat. Old Indian trick."

It makes Denorah do two extra dribbles, trying to keep music from playing in her ears.

Clunk.

"It's okay, it's okay," her dad says.

Rebound, reset.

Was that nineteen out of twenty or nineteen out of twenty-one? Shit. Twenty-one, then. Losing count never counts in favor of.

"I think it's Cassidy now," Denorah says, just because.

"Little Miss Cross Guns," her dad says back, which is what he calls her when she sounds like her mom to him.

"I see his new snag at Glacier Family Foods sometimes," Denorah says, saying the store's whole name because she likes the way it sounds.

"What do you know about snagging?" her dad asks, not patting his truck now.

"She's in produce now," Denorah says, a grin he can't see warming her lips.

Swish.

"Probably a vegetarian, yeah?" her dad says with a smile in his voice, and that tells Denorah all she needs to know about how much Jolene the Crow appreciates her dad coming out to do sweats with her man.

"Didn't know he had a lodge out there," Denorah says.

Dribble, dribble, find the valve hole by touch. Right when she rocks back to shoot, her dad touches his horn. Still: *swish.*

"Good, good," he calls out.

Returning to the line for number twenty-three, she sees a black hank of your hair breeze up from the bed of the truck and it stops her for a moment, the ball held at her stomach.

Her dad sees, leans out to look back, says, "What?"

"You're not supposed to be hunting," Denorah says, no joke to her voice about this of all things.

"What, you an officially deputized game warden now?" he says, settling back behind the wheel, leaning over to get at something from the floorboard.

Whatever he gets doesn't raise above the level of the door, but Denorah's pretty sure it's cold and beer-shaped.

And? Maybe he's not even lying.

Elk and deer don't have manes. Maybe he's a horse hunter now, she thinks, and has to turn around to the rim so he won't see the smile in her eyes. Just because she can't see him doesn't mean she can't hear a beer cracking open.

Noise, noise. The whole gym going wild.

"This the same sweat Nathan Yellow Tail's dad's making him go to?" she says.

Rattle, rattle, lucky roll.

"We're letting him in, yeah," her dad says like a challenge. Like she should try to call him on his kindness, his manners, on who this sweat is actually for, Nathan or the scrimmage.

"Coach says she sees you at the games," Denorah says, walking to the tall grass for the ball.

No answer.

She looks back to him.

"You look more like your mom every day," her dad says.

Trina Trigo, the grass dancer champion from high school, was even on a powwow calendar from back then. Except Denorah isn't sure if this is a compliment or if she only looks like her mom when she's saying stuff her dad doesn't want to hear.

Line up, draw a bead on those eighteen inches of circular orange up there. Remember that the higher you arc the ball, the more circular that rim gets.

The ball is a touch over nine inches wide. That leaves all kinds of room to play with. All kinds of lucky bounces.

But proper form is where it all starts. More you practice, the luckier you end up.

The ritual, the ceremony.

Dribble-dribble, thighs, extend, spin with the valve hole, hold that follow-through, hold it, hold it . . .

Swish.

Denorah smiles, is the deadliest Indian on the whole reservation.

"Do it again, it's worth twenty," her dad says behind her, low like an invitation he doesn't want just everyone tuning in to.

She turns back to him and he's pushed up against the back of the seat to dig in his front pocket. At least until whatever it is that's cold and beer-shaped and chocked between his legs spills forward.

"You've got twenty?" Denorah says back to him.

"I will tonight," her dad says. "When Officer Yellow Tail pays me."

"So it's like that," she says.

"Gratuity," her dad says. "He's very gracious."

"Double it if I go ten for ten," Denorah says.

Her dad raises his eyebrows, says, "You are my Finals Girl, aren't you?"

She smiles her smile that she knows is his, that she can't do anything about, and dribbles twice, *banks* it in just for show.

"Somebody give the little lady some dice," her dad says.

It's not luck, though. It's skill. It's practice. It's proper form.

"That's one," Denorah says, and turns her back on her dad again, imagines a gym all around her, wall-to-wall white people, all chanting for her to go home, go home.

She spins the ball toward her, dribbles twice, and lines up.

THE SUN CAME DOWN

Cassidy stands from his lawn chair to watch Gabe rattle over the cattle guard. The dogs, tongues hanging long from chasing that herd of elk off, swarm his truck before he's even got the door open, maybe thinking he's bringing those elk back, that he has them all in the bed of his truck.

Dogs are stupid.

"Ho, ho!" Cassidy calls out to them, slapping his thigh.

Gabe kicks his door open to scatter them but they just keep bawling their fool heads off. He wades out holding a rifle above his head like that's what the dogs are after.

"Don't you feed them?" he asks above the din.

"They like *red* meat," Cassidy calls back, making his way over.

"They're scratching the other scratches, man," Gabe says, pushing in between the dogs and the bed of his truck.

"What you got back there?" Cassidy asks.

"Not dog food," Gabe says back. "Unless—they eating spare tires now?"

Before Cassidy can get close enough, Ladybear nips at Gabe's left hand. In response Gabe pops the forestock down across the bridge of her nose. Then he steps into her space, driving her back, his lips thin like this could get serious.

Ladybear whimpers, backs off, the other two following.

"Shit," he says, shaking his hand, opening his door back up to see it with the dome light.

Cassidy leans over to see. Gabe's bleeding from the meat of his left palm. Two neat, welling punctures.

"Got rabies now," Gabe says, wiping the blood onto the saddle-blanket seat cover. "What, they on JoJo's side now? She even got the dogs turned against my ass?"

"Watch out for the horses is all I'm saying," Cassidy says, and returns to the lawn chair. The fire's down to coals mostly.

Gabe steps in, hunkers down into the other chair still holding his hand, settles the rifle across the top of his legs.

"Those rocks working out?" he asks.

"They're rocks," Cassidy says.

"Got water for the sweat?" Gabe asks.

"In there," Cassidy says, pointing to the lodge with his chin. "Got the money?"

"About that," Gabe says.

Cassidy chuckles and shakes his head, turns the bottle of water up and drains as much as he can without drowning. He's not thirsty, but he will be.

"She here?" Gabe asks, tilting his head over at the camper, its windows dark in the gathering dusk.

"Work."

"I've never done one at night," Gabe says then, leaning back in Jo's chair, the chair not quite bending. Yet.

"A sweat?" Cassidy says.

"There's nothing, like, *against* doing it at night, is there?" Gabe asks.

"Let me check the big Indian rule book," Cassidy says. "Oh

yeah. You can't do anything, according to it. You've got to do everything just like it's been done for two hundred years."

"Two thousand."

They laugh together.

Cassidy fishes a dripping bottle of water up from the cooler and spirals it across the fire to Gabe. The droplets spinning off it hiss against the embers, send up tiny geysers of steam.

"So what do you know about this kid?" Cassidy says.

"Nate Yellow Tail? You know. Twenty years ago, he's you and me. He's Ricky and Lewis."

"Half of us are dead, yeah?"

"Either that or one of us here is already half dead," Gabe says, and slings a dollop of water across the heat at Cassidy, to show this isn't completely serious. That it is, but he wants to get away with having said it.

"Maybe it'll be good for him, I mean," Cassidy says. "Help him out, like."

"Arrows are straight, but they have to bend, too," Gabe says, his voice dialed down to Wooden Indian to deliver Neesh's old line. It's what the old man used to always end his group sessions on. There was even a series of posters all along one wall of the substance abuse office, an arrow looking all bowed out at the moment of the string's release, like it's going to crack, shatter, blow up. But it doesn't. It's bent out to the side in the first poster, it snaps back a foot or two from the riser of the bow in the second poster, and then in the rest it's snapping back and following through, bending the arrow the *other* way now, and until the last possible instant before the bull's-eye, it's flopping back and forth through the air like that, trying to find true.

That's how they were supposed to be. It's what they, at fifteen, were supposed to have been doing. They'd been fired into adolescence and were swerving to each side now like crazy, trying to find the straight and narrow. If they did? Bull's-eye, man. Happy days.

If they didn't?

There were examples under every awning in town, drinking from paper-bagged bottles. White crosses along the side of all the roads. Sad moms everywhere.

"He'll sweat it out," Cassidy says. "Sing it out."

"Wish we had a drum," Gabe says.

"I got some tapes."

"Screw tapes, man. This is for Lewis, too, yeah? But don't tell Victor-Vector."

"Maybe don't call him that," Cassidy says.

"It's not bad, is it?"

"For Lewis," Cassidy says, holding his water up in salute.

Gabe lifts his the same, says, "He always was a stupid ass, wasn't he?"

"Smarter than you," Cassidy says. "He got out of here."

"But then he tried to come back," Gabe says, drinking once and swallowing hard. "They didn't shoot him until he tried to come back."

"He was just running for home base," Cassidy says. "They'd have shot him if he'd stayed where he was just the same."

"Why you think he did it?" Gabe says. "His wife, that Flathead girl?"

"She was Crow, man."

"Serious?"

"He probably couldn't have told you even while he was doing it, yeah?" Cassidy says, studying the clarity of this bottled water.

"Still," Gabe says, draining his, dropping it into the fire. The plastic shrivels even before the label flares up.

"Great," Cassidy says. "Pollute the rocks we're going to be breathing."

"Like I can fail the Breathalyzer any harder?" Gabe says back.

"So what's with the antique?" Cassidy says about the rifle across Gabe's lap.

"Old man finally parted with it," Gabe says, holding it across to Cassidy, around the heat of the fire.

Cassidy racks the bolt back, clears it, studies the long goofy stock.

"Think it's for NBA players," Gabe says. "Forestock's long like that so they don't have to bend their arms too much."

"It shoot straight?" Cassidy asks, shouldering it, training it out into the darkness, one eye shut.

"Like anybody still has shells for something that old?" Gabe says. "He only shot bird shot and rock salt through it, yeah?"

"The Great Mouse War," Cassidy says, fake-pulling the trigger. "I bet I got something that'd work. You know when Ricky—I went out to Williston to get his stuff."

"Oh yeah. What'd he have?"

"Nothing. His dad said he was supposed to have all their rifles, but his shit'd been cleaned out a crew or two ago."

"Tighty-whities."

"All that was left of the rifles was a bag of random-ass shells. Think they're still in the glove compartment, probably, with whatever kid book Lewis was reading back then."

Gabe leans forward to see the old Chevy up on blocks.

"Good you put that pony out to pasture," he says. "She got stuck everywhere, man."

Cassidy sets the gun back against the trash barrel, away from the fire.

"I'm going to fix it up," he says. "Body's still good, mostly. Just need to find a hood, and a bed. Maybe some fenders, too. An engine, some tires."

"Still hide your shit in it for safekeeping?"

Cassidy breathes in, looks over to the eye shine of one of the horses, watching them, its big ears probably catching every word, saving them for later.

"Can't even keep the ground squirrels out of it," Cassidy says what he knows is a moment too late.

Gabe knows about the thermos? How can he?

"Got *just* the gun for taking care of rodents," Gabe says back, nodding across to the Mauser. "Take it instead of the cash, for this?"

"You really think it still shoots?" Cassidy says.

"No reason it wouldn't."

"Gimme a minute," Cassidy says. "I'm weighing you giving me something that's old and broke and stolen against you *being* too broke to ever pay me money you owe me."

"Ha, ha, ha, ha," Gabe says, his mouth open wide enough for a laugh this slow and fake. "You can sell it for a hundred fifty, I bet. Maybe more if it's historical."

"And when your dad comes looking for it?"

"Sell it to him, he wants it again. But he gave it to me free and clear, Scout's honor."

Gabe rabbit-ears his first two fingers up but then lowers the index, turns his hand around slow to flip Cassidy off at close range.

"Sure, leave it, whatever," Cassidy says.

"Only if it's cool with JoJo, man," Gabe says.

"She doesn't like it when you call her that," Cassidy says for the fiftieth time this month.

"It's like 'yo-yo,' but with *J*'s, man," Gabe says, and Cassidy isn't sure whether Gabe's calling Jo a toy or whether he's talking about joints. Either way, he flips him back off, *both* hands, which is when headlights wash across both of them like a snapshot.

SHIRTS AND SKINS

It's not the same car you rode to the reservation in yesterday, but it's the same dad, the same son.

The dad is standing from the open door of the car, his headlights still splashing white across Gabriel and Cassidy, their hands up to protect their eyes, their shadows blasting back across the big pile of moldy laundry behind them and then the horse pens and all the darkness past that, where you're standing, the tips of your long hair lifting from the hot air the car was pushing in front of it.

"We surrender, we surrender!" Gabriel calls out, trying to duck away from all that brightness.

The dad reaches down, turns the lights off, and while he's leaned down, his son shakes his head a bit in disgust, says, "So these clowns are tradish?"

"It's not about the sweat," his dad says, not really using his lips, just his voice.

It's not about the sweat, you repeat, trying to keep your face perfectly still like that. It almost works, except you're pretty sure your eyes are grinning.

The night's about to start.

"Then what *is* it about?" the boy asks.

The dad sits back down into the car, clicks the middle console open like he forgot something. "Look at these two jokers," he says, face tilted down. "They were you, twenty years ago."

Cassidy is shooting a spurt of water between his teeth at Gabriel, and Gabriel, trying to avoid it, is collapsing one side of his chair, and Cassidy is trying to save the chair from folding down on itself.

The boy has to chuckle.

"They're *alive*," he says.

"There used to be four of them," the dad says.

The boy pops his door open, hangs a leg out, swings his hair back over his left shoulder.

"All of us'll fit in that thing?" he says about the mound of sleeping bags the sweat lodge is.

"Just the three of you," his dad says. "I'm handling the rocks, that's my part of it."

"How long?"

"Long enough."

They stand together, their doors closing at the same time, an accident of sound that makes the boy straighten his back, like it's bad luck.

Gabriel is already pushing up from his broken chair to greet them. His face is shiny with water from Cassidy's mouth. "To Officer Victor goes the spoils . . ." he's saying, wiping at his cheek with his sleeve.

"What does that even mean?" the boy says to his dad.

"He read it in some bullshit book," Cassidy says from his chair. "Ignore his ass."

"Gentlemen," the dad says, shaking the hand Gabriel is offering.

"Victor-Vector sounds like a cop even when he's off duty, man," Gabriel says with a halfway smile.

"I'm never off duty," the dad says back, nodding down to the cop car he's driving.

The boy doesn't look to the car with Gabriel and Cassidy, but to the camper. All the windows are dark.

"How long since you've sweated it out yourself?" Gabriel is saying to the dad.

"This is for him, not me," the dad says, and all eyes settle on the boy. "Nathan," he announces, the big introduction.

The boy keeps looking at the camper, like considering how to take it apart. Or—he can't see your reflection in a window, can he? Just your shape, your silhouette, your shadow? Your true face?

If the boy were to tilt his chin out at you for his dad right now, this instant, and if his dad leaned forward, peered through the darkness at the wild-haired woman just past the light, then this could all be over in a rush, couldn't it?

But it's better nobody sees you. Yet.

The boy finally drags his eyes away from the camper.

"You play basketball, yeah?" Cassidy says to him, about the scrimmage jersey the boy's wearing black side out.

"I played ball, I'd be skins," the boy says.

"Got a little court right over there," Cassidy says, hooking his chin to the left of the camper, back toward the road. "Maybe we could shoot to cool down later."

"Got glow-in-the-dark balls?" the boy asks right back.

"Son," his dad says.

"They call you Nate, right?" Gabriel says.

The boy shrugs one shoulder, says, "Gabe, *right*? Seen you around."

Gabriel purses his lips about this for a fraction of an instant.

"You drug him back from Shelby or what?" Cassidy says to the dad.

"Farther than we ever got," Gabriel says, making a show of turning to finally see what the boy had been staring at so hard. "He ever done a sweat?" he asks the dad, no eye contact.

"You can talk to me," the boy says.

"*You* ever done this?" Gabriel says, making a production of talking *to* the boy.

The boy shrugs.

Gabriel says, "The idea is it's a purification, like. Consider it a dishwasher, yeah? We're the dishes. It steams us up spick-and-span, man."

"That what your friend Lewis and Clark was coming back for?" the boy says. "Clean the spots off his soul?"

Gabriel smiles a tolerant smile, looks back to Cassidy, who wows his eyes out like what did they expect?

"This is about you," the dad says. "Not all that. Got it?"

The boy stares across the dying fire at the paint horse.

"Y'all know Lewis was coming this way, though?" the dad says to Gabriel and Cassidy.

"Never off duty . . ." Gabriel singsongs, just generally. "Always trying to solve some crime, put another Indian behind bars."

"Lewis left, he was a ghost," Cassidy says.

"White woman," Gabe adds, all the explanation necessary.

"And one postal worker," the dad adds, looking around the place. "She was Crow, wasn't she? Saw her picture in the paper. Bet the wife caught him creeping across to her tipi."

"Lewis wouldn't," Cassidy says.

"Wouldn't what?" the boy says. "Cheat on his wife, or kill two people?"

Gabriel touches a place on the side of his face, by his eye.

The dad's still looking around.

"Where's your dogs, Cass?" he finally says.

Cassidy looks around like just missing them.

"Cass's dogs, they're the criminal sort, I guess," Gabriel says, unbuttoning his cowboy shirt. "They see tribal PD, *pew*, they're gone for the hills, man. Anybody with a badge, I mean. Even that way with game wardens, right? They can't tell Denny Pease from PD. Stupid dogs."

Cassidy stands, is unbuttoning his shirt now as well.

"You haven't eaten today?" he says to the boy, the cadence of his words old and Indian, and mostly fake.

"Just water," his dad says for his son.

"Same," Gabriel says.

Cassidy nods that him, too, yeah.

"ESPN's on at eleven, yeah?" the boy says to his dad.

"It's on again at two," his dad says back.

"Speaking of numbers . . ." Gabriel says, squinting like it's painful to have to be bringing this up.

The dad passes him five bills. Cassidy tracks this money into the pocket of Gabriel's jeans, off and folded on the arm of the chair.

"You ever wonder where the term buck-naked comes from?" Gabriel says, down to his saggy boxers.

"Listen close," Cassidy warns, stepping out of his boots, "you're about to hear some good lies."

"Settlers moving into Indian territory used to call us *bucks*, back when," Gabriel says with authority, looking around for what to lean on while he one-legs it out of his boxers. "Because we were always horny, I guess, right? They could tell we were because we were naked, since Levi's hadn't been invented yet. So, you know, them Indians coming in to the trading post, *They're all naked again, Jim, what are we going to do? Look, look, hide the women, those bucks are naked, man, they're* buck-naked . . ."

"Told you," Cassidy says, folding his pants over the back of his chair.

"Isn't there usually singing or a drum or something?" the dad says, studying the mound the lodge is.

"Doesn't have to be," Gabriel says, balling his boxers up in his hands, being sure they're touching every last one of his fingers, the boy notes with distaste.

"I've got some tapes," Cassidy says, making like to go over to his camper.

"Don't worry about it," the dad says.

"It's just—" Cassidy says, but this dad does his right hand flat, palm down, and moves it from left to right, cutting this idea off. It's a hand signal the boy—you can smell it on him, can see it on his face—remembers from a picture book in elementary: how the old-time Blackfeet used to talk with sign language when they needed to.

He hates being from here. He loves it, but he also hates it so much.

"Just send him in when he's ready," Gabriel, naked, standing there like a dare, says to the dad, and holds the flap of the sweat lodge open for Cassidy to duck in. "Cool?"

The dad nods a curt nod and an ass flash later Gabriel is in the lodge as well, the Army coat flapping shut behind him.

"You're serious about this, really?" the boy says to his dad.

"He always has a bunch of dogs out here . . ." the dad says back like a question, shining his flashlight all around, holding it at his shoulder exactly like the cop he can't stop being even for one night.

The boy leans back against the car and peels out of the scrimmage jersey all at once, turning it inside out in the process, so it's shiny white now. He folds it neatly over his arm, all the same, like turning it inside out had been just what he wantd to do. The air prickles his skin. He rubs his arms with his hands, hisses air out through his clenched teeth.

"That horse is watching me," he says.

"Sounds like it's you watching the horse," his dad says back, still studying the night for the chance of dogs.

"So what am I supposed to do in there?"

"Figure it out."

"It's bullshit, you know."

"When I was fourteen I knew everything, too."

The boy shakes his head, kicks his shoes off, is already counting the seconds of this night.

THREE LITTLE INDIANS

"This lodge is *dank*, Nate," Gabe says when the shape of Nate finally darkens the flap. He's been saving that line special for the kid, just so he can hate on it. It's good to give them focus.

"It's Nathan," the kid says, settling in on the missing point of the triangle, the chipped-out little pit between them already disappearing in the darkness again now that the flap's shutting. Evidently Victor was holding it up for his son to enter. Probably making sure Gabe wasn't making it *actually* dank in here. It's a sweat lodge, not a human-sized bong.

"Welcome," Cass says, still playing the ancient Indian.

Gabe hits him in the chest with the back of his hand.

"First time I did this, I wore a swimsuit," Gabe says, trying to dial them all up to today instead of a hundred years ago.

"Thought it was supposed to be all hot in here or something," Nate says.

"You ready?" Cass asks.

"We can't see you nodding, man," Gabe says. "I mean, if you're nodding."

"Ready, yeah," Nate says.

"And this isn't some toughest-Indian-in-the-world thing," Cass says. "You're supposed to get hot, but not hot enough to pass out."

"Well, that's where the visions are," Gabe says. "But whatever."

"Think I'll be all right."

"You'll think this is stupid, me saying it now," Gabe says. "But the cool air, it'll be down near the ground. If you need a good breath."

"And it's about praying, too," Cass says. "Talking to whoever you need to talk to, all that."

"With my dad listening right outside," Nate says.

"Too many sleeping bags," Cass says. "This is just us in here."

"We'll be talking to a couple of our friends," Gabe says. "Just so you know."

"Which one?" Nate says. "The killer one, or the one who got killed?"

Gabe licks his lips, looks down into the darkness of his lap. It's just the same as the darkness everywhere else.

"When we were your age, doing these," he says. "Our . . . our counselor, this old dude, Neesh—"

"That's his granddad," Cass cuts in.

"You're nodding at Nate, I take it?" Gabe says.

"Nathan," Nate says.

"Neesh Yellow Tail was his granddad, yeah," Cass says.

"No shit?"

"No shit," Nate says.

"Anyway," Gabe says. "Neesh, Granddad, whatever, he told us that none of the old stories are ever about a war party attacking a sweat that's happening. That it wouldn't just be bad manners to do that, it would be the worst manners. You don't even jump somebody when they're done, are all staggering out, weak and pure and shit. It's a holy place, like. It means right here where we are, it's about the safest place in the Indian world."

Nate snickers, says, "Safest place in the *Indian* world? That means we're only eighty percent probably going to die here, not ninety percent?"

"Nobody ever dies in a sweat," Cass says. "Not even the elders. I've never heard of it, anyway."

"This where we eat the mushrooms?"

Gabe drops his head back to smile up into the idea of the domed roof muffling their voices, says, "Different tribe, man."

"Unless you ordered pizza," Cass chimes in, finally joining this century.

"I can do that?"

"After, sure," Gabe says. "I like meat lover's. That's real Indian pizza."

"Nobody says 'Indian' anymore," Nate says, voice somewhere between insult and disappointment.

Gabe closes his eyes, lilts out, "*One little, two little, three little Natives,*" lets it fall dead between them all, then says: "Doesn't really sound right, does it?"

"We grew up being Indian," Cass says, something about his delivery making it sound like his arms are crossed. "*Native*'s for you young bucks."

"And *indigenous* and *aboriginal* and—" Gabe says.

"This part of it?" Nate cuts in. "I supposed to be getting all sweaty from this history lesson?"

"You didn't wear deodorant, did you?" Cass says, not missing a beat.

Silence.

"Does that matter?" Gabe finally asks, kind of quieter.

Cass calls out a deep *ho* to Victor.

"We have to be sure to thank him each time he brings a rock in," Gabe says, back to normal volume. "Otherwise—this is what your granddad told us—otherwise, feeling all unappreciated, he might deliver a warmed-up buffalo patty in for us to pour water on, breathe into our lungs."

"Bullshit," Nate says.

"Exactly," Gabe says right back.

"Here," Cass says, reaching behind Gabe for . . . ah: the ceremonial golf club. Of course. He uses it to guide the flap open enough for Victor to step one leg in. A cool sigh of night air breezes in as well.

"Careful," Victor says, making sure his path is clear. When it is, he angles the shovel in. Balanced in it is a rock so hot there's lava worms crawling all over it.

"Thank you, firekeeper," Gabe over-enunciates.

In the splash of light coming in, Nate, pushed back from the pit, nods a quick thanks as well.

Victor rotates the shovel handle, spilling the rock into the pit, along with the embers and ashes he'd scooped up. A vortex of sparks trails up into the domed ceiling.

"Did you wet the sleeping bags and stuff?" Gabe leans over to ask Cass.

"It'd smell like dog if I did," Cass whispers back.

Gabe nods, checks the fabric all around them again.

"Does dog hair burn?" he says, just out loud.

"Thank you," Cass says up to Victor.

"Another coming," Victor says.

When the hot rocks are in the pit—there's room for maybe three more, total—and the flap's closed, their faces all underlit dull red, Gabe looks across to Nate, says, "Last chance, man."

Nate shakes his head no.

Cass reaches back, slides the cooler alongside. The dipper is an aluminum scoop, like for feed. Cass does a humming up and up then down again drumbeat in his chest, and Gabe gets the lope of it, falls in. When they were the kid's age, they always called drum circles circle jerks. And now here they are, carrying the beat.

Gabe shakes his head, amazed at it all, and ramps up his humming drumbeat, smiles a smile he can't help. There's five twenties in the right front pocket of his pants on the chair out there, and at least three of them are his—would be eighty dollars, but Denorah can shoot the hell out of free throws, can't she?

"Here we go," Cass says, breaking his own drumbeat for a moment, and scoops a dollop of water onto the two hot rocks.

The steam hisses up, boiling the air.

Gabe chances a look across at Nate, and for the first time there's a hint of uncertainty in the kid's eyes, and for an accidental flash Gabe is seeing himself in the side mirror of his truck, when D asked if he was hunting again, and he thought he saw black hair behind his reflection, lifting up from the bed of his truck.

Except that couldn't have been. And the dogs weren't smelling anything back there, either. They're just stupid dogs.

Gabe breathes the heat in deep and holds it, holds it, eyes shut.

DEATH, TOO,
FOR THE YELLOW TAIL

Victor plants the shovel into the ground by the fire after the next rock delivery—it takes two stabs to make sure it'll stand—then crosses to his car. Not to lean back against the fender until called again, but to settle down into the front seat, key the dashboard alive. He leans down to it, into it, comes up with a cassette tape. He holds it to the dome light to squint at, then flips it over to the side he wants, pushes it in.

Drumbeats well up from inside the car. Drumbeats and singing. It's been hot enough in the lodge for the last half hour that there hasn't been any singing, any talking, any anything. The last time he leaned in through that flap, he looked from face to sweaty face, gauging each of them in turn, then nodded, turned the shovel over, let the green jacket flop down.

Maybe it's working? Maybe this will have been a good thing?

Now he's looking at the green lights of the dashboard, is unhooking a handset from under the dash and clicking the connection open.

A crackling silence erupts from the top of the car, from a speaker up there. It's a loud nothing, full of emptiness and distance. Victor thumbs the sound away, pushes enough buttons or switches that the drumming and singing finally pour out of the top of that car all at once, making him flinch back from the suddenness. The sound swells, fills the night.

Inside the lodge one of them yips twice in celebration of this sound.

Victor nods with this, likes it.

He goes back to the fire, stirs it with the shovel, and notes the sparks drifting over to his son's scrimmage jersey. He saves it from the hundred airborne embers, folds it onto the broken chair set up by the lodge like an end table, so Nathan can find it first thing when he's done. Then he stirs the fire, watches the sparks spiral and climb even higher, like an invisible chimney, and then he leans the shovel against the trash barrel so he can inspect the rifle.

After making sure it's not loaded, he runs the bolt back and forth twice, swings it out like tracking something, and, of all the places in the night he could have pointed that barrel, he points it right at you, your head still turned to the side, your eye on the right side rolling back down that rifle at him.

Without even thinking about it—this is what you *do* when in a hunter's sights—you pull away.

Still, he sees . . . not you, but the motion of you. The idea of *some*thing.

He lowers the gun, stares out into the night.

"Jolene?" he calls out. "That you, girl?"

When you don't answer he flattens his lips and cuts a sharp whistle, slapping the leg of his jeans twice.

You're no dog, either.

Also: There are no dogs. Not anymore.

He settles the rifle back down into its place, watching the darkness the whole while. Moving mostly by feel, he pulls up three splits of wood from the pile, works them into the embers. Moments later one of them spurts a lick of flame up, and then all three are burning bright and orange and hot.

Victor stands in front of the fire, the dark silhouette of a hunter, still watching the darkness, the rifle in his hands again like a reflex, held crosswise down low.

From the lodge there's another *ho*, this one from Nathan—the first time he's been the one calling for more heat.

Victor considers the darkness, then finally turns away, trades the rifle for the shovel, and guides its blade under the burning logs, lifting a rock up and out. He shakes the shovel, ash and embers trailing off, and runs his gloved left hand up the handle, walks sideways to the lodge.

He taps at the door with his hip and it lifts up on a shiny silver strut, stays lifted.

Inside are three wet faces, each of them already spent. He deposits this next load of glowing rocks and just has his shovel clear of the lodge when one of the horses whinnies straight out of the heart of nowhere. Victor jerks hard enough to have dropped a burning rock if he'd still had any, but it's just a stupid horse.

Still Victor studies the night all around him, his eyes scanning and panning, trying to pick a shape out.

If he was smart, if he was listening to the horses, he'd already be gone.

You wouldn't leave, though, would you? You couldn't.

You stand over your calf until you can't stand, and then you try to fall such that your body can shield it. And then you come back ten years later and stand just outside the firelight, your soft hands opening and closing beside your legs, your eyes hardly blinking.

He can no more leave his calf than you could.

And now he's standing from his car a second time. With a beam of light to stab around.

You flatten against the ground, let that heat raze across your back.

But still, he knows. The way you can tell is the smell of the pistol at his hip. Its oily sick taste is in his hand now.

"Come on out!" he calls, his words rolling into the darkness, turning back to nothing.

The horses tell him about you some more, their warnings so clear, so urgent, so simple and articulate.

He had his chance, right? This is on him. He shouldn't have come out here.

Now his beam of light is disappearing behind the camper in hitches: walk two steps, shine the light all around, then jerk forward again, repeat.

When he's around the corner you can finally step out into the flickering light from the fire. The white and brown horse, the most clear-spoken of the three, stamps her feet, shakes her head back and forth.

You shake your head just the same back at her.

The two you want are just right there, in the lodge three steps away, naked and helpless. Gabriel Cross Guns, Cassidy Sees Elk. The only two left from that day in the snow.

But you don't want to get shot in the back again, either. You can still feel the pain from last time, don't need this dad to blow that hole open all over again before you're finished.

When he walks around the side of the camper, you follow, right in the scent-path still swirling in the air so clear you could close your eyes and not lose him. You know to stay far from the camper, though, so he can't pin you there in a sudden pool of yellow. A camper isn't a train screaming past, trapping you, but it might as well be.

When he edges up to the outhouse he's so sure you're behind, your leg muscles bunch so you can—

He brings the light around, freezes you in its glare, your mind losing itself in that brightness.

"What—*who*?" he says, running his pistol back into its holster at his waist. "You trying to give me a heart attack, Jolene?"

It's her shirt and pants you've stolen off the line.

"Jolene," you say, your voice creaky because your throat is new. You start to clear it but there's a sound intruding on this moment. You both look over to the road.

A truck grinding up the road?

"Wait, you're not—" Victor says, then leans in to see better. "You're that Crow from the newspaper, aren't you?" he says. "The one who . . . who—?" Then he's raising the fingertips of his left hand to his right forehead, to show what he's saying: "But what happened to your eye?"

I got shot there, you don't tell him. *Twice.*

He takes a step back all the same, says, "I thought you—that Lewis ki—didn't he . . . What are you doing up here?"

In answer, you bring your face back around to him, eyes wild, hair lifting all around, and say, "*This*," then rush forward, show him.

METAL AS HELL

Cassidy should have done this years ago. Sweats should be a regular thing. Just like Neesh told them back when, he guesses.

Back then, though, this would have been just one more thing to sit through, one more thing between the four of them and the weekend. A sweat was never a ritual, was always just an ordeal.

Cassidy nods to himself that, yes, he's going to keep this sweat lodge going, maybe even dial it back from sleeping bags to layers of actual hide. And maybe he'll petition Denny for hunting privileges again, right? Why not? Denny's settled down and married these days, is at all the basketball games, even. And, ten years has *got* to be enough punishment for nine elk. It's been a clean ten years, too. Well, come tomorrow it'll have been a clean ten years. Cassidy has hardly even shot any animals, just a mulie or three out on the flats, that one moose that was asking for it, and the odd whitetail. But that's more like herd management, he figures. Herd management and subsistence—that's his right as a tribal member, isn't it? How can slipping back into an elder section one time take that all away?

And if Denny says no, then, well. Once Cassidy and Jo are legal, she'll have hunting privileges, he's pretty sure. Or, if not from marrying in, then he's pretty sure she can transfer her Crow hunting stuff up here, if she gives it up back home. Then, as long as she's in the field with her tag whenever Cassidy pops an elk or whatever, Denny won't be able to say a thing. Or maybe she'll line up on a big bull herself.

Beside him, Gabe scooches back from the heat the rocks are throwing, shields his face for a moment with his forearm.

All you ever want this deep into a sweat, it's a bit of reprieve. But you've got to push *through* that.

"Good?" Cassidy says across to Nathan.

Nathan's sitting with his knees up, his head hanging down.

He sort of nods. Either that or a sluggish death rattle. A last spasm.

Cassidy angles the cooler up onto a corner and lays the scoop down flat, its nose in that corner, then tilts the cooler back the other way to get at the last bit of water.

"For Ricky," he says, tipping a sip out onto the ground before taking a drink himself. It's as hot as ten-minute-old coffee by now.

He offers it to Gabe, who takes it like each time, says, "For Lewis," spilling a bit out, but passes it on around without drinking. Because, he said early on, isn't that the dog food scoop?

Horses, Cassidy hadn't corrected. And just oats at that, because Jo's paint was raised to expect more than hay or cake. But Gabe doesn't really know horses, doesn't really know how inert oats are, that this scoop is probably as clean as any spoon at the diner in town.

Nathan takes the scoop, his hand trembling, his hair plastered to his face.

"For Tre," he says, a little water shaking out.

It's the first words he's volunteered in nearly an hour, by Cassidy's reckoning.

The kid's coming around. Breaking down. Playing along.

Good.

Tre is the high schooler the wake was for a couple of weeks ago—which, now that Cassidy thinks about it, is probably about when Nathan split town, ran away into the wilds of America. He only made it to some skunky trailer on the other side of Shelby, but that counts.

Tre, Tre, Tre. The wake was the first time Cassidy knew that was the way that name was spelled. He'd always assumed it had four letters, like what you carry food on in a cafeteria.

How had he died, even? Cassidy can't dredge it up, not with the heat turning his thoughts to syrup. Was he Grease's nephew, maybe? But that can't be right, Grease isn't old enough. Georgie, then? Somebody who was a senior when Cassidy was a freshman.

"Kill it," Gabe says to Nathan about the last of the water, and, after confirming with Cassidy—just eyes, no energy to spare— Nathan tips the scoop up, cashes it, holds it back across.

Cassidy takes it. Good thing about aluminum is it doesn't heat up in a sweat. What'd they use in the old days, wood? Horn? A bladder? The skullcap of a wolverine, because the old days were metal as hell?

Doesn't matter. This isn't the old days. Exhibit one establishing that: outside the lodge, Victor's tape goes silent, to the end of itself again, and then there's a few seconds of silence while the deck looks for the first song on the other side again.

"This one again?" Gabe musters the breath to say, because he thinks he's hilarious.

"Try going on a trip with him," Nathan says back, his chest shaking twice with what Cassidy thinks is a weak attempt at laughter. The weakest attempt at laughter.

Gabe is having to waver to stay sitting upright. But he can do it until sunup, too, Cassidy knows. Of the four of them, Gabe would always be the one still sitting on the toolbox in the bed of the truck after everybody else had slouched over, passed out. It was like he was waiting for something. Like he knew that if he gave in, shut his eyes, he was going to miss it, was going to get left behind.

Of the four of them—and Cassidy hates to say it—of the four of them, Gabe's the least likely to still be aboveground, too. He's always been the first to jump, whether it's off a cliff into some big water or into the face of some cowboy outside a bar.

"Like this," he's telling Nathan now, lowering his own mouth right almost to the ground and sucking air in, making a show of swelling his chest out because the air down there is so much cooler, so refreshing.

"Where a hundred asses have sat," Nathan's saying back.

"Don't forget the dog piss," Gabe says, just giving in and lying on down.

Cassidy smiles, greys out for a second, maybe two.

This is for Lewis, he's telling himself. Lewis, who was trying to come home.

It's funny, almost: Lewis runs for home, dies on the way. Ricky runs *away* from home, dies on the way. Gabe and himself stay right here, are perfectly fine.

"Hey," Cassidy says down to Gabe.

"Just resting my eyes," Gabe mumbles back.

Nathan lowers his face again, his long hair a wet curtain, the rest of him mostly a silhouette in the ashy, humid darkness.

"About Lewis," Cassidy says.

Gabe gets an arm under himself, cranks up to a sitting position, dirt sticking all to one side of him because he's so sweaty, and because the ground is thawing under them.

"We really out of water?" he asks.

"They said he had an elk calf with him, right?" Cassidy says.

Gabe fixes his unsteady eyes on Nathan, but Nathan's just still. Either not listening or listening and not caring.

"Serious," Gabe says, about Victor's tape. "I like drums as much as the next red-blooded, red-skinned, beer-drinking—"

"He was carrying an *elk* calf home," Cassidy insists.

"Wrong season," Gabe says, waving this off. "Must have been *slow* elk."

"Wrong for them, too," Cassidy says.

"Horse."

"You don't run with a foal. It's too heavy."

Gabe repositions himself, but even the air is hot.

"I never told you," Cassidy says.

Gabe stills, looks to Nathan again, then back to Cassidy.

"That last hunt," Cassidy says. "Thanksgiving Classic or whatever Ricky called it."

"Thought that was *me*," Gabe says.

"That little heifer elk Lewis shot," Cassidy goes on. "She had one in the oven."

"I thought I shot her . . ." Gabe says.

"Your brain's melted," Nathan tells him.

Gabe shrugs like the kid's right, says to Cassidy, "It was—it was Thanksgiving, man. Maybe that little elk just had a turkey in the oven, yeah?" He pats his own belly to show what oven he's trying to mean.

"It was the Saturday before Thanksgiving," Cassidy corrects.

"Tomorrow," Gabe says with a goofy grin, looking down to the watch he's not wearing and also doesn't wear, and wouldn't be wearing in a sweat anyway.

"Lewis *buried* it," Cassidy says. "That—that unborn calf, whatever."

This silences Gabe.

"This is that same scrawny pre-cow he made us drag all the way up the hill?" he finally says. "The one got us caught by Denny the man?"

"We were getting busted anyway."

"This is when y'all shot up that herd?" Nathan asks.

Cassidy and Gabe both look over to him.

"Denorah told me," he says, like challenged to answer.

"You told her?" Cassidy says to Gabe.

"Who *else* that was there might have told her?" Gabe says right back, then does his lips like he's going to spit onto the rocks but can't muster any spit, so ends up just leaning over like a drunk old man telling important secrets to the ground.

"Oh yeah," Cassidy says.

Denny. Denny Pease. Of course he would have told Denorah this story by now. Anything to make Gabe look worse than he already does.

"What are you saying?" Gabe says then to Cassidy, picking the idea of that elk calf back up. "That Lewis was all messed up? That all those elf books finally caught fire in his brain, made him kill two women and run around with an elk baby until the soldiers shot him down?"

"It wasn't the books," Cassidy says.

"Elves?" Nathan says, watching the two of them now.

"Breathe, breathe, you're hearing things," Gabe says.

"How much longer?" Nathan asks.

"You cured yet?" Gabe asks back.

"Of what?" Nathan says. "Being Indian?"

Gabe chuckles without really smiling, which is a sound Cassidy knows. He puts his fingertips to Gabe's chest to keep him there, says across to Nathan, "You can leave whenever you want, man."

"Once you've been *purified*," Gabe adds unhelpfully, and then leans over to cough a lung up. Maybe two.

After nearly a minute of it, Nathan says to Cassidy, "He going to be all right?"

Cassidy studies Gabe, on his hands and knees now, nearly puking. .

"One way or the other," he says.

Nathan shakes his head in amusement.

"My dad says he's busted him he doesn't know how many times," he says.

"White man's laws," Cassidy says. "Getting picked up, that just proves he's Indian."

"He says he busted you, too."

"Your dad's a good cop, mostly," Cassidy says. "Just messes up sometimes."

After a second or two, a grin crosses Nathan's face.

"He's standing out there like a cigar store Indian or something," he says.

"He called in sick on a Friday night for this," Cassidy says. "Because of being here, he's going to have to work shit detail for the next month, probably. He's doing this for you, man."

"He doesn't have to."

"Tell him."

"He doesn't understand anything."

"He was the first one into the Dickey house after that—Tina, with the gun?" Cassidy says, wincing from having to remember that. "He's scraped so many kids up off the asphalt he could prob-ably write the manual for how to do it best so they stay in one piece. He's had to carry stoned babies to grandmothers and he's had to walk out into the grass to find other grandmothers. Some of the drunks he shakes awake in the morning, they're stiff, and he remembers them from second-grade homeroom. His first week, he was the rookie cop they made drag Junior Big Plume in from the shallows, when his face was all . . . he sent my brother Arthur to prison, how about that? He doesn't want you to end up there, too."

"I'm not like him and Granddad," Nathan is already saying, his lower lip trembling hard enough he has to bite it in.

"He'll stand out there and keep that fire going for you for as long as you need. That's all I'm saying. Not every Indian dad's like that. You got one of the good ones, man."

"It'll turn into an old-time Indian story," Gabe chimes in, his voice weak and spent from the coughing. He plants a hand on Cassidy's shoulder to pull himself upright again. "It'll—it'll be the story of the dad who stands outside the lodge for seven days, having to go farther and farther out for wood to keep the fire going, and then he asks the beavers to bring him some, meaning he'll owe them a favor, and then when the fire almost dies out once, he needs some kindling, so he has to—has to call a *hawk* down to deliver him some dried moss, so he's going to owe him something, too, then, then it's something with a muskrat, then, then . . ." but he loses it to coughing again.

Cassidy shrugs to Nathan like, *Yeah, that.*

"Aren't we supposed to be singing and praying and all that?" Nathan says, looking from Cassidy to Gabe.

"We are," Cassidy says.

After that they all stare into the glowing rocks.

"We need more water," Gabe finally says. "Maybe if we had, like, water guns in here, right? Old-time Indians never thought of that, I bet."

He finger-shoots imaginary streams of cool, cool water at Cassidy, at Nathan, then into his own mouth, just drinking it up.

"You could have drank some from the cooler," Cassidy tells him.

"Got . . . standards," Gabe says.

"I'll ask my dad," Nathan says—any chance for escape—which is right when the flap pushes in the way it does when Victor is nudging it. Except, no Victor. Are the dogs back, then?

"Here," Gabe says to Cassidy, and hauls the cooler into his lap.

Gabe lies back for the sacred golf club, aims it for the flap, and pushes.

Outside, instead of Victor's thick legs, it's a woman's long, very nice ones.

Nathan, naked and fourteen, pushes back into the darkness with his heels.

"Holy shit," Gabe says to Nate, impressed. "You really order pizza?" Then, to Cassidy, "Town Pump delivers out this far? Also, Town Pump *delivers*?"

"I got this," Cassidy says, and sets the cooler to the side, stands up through the flap.

"How's it going in there?" Jo asks.

"Hot," Cassidy says, riffling his hair with his hand and looking down his front side. "Pretty naked, too, I guess."

Jo cringes back from the droplets of sweat Cassidy's hand is spraying from his head.

He stops, looks at his hand. It's still wet, like the whole rest of him. Then he looks past his hand. Usually if he's sweaty, the dogs are using him like a Popsicle. In this chill, though, the sweat won't be sweat for long. Couple minutes and it'll be pneumonia.

"See Victor when you pulled up?" he asks, looking around.

Jo turns to the darkness all around with him, says, "Thanks for bringing my clothes in."

Cassidy considers this, can't get it to track. Maybe he's that great a boyfriend, and he just forgot?

"Everything good at the store?" he asks, meaning: *Why are you here when you're supposed to be there?*

Jo gulps a swallow down, gathers her words in her mouth, is about to say whatever it is when Gabe calls a weak *Ho!* out from inside the lodge.

Cassidy keeps watching her face.

"This isn't your fault," she says at last. "I want to be clear on that. But—I called home on break, yeah?"

Cassidy nods, knows that that's when she talks to her sister, because no one watches the break room phone.

"You know your friend who . . . who got shot?"

"Which one?"

"Out by Shelby. Yesterday."

"Lewis."

"He killed his wife and that woman he worked with?"

Cassidy nods, not much liking this lead-up.

Jo hooks her right elbow into her left palm so she can hold her hand over her mouth, look away again. "That was—that one he worked with at the post office, I guess, she was my cousin Shaney. Shaney Holds. My sister just found out."

"Oh shit," Cassidy says. "Oh, shit."

Jo tries to shrug it off, can't. Cassidy goes to hug her but remembers at the last inch how gross he is right now.

"So . . . so what does this mean?" he asks.

"It means she's dead," Jo says, maybe about to cry. "My aunt, her mom, she's—Shaney was her last, yeah?"

"Of how many?"

"Last one to still be *alive*, I mean," Jo says, threading her hair out of her face, peering around it to see Cassidy's eyes for a moment.

"Shit," Cassidy says again. It's all he's got.

"I talked to Ross," Jo says. "He said I can have three days, starting an hour ago. One day to get there, one to be there, one to drive home."

"Don't worry about Ross," Cassidy says. "Gabe's been in the hole with him. Take all week if you need. Take two."

"I know you can't go—"

"I can—"

"Third week of a new job and you need some personal time?" Jo says, and lets that settle.

She's right.

"I wanted to just go straight there," she says. "When I didn't show up in the morning, though, I thought you might—"

"Thank you," Cassidy says. "I would have freaked out, kicked everybody in town's ass."

"Because that's how you are," Jo says with a smile.

"Gotta do what you gotta do," Cassidy says, happy to have made her forget her cousin for a moment.

Jo steps away from the lodge, bringing Cassidy with her.

"How's he doing in there?" she says.

"Nathan?"

"He's the freshman?"

"Eighth grade, maybe?" Cassidy says. "It's good, it's good. I wish—back when, I wish I would have paid attention when his granddad was doing all this for me, though. So I could, like, pass it on better."

"His granddad?"

"He was—don't worry about it. You need to go. You need some money, though."

"I can—"

"Take it," Cassidy says, turning to the truck on blocks, the thermos of cash in the crumbly glasspack. "That's why we've been saving it, right?"

He walks over, hooks his hands on the old grille guard to slide under but then stops at the last instant, remembering again how sweaty he is. And how naked. And how sharp all the hanging rust is down there.

Jo's right there beside him already, holding his arm. Pulling him to her.

They hug in spite of his sweat, her loose hair matting on his chest.

"You're going to need a shower now," he tells her.

"I like it," she says back.

"Let me get my coveralls," Cassidy says.

"I'm not completely useless, you know," Jo says. "I can get the money myself."

"It was my friend who killed her."

"Feed Cali?" Jo says, about the paint.

"I'm not going to call her that," Cassidy says.

"In your head you will," Jo says, and takes his face in her hands, pulls his mouth to hers, kisses him bye, and holds him there, her eyes shut.

"Careful," Cassidy says. "I am naked here."

She reaches down, doesn't help matters any.

"Two days," she says, backing away.

"Monday," Cassidy says back.

"I'll leave some towels by the fire," she says. "Boys always forget there's going to be an after."

Cassidy turns to the lodge, has to shrug. She's right. They were just going to drip-dry, maybe. In the freezing cold. Standing in the snow.

"You're good to drive?" he calls across to Jo. She's on the steps to the camper.

"It's not even that far," she calls back, then, about the drumming coming from Victor's car: "One of your tapes?"

Cassidy shakes his head no and then she's gone, inside, packing, the camper creaking and groaning, all the windows yellow now, which pretty much means their one light is on. But still, it looks alive in a way that pretty much makes all of everything worth it.

Out in the darkness the horses are stomping and blowing.

"Don't worry," Cassidy says to them. Then, more to himself: "I'll bring your scoop back, sheesh."

But where *is* Victor?

Cassidy studies the darkness for ten, twenty seconds, each colder than the one before, then whistles loud and hard to pull the dogs in.

Stupid dogs. Stupid horses. Stupid Victor.

On the way back to the lodge, walking faster the closer he gets, his breath chugging white before his face, he scoops up two dripping handfuls of snow then lifts the flap with his leg, slow-spins in, already holding those two cool handfuls of slush out.

"Coconut?" Gabe says, drunk on heat, taking his handful of cold and looking over to Nathan for the rest of the joke: "He knows I like coconut flavor for my Icee."

Nathan takes his, crushes it into his face, holds his hands there to try to get this coolness to last.

"Coco*nuts*," Cassidy says, shaking his own before sitting back down, and Gabe considers his handful of slush, considers it some more, then dollops it down onto the rocks. Steam billows up, dialing the heat in the lodge up an impossible degree or two more.

"Ho!" he calls out to Victor, but there's no Victor to say it to, just drums and darkness, horses and cars, and, standing right there, so close now, you.

Cassidy lets the flap shut them in again.

THIS IS HOW YOU LEARN TO BREAK-DANCE

The three things shuffling around for foot room in Gabe's head are:

1. a drink
2. a pee
3. Jo being out there now

What her being out there means is that staggering up and out into the cool air for the pee he desperately, desperately needs, even though he's drunk exactly nothing for this whole sweat, has to be deep in the negatives as far as fluids, really, what Jo being out there means is that . . . he needs a towel? A fig leaf? A Bible to cover himself with? Not one of the little green ones, but a big holy roller of a leatherbound book.

But—like there were never any naked dudes on the Crow rez?

Gabe chuckles to himself, slow-motions his fingertips up to feel his lips smiling, because his face isn't telling him anything at the moment.

"What?" Cass says.

Gabe just wobbles side to side, his wet head tracing secret fig-ure eights.

The kid has his mouth down right by the melting dirt, is suck-ing its vapory coolness.

Cass passes him the cooler. The kid tips it up like a giant cup, sluices the last-last memory of water down his gullet.

"Feel like I've heard this one somewhere before . . ." Gabe leans over to say to Cass about Victor's stupid drums.

"Shh," Cass says, his eyes closed like he's trying to be inside himself, is trying to really get into this sweat.

Sure, great.

Gabe closes his eyes too, swims through that powdery hot blackness and feels his shoulders melt down, his ribs sighing in when he breathes everything in him out, his fingertips bulbous and heavy now, his legs and feet somewhere else altogether.

Maybe this is how it works, he tells himself, at the same time trying to be quiet in his head, because talking to yourself is exactly how it *doesn't* work. The body slipping away is what allows the rest of you to float up, over, out. Maybe see some shit for once, yeah?

Except what Gabe settles on, it's not real, he knows. It can't be.

It's his father sitting in his chair in his living room on Death Row.

He's watching that same channel as always: that camera angled down onto the parking lot of the IGA.

On his rounded little screen there's nothing and nothing and then some more nothing on top of that, and then—and then a tall dog trots through on some dog mission or another.

Gabe's father grunts approval and Gabe looks over to him like, *What?* Like, *This is what passes for action?*

His father chins Gabe back to the television.

The same nothing, like bank robbers have looped the footage, are cracking into the IGA, stealing all the heads of lettuce they want, for their big salad enterprise.

Gabe snickers.

"Listen—" he says, making to go, to be anywhere else than this, there's got to be better visions, but now there's a flurry of motion on-screen.

Not dogs this time. Boys. Four of them.

The skin around Gabe's eyes draws in. Either in the sweat lodge or his father's living room, he doesn't know, and it doesn't matter.

They were twelve then. Him and Lewis, Cass and Ricky.

What they have between them is a single Walkman with that one tape Cass had stolen from his big brother Arthur.

Lewis is first.

He puts the headphones on, Cass holds the Walkman out, keeping the cable free, and Lewis nods with the synthesizer the way it starts out, and then he looks around at Gabe and Ricky and Cass, his face deadly serious, and the way his head is bobbing, he lets that infect the rest of his body.

When the beat finds his hand, his fingertips lift out to the side in some Egyptian pose that's already crinkling back up along his arm, hitting his neck, throwing his head to the side like he can't help it, and around him Cass and Ricky and Gabe are bouncing with it.

This is how you learn to break-dance.

Gabe smiles, watching the four of them all those years ago, Lewis already passing the headphones to the next popper-and-locker, holding the Walkman himself now, the music still in his head.

It always will be, Gabe remembers thinking. Knowing. Promising.

It always will be.

And beside him now, his father is looking past the television screen, to the walls of his living room, to his baseboards, which are . . . are crawling with—

Cass.

It's Cass sitting beside Gabe, not his father. They're in the sweat lodge.

Gabe breathes in deep, the hot air roiling in his chest, cooking him from the inside, and he tries to muster a smile because *they're* the turkeys in the oven now, aren't they? But his lips are traitors, are slugs, are so far from his face. When he looks across to check on the kid, make sure he hasn't passed out onto the rocks, he sees two more shapes sitting there, eyes boring down into the heat.

Ricky.

Lewis.

Except . . . except, Ricky, his face is leaking down, is beaten in, stomped in, and Lewis, he's starting to look up, and there's finger holes of light poking through his chest, and . . . and—

Gabe stumbles up into the ceiling of the lodge and dog hair rains down.

Some of it finds the rocks, hisses a bitter taste into the air.

"I've got—I've got to," he says, ducking now, his hand on Cass's shoulder, and Cass doesn't stop him from feeling his way around to the flap, birthing himself naked out into the night air.

A moment later, gasping the coolness in, Victor's drum loop filling all the empty spaces in the darkness, the cooler comes through the flap as well, for Gabe to fill. Because somehow this ordeal isn't over yet.

Gabe leans back, stares up into the wash of stars.

Let Jo walk up, look him up and down, shake her head. So he's not the toughest Indian in the world. He *is* the thirstiest, though, he's pretty sure. And not for stale water from Cass's tank.

He's got his own cooler just over there in the truck, right?

He finds the Mauser by the trash barrels, uses it like a cane for a few steps, leaves it against Victor's cruiser, pats the car's hood like thanking it for holding this for him. He leans on one of the chairs to steady himself and looks all around, taking everything in.

Except for the camper and the trucks, it could be two hundred years ago, he's pretty sure. Not a single electric light for miles in any direction. But he's glad it's not two hundred years ago, too. Two hundred years ago there wouldn't have been bottles of chilled beer in the cab of his truck.

When he shakes free of the chair to get some of that cold-cold beer, Cass's shirt tangles in his wet fingers. He holds it over his crotch in case Jo's about to jump up from behind Victor's car.

Speaking of: "Um, firekeeper?" Gabe says all around.

Nothing.

"Hunh," he says, and finally settles his eyes on the outhouse just back from the camper, nods about the hanging lantern in there, glowing yellow.

Victor's in the can.

Gabe grins a who-cares? grin, pushes off the side of the cruiser he's staggered into again somehow.

It's so cool out here. So perfect. The snow crunching under the soles of his feet is the best thing ever.

At his truck he stabs an arm in through the open passenger window, flips the cooler open, shoves his hand down into the water that used to be ice. There's still chunks in there, even.

He draws a beer out to himself, rubs the cold bottle all over his face, his chest, his arms. The hiss of it cracking open is amazing, the mist swirling up the best promise ever.

"I've been thinking about you," Gabe whispers into the mouth of the bottle, and tips it up, tries to go slow so he won't throw up.

While he's drinking, he left-hands a pee. Cass is always saying not to piss too close to the camper, either go on out to the trees or use the outhouse, that the whole place is going to start smelling yellow if everybody just splashes pee all over, but screw it. Victor's in there anyway, and Gabe can't wait.

Liquid in, liquid out.

With a gasp he finally breaks his long kiss with the beer, wipes his lips with Cass's shirt, oops, and manages a look down to what he's peeing on.

It's one of the dogs.

He angles his stream away, lets it sputter out, shakes off, and doesn't zip up since this isn't exactly a zipper situation.

He looks over to the camper, all its lights on. To the outhouse, hunkered down over its deep hole. To Victor's car, drumming its loud beat out into the night.

And the dog.

It's one of the two pups, not Miss Lefty, but . . . Dancer, yeah. Dancer the dead, dead, very dead dog.

Gabe squats down gingerly, unsurely, and touches the dog's matted coat.

"What stepped on you, girl?" he says, petting the dog's haunch.

Her guts are ballooned down into the interior skin of one of her back legs. Gabe's seen it happen before, to dogs that have been run over.

But this dog, it's been . . . *stomped*?

Her chest has been crushed, too, and because there was nowhere for the lungs and heart and liver to go, most of it's splashed out the mouth in what looks like a single chunky gout. The tongue is hanging, not swollen up yet.

"What the hell?" Gabe says, standing, looking out into the darkness instead of behind him, where you are, on the other side of the truck. If he just turned around, chanced a look into the passenger window, through the cab, there you'd be out the driver's side, watching him. Glaring hard at him, your five-fingered hands balled into fists.

He doesn't, though. And he won't. His whole life he's been looking in the wrong places. Why should tonight be any different?

"Cass," he says then, like trying it out, "one of your horses, man, it got out, I think. And it doesn't like your dogs."

He steps carefully around this dog, deeper out into the night.

Two slow steps later are the other two dogs.

Ladybear is dead, but Miss Lefty is still trying.

"Shit," Gabe says, dropping to a knee.

Miss Lefty whimpers.

"Shit shit shit," Gabe says, and sets his beer down in the snow, holds it there a moment to be sure he can let it go without it tipping over.

He feels around with his right hand for a rock, finds a good heavy one, then, with his left hand, makes sure where the dog's head is.

She's dead now.

He sets the rock back down, slumps on his thighs.

When he stands it's without his beer, without the shirt. When he looks back to his truck there's nobody there through the tunnel the windows make. Walking back, he runs the hair out of his eyes and smears blood all across his face.

That rock he used, or meant to use, it was the same one you used.

It's almost funny.

Back at the truck he grubs a rag up from under the seat, cleans his hands and face, then, with his other hand, liberates another beer, drinks it down all at once, and turns, does a running throw to sling the bottle out as far into the darkness as he can.

It doesn't land for seconds and seconds, and doesn't shatter when it does, just *thunks.*

Cass is *not* going to like this, he knows. Nobody likes all their dogs being dead at once. But it's not Gabe's fault, either. And if—if he leaves pretty soon after the sweat, then he won't even have to get involved in this, will he?

"You were never even here," he says to himself, looking around to make sure Jo's not suddenly standing there behind him, listening in.

Why would he even be thinking that?

"Getting jumpy in your old age," he mumbles, and hauls the cooler of still-cold water up through the window.

It'll be better than the water from Gabe's tank. And they'll need something better to dip it out with.

Gabe sloshes the cooler onto the hood, opens the passenger door and digs behind the seat, eyes staring straight up so his fingers can feel farther. Finally he comes up with some random metal thermos. He twists the cap off, dumps it into the floorboard then blows into the thermos once, hard, already turning his face to the side.

No mice skeletons or bug husks come back at him.

He holds it upside down, taps it against the front tire to break loose anything stubborn, and when nothing cakes out—it would just be coffee anyway, right?—he fixes its thin lip into his mouth, carries it like that, the cooler in both hands like the biggest, squarest, most refreshing fig leaf.

He's going to be a hero, bringing water back with actual chunks of ice still floating in it. And the dogs dead in the snow? They haven't even happened yet, aren't even real.

On the way back to the lodge he raises his voice, singing with the singers, walking with the drumbeat, Indian-style.

BLACKFEET INDIAN STORIES

Nathan remembers some stupid summer program years ago, where all the ten-year-olds were supposed to be learning traditional stuff. This was back when he had three braids, was still being groomed to be an All-Star Indian. Before he started being who he really was.

Tre had been there, too, his hair in traditional braids as well.

What they were learning for that week wasn't riding or archery or any of the cool stuff, but how to dry meat on a rack.

Sitting in the heat of this lodge, that's exactly what he feels like: one of those thin strips of meat on that rack of twigs, a slow fire burning under him, the sun baking him from above.

Except there's words cycling through his head, shaken loose by the steam. From when his granddad was taking him through the language. From when talking like that made sense.

Kuto'yiss.

Kuto'yisss"ko'maapii.

Po'noka.

Kuto'yiss is where his dad drove him back from yesterday, pretty much. The Sweetgrass Hills, but use it in a sentence: *I went out to Kuto'yiss maybe to die, Granddad. To be with Tre. But your stupid son dragged me back. I went there because you were always talking about the Sweetgrass money, do you remember? What America kept not paying us for the hills it stole?*

Use it in another sentence: *I'd rather die out in Kuto'yiss than under a car upside down in Cutbank Creek, like Tre.*

And what about Kuto'yisss"ko'maapii? It's not Sweetgrass Hills plus "ko'maapii," which was hard to wrap his head around back then. And also now.

What it means is Blood-Clot Boy, the hero kid born from a clot of blood, back when shit like that was always going down, at least according to his granddad, waving one more kid into the lodge for story time.

Nathan had never told anybody, but used to, second grade maybe, his dad braiding his hair before homeroom every day, he'd secretly known he was Kuto'yisss"ko'maapii. That he was here to save the people, then become a star in the sky. Then in seventh grade Mr. Massey had explained how every young Indian thinks he's Crazy Horse reborn.

Denorah Cross Guns had stabbed her hand in the air about this one, and Nathan sneaked a look back at her, like always.

"Not the girls," she said.

"You all think you're . . . *Sacajawea*," Mr. Massey told her with a shrug, his mouth tumbling down through all those syllables like the best joke.

Because Denorah Cross Guns didn't know enough of the old-time Indians to pick someone better, someone not a traitor, she'd saved it all up for the game that night, and fouled out, had to be dragged off the court for fighting, and her new dad had had to keep her real dad from crashing down onto the court as well.

Nathan had been there in the stands as well, yelling for her with the rest of the crowd, yelling that it wasn't her foul. But even if it had been, right?

Denorah Cross Guns isn't anybody's Sacajawea. And Nathan, he isn't any Crazy Horse *or* Blood-Clot Boy. He knows that now. Those three-braid days are over and done with. Never mind all this sweat lodge bullshit. Never mind his dad playing the drums out there.

When Gabriel offers the new cooler, Nathan takes it into his lap, uses the black metal thermos to scoop up some of that water that's so cold it almost hurts.

Cass nods at him to go on, that he's doing good.

Nathan dollops some of the water out onto the rocks and steam spits up between the three of them, stranding them in their own individual sweat lodges, almost.

Are the rocks even really supposed to be this hot?

Nathan doesn't think so.

No way could anybody stand this for more than an hour or two. Not without coming out cooked. A round or two ago Gabriel said he'd been *baked* before, sure, but this was another level.

There's still half the thermos of water left.

Nathan swishes it, swishes it again, and is about to drink when he remembers the rule: honor your ancestors. Which is what Cass told him. What Gabriel said was just to say somebody's name, somebody who might not be getting a drink otherwise, yeah?

"Granddad," Nathan says, loud enough for the two clowns through the steam to hear, and pours out half of what he was going to drink.

Across from him, he's pretty sure Cass nods that this is good, this is good. Now keep it going.

Back in his place in their triangley circle, Gabriel is next to get the cooler. The one he just delivered.

"Neesh," he says, like agreeing with Nathan, and tumps a splash down, doesn't take a drink himself. Meaning he probably drank his fill while he was out there.

"Think he's had enough?" Gabriel says just generally, passing the cooler to Cass.

Cass looks up, not following, so Gabriel explains: "His grand-dad, man. That's two drinks already. He's gonna have to go pee soon, think?"

He smiles after this, his mouth loose like his face is melting.

"What do you think ghost pee smells like anyway?" Gabriel's going on now. "You think it's like all around all the time?" He tries to haul his foot up to his nose to smell for ghost pee.

"Not hot enough for you?" Cass says back to him, then angles his face over to the flap, calls out a deep *ho* for another hot rock, even though the last one hasn't come yet.

Gabriel slumps in response, looks up into the ceiling like for something to save him, and big bad Officer Yellow Tail was right, Nathan kind of knows: Gabriel and Cass *are* him and Tre, twenty years down the road. Or, they would have been, if Tre were still around. Or, if he were over with Tre now.

This is all you really need, isn't it? Just one good friend. Some-body you can be stupid with. Somebody who'll peel you up off the ground, prop you against the wall.

Example fifty-eight, about: Gabriel has sharpened his hand into a blade, is touching Cass's shoulder with it, just enough to get a jolt of electricity from Cass, a jolt that can travel up his arm, cock his head over to the side in the stupidest, least ro-botic way.

"Shh, this is serious, man," Cass hisses to Gabriel, and Nathan shakes his head about the two of them, one grooving while sitting on his bare ass, one ceremonially dipping the new scoop into the water, holding it up like you have to look hard at it before tipping a little out for the dead to drink.

But then he doesn't tip any out.

He's still studying this black, onetime-pricey thermos.

"What?" Gabriel says, stopping his slow-motion serpentine groove. "I mean, I know it's not a dog food scoop, man, but some of us have higher—"

"Where'd you get this?" Cass asks, zero joking.

Gabriel shrugs, doesn't answer, goes back to his stoned swaying, and only looks around slow when Cass is up and gone through the flap, taking the black thermos with him.

"This mean it's over?" Nathan says to Gabe, and Gabe tunes back in, looks all around the lodge, finally settles on the cooler Cass let spill on his way out.

"Quick, kid," he says to Nathan about the spilling water, "say the names of all the dead Indians you know, be right back," and then he's gone just the same, and Nathan knows this was the plan all along: To strand him here alone with his thoughts, with his demons. With his granddad.

He shakes his head at the stupidity of it all.

What would Crazy Horse do? he asks himself. Probably stay in here all night, then stare everybody down when he walked out naked, all the rocks cool, outlasted.

Either that or he'd count to one hundred, be done with this Indian bullshit.

Highlights are on at eleven, he reminds his dad, out there somewhere.

How about we make them?

AND THEN THERE WAS ONE

Ten years and now you're here at last.

From the herd, you have the scent and the taste and the sound of Richard Boss Ribs getting beat to death in that parking lot in North Dakota, and you felt Lewis Clarke catching bullets with his chest, his body dancing against your own, his arms holding you like you were all that mattered, but this time you're going to *see* it happen.

It's going to be different. It's going to better. It's going to have been worth the wait.

Before, you were standing by the horse pens, close to the dogs. Now you're on the other side of the driveway, from walking back from the outhouse, your chin and mouth black with blood.

Neither of these last two know you're in the world at all. That day in the snow they shot you, to them it's just another day, another hunt.

That's why it has to be like this.

You could have taken them at any point over the last day, day and a half, but that's not even close to what they deserve. They need to feel what you felt. Their whole world has to be torn from their belly, shoved into a shallow hole.

The first one out of the lodge is the Sees Elk one, Cassidy. The name already leaves a bad taste in your mouth. He's standing in front of the lawn chair he left his clothes on. At first he'd grabbed the boy's bright white shirt when it was right there by the lodge, but he put it back, even trying to get it folded again, patting it into place. His own shirt isn't on the lawn chair anymore, but his pants are still there. He's trying to put them on but he's sweaty and they're tight and it's not working.

He grunts with frustration, sits in the chair and then straightens out in it, flattening his body to try to find less resistance. It's not the angle, though, it's the stickiness. The chair folds over, the left pair of hollow aluminum legs bending in.

He stands from the tangle, his pants halfway up, and slings the chair around and around, launches it as high and as far as he can, out past the horse pens.

It's because he watches it fall that he sees his shirt, a smear in the darkness over to the left of the trucks.

"Gonna shoot those dogs," he says, and takes up the black thermos, stalks out there.

A moment later the other one, Cross Guns—*Gabriel*, the first one to shoot his rifle into the herd that day in the snow—is standing naked in front of the lodge, watching his friend stalk off into the darkness.

For once he doesn't say anything.

Slowly, he becomes aware again of the lights in the camper still on, and of his own nakedness. He covers himself with his hands, darts to his own fallen-down bent-over chair, does the pants dance just the same as the other one.

"Victor?" he says all around, his voice deep like that can balance out his nakedness.

He rolls his shirt on sleeve by sleeve and you remember what the boy said before, about one team being shirts, the other skins.

"Guess the ceremony's over," Gabriel says, still watching Cassidy.

He's wrong. The ceremony's just starting.

Look over to the other one now.

Cassidy yanks his shirt up from the ground, tries to shove his right arm through the sleeve, but . . . it's wet, it's soaked something up, something more than just snow.

He peels back out of it, studies the spreading stain.

Blood.

It's then that he registers what he's standing in the middle of.

The dogs. *His* dogs.

All he came out here to do was shimmy under his truck, check the muffler, see if his black thermos is still there, if it was just bad luck that his friend had hauled an exactly matching thermos in from who-knew-where. Cassidy isn't trying to solve the big mystery of what happened to his dogs. Five seconds ago, there wasn't any big mystery. The dogs were just dogs, off doing dog things.

Like dying, evidently.

Like having their heads smashed in with . . . did the horses get free, stomp them? The dogs are forever harassing them. But still.

Cassidy looks over, the horses' eyes shining in the dull glow from the dying fire, nostrils wide from this death in the air. They're still in the pen, couldn't have done this.

So.

He comes back to the closest dog, sees the guilty rock. He edges over, lowers himself to his knees, the crust of snow sharp against the top of his feet. Right beside the blood-crusted rock is one of Gabriel's beers.

Cassidy is breathing hard now.

He looks over to the fire, to the lodge. To Gabriel, struggling to button his pants, having to hop on one leg so his other can be straight enough.

There's nothing funny about him right now.

You can read Cassidy's thoughts on his face, in the way his top lip is drawing up on one side: *Good-time Gabe. Dog-killer Gabe. Gabe the bank robber.*

Cassidy places his hand to the rock and, instead of hauling it up immediately, senses a presence the same way Victor Yellow Tail did. Not you this time, but—a pair of sudden and out-of-place eyes looking right at him from just a few yards away.

The Crow, the one who lives here, the one who leaves her scent everywhere, especially in her clothes. She's under the old truck just like she said she would be, one of her arms up in the chassis for that glasspack, but now she's motionless, doesn't know what this night is trying to turn into. "Is it there?" Cassidy says across to her, not loud enough for Gabriel to hear, and the Crow doesn't answer. "Never mind," he says, standing with the black thermos. "I already know."

With that he steps out, is standing by Gabriel's truck.

He pulls the passenger door open for the dome light.

Gabriel cocks his head over, says, "Cass?"

"Did you think I wouldn't notice?" Cassidy says.

Gabriel steps closer, eyes squinted.

He's heard his friend dial his tone down like this, but never for him, and not for years, probably not since . . . narrow your eyes so you can inhale it . . . not since Cassidy's big brother went to prison and Cassidy drank that whole bottle and broke into the high school at night, to wrench his brother's old locker door off, save it for him.

"Notice what?" Gabriel says, still edging in. "That I brought a lot of cold-ass water into that sorry excuse for a lodge, and then you spilled it all?"

Cassidy's body shudders with a sick laugh.

He punctuates it by slamming the thermos into the passenger side mirror of Gabriel's truck. The glass shatters, the frame swinging down on the lower part of the bracket still bolted to the door, the top arm scratching a raw arc into the paint.

"*What the hell!*" Gabriel says, in close now, leading with his chest.

Cassidy stands right into him for once, says, "Let me see your hand."

Gabriel backs up.

Cassidy reaches across, takes Gabriel's left hand in his own, turns it over for inspection. "She hardly even bit you," he says about the two punctures ringed with bruise.

"What are you—?"

"Is that how you justified it to yourself?" Cassidy goes on.

"The—" Gabe says, then sees it in Cassidy's eyes: "The dogs, no, yeah, I mean—that wasn't, I was going to—"

"Not the dogs," Cassidy says. "The money, Gabe. There was nine hundred *dollars* in there, man."

"In where?"

Cassidy spins the black thermos into Gabe's chest, says, "You *know* where."

Gabe fumble-catches the thermos, sets it purposefully onto the hood of his truck.

"You think I have nine hundred dollars on me?" he says, incredulous. "You think I've ever had nine hundred dollars to my name all at once?" To prove his innocence he shoves both hands into his pockets, rabbit-ears them back out all at once, five twenties fluttering out and down.

"I just got that from Victor," he says. "You saw, you were there, man."

"And that?" Cassidy says about his other hand, still wrapped in a fist, around whatever was in *that* pocket.

Gabriel looks down at that hand like he wants to know, too.

But he can feel it against his palm, too, can't he?

He steps back from Cassidy.

"I don't—this isn't mine," he says. "It wasn't here when I took those pants off."

"*What?*" Cassidy says, reaching in.

Gabriel steps back again. "Are these even mine?" he says, looking down to his pants.

"Show me," Cassidy says, his voice low and no bullshit.

Gabriel locks eyes with him, says, "Listen, I don't understand what's—" and holds his hand out between them, palm up, and opens his fingers, peek-looking at whatever he's holding.

It's the ring. The one Cassidy was keeping at the bottom of the thermos, for the Crow.

"*This* is how bad you don't want me with her?" Cassidy says, huffing a sort-of laugh out.

"No, wait, I don't—" Gabriel says, depositing the ring carefully on the hood of his truck to show how little he wants it. How little he stole it.

"And then you kill my dogs on *top* of that?" Cassidy says. "Did you catch whatever crazy Lewis had? I don't understand what's happening with you, Gabriel fucking Cross Guns. Tell me why you're doing this—no, no, don't even try. Just tell me where the money is."

"Listen, somebody's . . . I don't know what you're—" Gabriel starts, but then Cassidy cuts him off by one-handing the black thermos off the hood, spinning it in his hand to get the hold he wants, and slamming it into the windshield of Gabriel's truck, leaving a deep crater, the thermos in the white center like it's something that blazed down out of the sky for this truck and this truck only. Gabriel looks from the windshield to Cassidy then back to the windshield, his eyes flaring up at last.

"Right?" he says, matching Cassidy's rising tone, and steps in, wrenches his mirror the rest of the way off, holds it by the bracket and swings it into the rain gutter of the cab until the roof wedges in, making a deep, unfixable notch. "*C'mon*, man!" he urges. "Let's beat it to hell, yeah? Stupid truck, stupid truck, always getting stuck right when, right when . . ."

When Cassidy doesn't fall in, Gabriel slings the mirror out into the darkness, is facing Cassidy now, his chest heaving.

"But it's not the only truck that was always getting stuck, right?" Gabriel says, and brushes hard past Cassidy, is picking up speed by the time he pushes off from his own taillight, is already running before Cassidy can catch him.

"No!" Cassidy screams, diving, his fingers just hooking into Gabriel's right rear pocket.

For a moment Gabriel slows, but then the pocket rips away, shows ass.

"Gabe, Gabriel, *no!*" Cassidy screams from the ground, but it's too late.

If either of them looked just six feet into the darkness to the right, they'd see the white slash of your smile.

This is it. They're doing it.

Gabriel curls around to come at the old truck from the side and drives his shoulder into it with everything he's got.

He doesn't weigh much, but he weighs enough.

Cassidy is up and running already, but his pants aren't buttoned and are too long without boots and he doesn't get there in time, could never have gotten there in time.

The truck sways to the side, sways back, and Gabriel catches it in rhythm, pushes back hard enough that one of the cinder blocks under the front axle housing explodes, the driver's-side front lurching down like a horse taking a knee. No: like an elk that just got shot, doesn't understand, is crumbling down.

"No!" Cassidy screams, and hooks his fingers into the wheel well on the passenger side right as that cinder block comes down in stages as well, taking the two blocks under the rear axle with it.

For an impossible moment Cassidy holds the truck up, screaming, his mouth open as wide as he's ever had to open it, wide enough Gabriel even panics, wedges into Cassidy's foot space, hooks his hands in the wheel well like keeping this truck up is suddenly the most important thing in the whole world.

The truck doesn't know that, though. It hitches down farther through the cinder block, crushes down all at once.

Cassidy falls with it and goes lower, his face sideways to the snow in an instant, to look under, but there are no tires anymore, no wheels, even the brake drums are gone. The truck's sitting down on its frame. There's no seeing under it.

He hits the side of his fist into the ground over and over, and Gabriel's just standing there watching him.

"Hey, man, I got a good enough jack in the truck, we can—" Gabriel says, but Cassidy stands right into him, shoves him away.

Gabriel falls down, watches Cassidy from there.

Now Cassidy is . . . trying to force the hood open?

"Here," Gabriel says, pulling himself up and stepping in, but Cassidy elbows him away hard again.

"What's got into you?" Gabriel says.

Cassidy is crying now, sputtering, can't catch his breath.

Gabriel goes back, drives his elbow down into the mismatched hood once, twice, trying to remind the springs how they work.

The ancient catch releases and the hood pops up a few inches.

Cassidy pushes his hand in, forces the rust-frozen hook over to the right, and, with his other hand, lifts the hood in a screech of metal. He collapses back, covering his face from whatever's in there.

Gabriel looks from the ball of pain Cassidy is to the truck.

There's no engine, so he can see straight through to the ground.

It's the Crow. Part of her, anyway—her hair, matted deep in blood and brains, all of it soaking into a nice Hudson's Bay blanket. The crossmember at the back of the engine bay, right about where the front of the transmission would be, looks to have come down on her face, crushed her forehead in. And back out.

She was trying to ball up in the safety of the engine compartment, Gabriel can tell. She knew the truck was falling, she was scrambling ahead, pulling with anything she could grab on to.

It would have worked, too. It should have worked.

But they couldn't hold the truck up long enough. The truck that didn't even need to be falling in the first place, except to make a dumb-ass point. Except to get Cassidy back for bashing a windshield in, for some money and dogs Gabriel hadn't even had anything to do with it.

Still.

Gabriel covers his mouth with his hands, can't get his lungs to suck air in the right way anymore.

Now Cassidy's stalking back from the patrol car. With the Mauser.

Gabriel steps out into Cassidy's path, drops to his knees, offering himself, but Cassidy goes right around him, for the truck now sitting on the Crow.

He hauls the passenger door open and leans in, a great cloud of dust billowing up into the cab.

"Cass, man, I didn't—what was she—" Gabriel says.

And then he sees what his friend is doing. It's what Cassidy said earlier—that he probably had a stray shell that would fit the old gun. One of the ones from Ricky's foggy bag of stolen ammo.

Cassidy tries the first shell, and when it won't load he drops it, moves on to the next.

"You knew this is where I keep my money," he says to Gabriel like an explanation.

"Dude, *dude*," Gabriel says, standing, holding his hands out like they can fend off accusations, like they can stop bullets, like they can make all of this make sense.

Cassidy rams another shell in, works it back out, tosses it.

"Shut up," he says. "You're always talking. You never shut up. If you'd just listen for once in your life—"

"*I would never have hurt her!*" Gabriel screams.

They both hear it when the next cartridge slides in perfect, like made for this moment. Cassidy slams the bolt into place and steps out of the truck, the gun at port arms, his head loose like he's really getting ready to do this thing.

"We grew up together," he says, sort of crying, lips firm as he can get them. "I loved you, man. You saved my life so many times, and I saved yours back. But—but it was *her* now, don't you understand? I loved *her* now. *She* was saving my life. I was saving hers! Everything was *working* for once, don't you get it? And now . . . now . . ."

With that he shoulders the rifle, backs up enough to level the barrel dead-center on Gabriel's face.

Gabriel is breathing in spurts, shaking his head no, no.

When there's nowhere to go that Cassidy can't reach him with the rifle, he drops to his knees a second time. The rifle follows him, is tethered to the bridge of his nose.

"Do it, man," he says. "Fucking do it already. I don't deserve to— *Just do it!* Nobody will even know, nobody will even miss me, man! You're the only one who would, even. If—if you're . . . *Just do it!*"

To make it easy, he lifts his chin, stares straight up. A moment later he starts singing, kind of with the drums still bleeding out from the top of Victor's patrol car but kind of more, too. Something else.

"Shut up!" Cassidy yells down at him, stepping back from this, stepping back from *having* to do this.

But he keeps seeing the Crow, too, you know, the Crow through that engine compartment, under the truck *Gabriel* knocked over.

"What are you even doing!" he yells to Gabriel.

"My death song," Gabriel sputters. "Shh, this next verse is tricky."

"You're just making that up!" Cassidy tells him. "Everything that's Indian, you just make it up!"

"Shit, somebody's got to," Gabriel says, and goes back to the song.

It's not even words, is just that old-time sound, always rising higher and higher and then resetting, starting the climb again.

"I don't . . . I don't—" Cassidy says, lowering the gun, looking at his friend on his knees, tears coming down his traitor face, running down by his ears into his neck, into his shirt.

Cassidy is crying as well.

He wipes his tears away, raises the rifle back, can't hold it steady enough, but he's only ten feet away. It's how far Lewis was from you when he shot you the second time, in the head. And the third time.

It's the perfect distance. It's the distance they've earned.

Except this one is losing his resolve, is losing his anger, is falling into a grief hole inside himself. But he's on edge, too, the barrel of the rifle coming up like he means it, then dipping down again. His every nerve is frayed. What that means is that, when Cassidy sees a white flurry of motion directly behind Gabriel, he flinches back in response, startled, and tries to pull the rifle with him, ends up putting that jerking pull into a trigger he doesn't really know.

The sound is thunder, deep and bass and ragged. It splits the night in two, both halves falling neatly away, leaving Gabriel standing in the silence between them.

He looks down to his chest for the hole that should be there. And then he feels his face gingerly. Finally he pats the side of his head, comes away with blood.

His ear. His ear has a new notch in it.

He smiles with wonder, says, "Coup," and looks across to Cassidy, but Cassidy is dropping the rifle, is shaking his head no, his breath hitching in deep again. But this time it's with fear.

"What?" Gabriel says, unable yet to even hear his own voice, and looks behind him, to whatever's got Cassidy shaking his head no.

It's—Gabriel is trying to process it, trying to resist it—what he sees is what he's most terrified of ever having to see: the girl with the basketball, the Finals Girl. His daughter in her scrimmage-white jersey. Her name shapes itself on his lips a bit at a time, like trying to add up to her: *D, Den, Denorah.*

She's still standing, her hair spilled forward, her face angled down at the blood spreading over her bright white jersey like checking to see if this is really real, if this is really happening.

Gabriel falls back, unaware of his fingertips on the ground, unaware of anything except what's just happened, what can't be taken back, what can never get undone.

His little girl, she—earlier in the day, at the little pad of concrete behind her house, she'd toed up to that charity stripe, she'd used textbook form, and she swished forty dollars' worth of free throws through that net.

It was impossible, no kid could shoot like that. But she could. For forty dollars.

"I'll bring it to the scrimmage tomorrow," Gabriel'd said to her out the window of his truck, the engine already turning over to bring him here.

"It'll be gone by then," she'd said back, with her mother's mouth. "And, you can come to the gym again?"

"It's a scrimmage, not a game."

"If I'm playing, it's a game."

"I don't even have it yet," Gabriel told her, shrugging like this was the truth, the whole truth, and nothing but.

"Who's giving it to you?" she'd asked.

"Victor Yellow Tail," Gabriel said. "Tonight. Police money. That's the best kind, yeah?"

"For Nathan's sweat?"

Yes.

Denorah had logged that, he knows now and doesn't want to know, she'd logged it and weighed it and considered it, and now she'd caught a ride out here to collect before her loser-dad could spend what he owed. Before he could let it blow away across the snow.

Only, Cassidy shot her with a 7.62mm round before she could even announce herself, had shot her so clean that it hadn't even thrown her back into the lodge, had just blown a ragged plug of meat out behind her.

But she's not meat, she's my daughter, Gabriel says inside, screams inside, can't stop screaming about inside.

Exactly, you say back to him.

Gabriel slashes forward to catch her, but she tips forward onto her face before he's even two steps closer. He falls to his knees by his own truck, pushes his whole face into the ground, his lips right to the dirt all the tires have cleared of snow.

His girl, his baby girl. She was going to take the team to state, she was going to take the whole tribe into the pros, into legend. Everybody was going to quit painting buffalo and bear footprints on the side of their lodges, were going to have to learn to draw all the lines in a basketball. She was the one who could plant her feet, get the rim in her sights, and drain ten free throws in a row. Twenty. Fifty. A hundred.

She was going to make it out of here, like Gabriel never had. Like nobody ever did. Exhibition one: Ricky. Exhibition two: Lewis.

Had he really seen her earlier today at lunch, walking away from school in the cold in that same white jersey? Was seeing her like that supposed to have been a warning? Was it a vision? Is Trina parked down at the cattle guard? Did she hear the shot? Is she standing from the opened door of her car, listening with mom ears for the next shot? For footsteps running in the dark? For her ex, trying to come up with one more excuse?

Shit. Shit shit shit.

And: *no*.

There is no excuse. Not for this.

When Cassidy drops to his knees beside Gabriel like *What have we done here*, Gabriel pushes him hard enough that Cassidy falls and slides, hard enough that the recoil drives Gabriel over into the side of his truck.

"You *shot* her!" he screams, standing, his hands balled into fists. He's crying harder than he was, now. But at the same time he's mad, mad enough to reach around to his own cratered-in windshield, come back with the black thermos.

"And you—you pushed a truck onto Jo . . ." Cassidy says.

"Not on purpose!" Gabriel says, and then, just like he's supposed to, he steps out into the darkness after his best friend since forever, and when Cassidy crawls back, away from this thing trying to happen, Gabriel steps faster, finally comes down with his knees to either side of Cassidy's hips.

The thermos is alive in his right hand, is both completely weightless and the heaviest thing in the world. He rolls it for a better grip, for a final grip, for the best way to hold it when doing a thing like this.

"You shot her, man," he says, like he's pleading. Like he's trying to explain. "You shot *Denorah*. You shot my little girl . . ."

Cassidy is holding his hands over his face.

He nods that yes, yes, he did.

His body is hitching and jerking under Gabriel, and it's like a current is passing between them. Like they're kids again, learning to break-dance.

"I'm sorry," Gabriel says, and brings the butt of the thermos down with the weight of all their years of friendship.

Because he's holding it wrong, his pinkie finger is between it and Cassidy's eyebrow.

The thermos glances off and dives into the ground, its open mouth standing it up in the crusty snow.

Cassidy lowers his hands, blood sheeting down over his face.

He looks up through it to Gabriel, and they're both crying, neither can breathe right, neither wants to breathe ever again.

With an unsteady hand, Cassidy claps the snow for the thermos, finds it, passes it back up to Gabriel, and you have to cover your bloody mouth with your palm, because even in your most secret dreams you never would have guessed this part, would you have?

It's perfect, it's amazing.

Gabriel takes the thermos, their fingers touching over that black metal, and Gabriel remembers it all over again: D, earlier, turning back to him with that sharp smile, no-looking free throw number ten just like Jordan, and it hurts so bad that he closes his eyes, brings the thermos down again, with a crunch. The next crunch is wetter, the one after that deeper, punching through into a darker space.

The muscles closest to Cassidy's shin bone are the last to die.

Gabriel leans back, wavers, an insubstantial shape of a person.

Past Cassidy's head is a dead dog, and a beer, still standing.

Gabriel crawls over, trades the bloody thermos for the beer, and drains the bottle.

He still can't breathe. His right hand is slick with blood, and his face and shirt are spattered with it, and he doesn't know whether to laugh or die, really. Both seem reasonable.

He struggles out of the shirt and it'll hardly come, so he tears it, balls it up, stands to throw it as far as he can. It flutters, doesn't go anywhere. He kicks back through the snow for his truck and stumbles into the Mauser, is a skin now along with Cassidy, meaning they're on the same team, like always.

He looks at the rifle and then looks at it some more. His breath finally comes, washing through his head, leaving him dizzy.

The Mauser, yes, he decides. The Mauser, for the pest that he is. He can—he can be another statistic, he can make it so the pamphlets are right about Indian suicide rates, can't he? He can keep the numbers good, keep everyone from having to print up new pamphlets. He can go—he can go with Cassidy. Maybe even still catch up to him.

He picks up the Mauser, falls to the old truck, the one the Crow's dead under, and grubs cartridge after cartridge from Richard's foggy bag, only stops when he cues into an eye watching him.

"Jo," he says, like of course.

The hole he shot through Cassidy's floorboard all those years ago, it's pushed down now over the Crow's face, her eyeball bulging up through it. Gabriel turns away shaking his head no. His fingers are shaking too much to get a shell in right, though. He fumbles the one 7.62mm he finally finds into the snow. His chest shudders with laughter. He can't even do this right. He lets the rifle fall away, is looking back to the fire, squinting like to see it better. Or to see something over there better.

Denorah. Den. D.

He pushes away from the truck, makes himself take the walk to her. Just to hold her again. He wants to repeat her season average to her, and project what would have been her junior year on varsity, her senior year in the state tournament. He wants to tell her about all the games she would have won, all the posters they would have made with her on them. The line of shoes that would have been named after her.

Did you get the new Cross Guns yet?

They're so dope.

Do I look like her when I come up like this, on the toes?

And then he's there, stepping around the sparking fire.

"D?" he says.

Not because it's her. Because it's not. It never was.

Gabriel looks back to the mound in the snow his best friend is, then to not-Denorah again.

It's the—it's the *kid*? In the scrimmage jersey he has no reason to be wearing, that's black on the outside, bright white on the secret inside. His hair is down and everywhere, could be Denorah's hair, *was* D's hair.

"N-Nate?" Gabriel says. "*Nathan?*"

Where the Mauser caught him is low down in the left side. Not the kill zone, but close enough. The kind of shot where you just have to follow whatever you shot back into the trees, wait for it to collapse at the end of its blood trail.

But he's not dead yet. Not quite.

"Hard to kill, aren't you?" Gabriel says with an almost smile.

It shakes the boy awake, and, maybe because Gabriel's standing over him bloody-handed, bloody-faced, the boy jerks away, pushes with his heels, shaking his head no, no, and something else, the syllable and sounds coming fast, in a tumble, over and over.

Po'noka?

Gabe narrows his eyes, has to hunt deep in his head for this old word, then stand real still in his thoughts, wait for it to stand up from the snow, a brown form against all that white.

"*Elk?*" he says, and, following the boy's eyes, looks behind him, looks all around, but you're not there anymore.

When Gabriel comes back to the boy, the boy's still trying to get away, leaving more and more blood in the dirty snow.

"Wait, wait, let me get your dad," Gabriel says, going to his knees and holding his red hands up and out to show he's no threat.

It doesn't help.

The boy pushes back and back, past the lodge, under the lowest rail of the horse pen, leaving dark smears on the pipe.

"No, listen—" Gabe says, trying both to follow and not be scary, but he stops when the horses whinny in their panicked way about this intruder underfoot. "Shh, shh, guys," he says to them, stepping in, but the way he smells—they shy back, rear up, rise and fall in the darkness, and there can't be room in that pen for all four of them, can there? Their weight coming down shakes the ground and Gabriel looks away, kind of numb, finds himself just staring down at a clot of the blood the boy left behind in the slush. A clot of blood he probably needs, or would have needed, if the horses hadn't done their thing to him.

"Another fine job," Gabriel says, walking away from this, kicking snow with his bare feet, his hands running through his hair. He sits on Victor Yellow Tail's hood and stares into the fire, the drums beating, voices rising, his mind working, mouth muttering: Why did he think the kid was D, even? Why *would* he? It was . . . it was because the kid had been wearing a *black* jersey, right? And the last time Gabriel had seen the daughter, she'd been wearing white?

Still, was an inside-out jersey, plus long black hair, really enough for him to think Nate was Denorah? Was he not thinking right because Cass had just clipped his ear? Because Jo had just—Why was she even *under* that truck? Why was she home at all? Didn't she work most nights?

"What the hell is happening here tonight?" Gabriel says, pushing away from the car and looking all around.

"Po'noka?" he says at last, trying it out like it might be the key that opens everything up.

What would an elk have to do with this, though? How could an elk make them all kill each other? Why would an elk even care about two-leggeds, unless the two-leggeds were shooting at them?

And why is he even thinking like that? *Two-leggeds?* Has he fallen so far back into himself that he's sitting in Neesh's lodge again, listening to the old bullshit stories? If he's there again, though, then he's there with Cass and Lewis and Ricky, he figures. Back when there were four of them.

He rubs that spot beside his eye.

"One little, two little, three little Indians," he singsongs, and kind of laughs, kind of cries. It turns to coughing again, and when it won't stop he stumbles to the camper, tries the locked door, then feels his way around to the outhouse. All he needs is tissue, some toilet paper, something for his nose or he's going to suffocate.

When he swings the outhouse door open, Victor Yellow Tail is there, a bib of blood on his uniform shirt, his head lolling, his pistol in his hand like he had plans.

An elk mother will use her hooves when she can, but she'll bite if she needs to.

Gabriel closes his eyes, opens them again, and Victor Yellow Tail is still there, still dead.

"Then there was me," Gabriel mumbles, smiling a sloppy smile, and closes the door. It swings back open, so he shuts it again, and again and again and again, slamming it shut enough that none of this can even have happened.

But it did.

And he's the only one still standing knee-deep in it all, he knows. He's the one they're going to say did it, who cares why. Because he's an Indian with a Bad Track Record. Because a Tribal Police Officer Came Out. Because He Didn't Like His Other Friend's Fiancée. Because His Mind Boiled Out in a Sweat. Because His Murderer Friend Just Got *Shot*. Because the Great White Stepfather Stole All Their Land and Fed Them Bad Meat. Because the Game Warden Wouldn't Let Him Get His *Own* Meat. Because His Father Reported Him for Stealing a Rifle. Because the Rifle Was Haunted by War. Because because because. He did it for all those reasons and whatever else the newspapers can dream up.

Unless he runs.

Unless he runs to the mountains and lives there the old way, never comes back down, even for beer. But, maybe just to go to one of his daughter's games? Maybe just to stand by the Boss Ribs's grave fence? And wherever Cass gets buried? And Lewis?

He shuffles up to the fire, opens his palms to that wonderful heat. He's shivering, his teeth clanking against each other. He looks to the lodge, sprayed now with Nate's blood, hates himself for being thrilled it's not his *daughter's* blood, and then he studies the old truck, its frame on the ground. Finally his eyes settle on the mounds out in the snow.

He goes there, past the dogs, and drops to his knees beside his best friend.

"It's just you and me, man," he says down to him.

He sits down, the snow not even cold anymore, even though one ass-cheek of his pants is flapping. He works his legs under Cassidy's head, cradles his face, lowers his forehead to what's left of his friend's, and then he looks up fast, as far into the sky as he can.

"It wasn't her, man," he says, knocking his forehead into Cassidy's twice, kind of hard. Love taps. "It wasn't D, C."

Cassidy just stares. His eyes don't look the same direction anymore. In death, he's an iguana. Gabriel braces himself for Cassidy's mouth to open, for a great tongue to roll out, slap at something.

It wouldn't be the worst thing this night's had to offer.

"This is—this is goodbye, man," Gabriel says. "I'm going to—they're going to think it was me. And I guess it was, for Jo. And the kid, too. And you. Definitely you, man. You should have just—you should have pulled your shot an inch to the left, man."

He drills the pad of his middle finger into the dot of scar tissue by his right eye, the same place he's been touching since he was a kid.

"You always were a terrible shot, though," he says, then closes his eyes hard. "*But it wasn't D,*" he whispers, thrilled to be delivering this news. "It wasn't D. That's the main thing. She's all right. Now I'm . . . I'm going up to live with the—"

When he looks up to the snow crunching then not crunching, you're standing there, holding the Mauser across your hips, left hand ran all the way up the forestock, to the uneven checkering. It hurts to touch it, to even think about touching a *rifle*, but this is the only way now.

You can feel your eyes are the hazel and yellow that feels right, and that they're maybe a smidge or two bigger than makes sense for this face.

Gabe nods, says, "It was you who did all this, wasn't it? Lewis too, right?"

You don't owe him an answer. You don't owe him anything.

"Anybody ever tell you you've got eyes just like an elk?" he says. "Not the—the color. But . . . something, I don't know."

Down the slope the herd is already waiting for you, drifted in like ghosts, not even one of them bleating or calling. The ground under them is churned and dark and raw. The smell is so wonderful. You can't breathe it in deep enough.

"The kid saw you, didn't he?" Gabriel says, laughing it true. "P-Po'noka, right?"

"Ponokaotokaan*aakii*," you say down to him. Elk Head *Woman*.

Gabriel works through this, gets it enough, looks up to you and nods that he can see that, sure.

You hold the rifle out to him. An offering.

"Why?" he says, shying away from it, but finally having to catch it when you throw it sideways down at him.

He plants the Mauser's butt in the snow to prop himself up, says it again. "Why are you doing all this?"

If you tell him, he would get to die knowing it was all for a reason, that this has been a circle, closing. Which would be more than you ever got, that day in the snow.

You nod to the rifle he's holding, say in his bitter English, "Do it or I go after your calf for real."

He watches you for maybe five seconds here, and then he looks to this rifle.

When he racks the bolt back, the wet brass flashes in there for a bright instant.

"I dropped that shell in the snow," he says.

"It stinks," you tell him back, crinkling your nose.

"You'll really leave her alone?" he says, ramming the bolt home in a way that straightens your back. "You won't touch her? She's—you know she's going to get away from here, don't you? You can see that?"

It would be so easy for him to angle that barrel at you, wouldn't it?

But he's not thinking like a hunter right now. He's thinking like a father.

"Okay, okay," he says at last, and angles the awkward rifle around, the barrel chocking up under his chin, his head having to tilt up because the rifle's long. "Like this?"

His breath is fast and shallow like getting ready, and then he closes his eyes, pulls the trigger all at once.

Click.

"Oh shit," he says, flipping it back around with a halfway laugh, the barrel pointed right into you now, his finger still on the trigger, his thumb figuring the big obvious safety out.

"You promise you won't come after her?" he says one last time.

You shake your head no, so he'll get the rifle back under his chin. But then he stops, says, "Wait, does that mean you will or you won't go after her?"

He finally smiles when you're just drilling your eyes into him. He leans back a bit, says, "I always—I wanted it to be like those two Cheyenne I read about, yeah? I wanted to run my horse back and forth in front of all the soldiers, so it would be like . . . it would be heroic. Like the old days. Not like—not like this."

"Now," you tell him.

"Okay, okay, geez," he says, "at least let me—" and instead of using his own hand, he shoves his friend's dead index finger through the trigger guard.

"I killed his almost-wife," he explains, getting the finger positioned just right. "This is—he's avenging her, like. It's an Indian thing. You'd understand if you were, you know, a *person*."

He opens his mouth, swallows the barrel in deep enough that his eyes fill with tears. The metal rattles against his teeth. His breath is fast and shallow, like it matters anymore how much air he has.

"D, D, D," he says around the barrel, and nods once to himself for rhythm, then again to be sure, and on the third time he raises his fingers over his friend's hand and gallops them down one after the other, until the last one, the one that makes the trigger pull, and right as the sound blasts a fist-sized hole through the top of his head you realize that he's made his fingertips into horse hooves, that it's still the cavalry taking a shot at him, and finally getting lucky.

The rifle is angled away from you but the red mist of him rises, plumes over, coats your face.

You wipe it off, don't lick it away, then look back down the idea of the road, to the cattle guard out there in the deep dark.

Now there's only one left, one you just promised you wouldn't go after.

Killing a calf is the worst of the worst, you know.

Beside it, breaking a promise is nothing, really.

Nothing at all.

MOCASSIN TELEGRAPH

Say we're all in a John Wayne movie. Say your trusty reporter here has his ear pressed right down to the railroad tracks, so he can listen up the future.

What am I hearing? you ask.

The bus tires of Havre leaving their parking lot for tonight's big scrimmage with the girls, sure. But you don't need to be a real Indian to know the Blue Ponies are coming to town for a grudge match, to prove that last year's tournament win was due to skill, not injury.

No, you come to this column for the real dirt, don't you? Let me dish. And, as always, you didn't hear it from me.

Word has it that a certain big-time college scout has been seen in blaze orange down at the diner. Further word has it that, once his lunch schedule was established, it's completely possible that a certain coach may or may not have sidled up to his table, mentioned where the elk have been the last week or two, and kind of, shall we say, bun-dled that tidbit in with a certain scrimmage going down tonight.

Supposedly, the trade was that if a certain scout bagged his trophy early enough in the day, well, that would leave his evening free, wouldn't it?

And, if he was free enough, why not come for the junior high game as well, right? You thought I was talking about a certain high school coach, not the coach who wears her hair in two pigtails?

For shame.

Junior high coaches know as well as high school coaches that all the elk are bunched together up towards Duck Lake all this week. The game wardens have been trying to scare them back over to the park, or onto the old folks' happy hunting grounds, but elk are elk, right?

This isn't the Fish & Game column, though. This is what you won't hear anywhere else, unless you have your ear to the rail like me. Trust me, a certain junior high coach either has or hasn't got a big-time college scout out to watch her star player. You know the one. You've seen her after practice, smoking the varsity girls and boys both? We've never had a player like her, niiksookowaks. This is history in the making. I'll be there, trying to read over the scout's shoulder.

And, remember, you didn't hear this from me.

IT CAME FROM THE REZ

SATURDAY

Denorah can tell the order the sweat lodgers got there last night.

Cassidy was first, of course. It's his place. He didn't so much get there as just never leave. Her dad was next, front tires cocked at what he tells her is a rakish angle, like his truck just stopped there in the middle of some crazy slide and he had to wait for all the dust to settle before kicking the door open, stepping down, peeling out of his sunglasses one side at a time. After that dramatic or *not* dramatic entrance had been Victor and Nathan Yellow Tail, the cop car nosed right up to the fire like claiming it for its own, its tracks in the snow showing where it had to step off the sort-of road to get around all the trucks that thought they were so important.

Last, this morning probably, when her shift was over, was Jolene, pulled up right behind Cassidy's old truck that used to be up on blocks but is flat to the ground now, like embarrassed about something it's done.

None of the sweat lodgers are up and about yet. Denorah would think it was for the usual reason—beer-after-ceremony, which her dad would have claimed was "rehydration"—but then Victor's car wouldn't still be there. And no way would *Officer* Yellow Tail let *minor* Nathan drink with her dad and Cassidy, even if Cassidy's finally settling down a little, according to Denorah's mom.

"Hello?" Denorah says, still a good walk away from it all. She could have screamed it if she wanted, she supposes. Near as she can tell, the place is dead times two. Even the sweat lodge is collapsed, smoke and the lines of heat blurring the air above it, the blankets and whatever smoldering, meaning it's a trash pit now. Next time, the sweat will be somewhere else.

It's good Denorah's mom didn't see that smoke.

"I'm only letting you do this because that Crow from the grocery is there," she'd just told Denorah at the cattle guard, after confirming the presence of Jolene's truck. "I'll be back in one hour, got it? You're lucky I had to return this to Mona."

"This" was Mona's casserole dish that had wended its way from Tre's house to Denorah's, because Trina is always coming up this way to smoke cigarettes with Mona in her new trailer. There's an old bear that gets after the berry bushes just down from the trailer in the spring, and Denorah's mom is forever talking about that—as she calls it—silly old bear. Silly or not, that bear is her mom's excuse for one more cigarette, one more pack, one more carton. It's like she's the most willing prisoner in Mona's little window nook thing, that Denorah thinks looks like the cockpit of a spaceship, like the two of them are plotting some big escape, once Denorah's out of the house.

"One hour, right here," Denorah said back to her mom.

It feels military, repeating commands back so there's no confusion, but it seems to result in less grief, so Denorah plays along.

Still standing at the cattle guard, Cassidy's place either a ghost town or a junkyard, no dogs even—no dogs?—Denorah looks back to the road for her mom's car. Past where the road ducks down to the right there's just snow and snow and more snow, though, and then the shimmer of the lake where her dad told her one of his running buddies died, way back.

But her dad's got a story for every place on the reservation, doesn't he? If not someone he used to run with in high school, then a coulee where he popped a blacktail once, a ridge where he found a little pyramid of brass shells for a buffalo gun, a place he once saw a badger humping it across the grass, an eagle dive-bombing it like it thought this was the biggest prairie dog ever.

When she was a girl Denorah had soaked every one of those stories up, and then, later, her mom told her to be careful what she took for gospel. The dead friend stories, though, Denorah still kind of believes those. Because it would be bad luck to lie about that, she thinks, and her dad's the kind of superstitious he thinks nobody notices. Case in point: That day him and Ricky and Cassidy and Lewis popped all those elk back in that section they weren't supposed to be in, the one down by the lake? He's never mentioned that to her even once, even in defense, even to give her the rest of the story, how it wasn't like it sounds, the story her stepdad told her isn't the real story, isn't the one with feet on the ground and smoke in the air, bang bang bang. And the reason he hasn't said anything to her about it, she's pretty sure, it's because to talk about it out loud would throw his sights off next time he was lining up on an illegal elk, those being the only kind he can shoot anymore.

The same as he's never told her his version of that elk massacre, he's also never told her *how* his friend died at the lake. Just that that's where his body was. To talk about what actually happened might get him in Death's crosshairs, the way he thinks. So, because he won't speak directly about that story, she kind of believes it, in spite of her mother's warning. But still, her dad's got to *think* of that dead friend still, doesn't he, even if he won't talk about him out loud? How could he not? Every time he's out here to see Cassidy he probably stops halfway across the cattle guard and looks back to Duck Lake. He says that when his *other* dead friend Ricky found his *lake*-dead friend, Ricky got hauled into jail himself. Not because he did it—everybody knew who did it—but because he'd had to break into one of the summer people's lake houses over there to call in about that body, and the cops couldn't look past a breaking-and-entering, not when there was property damage.

It was all part of a lesson her dad was trying to impart, Deno-rah's pretty sure, which was why he was even talking about the whole thing at all, but she isn't sure if it was a warning against calling the cops or against finding a dead body. Maybe both at once? Probably the idea was that when you see somebody dead and floating like that, you just keep walking, let somebody else find it, or nobody.

She knows the joke about how Indians are crabs in a bucket, always pulling down the one that's about to crawl out, but she thinks it's more like they're old-time plow horses, all just walking straight down their own row, trying not to see what's going on right next to them.

Speaking of horses: Cassidy's?

Last time she was out, her dad had let her sit up on that paint horse, the one Jolene calls Calico, like a cat, but that was . . . was it last summer? Was Jolene living here by then? Yeah, she was. That was when her dad was still calling her Dolly, like the best joke ever, and Cassidy had even played along at first, faking like he had a beard—like, if his girlfriend was Dolly, that meant he was Kenny, ha ha ha. It had been so stupid that it had been hard for Denorah not to smile about it. The way they were fooling around so natural made her kind of see her dad and Cassidy twenty years ago. It had been a good day. But now the pens are empty, the gate flapping. Cassidy wouldn't have sold his Indian ponies, though. They're probably grazing in some meadow, won't trail back to the barn until dark.

Also: Who cares?

Denorah's here for forty dollars, not to conduct the Big Horse Poll and Headcount.

She nods to herself about this and leans up the road, follows its loop around and down, keeping to the ruts because the snow's crusted hard and she doesn't need to hyperextend a knee before tonight's game.

She's almost to Jolene's truck when the driver's door opens and Jolene cocks her right foot up on the duct-taped armrest, to tie a high-top tighter.

Her long hair blows out over her knee.

"Hey," Denorah calls ahead, to keep from getting shot.

Jolene flinches around, clears her hair from her face, from her blown-red right eye, and she *isn't* Jolene.

"Whoah," Denorah says, stopping hard, looking around at everything all at once, to be sure this is still Cassidy's place.

Not-Jo snickers, keeps tying her laces.

"Who are you?" Denorah asks.

"Don't worry," Not-Jo says, "this isn't a raiding party, little girl."

"*Little girl?*"

"Young lady?" Not-Jo stands from the truck, sways her back in, extending her arms to either side, wrists up, stretching. It's a full-body yawn. She's wearing black gym shorts and a faded yellow T-shirt with the arms scissored off, the neck cut out, maroon sports bra.

"Where's Jolene?" Denorah says, not even trying to reel the accusation in.

"You're Gabriel's girl," this woman says, angling her head over to study Denorah. "You *do* look like him. That's not an insult."

"You're Crow, aren't you?" Denorah says.

"Your dad would have been pretty—I mean, if he was a girl," the woman says. "I'm Shaney, Shaney Holds. Jolene's best cousin. Maybe the best cousin of all time, jury's still out on that one."

"What are you doing here?"

"Getting interrogated by a kid?" this "Shaney" says with a smile, then reaches importantly back into Jo's truck and hauls out a basketball, claps it in front of her like something's starting.

"You play, right?" she says, passing the ball across to Denorah. "Your dad says you're pretty good."

"Where is he, do you know?" Denorah says, casting around Cassidy's place a third time.

"Good luck with that," Shaney says with a smile.

"What do you mean?"

"That kid . . . Nate?"

"Nathan Yellow Tail."

"He heard the dogs bawling after something down that way," Shaney says, hooking her chin downhill to where the trees start. "His dad, that big cop guy, he thought it would be all super-Indian if they rode horses down to check it out."

"My dad can *ride*?" Denorah says.

"Just glad they're gone," Shaney says. "I can't shoot the ball when the horses are in their pen. I think one of them's gun shy or something, I don't know. Gets them all riled up. But, now that they're gone . . ."

She opens her hand for the ball and Denorah underhands it back to her.

"Why's the sweat lodge burning?" Denorah asks.

"They used plastic for the frame," Shaney says with an amused shake of the head. "It kind of melts in the heat, I guess? Whole thing collapsed in on the rocks. They told me to watch it, make sure it doesn't catch the grass."

Denorah nods. That sounds about right for her dad.

"Even Nathan rode a horse?" she says, incredulous. "He's always being all gangster."

"Two hundred years ago, the gangsters rode war ponies,"

Shaney says, and shuts the truck door with her hip. "Twenty-one till they get back? I want to see if your dad was lying about what you can do out there."

Denorah looks to the goal poking up from the grass maybe fifteen yards to the left of them, over from the outhouse. It's a square, rotting-away backboard nailed flush to a tribal utility pole—the kind of court where if you don't slash in from the baseline for a layup, then where you come down, it's into a rake of creosote splinters.

"Got a game this afternoon," Denorah says.

Shaney nods, looks out to the grey trees, like for the men.

"You can hang in the camper if you're cold," she says. "Or sit in the truck. I think they broke all the lawn chairs around the fire last night."

She's right: the chair by the dead fire is folded over, the one by the lodge is bent in on itself, and the other is sideways, thrown out in the grass and snow.

"Did you play?" Denorah asks. "In high school, I mean."

"I used to *eat* basketballs, little girl," Shaney says, clapping the ball hard in her hands, and Denorah knows right then she's not sitting in any camper, she's not sitting behind the steering wheel of any truck.

"Maybe just twenty-one," she says to Shaney. "Until they get back."

"Sure your coach won't mind?"

"Not if I play like I play, she won't."

"How old are you?" Shaney asks, the crow's feet around her eyes crinkling with amusement.

"How old are *you*?" Denorah says right back.

Shaney hooks her head for Denorah to fall in. Denorah does, and, turning away from the drive, she sees that her dad's windshield is caved in on the passenger side. It stops her for a moment, but that could be anything. Knowing him, he's got six different stories cooking already for what happened, each more epic and unbelievable than the last, none of them involving him being at fault.

The seventh story will probably be about how he needs this forty dollars toward a new windshield. Does his Finals Girl want him to freeze, come January?

Denorah follows the path Shaney's picking through the hard snow. It's rocks and dry patches, but it gets them there without wet feet or bleeding shins.

Shaney bounces the ball high off the concrete and tracks it while working a hair tie off her wrist, gathering her hair behind her neck. On the ball's third bounce she rabbits forward, snatches it on the way to the bucket, then stops on a dime, fakes once, and goes up, executing a neat little fadeaway that banks in like money.

"Your coach let you Reggie Miller your left foot out like that?" Denorah says, down on one knee to retie her right shoe.

"Crow ball," Shaney says. "What do y'all play up here? Big on the fundamentals, all that boring-ass stuff?"

Denorah switches to her other shoe, battens it down tight, making sure the bows are even. Not because she's superstitious but because it makes sense to have them both the same.

"Done stalling yet?" Shaney says from the pole, and snaps a bounce pass across.

Denorah has to stand fast to catch it at her stomach, keep it from slamming into her face.

Shaney has her by six inches, she guesses. But tall girls are never the ball handlers, at least not in small schools—not in reservation schools. Tall girls get trained on boxing out, on rebounding, on posting up and setting screens, using their hips and elbows. All of which a team needs to win, for sure. None of which are much use one-on-one, which is a game of slashing, of stopping and popping.

Denorah dribbles once to get the right feel for this ball, this court.

"Warm up?" Shaney says, bouncing in place.

Denorah snaps the ball back to her, says, "So you can clock my dominant hand, my favorite place at the top of the key?"

Shaney chuckles, says, "There's no key out here, little girl. Just you and me."

"The Blackfeet and the Crow . . ." Denorah says.

"If that's how you want to look at it," Shaney says, stepping out to what would be the free-throw line and waiting for Denorah to step into position in front of her.

Denorah takes her time, won't be rushed.

"Don't want to wear you out for your big game or anything," Shaney says with a little bit of bite, dropping the ball in front of her for Denorah to check.

Denorah takes the ball in both hands, spins it back toward herself, and makes a show of looking around, says, "What, there another baller out here I'm not seeing?"

"Cocky, I like that," Shaney says, taking the ball back. "Just like your dad."

"Done stalling yet?" Denorah says, getting down in the stance, palms up, tapping her forearms on the outside of her knees twice like activating Defense Mode.

Shaney dribbles once, high by her right hip, and then turns around, giving Denorah her ass, backing her down already, which is what you do when you have a size advantage.

When you're on the wrong end of that size game, though, then you can time it out, stab an arm in, slap the ball away.

Denorah gives ground like she's falling for this, then, the next time Shaney goes for a bounce-against, the round of her back to Denorah's chest, Denorah steps back—pulling the chair out, Coach calls it—comes around with her right hand, reaching in for that blur of orange leather.

Except Shaney wasn't backing her down. She was baiting the trap.

What she does now is peel around the other way, her long legs giving her what feels like an illegal first step, and by the time she's done with that step, throwing the ball ahead of her in a dribble she'll have to chase down, Denorah's already out of position, can just watch.

She's never been spun on like this.

To make it worse, Shaney doesn't just lay it in, either. She catches her dribble in both hands, rocks her elbow out hard to the right, and plants one high-top on the pole about chest-level and uses that to push higher, twisting in the air to come around the right side, having to guide the ball *around* the net on the way, like having to fight through the trees to get to the bucket.

She lays it in gentle with both hands, lands already jogging backward.

Fucking-A, Denorah knows her face has to be saying.

This might be a game.

THANKSGIVING CLASSIC

15–15, and Denorah isn't having to run her flyaway hair out of her face anymore. Now it's pasting to her skull with sweat.

She dribbles in hard to the left, Shaney bodying right up to her but not tangling their feet somehow, and stops, makes to rise up, getting Shaney's long body into the air. It's one of the only two strategies she's found that are worth anything against this tall, slashy defender. Trick is, long bodies stretched out, they take longer to recoil back down, go a different direction.

Instead of letting her feet leave the ground, Denorah reels the ball back, both hands because Shaney will slap it out into the snow again, and leans over to the right, ducking ahead under Shaney's already-coming-down arm.

Position, yes. When you're outgunned, all you can do is whatever you have to for position. Not that there's a ref to blow a whistle, but even a Crow knows that bringing an elbow down into the neck and shoulder of a player in the motion of shooting, that's a do-over.

Now Denorah lets her feet leave the ground, still exploding forward under Shaney's wingspan, and she teardrops the ball up and over, *in*, just enough soft touch, because this bullshit plywood backboard isn't trustworthy, not for someone who hasn't killed a thousand sundowns out here, the clock always ticking its last three seconds down.

"Cheap . . ." Shaney calls out, just generally.

"Sixteen," Denorah says back, collecting the rebound before the ball can get slick in the snow.

She dribbles it slow back to the top edge of the court, bounces it across to Shaney, who, Denorah's satisfied to see, is finally breathing hard as well, her mouth moving like she's the kind of player who's used to having a piece of gum in her mouth. Or used to chewing cud, ha.

"How long you been playing?" Shaney asks. "Your dad never said."

"Was born on a court," Denorah says, Shaney lowering the ball right to the concrete, rolling it slow between them, giving her time to crowd in.

"So this is what's most important to you, right?" Shaney says. "Basketball? Matters more than anything to you?"

Denorah fixes Shaney in her eyes for a moment, like taking stock. "And you think you can take it away from me?" she finally says. "That you can break my pride before the game tonight? You a Blue Pony in disguise?"

"Home court advantage, little girl."

"You're *far* from home," Denorah tells her, lowering into triple-threat, leading with her face. In practice, Coach will put a big hand on Denorah's forehead while she slashes the ball back and forth and all around to pass, to shoot, to dribble. Now Shaney does the same thing, her rough palm right between Denorah's eyebrows. It's a violation, would be a foul in any game with a whistle, but, too, it slows the whole world down, lets Denorah sort of see this not from her triple-threat position, but from the side, in ledger art, like this battle between the two of them is so epic that it's been painted on the side of a lodge, and inside that lodge, an old man with stubby-thin braids is recounting the story of that one time the Girl played a game for the whole tribe. How each dribble shook the ground so hard that over in the Park great mountainsides of snow were calving off, rumbling down, shaving the foothills of trees. How each time the ball arced up into the sky it was merging with the sun, so that when it came down it was a comet almost, cutting through that orange circle of a rim. How each juke was so convincing that the wind would come in to take that player's place but then would get all scrunched up because the player was already back in that space, cutting the other way, her path as jagged and fast as a bolt of lightning.

This win isn't just for pride, Denorah tells herself, in order to push harder, be faster, jump higher. It's for her tribe, her people, it's for every Blackfeet from before, and after. "You don't win today," she says, speaking right into Shaney's wrist.

"And you do?" Shaney says back, getting light on her feet for what she must think Denorah's move is about to be.

"I am," Denorah says, and pushes hard with her forehead, nudging Shaney back just enough to clear some space.

She uses it to launch up and back, up and back. It's improper form, is even poor practice, as it's nearly impossible to replicate all the variables of a fallaway like this, but you can't always go by the textbook, either. Some games, you are Reggie Miller. And, if you're really good, you're maybe even Cheryl.

Denorah rises and rises, falling back at the same time, Shaney lowering her arms to swing them up together, extend enough to block this shot, but that smidge of time it takes to lower and jump, gather and push up, it gives Denorah just enough window to release the ball through.

Still, because of Shaney's length, Denorah has to adjust at the last instant, arc the shot even higher than she'd wanted, make it even more of a prayer.

It *just* clears Shaney's fingertips.

Denorah lands on her ass in the snow a full second before the ball catches the front of the rim, shudders the whole thing, and then—bounce, bounce, jiggle-jaggle—it drops through. Denorah rolls over three times in celebration, snow and dry grass all over her. She's spent more hours on the court than off, she'd bet, and played against girls her age and older, guys, too, on Sunday nights when the gym's open, she's even had the ball at the end of the game more than anybody on her team, but still, this shot, this one lucky roll, it's better than any of the rest of them.

"*Two*," she calls out, because that's how they've been playing, and Shaney's pissed enough she rips the hair tie from her ponytail, runs to the edge of the concrete to throw it as far as she can. It's a scrunchy, though. Too much air resistance. It flutters, dies, doesn't go anywhere.

"You can't beat me," she says—*growls*, really.

"Eighteen," Denorah says, standing, keeping a close eye on Shaney.

Riled up like she is, there's something almost animal about her. In a game it's the kind of thing Denorah could use to get to the free-throw line. Out here miles from anybody, it's more likely to earn her an elbow in her ribs.

It'll just mean she's winning the real game, though.

Shaney gives her the ball, bodies up close enough that Denorah gets a bug's-eye view of her knitted-together, scarred-up forehead, and Denorah fakes back like to repeat that Hail Mary fallaway but Shaney doesn't take the bait, is all over her when she puts the ball down to drive.

Still, Denorah gets the step—you can always get the step, if you want it bad enough—runs the ball as far out in front of her as she can to flip it up at the last possible moment before her next foot touches the ground.

It's pretty, and it's on target, but Shaney's been *on* this ball since the moment she checked it. She doesn't just slap it down, either, she smothers it, she collects it, she wraps around it like a fullback, falls hard enough back into the pole that rotted wood from the backboard rains down over her.

She waves it away from her face, shakes the pain off, her hair almost completely hiding her face now, her teeth flashing in that black shroud.

"You all right?" Denorah says.

"Check," Shaney says, leaving the ball behind her as if disgusted by it.

Denorah uses the toe of her right shoe to flip it up to her hands, a move Coach would be all over—hands, hands, basketball players use *hands*—and, on the way to the top of the key she chances a look back to the dead fire, the smoldering lodge, the horse pens, all the empty trucks. The camper, the outhouse. The whole reservation as backdrop.

"Where are they?" she says, kind of just out loud.

"They're not going to save you, little girl," Shaney says, already in her place.

Not even a *dog* has made it back, though? And what happened to the windshield of her dad's truck?

"I'm not a little girl," Denorah says.

Shaney starts to say something about this but swallows it.

"My mom's coming back by in about fifteen minutes," Denorah adds.

"She can play winner, then," Shaney says, clapping twice for the ball.

Denorah rolls the ball to her slow enough that the lines don't even blur.

Shaney snatches it up the moment it's close enough, follows through on that forward dip of her body, twitches ahead like enough with the bullshit finesse, this time she's going *through* Denorah.

Because she can't get too banged up for the *other* game she's playing today, Denorah flinches back, ready to give ground, sacrifice a point to save her body, but then at the last moment Shaney breaks right, the exact same move Denorah just used on her: get the first step, then stretch out, flip it in.

The reason it didn't work for Denorah was Shaney's length, which Denorah doesn't have.

One blurry dribble and then Shaney's flipping the ball up.

It catches the backboard high and comes down slow, flushes down and through, the net popping up behind it exactly like an old man's lips after he's leaned over to spit.

"Good one," Denorah says, chocking the ball under her arm.

Her legs are trembling, spent, her lungs raw, her heart beating in her temples. This is no way to prepare for tonight's game. Still, if her mom's car crests over the cattle guard, she's going to hold her hand out, tell her to wait, she's got to finish this.

Forty dollars or not, right here's where the real money is.

"Sixteen–eighteen," Shaney says.

"Give up now, you want," Denorah says back. "There's no shame. I'm younger, faster, play every day. You've taken this farther than anyone else would have."

Shaney laughs at this.

"You should probably be asleep now anyway," Denorah says, "right? Or you and Jo on different schedules or something?"

"I slept for ten years," Shaney says back.

After a breath to make sense of this, to *not* make sense of this—she didn't step on a court for a whole decade, and can still play like this?—Denorah bounces the ball across.

Because she's winded, Shaney takes the ball up into the Crow version of triple-threat, which more and more Denorah's thinking might actually be some sort of quadruple-threat, and turns around to back her defender down, probably muscle her back at the end of that, fall back on one leg, bank it in. Not showy, but, if there's no three-second violations, generally effective in one-on-one like this.

But now Denorah knows not to try to reach around, slap the ball. That's what Shaney's waiting for. Probably she's just *acting* spent, is really ready to spin off Denorah, go up and under, lay it in.

Denorah thins her lips, shows her teeth where Shaney can't see, and shakes her head no to the chance of that happening. Not on *this* defender. Not in *this* game.

Still, when Shaney bounces back into her, she can't help but give six inches, a foot.

Again, again.

Denorah steps in to regain ground, leading with her hips now because Coach says that's where women are most solid, and when Shaney's hair is in her mouth she spits it out but doesn't raise a hand to guide the strands out, because being grossed out doesn't matter, not when a point's at stake.

Except—

There's something *wet* on Denorah's chin?

Now she does raise the back of her hand, to wipe at it.

Blood?

Did she bite her tongue? Bust a lip?

No.

She backs off a full foot, to study Shaney's back.

"Hey," she says, stopping the game. "You're bleeding."

The whole back of Shaney's pale yellow shirt is red and dripping, her hair all matted in it.

"When you hit the pole that last time," Denorah adds.

Shaney keeps dribbling, the ball a metronome. Her face shrouded under her everywhere hair.

"We're playing," she says.

"But—"

Shaney spins against nothing, is playing mad now, is up against an imaginary defender.

She slashes past Denorah, is already pulling the ball up under her arm and behind her like protecting it for a bust-through, and, because she can, because she hasn't been in this backed-off of a position yet, Denorah reaches an easy hand out, slaps the ball from around Shaney's back, doesn't even have to shift her feet.

It's not a defensive move, it's a time-out.

The ball rolls off Shaney's knee, out into the crunchy grass and snow. Shaney, her momentum already gathered, has no choice but to keep surging forward. For the second time in as many plays she slams into the utility pole, shaking the janky backboard, more splinters and bird-nest trash sifting down. Denorah steps out of that bad rain, clocks Shaney coming down hard and awkward right *on* her back, like somebody cut her legs out from under while she was up there walking on air.

She flips over fast, onto her palms and toes, and then she rolls her shoulders slow, her hair all around her face, and screams straight down into the concrete, screams for longer than her lungs should have air for.

Denorah turns her head, like studying this from a slightly different angle can make it make sense.

"Hey, hey, are you all—" she tries, leaning ahead with her hand open like to help, but now Shaney is standing in her easy, athletic way, her body loose and dangerous again.

She guides the hair out of her face and . . . her eyes. They're different. They're yellow now, with hazel striations radiating out from the deep black hole of a pupil. Worse, her eyes are too big for her face now.

Denorah falls back, sits on the concrete with maybe half her weight, the rest on her fingertips.

She's not making the game tonight, she knows.

"What—what are you?" she says, breathing hard from fear now, not exertion.

"I'm the end of the game, little girl," Shaney says, then twitches her head around, stares hard at Cassidy's camper.

Dad? Denorah says deep inside, her heart fluttering with hope.

She looks to the right, trying to will three or four horsemen up from the grey trees, dogs weaving ahead of them.

There's nothing.

"The end of *your* game, anyway," Shaney goes on.

"Why are you doing this?" Denorah says, her voice getting more shaky at the end than she planned.

"Ask your father," Shaney says right back, still watching whatever she's watching over at the camper, or the sweat, or the cop car.

"My dad? Why? What did he do? He doesn't even know you."

"We met ten years ago. He had a gun. I didn't."

To prove it she whips her hair away from her melty forehead, leans forward so Denorah can take a long look.

"He . . . he wouldn't—"

"Shouldn't, wouldn't," Shaney says. "*Did.*"

"Just—just let me go," Denorah says. "You win, okay? We can . . . this is between you and him, then, right? Why do you even need me?"

Shaney settles her weird eyes back on Denorah.

"You're his calf," she says, like that explains anything.

"You're not really Crow, are you?" Denorah says.

"Elk," Shaney says back with a grin.

"My mom's on the way," Denorah says.

"Good," Shaney says back.

Denorah stares at her about this.

"What if I win?" she finally says.

"You won't," Shaney says. "You can't."

"I was," Denorah says. "I am. Eighteen-sixteen."

Denorah stands, staring into Shaney's nightmare face the whole time.

"I don't care what you are," she says. "When you're on this court, you're mine."

"And that's precisely what I'm here to take away from you," Shaney says back. "Before I take everything else."

Denorah gives Shaney her back, steps out into the snow to collect the ball, comes back to the pad of concrete, and cleans the soles of her shoes on the opposite legs of her shorts.

"My ball, right?" she says.

Shaney doesn't say yes and doesn't say no, just takes the check pass.

Denorah walks to her place facing the goal, says, "It's my ball, and"—pointing with her lips—"I'm putting it right there, and there's not one single thing you can do about it."

This is word for word what her dad used to tell her when she was a kid and they were playing in her granddad's driveway, when she could hardly even hold the ball, when he would have to scoop her up under the arms at the last moment of the layup, hold her up to the basket.

But sometimes he'd set her up in defensive position, get loose in the shoulders, his head rocking back and forth, and look up to the goal, tell her he was going to put it right there, and there's nothing Denorah can do about it.

Which is where it all started, she knows.

"What's wrong with your back?" Denorah says, catching Shaney's rolled ball under the sole of her right shoe.

"I'm dying," Shaney says, easy and obvious as anything.

"Serious?"

"But not yet, don't worry."

Denorah isn't sure what to make of this so she just looks to the opposite two corners of the court like confirming with her team- mates, and then she feels her mouth curl into her dad's reckless smile. Whatever this is, it's about to happen.

Shaney, whatever *she* is—some Indian demon from way back, some monster her dad found buried on some hill out here, a ghost woman he left in a rolled-over car—she steps in, gets down into defensive stance, her long fingers ready, her teeth showing.

Denorah turns to the side, dribbling with her left and taking stock, and in her head she says a silent apology to Coach, for the move she's about to try.

One thing about Coach, she *does* believe in the fundamentals. Nothing fancy, nothing showy. Three times already this season Denorah's been benched for showing off. Once it was for circling the ball around her waist before a layup on a breakaway, never mind that the crowd all came to their feet for that. Another time it was for passing between a defender's legs, which made that girl mad enough that she ended up getting kicked out a quarter later.

The third time Coach benched Denorah, it was for dribbling behind her back when there was no advantage to do it. Coach had been right, too—it *had* been completely for show, for joy, had been one hundred percent because Denorah *could*.

Never mind that she almost lost the ball, had to step long to keep up with it.

Alone on the little court at her house, though, she's been prac- ticing a new move.

About a third of the time, with no defender, when she's holding her mouth just right and the wind's in her favor, she can stick it.

Okay, one time so far she's sort of nailed it. Everything but the actual shot at the end.

Still, "Bet they didn't teach this at elk school," she says, and then, before Shaney can react—*using* that moment of confusion—she flips the ball around her left hip with her right hand, more a bullet pass than a real dribble, one she has to hula her hips forward a smidge to allow.

The ball bounces once with her serious English on it and then it's beelining for the right corner of the concrete pad and Denorah is already in motion, diving for it, her body blocking Shaney out behind her. Two out of every three times she's done this at home—okay, nineteen out of twenty—she can't catch the ball, has to run her effort off in the grass and snow. It's nearly impossible to catch it, much less turn it back toward the bucket. It's a move Coach would have outlawed for sure if she'd ever seen it. It's a move the crowd would shake the roof off the gym for, if they ever saw it. More important, it's a move that'll break the heart of any defender, Denorah knows. More important than *that*, it's the very last arrow in her quiver, and it's already slashing across the court, is going to bounce out of bounds if Denorah doesn't—

She *just* gets her fingertips to that spinning-away leather, Shaney so close that her hair is coming around Denorah's own face. Committing all of her weight and muscle and hope, cashing in every hour she's spent sweating it out in practice, Denorah pulls that ball tight to her ribs, hands clamped hard to each side so it can't be poked away, and turns on the ball of her left foot, her right shoe already coming up, and up.

She's too close already, though. This court is so small. The burst of speed she needed to catch up with the ball, it's left her already under the basket, where the only thing she can do is the first thing Shaney did to *her*: plant that right shoe as high up the utility pole as she can, wait for her weight to collect behind it. It gives her enough grip for the sole of her shoe to stick when she pushes off, when she forces her already-twisting body up into the air, the net scratchy against her face, her mouth open not in a scream but a war cry, her face full of Shaney's hair because she's right there, is coming up with Denorah, is going to slap this one down no matter how high Denorah climbs.

The only thing Denorah can do, her only hope, it's to extend the ball as far from her body as possible now, *around* Shaney's side where any defender would least expect it, meaning Denorah's one-handing it now, has just enough grip to spin it up, kiss it soft off the other side of the board, and then she's falling away, is falling for miles, back into legend.

The concrete jars her from tailbone to neck, leaves her spitting tongue blood and cheek gristle, but still she sees the ball slip through neat as anything, a pretty little reverse by a player who shouldn't even have that kind of reach, that kind of vertical, that kind of English.

It's about heart, though, Coach is always saying.

When Denorah smiles, she's sure her teeth are red.

"Nineteen," she says, chocking her face up like does Shaney have anything to say about *that*, and then she cringes away all at once, from . . . from—

From splinters in the air?

And sound. Her head is full of it.

A *gun*shot.

She looks up to where Shaney is glaring.

Cassidy's camper.

No, the outhouse.

Victor Yellow Tail is wavering a few feet from it, the door open behind him, his whole front soaked in blood, a pistol flashing in his right hand.

What he just shot was the utility pole.

The backboard is raining more of its rotted wood down.

Shaney bares her teeth, her whole body quivering.

"I killed you," she says across to Victor.

"*Where's my son!*" Victor loud-whispers back—no throat to speak with—and loosely aims the pistol again, shoots.

This time the concrete in front of Shaney chips up. Her leg snaps back and away and Denorah can tell she wants to explode away from this spot, run and run, be miles away.

Now Victor's falling to his knees with the effort of shooting, of screaming, of bleeding so much. But he's still pointing that drooping pistol ahead of him.

Shaney turns her head to the side like he better not, but he pulls the trigger.

This shot catches her in the right shoulder, flings her off the court, into the frozen grass and snow.

Denorah stands, doesn't know what to do.

Instead of just lying there and hurting, like would make sense, Shaney is flopping and writhing in the snow, screaming from the pain, the fingers of her left hand digging into her shoulder, and . . . and: no.

Her face.

Her head.

She arches back, her fingers deep in the meat and muscle of her shoulder, and her face is *elongating* from the strain.

Her cheeks and chin tear with a wet sound and the bones crunch, resettling.

At the end of it her long hair is blowing away from her, isn't connected to her scalp anymore, and her face, it's, she's, her face is—

Not a horse, which is what Denorah thinks at first.

Not a horse, an *elk*.

Elk Head Woman.

Denorah falls away, stands again, knows only to run, to leave, to not be here for whatever's next.

Where she runs is straight to Victor, the cop, the one with the gun.

She slides to her knees in front of him, grabs on to him, and his right hand falls across her back, the pistol hot at the base of her spine.

"Na-Na-Nate," he manages to say.

"*What is she?*" Denorah says, crying, holding so tight on to his bloody shirt, but then with his left hand he guides her away from him, pushes her behind.

Elk Head Woman is standing, is walking this way, her ungainly head turned to the side to better see them with her right eye.

"Go," Victor whispers to Denorah, "*run*," and she does, on all fours mostly, and when Victor's pistol fires again she falls ahead from just the massive cracking *sound* of it, and where she falls, it's into the smoldering lodge.

It's a pit of bodies.

The first she sees is a dog, mouth open, eyes staring at nothing.

She pushes away, trying to climb out, and then it's Cassidy, his face caved in and half burned away.

Denorah screams, can't breathe, can't do anything.

The hair in her hand is, it's—this is her dad.

She opens her mouth, doesn't have any sound left.

Behind her and all around her then, Victor screams through his bloody throat from whatever Elk Head Woman is doing to him.

Denorah rolls around, sees just grey sky above her, and then her right palm finds an ember. She snaps her hand back, holds it to her chest, and, just working on automatic, on instinct, fights her way up from the lodge, her knees and clothes and all of her sticky with ash and gore.

On the other side of the lodge, through the scrim of smoke but staring right through it, is Elk Head Woman.

"*You killed them all!*" Denorah screams through the smoke, holding her right hand with her left. "You killed my . . . my—"

Instead of answering—her mouth is no longer shaped for human words—Elk Head Woman steps forward, over what's left of Victor's broken body, his head just hanging by tendons now, and the way she looks down to get her feet in the right place, it's because her eyes are on either side of her head now.

Denorah steps back, falls down, comes up already running.

In practice, there's a drill Coach makes the team do maybe once a week. More than that would leave them too beat-up to play. But, once a week, she'll line all the girls up on the baseline and step in among them, the ball rocked back to let loose.

The whole time before she blows that whistle, too, she's yelling to them like a drill sergeant, asking, *How bad do you want it? How bad do you want it?*

At the end of a game or the beginning, it doesn't matter, it's never the fastest or the strongest player who gets that rolling-away ball. It's whatever girl dives the hardest. Whoever fights for it. Whoever doesn't let anybody take it away. Whoever doesn't care about their precious hair or skin or teeth. It's all about whoever wants it the worst.

This is the drill Denorah's running right now.

Only, this time it's no drill.

ONE LITTLE INDIAN

The first place Denorah's going is Mona's. If she can cut across, find the road, it'll take her right to Mona's trailer, and, and, and maybe that old bear will be there, maybe he never went to sleep this winter, maybe he'll smell an elk and stand up, forget about waiting those berries out.

It's a good plan—it's a stupid, stupid plan—until, maybe a mile out, her lungs raw, her shins bloody from how hard the snow's crusted, her feet soaking wet and numb forever, Denorah runs right up to the lip of a drop that's probably a hundred rocky feet down.

The wind coming up it pushes her back, saves her life.

Down at the way-bottom there's an old broken-down corral and a stone something, but nobody's lived down there for eighty, a hundred years. This is one of those far-out allotments that some-body tried to make work, but Blackfeet aren't farmers, Blackfeet aren't ranchers.

"*Nooo!*" Denorah screams down into that big empty space she can't cross.

She looks back like she's been telling herself not to, and maybe a quarter mile back, not even that, it looks at first like a horse is cresting the rise. Her heart swells, thinking it's her dad on one of Cassidy's horses, but that was a lie, the men never went down to see what the dogs had tied into, the dogs were already dead by then, they all were.

And this still isn't a horse head.

It's Shaney, whatever she is. Elk Head Woman.

She walks forward, has human shoulders, a woman's arms, one arm red from all the blood coming down from her shoulder. Long gym shorts and tall socks. Wide eyes fixed right on Denorah.

"*Why won't you even run!*" Denorah yells back.

Elk Head Woman just keeps coming.

Denorah bounces on her feet, her back to that long drop, and looks left and right, her only two choices.

Right is more of the same: what looks like snow forever, all the way over to the Park, but with coulees gouging through deep and sudden. Left is that same snow for maybe a mile, but then there's the lake out that way. The one her dad's friend drowned in, or whatever. And—and that's where his other friend got arrested—

"Oh yeah," Denorah says.

There's *lake houses* over on the shore. Cabins with steep roofs and canoes lashed to the porch across the door, like that would really keep anybody out. Her dad's friend Ricky had broken into one to report that body floating facedown in the water—to *call* and report a body in the water.

Meaning: a phone.

Denorah can call Mona's from there, she can call the Game Office, her dad, her new dad, she can report the massacre, report an officer very, very down.

Denorah looks back to Elk Head Woman, gone again, slogging through the deep snow that's between the rises—she only walks in a straight line, like following a ridge would be undignified, would be the land making her do something.

Picking her foot placements as carefully as she can, so she's walking where the wind's scoured the snow away, Denorah goes left and keeps low, always picking the side of the brush that threatens to drop her down the cliff instead of the other side, where she can be seen.

She remembers some grade school teacher talking about how Indians having long hair—it was Miss Grace, a French-accented blond woman from Canada—how long hair, it helped with hunting. When hanging down, it wavered back and forth like grass, and it hid the recognizable human features.

It had been bullshit, of course—hair isn't grass, faces are faces—but Denorah has never forgotten it, either.

Running now, half certain that Elk Head Woman anticipated which direction her prey would take, is cutting across on the diagonal, is about to step over this rise *right here*, Denorah hooks a finger into the tail of her braid and runs the band off, combs the rest free with her fingers.

In her head she's practicing what she's going to say into the phone.

My dad, Cassidy, she killed them all, you've got to—

No. Start with Victor.

Your . . . your cop, your officer, Nathan Yellow Tail's dad, he tried shooting her but she . . . she—

Also: *Her back is already hurt. That's where you need to shoot her. If you shoot her from the front she'll just pull the bullet back out.*

Like she's even going to get to call. Like she can even make it the two more miles to the lake. Like she's not going to fall one too many times, roll over, find Elk Head Woman standing over her.

Why would a phone even still be hooked up out there over winter?

But where else is there to go?

Denorah shakes her head, her hair down now.

She imagines Coach behind her, blowing the whistle.

The next time she looks back, there's no elk head cresting. It doesn't mean anything, she tells herself.

Run, run.

She does, harder, and it's good that she does. This time when she looks back, Elk Head Woman is right there, maybe forty yards back.

She stops, turns her head to the side to get Denorah in one of her big eyes.

"I beat you," Denorah mutters, not even close to loud enough, and forces herself up a steep rise, comes over it at a shamble, into . . . into—

Somebody's old place. A ramshackle house low to the ground, all the windows gone, the walls peeling away. Two old hatchbacks left where they died, it looks like. A barn or shop that's blown over except one corner. The only structures still standing, not caring about the wind and snow and loneliness, are three rusted-purple boxcars parked nose to tail, the kind people use for storage, the kind her stepdad recommends all around the reservation since they're about the only thing bears can't get into. Whoever set them up here was playing with a big train set, it looks like—no, of course: they were trying to get a snowbreak going. Giving the snow something to drift against, to keep it from building up against the house, but these boxcars are train-tall, too, are sitting on blocks or actual wheels or something.

"*Hello!*" Denorah yells down to the place, but it's obviously abandoned.

And, is she hearing footsteps behind her? *Hoof*steps?

She surges downhill, sliding on her butt and the heels of her hands over and over. When she looks back this time, Elk Head Woman is walking a straight line down the hill, not slipping even once, because elk always know where the foothold will be.

Denorah turns, frantic, considers trying to stay on the other side of one of the hatchbacks, always moving to the front when Elk Head

Woman comes around the back, but all it takes to lose that game is one good trip. And the house, going in there she'd just dead-end in a bedroom, die there when Elk Head Woman filled the doorway.

Denorah shakes her head no, there's nothing for her here. This is just a place to run through. She does, deciding at the last moment that burrowing down *under* the middle boxcar might slow Elk Head Woman down. If she only walks in straight lines, maybe she doesn't bend over to go under things, either, right?

It's as good a guess as anything.

Denorah slashes forward, Elk Head Woman only two car lengths behind now, and forces herself through the wind-scoured crust of snow packed between . . . probably not wheels, but it doesn't matter.

Immediately she regrets closing herself in like this, and panics hard, digging with her hands, kicking with her legs until she surges ahead into . . . a dry cave under this boxcar. A magic kind of place. So quiet but not quite dark: the sunlight's seeped in through the thousand-million crystals of snow packed all around her, making the walls glow blue like ice.

Not a cave, she tells herself, though. A tomb. A grave.

She gathers her will and pushes into the far blue wall, takes a deep breath to push through, but then each sidearm of snow she sweeps away, ready to break into open air, there's just more snow, and more snow. She gasps her lungs empty, tries to suck a breath in but there's only snow everywhere, in her mouth. She gags, bucks, gets her feet under her as best she can and just *pushes*. Into more snow.

But her hand, it's through, it's out there.

She's swimming now, swimming up through a slushee, not quite surfacing but pulling enough crusted snow down that a sort of sinkhole to the sky opens above her. Her mouth at the bottom of that funnel, she draws as much air in as she can. And again.

Like whoever put the box cars there planned, and like she didn't think to anticipate, the snow on this is drifted deep-deep, and sloped out for probably thirty feet.

Denorah trudges out through it, the crust cutting into her neck, then her chest, then her stomach, thighs, shins.

On level ground at last, she lowers herself to her fingertips, shakes her hair out of the way, and looks back through the chasm she just made, that'll probably hold for a few more minutes yet.

Elk Head Woman's high-tops and tall socks are there through the opening, blurry through the wall of icy snow on the other side. But they're not moving. For the first time, they're not moving.

What? Denorah says to herself.

She stands, ready to run, but then doesn't. She looks through again.

Elk Head Woman's high-tops and socks again. Still.

"What the hell?" Denorah says, looking left and right to be sure this isn't a trick, that Elk Head Woman isn't coming around either side.

Has . . . did Shaney get stupid when her head went all elk? Is she *acting* like an elk now more than a person?

Denorah stares at the backside of the drift she just clawed up from.

Jutting up from it, to the roof of the boxcar, is the built-on ladder.

This is what smart girls do, she tells herself. When the killer's right on their tail, they run up to someplace they can't get down from.

But she has to see. She has to know.

Denorah nods to herself, nods again, then backs up and rushes ahead, to run up the side of the icy drift, clamp a hand onto that lowest rung.

She makes it three steps before the drift swallows her whole for a second time.

Ten sputtery seconds later she bursts up through the drift partway up the ladder, stabs a hand up one rung higher, pulls herself free.

She clambers to the top, hooks a leg over, and just tries to breathe.

She's wet head to toe now, which isn't wonderful.

It's windier up here, too. Of course.

Denorah hugs herself and inches forward, being sure of each footstep before giving it the rest of her weight. She does *not* need to fall through, into whatever got left behind in this boxcar.

The last four feet are on her stomach, her hair coiled in her hand so it doesn't blow over the edge ahead of her, give her away.

Elk Head Woman is just standing there, her ungainly head cocked a bit to the side, the boxcar locked in her glare.

Denorah smiles.

You're afraid of trains, she doesn't say out loud.

But it's true.

Elk, which is what Elk Head Woman must be in there somewhere, maybe more and more with each step, they're train-shy. Her dad told her this. It was a story from one of his great-uncles, about how once all the men in town had backed a herd of elk up to the tracks, blown them away when the train came. They hadn't meant to use the train as a fence, were just using it for sound cover since they weren't supposed to be shooting in town, but it had turned into a fence all the same. The one or two elk that got away, her dad said, had told the rest the Truth About Trains, and that was that, no more using train tracks to hunt.

Evidently trains themselves are even scarier, never mind that there's no wheels on this train. Never mind that the cars aren't even actually connected. Never mind that there aren't any *tracks*.

Elk might be tough and fast, Denorah figures, but they don't seem to be the best problem-solvers. Still, it's not going to take

Elk Head Woman forever to figure out that this train is only three cars long, and not making any sparks, not filling the world with sound.

Elk Head Woman opens her mouth and a low bleat eeks out, like testing this situation, like announcing her uncertainty, like asking the herd for help here. When none comes she steps back, like the train trance she's in is losing a bit of its grip.

Denorah turns, crawls back to the ladder, hand-over-hands it down into the drift, kicks out the other side, walking her previous churned-up path.

Still no Elk Head Woman.

"Choo-choo, crazy lady," Denorah says, tossing a middle-finger salute off her forehead—another thing she learned from her dad, every time they'd just passed a cop.

The lake, she's saying inside.

She can make Duck Lake now.

The one time she looks back, there's no Elk Head Woman rounding either side of the boxcars. But there will be, she knows. There will be.

Denorah quickens her pace.

BLOOD-CLOT BOY

She should have crossed the dirt road she needs ten minutes ago, Denorah knows. Twenty.

It's like—it's like all the roads are gone. Like the reservation's dialed back a hundred years, to before cars. Like that broken-down corral back there, like it's probably still standing, has a stone house beside it now, smoke curling up from the chimney.

Either that or Denorah's a town girl, knows every inch of the basketball court, but the ungreat outdoors? Not so much.

One tree is the same as the next. All this snow looks like all the rest of the snow.

The lake, though.

Every few hundred yards she'll work her way up a rise, see it shimmering in the distance.

What time does it get dark? Four?

Coach is going to flip when her star player doesn't show up an hour before the game. But that's good. Wait, no: That doesn't matter. By then Denorah's mom will have called in the National Guard, probably. She'll have walked up Cassidy's long driveway, found all the bodies burning, seen the blood splashed on the court, found Victor Yellow Tail over by the outhouse, killed twice.

And . . . and there are tracks in the snow, aren't there? Denorah looks behind her to be sure.

Hopefully Elk Head Woman is still stuck on the wrong side of that ghost train. But: Don't count on that, Denorah tells herself. Elk Head Woman's got to be close already. Don't look now. Okay, don't look again, and again.

Denorah sags to her knees, makes herself push up, push on.

Her first wind was spent before Cassidy's camper was even out of sight. Her second wind didn't even register. She's going on pure need to survive now. Need to survive and the conditioning Coach is always saying can decide a game.

All that plus a little hope: the lake houses.

Maybe some crazy hermit of an ice fisherman is down there, snowed in. Maybe some of the high schoolers have broken into one of the cabins like always, are partying this weekend. Denorah can . . . she can take one of their snowmobiles, tear out of there, run for Canada.

Are there any train tracks between here and there? She's not counting on bears to save her anymore, but train tracks.

Run, run, she tells herself.

Last three seconds of the game, *push*. And again.

Her lungs aren't burning anymore, they're cold, and there's blood in the back of her throat she's pretty sure, what Coach calls lung cheese. But Denorah should have blown all that out two months ago, when practice started. And she's lactose-intolerant anyway, she says about the lung cheese, trying to make a joke out of it. The saddest, most-alone joke.

She breathes a long hair down her throat and has to stop to cough it up, puke a little besides.

She's not going to make it.

Is the lake still the same distance away? It can't be.

Denorah closes her eyes tight to reset, to find herself in all this pain, all this cold. Distantly, like it's someone else, she's aware that she's on her knees, that her hands are over her face.

The road has to be up here somewhere. It *will* be up here.

What's happening is that she's thinking like she's in a car, where distance goes by fast. But she's on frozen wet feet, and not taking anything even close to a straight line, and the road *does* swerve the other way anyway. She's probably just coming in above that bend, meaning the road will be farther.

Don't panic, girl. Gather the ball, collect your wits, and check the time clock.

Denorah lowers her hands, looks up to the hazy sun.

At least three hours left, she decides. Three hours before Elk Head Woman is stepping out of every darkness, which will be all there is.

But you'll be dead long before then, she reminds herself, and brings her face down. To a long brown face watching her from maybe twenty feet ahead, just behind the next rise.

Denorah knows not to flinch back, knows better than to scream, but still, inside, all her big plans are falling off their flimsy metal shelves, clattering down into the pit of her stomach.

This is it, then.

Denorah stands, her hair lifting all around her, her hands opening and closing by her thigh because she's about to be gouging eyes and tearing ears, whatever she can get to—you come at a reservation girl, bring a box of Band-Aids—but then . . . then—

It's a buck. A mule deer. She knows because back when she still had to stand up on the seat of her dad's truck to see over the dash, he taught her how to tell mulies and whitetails apart. It's size, sure, and how their racks are shaped if they're male, but before all that, it's color. Mule deer are dusky brown for life out here on the flats, and they don't have white rings around their mouth and nose so much, and, according to her dad, they taste better, but color's an easier tell than running one down, taking a bite.

This one's just staring at Denorah with his big black-marble eyes. Waiting to see what she is, his tail twitching out the seconds.

And then he's looking past her. Behind her.

"No . . ." Denorah says, turning around all the same.

Elk Head Woman, plowing ahead, leading with her wide forehead.

Denorah turns back to tell the mulie to go but he's already gone, running down some frozen creek bed, probably—yep: on cue he bounds up like off a hidden trampoline, floats for a span of ground Denorah is completely jealous of. The instant he hits the ground his hooves are already digging in, churning him forward.

"Hit it, brother," she says, and urges her own self forward as well.

She's got maybe a quarter mile of space between her and Elk Head Woman, if that.

What was that little kid story her dad told her about whitetails? He said he'd heard it from his granddad, but Denorah learned later that he never knew her great-granddad—their years hadn't overlapped, quite. He's heard it from *some* granddad, then. Either way, when he told it, it was so real. The way whitetails got that white ring around their mouth and nose, according to him, it was because they were always sneaking into Browning to drink from the bowls of milk everybody used to leave out, from back when there weren't any reservation dogs, only reservation cats. That was why the whitetail could come into town like that: no barking. But the cats were too good, they got the mice all so scared that the mice got smart, started living so deep in the walls of the houses that the cats couldn't get to them, so one day all the cats just left. It was two, maybe three days after that the first dog trotted into town with a stupid grin on its face, looked around for what it could pee on.

Denorah hates that she'd believed that, once upon a time. And she wants to cry for not getting to believe it anymore.

Yes, the deer drank milk, and that left their mouths ringed white. Fuck it.

Run, run.

She tells herself not to, but she looks back all the same.

No Elk Head Woman. Meaning—meaning Denorah can wait for her to slog up from whatever dip she's in, or she can cover some more ground.

She moves, moves.

What she's counting on now, since the lake isn't getting any closer, is finding the road before her mom's driven down it, finding it and then flagging her mom down, not even letting her stop the car all the way, just climbing in, locking all four doors, and waving her ahead, Go, go, faster, I'll explain later, *go*.

Denorah falls, gets up, falls again, gets up again, and now the horizon is wavery. Not from heat, but from exhaustion. From cold. From no more adrenaline. From too many last three seconds.

But then . . . then: Elk Head Woman *is* Crow, right?

Denorah stands, pushes on, making herself run again.

No way does a Crow win. Not today, not here.

Even *if* the world's blurry. Even *if* Denorah's lungs aren't working. Even *if* she can't feel her legs at all. Even if she's seeing ledger art come to life now in front of her.

She slows, shakes her head, tries to clear her eyes.

The ledger art remains. It's there not fifteen feet in front of her.

A dying Indian slumped forward on a horse, what she sees at every booth at every powwow: The End of the Trail. The only difference is that the tired war pony, it's usually either in silhouette or just white, to better see the dying Indian's bare leg on that side.

This horse is a paint.

It raises its head, gives an obligatory whinny about Denorah.

"Calico?" Denorah says weakly, sure this must be a death vision.

Calico whickers, blowing her lips at the end, and Denorah tracks up the horse's neck.

Tangled in its long mane, tangled tight, are fingers. Behind them, on Calico's back, his blood coating down her side, is the dying rider—

"*Nathan!*" Denorah screams, running to him, hardly registering that they're up on the hard hump of exactly the road she's been trying for.

She gets to his left leg with her hand, and that nudges him awake. He looks around, then down to her.

"D," he says back with half a messed-up smile.

"Are you—what—let me," Denorah says, no clue where to start or how to start it.

At which point Calico dances to the side, away from Denorah.

"Po'noka," Nathan says, sitting all the way up now.

Denorah tracks from his eyes to where he's looking: behind her.

She turns around already shaking her head no.

Elk Head Woman.

So close.

Two free throws away and walking in a straight, pissed-off line. Probably because Nathan's supposed to be dead, not still dying.

"*Go, go, go!*" Denorah says up to him.

He reaches an arm down for her, to haul her up onto Calico with him, but the effort nearly tilts him off, and grabbing onto her looks like it would rip him in half anyway. *More* in half. Denorah pushes him onto Calico's back with both hands, holds him there.

"*No,*" she says, "I'll—I'll lead her to the lake. Tell my—go to town, can you do that? Ride right the fuck into town and tell them, tell them all . . . You know where the Game Office is? Just . . . find my dad, tell him I'm headed to the lake, the one that Junior guy died in, Duck Lake, tell him—"

"Your . . . your dad," Nathan manages to say. "He's—isn't he dead?"

"My *other* dad!" Denorah yells, then grabs on to Calico's head, hauls her around, and slaps her hard on the ass, screaming at the same time.

Calico explodes ahead hard, even wheelie-ing up at first, which Denorah knows is called something else when it's a horse, but there's no time, there's no time.

Elk Head Woman steps up onto the packed dirt of the road.

She's looking at Nathan and Calico, is considering them.

"*Hey, you!*" Denorah says, bringing that long elk face around to her. "Nineteen-sixteen," she says, touching her own chest, then pointing across to Elk Head Woman. "Thought we had a game to finish here."

Denorah gets the full attention of one of those big yellow eyes, and she doesn't wait, she's already running.

This isn't a second or even a fourteenth wind, she knows. This is running on hardpack with feet she can't even feel. This is running downhill, toward water.

This is the real last three seconds.

WHERE THE OLD ONES GO

It doesn't make sense that Denorah is just now getting down to where the lake sort of is. She's been running for *years*, she knows. For her whole life, maybe. And not running the whole time, either. At least three times now she's crashed and burned, just flattened out there, ready to give up. Her chin is raw from scraping, the palms of her hands are bleeding, and she's not thankful that she can feel her feet again. Her feet are full of needles.

In her head she mumbles apology to Coach. Players are supposed to save their legs for game day. Denorah's not going to be able to walk for a week, she knows. If then.

But first she's got to live.

The last time she fell and decided to just rest her eyes for a moment, for a breath, for two, okay, the hard dirt so right and perfect against the side of her face, she came to all at once, instant panic, and rolled over to see Elk Head Woman just two fence posts behind her.

She's walking on the road now too, even though the road curves and twists, banks and falls away in places. If this were fair, if Elk Head Woman were sticking to her own rules, she'd still be taking a straight line, wouldn't she? She'd be getting bogged down out there in the deep stuff. Even elk bog down, right?

But elk walk the road, too, Denorah knows. She's seen them doing it, all in a long line, heads drooping like it's the Elk Dust Bowl, the Great Elk Depression.

"*What do you want?*" Denorah screams back, standing her ground, leaning forward from screaming so hard. "What did I ever do to you?"

For the first time Elk Head Woman's pace steps up.

Denorah falls back, turns it into another push, another run.

What's it going to take? What else can she do? And, how has she missed her mom? There isn't some other road out here, is there? If only her dad were here. He knew all the old poacher cut-acrosses, all the shortcuts if you had four-wheel drive.

Denorah falls again, leaves even more hand-meat and knee-meat and mouth-blood on the road, and then she rises again, not running anymore, just stumbling.

She's not going to make the lake by dark. She's not ever making the lake.

And—and Nathan, he probably fell off Calico a hundred yards out. He doesn't know horses any better than Denorah, and he was already half dead anyway.

It's just Denorah and Elk Head Woman, then. One-on-one.

Denorah walks backward a few steps, sees that distinctive head crest over the road, ears pasted back.

She shakes her head no, no, please, and nearly falls down again, has to catch herself on one of her raw hands, push up hard before it turns into a dirt nap.

Ten, twenty steps later, there's a break in the grey trees beside her. A—a *gate*.

Denorah looks back and Elk Head Woman is gone for a moment, so, no time to think about it, she steps out onto one side of the big corrugated silver tube that runs under this offshoot road, and she jumps from it to the top wire of the gate and flops immediately and unintentionally over the top strand, praying she's not leaving tracks. Without looking back, she staggers ahead, her right hand patting numbly at the point of pain in her belly, probably from a barb. But that hardly even matters at this point. The road's a two-track, but one that hasn't been used for however many snows there's been so far this year.

She tries to keep to the hard hump in the middle but loses the road almost immediately, is in some trees now, is using them to stand, to pull ahead.

Don't look back, don't look back.

Just forward, go, keep moving.

Maybe there'll be a phone booth right up here, she tells herself, her thoughts getting loopy now, the trees smearing into a wall of upright logs. Denorah hand-over-hands down along that wall, feeling for the opening. When she finds it, she was so sure the wall was going to go forever that she falls straight through, is sliding downhill, scraping and rolling through rocks and bushes and dead wood.

She lands in a ball of pain maybe ten seconds later, looks back up.

Oh. She was on a ridge. The road must have hooked back to the right to keep this from happening to all the trucks. But, unlike a truck, she'd gone straight.

Denorah pulls herself up with a bush that scratches every part of her face, even her lips—is this what her dad calls *buckbrush*? Or, used to call, she reminds herself.

"But I'm a doe," she says, drunk with the pain of it all, and puts one foot in front of the other, and then repeats that complicated process, and a court length or two into that she realizes that this is what it's like to die, isn't it?

You hurt and you hurt, and then you don't.

It's soft at the end. Not just the pain, but the world.

And at least she'll die with that, she knows: The *world* killing her. Not the Crow. Not Elk Head Woman. Not the thing that got her dad.

"I'm sorry," she says to the idea of him.

Not because he died however he died, but because she never told anybody to let him stay when they were dragging him from the gym. Because she pretended she didn't know him. Because she was embarrassed. Because—because she's still that girl standing on the bench seat of his truck beside him while he's driving, her hand on his shoulder, the cab full of his stories that were all true, she knows.

Because because because.

Her breath hitches deep and she stops, her hand on an aspen, a birch, she doesn't know stupid trees, trees are only good to make basketball courts from. This tree holds her up just the same. She pats it in thanks and looks past it, to where she's going to die.

It's a field of . . . not spikes of snow, no, there's no such thing.

Bones.

"What?" she says.

She—she can't be *that* far out, can she? Marias, that massacre or whatever? The bones from that wouldn't still just be lying out there, would they?

Bones don't last that long.

Unless. Unless she already died a few steps back, and is walking forward through her people's past now, maybe. Is that how dying works?

She looks behind her—nothing calling her back—and steps forward gingerly, to crack this last Big Indian Mystery.

It is another world, the kind she wants to hold her breath in. Not to keep it from getting into her lungs, but because it's sacred. There's skeletons all around her. Not Indians, she can tell that now, not people at all, but . . . cattle? Her new dad's told her about grizzly stashes, but those are always back in the trees, not out in the open like this.

No, this is something different, something worse.

Elk.

Denorah nods to herself, puzzling the bones together in her head.

Elk, definitely.

There's one side of a rack tilted up over there, even, unbleached and frozen, and—she looks around faster now, more desperate.

This can't be *that* place, can it? The place her dad would never tell her about, where him and his friends blasted all those elk ten years ago?

But it is that place.

Denorah swallows, settles down to her knees, her hand tracing the gentle curve of a weather-smoothed rib until it's shattered about halfway down. And the rib beside it as well, just the same. From a gunshot. Maybe even from a bullet shot from her dad's gun.

Denorah looks up the steep slope, can almost hear the rifles, can almost see her dad and Cassidy, Ricky and Lewis, so proud, so thrilled with their luck, with what great hunters they were.

Her heart beats once, seems to stop in her chest.

"Dad," she says.

This is where it happened.

Part of her new dad's story of this, too, the end of it, was her real dad and his buds throwing their caped-out trophy bull back down the slope, after trying to bargain to keep just it, please, not even any of the meat, not even the young little elk.

That's when she knew it was a true story. Because that's exactly the kind of thing her real dad would have asked for: the horns.

But, that story being true, it also means—it means her dad really and truly did this, doesn't it? Instead of being the one down in the encampment, bullets raining down all around, punching through the hide walls of the lodges like she knows happened to the Blackfeet, to Indians all over, her dad was the one *slinging* bullets, probably laughing from the craziness of it all, from how, this far out, they could do anything, it didn't even matter.

"I'm sorry," Denorah says to the elk rib she's touching, and closes her eyes.

This is a good place, she tells herself. A good enough place. She can lie down here with them, can't she? If they'll have her.

When she opens her eyes ten, twenty seconds later, it's from snow crunching behind her.

She sways her back in but she has to do it, has to look around.

Elk Head Woman.

This close, her head is even more wrong.

But she's not looking at Denorah, has forgotten all about Denorah.

Elk Head Woman falls to her knees too, her human hands to these elk bones, her nose dipping down to touch a skull, and staying there.

Denorah is breathing heavy, can't move.

All at once Elk Head Woman thrusts up, casts her long head around, looking for, looking for—

There.

Just an icy patch of grass like all the rest.

But not to her.

She makes her way over, falls to both human knees over it, lowers her head.

"You—you were here that day, weren't you?" Denorah says, and Elk Head Woman snaps her face over, her eyes hot and fierce.

Denorah starts to reach a hand across, like the daughter of Elk Head Woman's murderer can do anything good here at all, but then she remembers Victor Yellow Tail's broken body. And

Cassidy's, and Jolene's. Her dad's. She pulls her hand back to her chest, holds it there.

Now Elk Head Woman is leaning forward on her right arm, her palm to the bare dirt, like she can feel something down there.

Denorah can feel it writhing around down there, too.

"What is it?" she asks without thinking, but Elk Head Woman is already digging, frantic, her elk mouth making desperate little chirping sounds.

Denorah, shaking her head no, leans in just enough to see the birth: a fragile brown leg kicking up from the dirt, ten years after it should have rotted away, and then a thin little flank there under a swipe of earth, and now Elk Head Woman is digging faster, more desperate.

An elk calf, still wet, shivering.

She pulls it up to her human chest, its neck too weak to hold its head up, its chin on her shoulder.

Elk Head Woman's whole body hitches up, and then sighs with the perfectness of this skin-to-fur contact.

Which is when the rifle shot opens up the world.

Just past Elk Head Woman, a spurt of snow geysers up, the powder hanging there while the sound rolls away. Denorah looks back up the long slope, to . . . to—

"You made it," she says, in wonder.

It's her new dad, in his Fish & Game shirt.

Meaning—meaning *Nathan* made it. He Paul Revere into Browning with half his blood gone, must have ridden right up to the Game Office, stayed conscious long enough to tell Denny Pease that his new daughter, stepdaughter, she was out by the lake, Duck Lake, and there's a . . . there's a monster—

Her new dad knew just where to go, and just how to get there. There's only one spot Gabriel Cross Gun's daughter would end up. Where *his* daughter would be.

His next shot slaps into the ground just in *front* of Elk Head Woman, like showing her he can shoot past her, and he can shoot just short of her. Translation: she's next.

Elk Head Woman understands this, resists all her instincts to run, instead turning to curl around her calf, give her back to the slope, hoping her body can be thick enough to keep her calf safe. Because that's what an elk mother does, isn't it? That's the only thing you've ever really wanted to do this whole time, ever since you found yourself suddenly back in the world. Just—your anger, your hate, it was coursing through you so hot, and you got lost in it, and—

Denorah looks up that long hill, into the winking scope and dead eye of her new dad, and then she looks to Elk Head Woman, to the calf, and she sees now that both her fathers have stood at the top of this slope behind a rifle, and the elk have *always* been down here, and it can stop . . . it *has* to stop, the old man telling this in the star lodge says to the children sitting all around him. It *has* to stop, he says, brushing his stubby braids out of the way, and the Girl, she knows this, she can feel it. She can see her real dad dead in that burned-down sweat lodge, the back of his head gone, but she can also see him up the slope ten years ago, shooting into a herd of elk that weren't his to shoot at, and she hates that he's dead, she loved him, she *is* him in every way that counts, but her new dad shooting the elk beside her isn't going to bring him back, and as long as she keeps dribbling behind her back when she doesn't have to, then her real dad won't even really be gone, will he? He'll still be there in her reckless smile. Because nobody can kill that.

So—this is where the old man looks from face to face of the children in the lodge with him, a blanket of stars spread out around them, this is where he says to all the children gathered around the fire that what the Girl does here, for Po'noka but also for her whole tribe, what she does is slide forward on her bloody knees, placing her small body between that rifle and the elk that killed her dad.

She holds her right hand up the slope, palm out, fingers spread—the old man demonstrates—and she says it clear in that cold air: *No, Dad! No!*

Is it the first time she's called him that?

"It is," the old man says. It is.

By slow degrees, the rifle raises, its butt settling down onto Denny Pease's right hip. He's just a silhouette all the way up there. Just another hunter.

For a long moment Elk Head Woman doesn't move, is just hunched there around her calf, but then her long head wrenches around, ready to flinch from that next shot boring into her back, to take her legs away again, to start this whole cycle all over.

Instead, the man-shape up there, he's sliding his right hand sideways, palm down, left to right like this, the old man says.

It's the Indian way of saying a thing is over. It's what he used to end every meeting with, when he was trying to pull Gabe and Cass and Ricky and Lewis back, keep them alive. It's what he would have told his grandson, if he could have.

It's over, enough, it can stop here if you really want it to stop.

The Girl nods about this, knows what this hand signal means. She turns back to Elk Head Woman beside her, but Elk Head Woman is jerking now without even being shot, is falling over onto her side, still holding on to her calf, protecting it from whatever this next thing is.

It's her collapsing into the snow, her legs and arms kicking and reaching, twisting and creaking. Finally her right leg kicks through its human skin, is coarse brown hair underneath. Then an arm pushes through, has a clean black hoof at the end of it.

An elk cow stands up from the snow and lowers her face to her calf, licks its face until it wobbles up, finds it feet, and that's the last anyone ever sees of those two, walking off into the grass, mother and calf, the herd out there waiting to fold them back in, walk with them through the seasons.

Because it's the end of the story, the old man holds his right hand up again, like the Girl did that day, and all the children do as well, and then, just like the Girl does four years later, when her team loses State in double-overtime, he balls that hand into an upraised fist. What the Girl will be doing with that held-up fist at the end of that forever game, it's honoring the Crow team that finally figured out how to shut her down—the first defense to ever do that, and one of the last.

That show of sportsmanship, of respect, of honor, it's what gets silhouetted on thousands of posters all through high school sports, all across the land that used to be hers.

It's not the end of the trail, the headlines will all say, it never was the end of the trail.

It's the beginning.

ACKNOWLEDGMENTS

I don't write this novel without Ellen Datlow—not sure how I would write horror at all without her being her—so, thanks, Ellen, always. Not sure how I'd write *this* novel without how Louise Erdrich's *The Antelope Wife* lodged in me, either. But that's everything she writes. Her stories and characters and scenes are shattered all through my heart. Remove any one of them and I bleed out fast. Too, there's Elizabeth LaPensée's *Deer Woman: A Vignette*, which I picked up at the first Indigenous Comic Con. Or, I think Lee Francis IV maybe gave me a copy when I was there? I don't remember for sure, but somehow I ended up curled around that comic book, and couldn't stop thinking about it. And I'd be lying if I didn't also cite the seventh episode of season one of *Masters of Horror*, "Deer Woman" by John Landis. I really liked how that woman kicked whoever needed kicking in that story. I want that for all Indian women. I also want them all to live, too, please. Some of them are my sisters, my nieces, and all of them are my cousins, my aunts. And Joe Lansdale is always kind of my model for how to write, how to get heart and laughs and action and everything good on the page, in whatever genre. And . . . somehow James Dickey's poem "A Birth" is either really deep in the grass of this story or it's part of my writer DNA in a way I can't shake. The timid steps that new horse takes in that poem, into my world, that's the way the elk walk for me in this novel. They're looking at me while they graze, I mean, and if I don't do them right, then they're coming for me. I probably won't hear them either, since my music's always blasting. Example: when very first starting this novel, the song I had on repeat is D-A-D's song "Trucker." Rounding the corner to done, though, I needed people, not music. I think Matthew Pridham and Krista Davis were the first to read it, but Matthieu Lagrenade and Reed Underwood and Bree Pye and Jesse Lawrence and Dave Buchanan were close behind. Thank you all. Hope I'm remembering everybody there. If not, then write your name here: _____. Thank you, _____. Thanks as well to Alexandra Neumeister and David Tromblay and Theo Van Alst, and Billy J. Stratton, none of whom read this while I was writing it, but talking about different things with each of you nudged me this way in the story instead of that way. Thank you for those talks. And, talking about talking, I don't speak Blackfeet, but Robert Hall and Sterling HolyWhiteMountain were able to make that all right and proper for me, along with a lot of Browning and Blackfeet Reservation details, which I didn't know, as I didn't grow up there. Which isn't to say I didn't still manage to jack it all up. But, if so, it's on me, not them. Thanks, Robert and Sterling. And, okay, Sylvester Yellow Calf too, you're on every page. And Pat Calf Looking, my great-uncle, you're maybe on some as well. While finishing this, I taught a grad seminar on the haunted house, which was so helpful. Won't list all the students in there, as I'm probably already getting the staredown (I'm really stretching "one paragraph" here . . .), but our discussions in that classroom were so vital to me, getting this novel together, as were some old and not-so-old haunted house discussions with Nick Kimbro. Too, thanks to my brother-in-law

Oliver Smith, for doing some last-moment eyeshine research deep into a writing night. And thanks to Migizi Pensoneau for helping me get some Great Falls facts . . . I won't say "right," as I tend to change things in the writing of them, but "less wrong," anyway. I hope. Maybe. Thanks also to Jill Essbaum, who doesn't yet know I smuggled the opening line from her *Hausfrau* up onto the rez, with some slight liberties taken. Really, though, thanks for always being my lifeline on the mountain, Jill. I can't write if I don't come back down, right? And, talking about making it down the mountain mostly in one piece: thanks to my dad, Dennis Jones, for taking me out before dawn morning after morning, when it's so dark that everything's sort of glowing light blue, and you can hear the elk so close that you're pretty sure you can reach your hand out, touch them. Only, they're ghosts, aren't they? They're so much smarter than I'll ever be. Mostly what I come back with are stories. But stories last longer than meat, I say. One of those stories is from my great-uncle Gerry Calf Looking. It's about how one time a herd of elk came to Browning, and how the train came through right when it needed to. And I bet either some of John Calf Looking's actual stories are in here, or I stole the way he tells those stories. But I also stole the way Delwin Calf Looking said "Tasco" once when we were out after deer, so, you know: stealing's what I do, yeah? Or: I'm always listening, anyway. And, the next-to-final reader this time was Mackenzie Kiera, who didn't just give this novel a pass, she stepped inside it, looked out from it, and guided me back in, walked me through all the story's rooms, one of which is the living room of the house I'm currently renting. It has this high, slanted ceiling, I mean, and this crazy-eyed light that doesn't know what to do with electricity. So, thanks, ridiculous, probably-haunted light. I'd have never looked down through the blades of my ceiling fan without you. Thanks also—and this is maybe the first time I've done this, and I might still delete it, because who can believe me—to the dog that grew up with my kids, Rane and Kinsey. You were Harley here, Grace. You were the best girl. And? Thanks also to a dude I worked with at a warehouse a long time back. Butch. I hijacked you, man, renamed you Jerry. But it's just because I miss you. You're also in my "Discovering America" story from better than twenty years ago, I guess. I can't quit writing about you. And, after all of that to get the novel together, after all the hijackings, all the lifelines and late-night texts, all the people talking me down from the many ledges ringed around each and every novel, thanks to BJ Robbins, first for making it better, for asking the good questions that I was kind of hoping nobody would think to ask, and second for believing in it enough to get it on the right editor's desk. That editor was and is Joe Monti. Since it's my name on the cover, you maybe can't see the impression his hands have left in this story, this book, but, really, this novel didn't find its final form, the one you're holding in *your* hands, until he kind of shrugged and asked, "What if it was this way instead of that way, yeah?" It's this way now. The other way—what was I even thinking? But sometimes, some books, it takes the right editor to push it those last few steps, into what it *can* be. Thanks, Joe, thanks, BJ, thanks, Lauren Jackson, best publicist ever, thanks to Madison Penico for helping me get this one right line-by-important-line, thanks, first, second, last, and current readers, thanks to everybody, especially the people I'm forgetting, the animals I'm keeping secret, but mostly, as always, thanks to my beautiful wonderful smart and perfect wife, Nancy, for putting herself between me and the world time and again, leaving me little pockets out of the wind where I can sometimes write a book or three. I write nothing without you shielding me like that. But, really, thanks just for seeing me across a wash of sand when we were both nineteen, and holding my eyes that one little moment longer, a moment that's lasted and lasted for us, and still has a lifetime to go.

ABOUT THE AUTHOR

© Gary Isaacs

Stephen Graham Jones has been an NEA fellowship recipient;
has won the Jesse Jones Award for Best Work of Fiction from
the Texas Institute of Letters, the Independent Publishers Award
for Multicultural Fiction, a Bram Stoker Award, four This Is
Horror Awards; and has been a finalist for the Shirley Jackson
Award and the World Fantasy Award. He is the Ivena Baldwin
Professor of English at the University of Colorado Boulder.

THE
ONLY
GOOD
INDIANS

BY STEPHEN GRAHAM JONES

This reading group guide for *The Only Good Indians* includes an introduction, discussion questions, and ideas for enhancing your book club. The suggested questions are intended to help your reading group find new and interesting angles and topics for your discussion. We hope that these ideas will enrich your conversation and increase your enjoyment of the book.

INTRODUCTION

Four American Indian men from the Blackfeet Nation, who were childhood friends, find themselves in a desperate struggle for their lives against an entity that wants to exact revenge upon them for what they did during an elk hunt ten years earlier. Not just them, either, but their families and friends.

TOPICS & QUESTIONS
FOR DISCUSSION

1. The story opens with a dark and pessimistic headline that rings true by the end of the prologue. What do you think of Ricky's prediction? Do you believe what he saw that night or was it a figment of his imagination? Could he have changed the outcome?

2. Lewis is haunted by one afternoon of hunting with his friends, instead of all the other hunts he's been a part of. What about that hunting trip was unique? Do you understand how it was a violation?

3. Lewis believes he is being pursued by the spirit of the young mother elk he killed. How does his recent string of bad luck chip away at his sanity? Discuss the combination of factors that push him over the edge. Would these circumstances have driven you insane?

4. Lewis convinces himself that Elk Head Woman has infiltrated his life first as Shaney, then as Peta. What convinced him each

time? Did you expect him to kill them both? If so, at what point did you realize he would go that far?

5. As the story unfolds, it seems less and less likely that Ricky and Lewis were imagining that an elk was after them. What did you believe was happening at this point in the narrative?

6. Discuss reading the chapters as told through the Elk's voice. Why do you think the author chose to include this point of view? What fresh insight does it provide? How did it change your understanding of the first few chapters?

7. Having already heard Lewis's account of that night ten years ago, how does hearing the Elk's version of events change your perspective? Does it make the revenge justified?

8. We've had little access into the world Lewis and Ricky left until the novel moves to Blackfeet Nation. Discuss what you learn about the reservation from Gabe and Cassidy.

9. As the story progresses, the chapters continue to be told from the perspective of a pivotal character but with one significant difference: moments from the Elk's perspective are now interspersed. Discuss this narrative choice. How does it affect your view on the unfolding events?

10. On page 156, Gabe mentions emptying his dad's freezer of the last bit of elk meat from their hunt ten years ago and feeding it to the dogs. How does this moment tie into what is happening now? How does it foreshadow what is to come?

11. Describe Denorah. Discuss how she ties into what is happening with Elk Head Woman and why she is involved in what happened ten years ago.

12. As the three embark on the sweat, they reveal more about life as a Blackfeet, both past and present. Discuss the challenges they face and their differing methods of dealing with them. What, if anything, was surprising or unexpected about their experiences and their conversation during the sweat?

13. When Elk Head Woman sets her plan in motion, things unravel quickly for Cassidy and Gabe. She has no remorse for anyone who gets caught in the crossfire. Discuss how everything from one angle implodes, but from her standpoint goes like clockwork. Do you think this level of violence is warranted? Is this revenge or overkill?

14. Denorah and Elk Head Woman go toe-to-toe on the basketball court long before the true game begins. Why does Elk Head Woman draw this out? Imagine you are in Denorah's shoes. Could you run, fight, and endure for as long as she does?

15. Denorah brings the saga full circle in the end. How does she influence the outcome? How do her actions compare to the choices her father made? What is the relationship with her parents? The day's events have a significant ripple effect in the tribe. Why is Denorah's story passed on?

ENHANCE YOUR BOOK CLUB

1. Stories, lessons, and legends are passed down from one generation to the next. Discuss the theme of generational knowledge and how it is an undercurrent for each character and influences their decisions. How has the history you've inherited influenced your life? How does it influence these characters'?

2. Everything that happens throughout the novel is possible only because, on some level, the characters believe that it is. Did their culture and upbringing influence that? Discuss what you believe about spirits, the afterlife, and what is possible or impossible. How easy or difficult was it for you to suspend your disbelief?

3. Early on, we learn the full meaning behind the title: *The Only Good Indians*. Discuss the meaning behind this insult and the author's choice to use it as the title. Does it give power to the saying or does it take it away? Discuss its significance in the context of this novel, as well as in the world you live in.